THE LIBRARY
ST. MARY'S COLLEGE OF MARYLAND
ST. MARY'S CITY, MARYLAND 20686

Biology of RNA

BIOLOGY OF RNA

J. L. Sirlin
Department of Anatomy
Cornell University Medical College
New York, New York

ACADEMIC PRESS   New York and London   1972

COPYRIGHT © 1972, BY ACADEMIC PRESS, INC.
ALL RIGHTS RESERVED
NO PART OF THIS BOOK MAY BE REPRODUCED IN ANY FORM,
BY PHOTOSTAT, MICROFILM, RETRIEVAL SYSTEM, OR ANY
OTHER MEANS, WITHOUT WRITTEN PERMISSION FROM
THE PUBLISHERS.

ACADEMIC PRESS, INC.
111 Fifth Avenue, New York, New York 10003

United Kingdom Edition published by
ACADEMIC PRESS, INC. (LONDON) LTD.
24/28 Oval Road, London NW1

LIBRARY OF CONGRESS CATALOG CARD NUMBER: 71-127703

PRINTED IN THE UNITED STATES OF AMERICA

To Raquel, Jordanna, and Anton

Contents

Preface   xiii

Part I   MOLECULAR ASPECTS OF RNA

Chapter 1   RNA Molecules as a Class

The Impact of Molecular Biology	3
Historical Profile	5
Informational Macromolecules	9
General Structure of Nucleic Acids	13
Structure of DNA	16
Structure of RNA	19
Mechanisms of Nucleic Acid Synthesis	20
Replication of DNA	20
Transcription of DNA	29
Replication of RNA	31
General Considerations on the Synthesis of Nucleic Acids	32
The Genetic Code	33
Wobble Theory	37
Particular Aspects of Translation	39
Initiation of Translation	39
Termination of Translation and Its Informational Suppression	42
Mistranslation	44

Basic Nature and Evolution of the Code	46
Evolution of Decoding	49

Chapter 2 Molecular Species of RNA

Messenger RNA	51
Characterization of Messenger RNA	52
Some Metabolic Characteristics of Messenger RNA	55
The Informosome	57
Eukaryotic Labile RNA	58
Mitochondrial Labile RNA	61
Ribosomal RNA	61
Major Species of Ribosomal RNA	62
Minor Species of Ribosomal RNA	65
Mitochondrial and Chloroplastal Ribosomal RNA	69
Transfer RNA	70
Other Low Molecular Weight RNAs	79
Chromosomal (Derepressor) RNA	79
4 to 7 S Nuclear (Nucleolar) RNA	80
Bacterial 7 S RNA	82
Homopolymers	82

Part II FUNCTIONAL ASPECTS OF RNA

Chapter 3 Transcription of DNA-Like RNA: Messenger RNA and Eukaryotic Labile RNA

General Structural Basis of Transcription	88
General Basis of Regulation in the Prokaryote and Eukaryote	93
Prokaryotic Messenger RNA	95
Specific Regulation in the Prokaryote: The Operon Model	95
Less Specific Regulation in the Prokaryote	99
General Aspects Related to Bacterial Messenger RNA	100
General Regulation of Bacterial Messenger RNA	106
Viral Messenger RNA: A Survey	109
Extent of Genomic DNA Transcribed in the Prokaryote	112
Conclusions on the General Regulation of Prokaryotic Messenger RNA	114
Eukaryotic Regulation: The Problem at the Cellular Level	117
Eukaryotic Messenger RNA: Participants in Its Regulation	119
Heterochromatin: A State of Chromatin	120
Histones: Not Exclusively Repressor Molecules	122
Stable Regulation of Messenger RNA	127
The Format of the Stable Regulation of Messenger RNA: The Concept of Transcriptional Program	139
Molecular Redundancy of Eukaryotic DNA	141
Extent of Genome Transcribed in the Higher Eukaryote	149
Hypotheses in the Eukaryote: Models of Genetic Organization and Regulation	152

Contents ix

    Cytological Evidence Bearing on Eukaryotic Messenger RNA   158
Eukaryotic Labile RNA   169
    Biological and Molecular Aspects of Labile RNA Synthesis   169
    Function of Labile RNA: Precursor to Messenger RNA or Regulatory
        RNA?   172
Conclusions on the Regulation of Eukaryotic Messenger RNA   179
Adaptive Regulation of Eukaryotic Messenger RNA: Posttranscriptional
    Modulation?   184

Chapter 4   Transcription of Stable RNA: Ribosomal and Transfer RNAs

Major Ribosomal RNAs   188
    Cistrons in the Eukaryote   188
    Cistrons in the Prokaryote   193
    Pathways of Synthesis in the Eukaryote   194
    Pathways of Synthesis in the Prokaryote   197
    Regulation of Synthesis in the Eukaryote   198
    Regulation of Synthesis in the Prokaryote   211
    DNA, Ribosomal RNA, and Informational Aspects in Mitochondria
        and Chloroplasts   218
5 S Ribosomal RNA   222
    Cistrons in the Eukaryote   222
    Cistrons in the Prokaryote   223
    Pathways of Synthesis   223
    Regulation of Synthesis   224
Transfer RNA   226
    Cistrons in the Eukaryote   226
    Cistrons in the Prokaryote   229
    Pathways of Synthesis   230
    Regulation of Synthesis   231
    Mitochondrial and Chloroplastal Transfer RNA and Synthetase   241

Chapter 5   Translation

Basic Mechanism of Translation   245
    Prepolysomal Events   245
    Synthetases   248
    The Ribosome   249
    Polysomal Events   252
    Some Characteristics of Eukaryotic Translation   256
Special Topics in Translation   257
    Models of Structure-Function Relationships   257
    Quantitative Aspects of Transfer RNA during Translation   261
    Effects of Antibiotics on Translation   263
    Nontranslational Oligopeptide Synthesis   264
Regulation of Translation   266
    Regulation in the Prokaryote   266
    Regulation in the Eukaryote   272

	Contents

Transfer RNA as a Regulator of Translation: A Review 288
The Hypothesis: Modulation of Translation 288
Evaluation 302

Part III BIOLOGICAL AND EVOLUTIONARY ASPECTS OF RNA

Chapter 6 Topics of Interest to Cell and Developmental Biology of the Eukaryote

The Nucleolus 308
The Ribosomal RNA Cistrons: Natural History 309
Nonribosomal RNA Species in the Nucleolus 314
Nucleolar Protein 316
Ultrastructure and Organization of the Nucleolus 319
Nucleolus-Associated Chromatin 324
The Eukaryotic Cell Cycle 327
Introduction 327
The G_1 Period 331
The S Period 333
The G_2 Period 337
Mitosis or M Period 338
Some General Points Concerning the Cell Cycle 341
Embryonic Development: A Review 344
General Aspects of Development 344
Replication during Development 353
Transcription during Development 357
Translation during Development 382
Transfer RNA and Related Enzymes in the Modulation of Developmental Translation 390
Embryonic Induction 394
A Cell's Life: An Overview 400

Chapter 7 Evolution of RNA: An Evaluation and Epilogue

The First Genetic Substance 404
The Dawn of the Cell Nucleus 405
DNA 410
Structural Complexity of Eukaryotic DNA and Some of Its Biological Implications 410
Evolutionary Parameters of DNA 422
Macromolecular Associations of Eukaryotic DNA 431
Eukaryotic Cytoplasmic DNA 435
RNA 437
Transfer RNA 440
Ribosomal RNA 447
Informational RNA 457
Eukaryotic Regulatory RNA 466

Contents xi

Appendix

Resume of Techniques 475
List of Abbreviations 477
Glossary 477

Bibliography 481

Index 517

Preface

My purpose in writing this book is to present a unified account of the biology of RNA. My specific aim is to compare the prokaryote with the eukaryote, with emphasis on the eukaryote and particularly its regulatory organization and evolution.

The book is divided into three parts in which the form and function of the molecular species of RNA and how they relate to one another and to the needs of the cell are discussed. Part I is comprised of two chapters—Chapter 1, an introduction to the study of the molecules, and Chapter 2, a description of the individual RNA species. Chemical properties of the molecules are considered only insofar as it is necessary to understand function. Chapters 3 through 5 make up Part II. The regulation of transcription is discussed in Chapters 3 and 4. Chapter 5 deals with the role of some of the species in translation. Finally, Part III offers a review of topics in the cell biology of the eukaryote (Chapter 6) and an evolutionary appraisal of RNAs grouped by categories (Chapter 7).

I wish to thank those of my colleagues and students who, unknown to them, helped my thinking in the course of discussions and lectures. Among my Cornell students, I owe a debt of gratitude to Barry Kaplan and Stuart Bergman, who criticized the last drafts of the book. I am obliged to the staff of Academic Press for their patient cooperation. Last but not least, I thank my wife for a warm mixture of forbearance and encouragement.

J. L. SIRLIN

Part I
Molecular Aspects of RNA

Chapters 1 and 2 focus, respectively, on the general structural characteristics of RNAs that permit us to group them together as a family of molecules, and on the particular structural and functional characteristics that enable us to recognize them as distinct molecular species. The basic considerations are the principles underlying the structure of the molecules, their mechanism of assembly, their status with respect to the genetic code, and their relationship to the processes that intervene in decoding. Aspects of each RNA species to be discussed include their individual structure, metabolic behavior, and interrelationship with other molecules, which together determine the function of the species. These chapters place the RNA species in their roles in relation to the genetic function of the cell, each species representing a different facet of the information contained in DNA and ultimately designed for protein synthesis.

CHAPTER 1

RNA Molecules as a Class

THE IMPACT OF MOLECULAR BIOLOGY

Much of the mystery surrounding the status of complex biological molecules was dispelled when it was first shown for bovine insulin (Sanger and Thompson, 1953) that the protein could in principle be defined by its amino acid sequence. A few years later, the discovery that the severely abnormal sickle cell hemoglobin differed from normal in a single amino acid (Ingram, 1957) had much the same effect as the pioneer work on insulin. This meant that macromolecules did not escape classical chemical laws, that is, both the structure and function of a macromolecule are uniquely determined by properties of the submolecular structure (although enzymes may affect the final structure). In addition, the implication of these findings was to corroborate the "one gene-one enzyme" hypothesis advanced previously from biochemical genetics (Beadle and Tatum, 1941), which now became "one gene-one protein." More generally, the macromolecular endowment of any organism was chemically accountable for in the vast potential information in the DNA. Since that time the molecular approach to biology and genetics has grown rapidly and led to present ideas, some of them fairly good, on how genes act to direct their product. By 1966, with the unraveling of the nature of the genetic code, biology reached a state similar to that of physics in the 1920's when the understanding of the nature of the atom was achieved.

The appeal of these molecular studies undoubtedly comes from the satisfaction of being able to explain complex phenomena at a fundamental level. At present, molecular analysis attempts to explain the most intricate processes in cell, organism, and mind; and it is not a too distant hope that we may soon understand the bacterial cell, if not fully, at least adequately. Concurrently with these advances, machines have taken over almost to a qualitative degree the handling of the complicated space-time interrelationships involved in biological processes. In effect, the computer is filling the gaps between the facts of the reductionist experimental scientist. Because understanding of these relationships might well be limiting for new levels of understanding, the ultimate consequences of this mechanization could be momentous.

Over the last 20 years a new attitude has been incorporated into biological thinking from the more deductive branches of science and particularly physics (Platt, 1964), with biochemistry providing mainly the facts and techniques. This attitude has unwittingly resulted in a loss of feeling toward the more biochemical aspects, which, to the more traditional mind, appears to be almost an antichemical attitude. Thus, according to Chargaff (1968), molecular biologists not only have to live with "black boxes" but these are filled with "unobservable" substances. Such, however, is the price of progress. To cite an example, the first artificial synthesis of a transfer RNA (tRNA) gene, although not the complete one coding for the precursor tRNA, has been attained (Agarwal *et al.*, 1970). Comments on the various phases and motivating goals in molecular biology can be found in Stent (1968). The reader, however, need not worry unduly about Stent's epitaphic quality concerning the coming to an end of progress. In better spirits, Hess (1970) has also written on this subject.

A basic hypothesis in molecular biology is the "central dogma" proposed by Crick (1958), which states that genetic information is passed in one direction only, namely, from nucleic acid to protein[1]:

$$\text{DNA} \longrightarrow \text{RNA} \longrightarrow \text{Protein}$$

As far as is known, this direction is not contradicted in any biological material.[2] At the molecular level the hypothesis is distinctly Darwinian: the

---

[1] The order and direction of this flow is intuitively visualized from the gross chemical composition in *E. coli:* 1% DNA, 6% RNA, and 15% protein.

[2] At the present time, it is clear that during the life cycle of single-stranded oncogenic RNA viruses, genetic information flows in the direction RNA → DNA → RNA (Temin and Mizutani, 1970). By using more than one novel type of polymerase, principal among which is the RNA-primed DNA polymerase or reverse transcriptase, the initial RNA template forms a hybrid with its product DNA which then by complementarity gives rise to helical DNA for integration into the host genome; the integrated DNA produces viral transcripts. For these particular viruses, the genome may be said to alternate between a RNA and a DNA form. While the newly found direction of flow of information reverses the normal direction of flow, this does not however violate the spirit of the dogma which

opposite flow of information would be predicted on a Lamarckian basis. The point in mentioning it here is to focus attention on RNA as the conveyor of genetic information.

Much of the strong deductive element behind the dogma had its origin in general considerations on coding for protein synthesis. These were prompted by the elucidation, a few years before, of the structure of DNA (Watson and Crick, 1953), by then recognized as the genetic material. The structure by itself showed that unique base sequences could be converted into amino acid sequences in protein. This view was reinforced by fine genetic analysis in phage (Benzer, 1959, 1961) that revealed a sequential topology and linear topography of cistrons down to the single base level. The obvious intermediate was RNA whose similar sequences suggested their origin in the nuclear DNA. As far as RNA synthesis is concerned, this meant a central role for the cell nucleus. The lasting conceptual influence of the dogma will become clear in what follows.

HISTORICAL PROFILE

A profile of events leading to the concept of separate RNA species should begin with the 1940's. At that time a general belief that genes might contain or consist of protein was still widespread. It was also believed that the nucleus was an important center of protein synthesis. In 1941, Caspersson proposed, on the basis of ultraviolet cytophotometry, that RNA was in some way involved in protein synthesis. Brachet (1942) had marshaled enough cytochemical evidence on cell basophilia to arrive independently at the same conclusion. In keeping with then current ideas, Caspersson (1941) thought that the actual carrier of information from the nucleus was protein, and in turn protein synthesis in the cytoplasm was activated by RNA in a manner as yet unknown. In retrospect, these contributions elicited the first interest in RNA. Soon after came the identification of DNA as the bacterial transforming agent (Avery *et al.*, 1944). Within a decade DNA was shown to be the genetic ma-

---

remains that protein is the dead end of genetic information. Perhaps then, as proposed recently by Crick (1970), the present status of the dogma, as far as normal cells are concerned, is indicated in the following diagram (broken lines represent rare transfers of information).

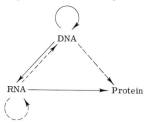

terial in viruses (Hershey and Chase, 1952). This important contribution, the first of a long series from the work on phages (Cairns *et al.,* 1966), had a greater effect than the previous contribution by Avery because by this time the genetically monotonous "tetranucleotide hypothesis" of DNA structure had been abandoned. Later bacterial transduction experiments pointed in the same direction. The corresponding genetic role for viral RNA was shown in 1956 (Gierer and Schramm, 1956), in fact, with purified viral RNA as the sole infective agent.

The Watson-Crick model of DNA was published in 1953, 1 year after Hershey's experiments. Crucial as this model has been for guiding later work, however, it should be remembered that it (as well as the preceding model by Pauling of the protein α-helix) came as the culmination of the X-ray crystallography school of Atsbury, Bernal, Wilkins, etc. A highly personalized account of the making of the model is written by Watson (1968). The model suggested immediately how RNA was formed. Intensive work in this area resulted by early the next decade in first-hand knowledge of the mechanism of synthesis by RNA (and DNA) polymerase.

By the time the dogma was announced in 1958 it was becoming clear, after some initial debate, that most RNA originated in the nucleus. The evidence was biological (Brachet, 1957; Hämmerling, 1953) and came mainly from autoradiography (reviewed by Prescott, 1964). That the RNAs of higher organisms were made on DNA templates was shown using actinomycin in the late 1950's (reviewed by Reich and Goldberg, 1964). However, rigorous proof of template DNA for each RNA species had to await the development of molecular hybridization. That the template was indeed a nuclear template became unequivocal after further work.

A general belief persisted that the RNA which carried the nuclear information was ribosomal RNA (rRNA), that is, this RNA was the template for protein synthesis in the cytoplasm (Crick, 1958). The reason for this belief was that this type of RNA was the most abundant and, furthermore, no other one was known for this role. However, skepticism began to mount when no sound experimental or even a theoretical basis could be advanced in support of rRNA being the template. In particular, rRNA did not have the template instability (Davern and Meselson, 1960) predicted for rapid bacterial enzyme induction, and its composition and size were similar enough in different bacteria, irrespective of the DNA composition (Belozersky and Spirin, 1960). Many mechanisms of protein synthesis that were proposed before the dogma and before the details of protein synthesis became known as we know them today, are reviewed by Zamecnik (1969).

Finally, the postulation on genetic grounds of messenger RNA (mRNA) by Jacob and Monod (1961), based on the interpretation of phage lysogeny-induction, on the one hand, and of bacterial enzyme induction-repression, on the other hand, explained the cytoplasmic template as it has now come to be accepted. In retrospect, the first to have observed mRNA were Volkin and Astrachan (1956) in phage-infected bacteria. Considering first what was known of fine genetic structure (Benzer's work and intragenic recombination), the operon model by Jacob and Monod was indeed a remarkable synthesis—to this day the signal one of molecular genetics. In addition, the model made DNA and ribosome and tRNA fit together into a working pattern. Shortly after, mRNA was identified biochemically. Brenner *et al.* (1961) went on to show that mRNA formed after phage infection became associated with "old" bacterial ribosomes to make "new" phage proteins, thus confirming the concept of mRNA. The assumed complementarity between phage DNA and its mRNA was confirmed by hybridization (Hall and Spiegelman, 1961). Messenger was found in noninfected cells (Gros *et al.*, 1961) and incorporation into protein in a cell-free system was shown to respond to added mRNA (Matthei and Nirenberg, 1961). Soon the code was broken with the same system when polyphenylalanine was obtained under the direction of poly U, the first artificial mRNA (Nirenberg and Matthei, 1961). The discovery of the polysome (Warner *et al.*, 1962; Wettstein *et al.*, 1963) allowed one to visualize, this time in eukaryotic cells, how mRNA went about its task. Basic understanding of protein synthesis was now well under way.

Because of its bulk ribosomal RNA was the first RNA to be observed in the cytoplasm. By the early 1950's, the combined techniques of cell fractionation and electron microscopy had established a correlation between the RNA (as part of the ribosome) and protein synthesis. By contrast, nucleolar rRNA, observed as early as was rRNA in the cytoplasm, had to await characterization until 1962 (Perry, 1962). This study thus established the origin of rRNA in the cell. However, as mentioned above, even as early as 1958 it was thought that rRNA was the direct template for protein synthesis. The 1961 experiments demonstrating the existence of mRNA also made clear that rRNA assisted in rather than directed protein synthesis. Surprisingly enough, only recently we have begun to understand the function of this RNA in the ribosome.

Among the first to find what was later to be transfer RNA in the cell supernate were Brachet and Jeener (1944). The original mention of an adaptor RNA (or oligonucleotide) was made on theoretical grounds in a note by Crick (1957) to obviate the lack of steric matching between polynucleotide and polypeptide. Hultin (1956) had found bound amino acid in the cell supernate that behaved as an intermediate in protein synthesis. Soon after, Hoagland and associates (1957) reported the acceptance of activated amino acid by tRNA and its subsequent appearance in polypeptide. That the activation energy was derived from phosphorylation had been predicted as early as 1941 (Lipmann, 1941). Hoagland's discovery gave great impetus to the work in the area of protein synthesis. As soon as the molecular parameters (size, unusual bases, multiplicity) became known, it was clear that tRNA was one of the most favorable RNA species for sequencing. However, it was not until almost a decade later that Holley *et al.* (1965) published the sequence of yeast alanine tRNA, the first for any nucleic acid. Fortunately, it turned out that tRNA has an elaborate tertiary structure, and important advances in relation to general RNA function are expected from study of this molecule.

Ribosomal 5 S RNA was discovered in 1964 (Rosset et al., 1964), and the first sequence was completed in 1967 (Brownlee et al., 1967). The first to propose labile (nuclear) RNA was Harris (1959). His arguments were strongly resisted mainly because no one could see a reason for such lability in a nuclear RNA. It was indeed the forerunner of eukaryotic complications to come. The RNA has since been confirmed, the function remaining hypothetical. Derepressor RNA was postulated by Frenster (1965) in order to activate DNA for RNA synthesis in the eukaryote, and possible but still disputed evidence in favor of it is now available. Regulator (messenger) RNA, inferred to code for bacterial and viral repressor, has been observed.

It can be seen from this profile of RNA species that one of them, 5 S RNA, was found empirically. Almost the opposite is true of mRNA and tRNA which were predicted by deduction. Today, several other RNA species are known, of low molecular weight, in particular, which are described in the text. Significantly, while the whole bacterial repertoire of RNAs has probably been known since 1967 with the discovery of a 7 S RNA (Hindley, 1967), in all likelihood new eukaryotic RNAs have yet to be discovered.

In considering cell function, if one were to evaluate what has been gained in these years, then in the prokaryote the mechanism of macromolecular syntheses is understood, but its basic regulation is not. This is because the bacterium is an exquisitely poised cell and not a random assortment of mechanisms; to learn about the overall regulatory integration of these mechanisms in relationship to one another and ultimately to cell metabolism remains the outstanding challenge. In the eukaryote, on the other hand, seldom accessible to the molecular geneticist, progress is in the hands of the molecular biologist. We have just begun to learn what goes on molecularly without having to borrow second-hand from the bacterium and, contrary to first shallow impressions, the mechanisms in the eukaryote appear to be extremely more complex. Judging from the larger number of macromolecular species involved, which applies very much to RNA, there will surely be unsuspected regulatory mechanisms. In fact, extrapolating from the gradual increase in complexity of bacterial regulation as the knowledge of it has progressed, it can be positively predicted that in a few years eukaryotic regulation will bear little resemblance to what we know of it today. The nucleated cell is a subtly balanced system within another subtly balanced system, the organism, and each has its own laws and mechanisms of regulation (although, obviously, these mechanisms are congruous with one another). Nevertheless, the molecular study of eukaryotic cell function and differentiation has progressed substantially during the last few years. Two uncharted areas, probably the most challenging conceptually, are those of immunological and neuronal function.

From what was just said, it is no understatement that at this time we do not understand how regulation is integrated at the cellular level and, hence, how the cell is organized. Thus we may not have found which are the higher principles of regulation involved. Two aspects stressed throughout the text in relation to this question are the dependence of macromolecular syntheses on cell metabolism and of the synthetic machinery on cell infrastructure. The broader aim is, however, to delve into the structure of cell function itself as this concerns RNA, and in so doing to emphasize the self-organization of biological function.

INFORMATIONAL MACROMOLECULES

By way of definition, informational macromolecules are those macromolecules through which genetic information actually flows. Genetic information may be defined as that primarily required to assemble a protein, hence ultimately required to perpetuate biological order, but the definition applies also to the information for RNAs which are not translated into protein. Between cell and organismal generations the informational molecule is DNA, which functions as the repository for information or code, and based on present knowledge accounts for most, and probably all, of the genetic information. Within the cell, the informational molecule is mRNA (Fig. 1). Althought all RNA species function by virtue of the information they embody, only mRNA transmits the information or is a carrier of information for protein. (That RNAs, including mRNA, may be transferred between cells remains however a possibility.) DNA directs its own replication and, through RNA, the replication of the cell containing them. DNA has, therefore, a dual function, replication and transcription, and the two are by essentially similar mechanisms.

It was first pointed out by Dounce (1952) that genetic information is represented by linear sequences. That plants and higher animals have similar DNA base ratios in itself shows that only base sequences are responsible for the genetic differences. From the previous definition, these linear sequences will reappear as corresponding linear sequences in protein.

The information in DNA is contained in coding sequences ultimately destined to appear either in RNA or protein, and in noncoding sequences which can act as signals, e.g., initiation and recognition (or have other functions, e.g., mechanical). Coding sequences are also referred to as structural.[3] The reason for the two types of structural sequences is clearly because DNA contains the information both for proteins and for RNAs with which to make them. Structural sequences are therefore either informational or noninformational, according to whether or not they code for protein. Hence informational sequences are transmitted via mRNA for translation into protein. The noninformational sequences are transcribed into RNA which is not translated but is thought to function as

---

[3] As used in the field, the term "structural" has therefore a coding connotation only. The structure alluded to is that of the product RNA or protein, not that of the DNA itself. However, there may also be mechanically structural sequences in DNA, but in the text the term "structural" is used sparingly in this noncoding sense, and only when the context obviates any confusion with coding sequences.

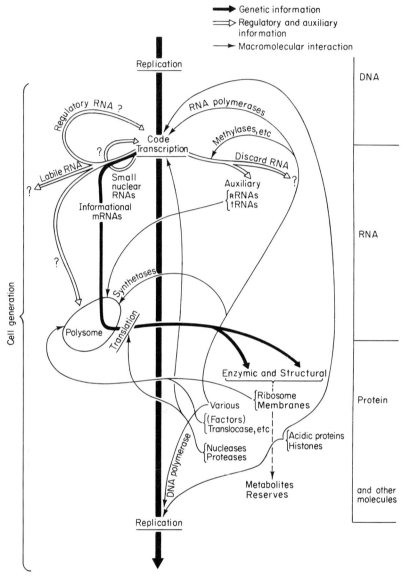

Fig. 1. Diagram of the flow of genetic information in the eukaryote illustrating the specification within the cell of the macromolecules discussed in the text. It has been simplified by omitting factors and other regulatory proteins that act on transcription and replication, and feedback loops from metabolites to almost all the functions represented.

either regulatory or auxiliary RNA (Fig. 1). Noninformational RNA might be particularly significant in the eukaryote for regulation of the genome. However, since some of the protein has similar regulatory functions, the final function of some informational RNA may also be regulatory.

Structurally speaking, if possibly not phylogenetically, all RNA molecules are patterned on DNA. Also, on present knowledge, all cellular RNA is made on DNA. However, the evidence from experiments with actinomycin supporting this statement is of a negative character, and therefore cannot logically prove it. This then leaves open the possibility that RNA may form on a RNA template (Spiegelman, 1969). Evidence in this direction might be the helical RNA that has been extracted from vertebrate cells (Colby and Duesberg, 1969), although it remains to be shown that this RNA is helical in the cell and the product of self-replication. Similarly, the reversed transcription from RNA to DNA which has been authenticated only in oncogenic RNA viruses (see footnote 2) could conceivably have wider biological implications in the eukaryotic cell. Some reports have appeared in this direction, but are still unconfirmed.

Strictly considered, the genetic code in DNA is only in relation to information destined for protein and presupposes the machinery to use the information for making protein. For this reason, the design of the code has been compared with a computer program wherein the structure and interconnections of the computer (the cell) are presupposed (Dayhoff, 1969). The meaning of this comparison is clear when one considers decoding of DNA at translation. However, two molecules in particular, tRNA and tRNA synthetase, function as a "code within a code" by virtue of the fact that their mutual recognition in connection with translation is what ultimately determines which amino acid enters the polypeptide. That DNA can function as a code is because its base sequence is largely irrelevant in determining the structure of the molecule. "It is this physical indeterminacy that produces the improbability of occurrence of any particular sequence and enables it to have [informational] meaning" (Polanyi, 1968). By contrast, the pattern of atoms in a crystal is an instance of complex order without appreciable information content. However, intramolecular electrostatic forces in DNA may somewhat influence the sequence (De Voe and Tinoco, 1962) and along these lines there appears to be a certain rhythmical periodicity of sequence in yeast DNA (Fiers, 1966). Nonrandomness of DNA sequences will be further discussed in Chapter 7. (At the other extreme, it needs mention that protein sequences have a structure which is in fact the basis of their function.) To give an idea of the enormous potential information in DNA, a molecule with 10^3 bases might have any of 10^{602} (4^{1000}) possible sequences. The combined weight of this number of sequences is enormously larger than the weight of the Earth. Each of these unique sequences has a complexity of 10^3 by definition (Wetmur and Davidson, 1968). For comparison, a random polypeptide consisting of 333 amino acids (equivalent to 10^3 bases in DNA) has a possible 10^{433} sequences.

Informationally speaking, DNA and protein, representing a colinear translation, are one-dimensional, that is, their linear primary structure determines all other structure.[4] DNA and mRNA function at a level somewhere between primary and secondary structure, although for mRNA a low-order tertiary structure remains a possibility. Metabolically speaking, all steps leading up to translation are energetically reasonably simple when compared with intermediate metabolism (Lipmann, 1965). These steps are also mathematically simple. However, stereochemically they are complex (Dayhoff, 1969). On the other hand, translation itself is complex in all respects since it requires the intervention of tertiary and quaternary structures, e.g., tRNA, enzymes, and ribosomes in order to achieve specificity and plasticity; one may thus speak of the quinternary structure of the translational apparatus. Translation is also, of course, the meeting ground for informational and auxiliary RNAs (Fig. 1). Compared with one-dimensional molecules, the study of structure-function relationships in translational molecules is just beginning. Related to this question, a chronicle of structural versus informational ideas in molecular biology is given by Stent (1968).

In regard to the eukaryotic cell it can be seen in Fig. 1 that there are two major feedback pathways resulting from the flow of genetic information, one to the code or its transcripts and the other to the polysome. In this case the feedback macromolecules are the more remote products. This means that in each cell these molecules (RNA polymerases, synthetases, methylases, factors, etc.) must be present from the beginning of the cell cycle, that is, they are either received or made on programmed polysomes received in the cytoplasm from the parent cell. Other obligatory macromolecules (histone, acidic protein) come along with the parental chromosomes — a reminder that chromosomes and not just DNA are handed down to a cell. (Clearly, obligation is not at all confined to macromolecules.) Eventually, and especially if progeny cells are to follow, the molecules or the mechanism for making them have to be replenished by the present cell. Not fortuitously, it is these two feedback

---

[4] Orders of structure in nucleic acids are described later, so that only a word about protein structure need be said here. The primary structure of a protein is a simple one with amino acids joined by a recurrent peptide linkage to form a right-handed α-helix. This helix is not universal in protein, but, because it resembles the DNA helix considerably, it may be briefly described. Hydrogen bonds occur within and parallel to the peptide bonded chain, their structural role probably being secondary to that of the nonpolar groups in the amino acid residues. Nonpolar bonds, H bonds, and even covalent bonds (-SS-) between these residues determine the higher order structure of the protein. On the other hand, the interior of many globular proteins (e.g., myoglobins) is hydrophobic, which, electrostatically speaking, helps make the molecule a compact one; the outside of globular proteins is polar. The standard amino acids are all derivatives of the smallest one, glycine, with the α-carbon H substituted by as many side groups. Conceivably, the exclusive L-isomerism of amino acids is mainly for reasons of secondary structure (Doty and Lundberg, 1957).

pathways that most need elucidation. However, a third feedback pathway is that of the code on itself; this pathway will be discussed as we proceed, but from the number of question marks attached to it in Fig. 1 it is evident that this pathway is in equal or even greater need of explanation. A glance at the figure further suggests that, even if the mechanism of encoding information is one, the mechanisms of regulation (of which the three pathways are a part) must be many. What was said here means, therefore, that although the genetic information comes from DNA, the cell still comes from a cell. One may think of the cell as a self-perpetuating system whose information is continuously replenished from DNA.

Separate DNA-dependent genetic systems are present in cytoplasmic organelles, such as mitochondria. These systems partly depend on information derived from, and to some extent are integrated with, nuclear DNA (review by Beale, 1969). The organelles are important physiologically, but they also may have an informational role for the cell which is as yet undetermined. This could mean not just an adjunct genome but an extension of it, rather analogous to the bacterial plasmids which will be mentioned later.

At this time it may not be possible, however, to say that all information destined for macromolecules is in DNA or is linear. This nonlinear or pattern information may reside in cell structures such as cortices and perhaps membranes, and will be briefly discussed in connection with development (Chapter 6). In this case obligatory priming by preexistent structures may be involved. The question is how this potential information relates to that in DNA. For this reason, at the moment, it might be better to consider this information as being nongenetic information. Nongenetic information derived from molecules other than nucleic acids would mean no violation of the central dogma, which negates this derivation for genetic information only. Moreover, such information is as likely to be carried by molecules as by the interaction between them, in this case being second-order with respect to the genetic information in DNA.

GENERAL STRUCTURE OF NUCLEIC ACIDS

Some general traits of RNA are presented here before introducing the traits of individual species in Chapter 2. Because the traits are similar in DNA, but more regular, this will be discussed as a prototype of nucleic acids at some length. Structurally and functionally, RNA is basically understandable in relation to DNA. However, it must be stressed from the beginning that the detailed native molecular structure of DNA is not known with absolute certainty.

All RNA molecules are basically similar heteropolymers (polynucleotides) formed of four monomers in a linear, nonbranched sequence (Fig. 2). The essential difference between the various RNAs is their sequence which ultimately determines their higher order of structure[5] and, thus, function. Each monomer nucleotide in the polymer consists (1) of one purine or pyrimidine base bonded to ribose (a five-membered D-furanose ring) to form a nucleoside, and (2) the ribose in turn is linked to inorganic orthophosphate to form the nucleotide. The bond with ribose is weaker in the case of purines. The purines are adenine and guanine; the pyrimidines are cytosine and uracil. The usual tautomeric forms of the bases are the amino (adenine, cytosine) and keto (guanine, uracil) in preference to the imino and enol. After the primary sequence of the polynucleotide has been laid down, a proportion of the bases in some of the RNA species are enzymically modified by methylation, thiolation, etc. There are some 60 or more "odd" nucleosides, including pseudouridine and inosine (derived from the purine hypoxanthine), and it is probable that all arise by enzymic modification at the polymer level (Hall, 1970a). By influencing the H-bonding pattern the effect of these modifications on tertiary structure can be considerable.

Adjacent nucleotides are linked by 5'- to 3'-phosphodiester bridges between the ribose units, the intranucleotide linkage of each ribose being the 5' as in the precursor nucleoside triphosphates. The integrity of the polynucleotide depends on this backbone with a definite chemical polarity. The recurrent negatively charged phosphate groups in the backbone give the molecule a polyanionic character, the negative charge being on one oxygen atom per phosphate group. These groups usually are electrostatically neutralized by oppositely charged divalent cations and perhaps by polyamines. In the aqueous cell environment

---

[5] Primary structure of RNA is given by the covalently bonded base sequence. Secondary structure refers to the relative position of adjacent nucleotides that is influenced by electrostatic (van der Waals) forces. Tertiary structure involves the spatial arrangement of the macromolecule in which folding is important; the structure depends on noncovalent bonds (except for possible covalent -SS- bonds in the case of tRNA) between distant nucleotides along the molecule; see below for further definition of tertiary structure. Quaternary structure results from chain association; the term is also nominally used for multimolecular associations of RNA-RNA and RNA-protein as, for example, in the polysome. Primary structure of a molecule is obligatory, but, as the structure rises in order, it becomes progressively more facultative because it becomes progressively more sensitive to the environment.

The above definition of tertiary structure requires amplification. First, there is a low order tertiary structure in helical regions of RNA molecules resulting from folding and primary base pairing within the same strand. Second, there is a high order tertiary structure derived from folding of the helical regions onto other helical regions, which is associated with secondary base pairing (between unpaired bases in loops) and to a certain extent is conditional on the primary base pairing. These two suborders are well illustrated in tRNA. Finally, it is tautology to note that all matter is three-dimensional, yet this does not necessarily mean tertiary structure. For instance, the reason why helical DNA as a linear molecule, although base paired, usually has no tertiary structure, is because the base pairs are between two different strands; yet it cannot be excluded that DNA possesses, perhaps restrictedly, some tertiary structure, but so far this has not been demonstrated.

Structure of Nucleic Acids

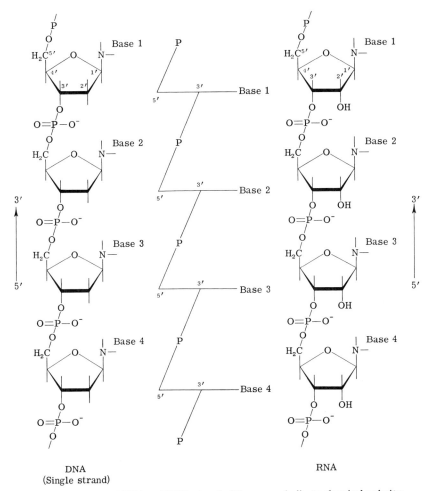

Fig. 2. The structure of DNA and RNA strands. The arrows indicate chemical polarity.

these ionic bonds are not believed to determine the shape of the RNA molecule. However, if the bonds are to cationic residues of protein, it is possible that the shape of the RNA molecule could be affected.

The similarity in the basic structure of RNA compared with DNA is shown in Fig. 2. One general difference in composition is that DNA contains thymine instead of uracil, though there is some thymine (as methylated uracil) in RNA. Certain DNAs contain 5-methylcytosine, and in certain phages 5-hydroxymethylcytosine and 5-hydroxymethyluracil in place of cytosine and thymine; the hydroxymethyl group can be glucosylated. A second difference is that DNA contains 2-deoxyribose instead of ribose. Interestingly, the thymine substitution could reflect the general lack of appropriate kinases to phosphorylate dUMP in a separate DNA precursor pool, since certain viral DNA contains uracil (see Lesk, 1969, for other explanations). On the other hand, the reason for the sugar replacement is not immediately obvious since RNA helices do occur (and DNA acts as mRNA in the presence of antibiotics), but it might explain the dual function of DNA in replication and transcription as described later. The deoxyribose also confers a greater chemical stability to each DNA chain, but it is not clear if this effect of the sugar on the chain is so important as to necessitate its presence.

In summary, it would seem that the thymine substitution is trivial in explaining the need for DNA in the cell but the sugar substitution is relevant. Hence DNA combines the intrinsic stability of a helical molecule (which is more than just the stability of each chain) with maintaining separate the replication and transcription of the code, and this may be possible because of the deoxyribose. In addition, only a helix can be semiconservatively replicated, as discussed later, and the result that one parent helix always produces two daughter helices to select from may well be an evolutionary advantage. However, to anticipate a different consequence of this type of replication, it is molecularly so complex that ease of replication is certainly not a reason for DNA in the cell.

Structure of DNA

With exceptions, DNA usually forms a perfect double helix composed of two individual right-handed helices. The helical structure initially proposed by Watson and Crick (1953) was later confirmed in all its major traits (L.D. Hamilton, 1968), and has now been directly visualized with the electron microscope (Ottensmeyer, 1969). The structure is that of bases in one chain paired with the complementary bases in the other chain. The base pairs are stacked at the center of the molecule as planar structures, and the two backbones run on the outside (Fig. 3). The role of water intimately bound to the structure is not understood. In the native condition, referred to as the paracrystalline form B, the bases project from the backbone perpendicularly to the long axis. By association with histone in the eukaryotic chromosome (histone is absent in the prokaryote) the bases lose their perpendicularity. As will be described in Chapter 3, also other physical properties of the helix can be altered in this case; the same may be true of the helix within the phage capsid. Electrostatically, considering the associated electron clouds or van der Waals spheres, the bases nearly fill the inside of the helix, and in this sense the molecule is a solid rod. The helix is marked by a high degree of stereochemical order and looks the same from either end, except for the antipolarity of the component chains (see Watson, 1970). Because of base pairing and simplicity of design any sequence tells more about the local structure of DNA than any protein sequence tells about the structure of protein.

Structure of Nucleic Acids

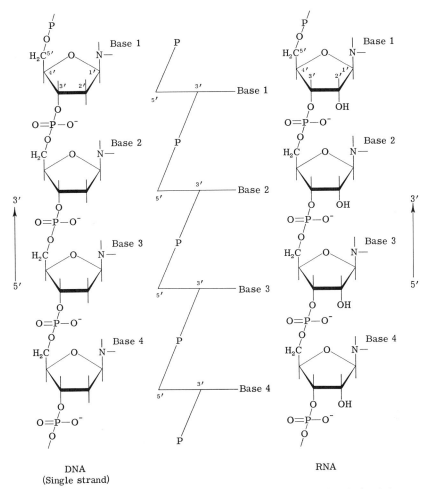

Fig. 2. The structure of DNA and RNA strands. The arrows indicate chemical polarity.

these ionic bonds are not believed to determine the shape of the RNA molecule. However, if the bonds are to cationic residues of protein, it is possible that the shape of the RNA molecule could be affected.

The similarity in the basic structure of RNA compared with DNA is shown in Fig. 2. One general difference in composition is that DNA contains thymine instead of uracil, though there is some thymine (as methylated uracil) in RNA. Certain DNAs contain 5-methylcytosine, and in certain phages 5-hydroxymethylcytosine and 5-hydroxymethyluracil in place of cytosine and thymine; the hydroxymethyl group can be glucosylated. A second difference is that DNA contains 2-deoxyribose instead of ribose. Interestingly, the thymine substitution could reflect the general lack of appropriate kinases to phosphorylate dUMP in a separate DNA precursor pool, since certain viral DNA contains uracil (see Lesk, 1969, for other explanations). On the other hand, the reason for the sugar replacement is not immediately obvious since RNA helices do occur (and DNA acts as mRNA in the presence of antibiotics), but it might explain the dual function of DNA in replication and transcription as described later. The deoxyribose also confers a greater chemical stability to each DNA chain, but it is not clear if this effect of the sugar on the chain is so important as to necessitate its presence.

In summary, it would seem that the thymine substitution is trivial in explaining the need for DNA in the cell but the sugar substitution is relevant. Hence DNA combines the intrinsic stability of a helical molecule (which is more than just the stability of each chain) with maintaining separate the replication and transcription of the code, and this may be possible because of the deoxyribose. In addition, only a helix can be semiconservatively replicated, as discussed later, and the result that one parent helix always produces two daughter helices to select from may well be an evolutionary advantage. However, to anticipate a different consequence of this type of replication, it is molecularly so complex that ease of replication is certainly not a reason for DNA in the cell.

Structure of DNA

With exceptions, DNA usually forms a perfect double helix composed of two individual right-handed helices. The helical structure initially proposed by Watson and Crick (1953) was later confirmed in all its major traits (L.D. Hamilton, 1968), and has now been directly visualized with the electron microscope (Ottensmeyer, 1969). The structure is that of bases in one chain paired with the complementary bases in the other chain. The base pairs are stacked at the center of the molecule as planar structures, and the two backbones run on the outside (Fig. 3). The role of water intimately bound to the structure is not understood. In the native condition, referred to as the paracrystalline form B, the bases project from the backbone perpendicularly to the long axis. By association with histone in the eukaryotic chromosome (histone is absent in the prokaryote) the bases lose their perpendicularity. As will be described in Chapter 3, also other physical properties of the helix can be altered in this case; the same may be true of the helix within the phage capsid. Electrostatically, considering the associated electron clouds or van der Waals spheres, the bases nearly fill the inside of the helix, and in this sense the molecule is a solid rod. The helix is marked by a high degree of stereochemical order and looks the same from either end, except for the antipolarity of the component chains (see Watson, 1970). Because of base pairing and simplicity of design any sequence tells more about the local structure of DNA than any protein sequence tells about the structure of protein.

Structure of Nucleic Acids

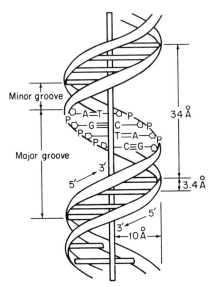

Fig. 3. Diagrammatic representation of the DNA helix. The rungs between bases indicate H bonds. The arrows along the strands indicate their chemical polarity.

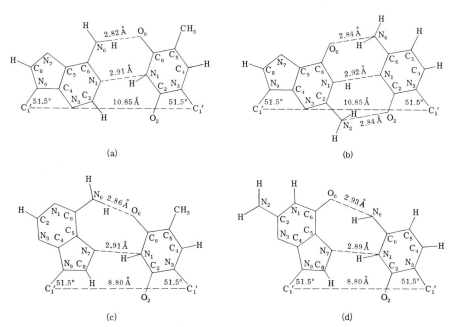

Fig. 4. Base pairing in DNA according to Watson and Crick (1953) and Hoogsteen (1963). Pairs (a) and (b) represent adenine-thymine and guanine-cytosine, respectively, according to Watson and Crick; (c) and (d) the same pairs according to Hoogsteen. Taken from Arnott *et al.* (1965).

Guanine-cytosine and adenine-thymine are the standard complementary or Watson-Crick pairs. The base ratios resulting from these pairs were discovered by Chargaff et al. (1951). The pairs form spontaneously between synthetic polynucleotide chains, indicating that the stereochemistry is right; they require the correct tautomeric form of the bases. From the Watson-Crick base pairing it follows that whatever the sequence of bases, or the preponderance of any one pair, the result is a one-to-one ratio of purines to pyrimidines and of keto to amino bases. The two purine-pyrimidine pairs are of equal size and permit a uniform spacing between the two backbones of the molecule. By contrast, purine-purine or pyrimidine-pyrimidine pairs have very unequal sizes, the first being larger, and would distort the molecule. The pairs are determined by H bonds between certain atom groups in each base: guanine and cytosine share three H bonds, while adenine and thymine share two. These specific bonds constitute the basis of complementarity. [H bonds are highly directional bonds between molecules with a polarity of charge (Watson, 1970). They are weak bonds (3-7 kcal/mole) that require no enzyme to make or break them; hence these bonds are suitable for interaction between molecules. For comparison, covalent bonds are strong (50-110 kcal/mole) and in biological systems their formation is catalyzed by enzymes; these intramolecular bonds have a structural function.]

However, the type of pairing just described is not unequivocally established over that put forward by Hoogsteen (1963). The two types of pairing are compared in Fig. 4. Regardless of which pairing is involved, each turn of the helix (or pitch) contains ten base pairs. Groups of H bonds occur regularly every 3.4 Å which is the distance between nucleotides along the axis of the helix. Each turn therefore spans 34 Å (Fig. 3) as has now been directly observed by electron microscopy (Ottensmeyer, 1969). An average bacterial DNA molecule of 10^9 daltons contains about 4×10^6 H bonds. These bonds are strongly responsible for the replication and transfer of information from the helix. Also, their collective sum constitutes a force that holds the helix together. However, for this an energetically more important part is played by the water repulsion of the internal vertically stacked bases, which are hydrophobic compared with the exposed hydrophilic backbone. This nonpolar stacking of bases creates mutually attracting van der Waals (or London) forces and electrostatic forces (Pullman and Pullman, 1969) which stabilize the helix. The greater stabilization is by stacking and not by H bonds because, thermodynamically speaking, H bonding between bases merely replaces H bonding between bases and water without a net gain in free energy. By partly contraarresting the hydrophilia of the backbone, divalent cations also contribute in stabilizing the helix. A similar effect may result from association with polyamines.

DNA in solution is therefore a rigid and thermodynamically stable molecule that never breaks apart, except reversibly for a few terminal nucleotides. It was assumed that at any time a few pairs along the molecule are transiently opened and the molecule was said to "breathe" (Printz and von Hippel, 1965). At present it is favored that under physiological conditions the molecule may only be transiently distorted rather than opened (Vinograd, 1970), which still reveals the dynamic nature of the H bond. However, in the eukaryote even extensive open regions are a possibility. In any event, for total breakage of the molecule to occur extreme conditions of pH, temperature, ionic strength, etc., are required. These are rarely, if ever, met under physiological conditions. When breaking into two randomly coiled strands occurs it is in a highly coordinated or cooperative manner attesting to the regular H bonding, and the molecule is said to "melt." The higher the proportion of guanosine-cytosine pairs, the more stable the molecule. Somewhat surprisingly, DNA is very susceptible to mechanical shearing which involves the rupture of phosphodiester bonds.

DNA molecules can be very long. The single molecule in the *Escherichia coli* chromosome is 2.8×10^9 daltons, or 4×10^6 nucleotide pairs, or 1.1 mm extended length. If individual human chromosomes consisted of a single molecule, for which the evidence is not compelling, this could weigh as much as 10^{11} daltons for the largest chromosome (DuPraw, 1968). These dimensions make DNA longer by several orders of magnitude than any other biological molecule, including homopolymers such as glycogen (5×10^6 daltons). The density of DNA (about 1.8 g/cm^3) is higher than that of protein (1.4 g/cm^3).

Structure of RNA

With few if any exceptions, RNA molecules are single linear chains with extensive helical segments present in varying degrees in all species. These helical regions adopt *in vivo* the structure corresponding to the crystalline A form of DNA with the bases tilted at 20 degrees. Because the Watson-Crick pairing is irregular and incomplete, the structure can be more elaborate spatially than the relatively uniform DNA. Some rare RNA viruses and replicative forms of RNA viruses, in general, are regular helices with 11 base pairs per turn. Studies with synthetic helices indicate a transition to 12 base pairs per turn at higher ionic strength, which may be functionally significant (Arnott *et al.*, 1968). The data also show that the helices can adopt several conformations all similar to DNA-A, the 2'-hydroxyl group possibly being responsible for this constancy.

In certain nonhelical regions of the RNA molecule, as for example in rRNA, the backbone can coil on itself permitting bases a few nucleotides apart to approach one another and form stacks held together by van der Waals forces. These forces stabilize the local secondary structure of the coil, which is further stabilized by salt bridges between phosphate groups. Also, compared to DNA, there are extra H bonds possible with the 2'-hydroxyl group. In the particular case of nonhelical phage RNA within the virion, this has a compact secondary structure due as much, or more, to bonding with capsid protein as to intramolecular H bonding.

Thus in nucleic acids in general, as in proteins, the integrity of macromolecular conformation depends both on internal binding forces and environmental factors. The analogy is very striking in the case of tRNA with a small size conducive to a distinctive conformation with elements of both secondary and tertiary structure (Adams et al., 1967). Generally however, because of the strong environmental dependence, RNA molecules free in solution may be deemed not to possess a unique structure.

Compared with DNA, RNA molecules free in solution are much more labile and less cooperative in structure as seen during melting, and therefore are prone to spontaneous degradation or hydrolysis. On the other hand, it is becoming increasingly evident that much of the RNA in the cell is associated noncovalently with protein or protein-containing structures. All RNAs, except for tRNA, are in this way more extensively proteinated in the eukaryote than the prokaryote. For this reason the molecules may adopt a tertiary structure partly imposed and protected by protein. It could well be that most RNA in the cell possesses an intrinsic or imposed higher order structure, in either case, a stable native structure. Moreover, because of proteination, it could well be that there are no exclusive RNA-RNA interactions in the cell, in the manner that there seem to be protein-protein interactions. In general, RNA is shorter and denser than DNA.

MECHANISMS OF NUCLEIC ACID SYNTHESIS

RNA resembles DNA in the manner of assembly, and for this reason it is advantageous to look first at DNA.

Replication of DNA

Replication of Prokaryotic DNA

It is a current assumption that during DNA replication the H bonds are broken and the two strands of the helix pull apart (see Ingram, 1966; Watson, 1970). This denaturation, and perhaps the unwinding of the helix, may be assisted by specific proteins (Alberts and Frey, 1970). In the cell, as the strands become free they are reproduced by a polymerase simultaneously and in parallel into complementary replicas, i.e., replication is symmetric. Each nascent strand remains attached to its template forming a helix that is now half old and half new, in other words, barring crossover the single strand is the conserved unit. This is the semiconservative type of replication most likely to apply to replication *in vivo* (Meselson and Stahl, 1958; Taylor et al., 1957). In addition, evidence in bacteria (Yudelevich et al., 1968) indicates that replication is by short segments (known as "Okazaki fragments" after their discoverer), and that later these segments are linked by a ligase as discussed below. However, given the uncertainty also discussed below concerning which DNA polymerase is functional in the cell, the mechanism of replication cannot be considered as established.

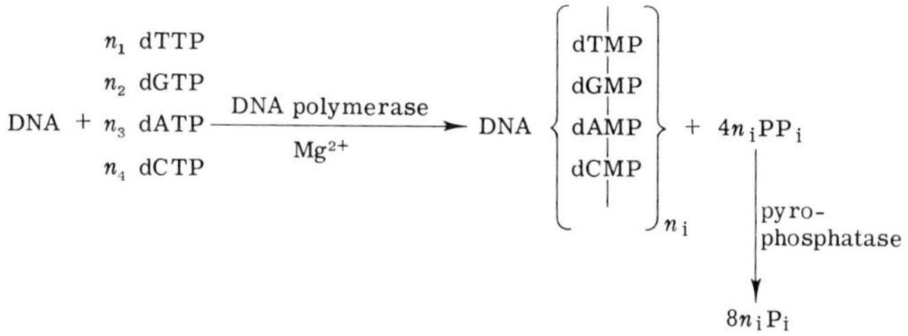

Nucleic Acid Synthesis

$$\text{DNA} + \begin{matrix} n_1 \text{ dTTP} \\ n_2 \text{ dGTP} \\ n_3 \text{ dATP} \\ n_4 \text{ dCTP} \end{matrix} \xrightarrow[\text{Mg}^{2+}]{\text{DNA polymerase}} \text{DNA} \left\{ \begin{matrix} \text{dTMP} \\ \text{dGMP} \\ \text{dAMP} \\ \text{dCMP} \end{matrix} \right\}_{n_i} + 4n_i \text{PP}_i \xrightarrow{\text{pyro-phosphatase}} 8n_i \text{P}_i$$

Chart 1

Despite this uncertainty, it is possible to infer very generally the behavior of the polymerase from *in vitro* studies with the Kornberg enzyme obtained from *E. coli*, which, as we shall see, is probably not the functional polymerase. Based on this inference, the polymerase moves along the parental helix causing the stepwise addition to the nascent chain of 5'-nucleotides derived from triphosphate nucleosides, believed to be the precursors which are combined with Mg^{2+} in the cell pool (Chart 1). The direction of chain growth is a function of the chemical polarity in the precursors. The polymerization reaction occurs by a nucleophilic attack by the terminal 3'-OH in the chain on the innermost α-phosphate of the incoming 5'-triphosphate nucleoside (Kornberg, 1969). The chemical direction of growth is therefore 5' to 3'. By itself the polymerization reaction is thermodynamically unfavorable (about -0.5 kcal/mole), but the energy liberated by the pyrophosphate split (-7 kcal/mole) drives the overall reaction toward synthesis. Since pyrophosphatase is always present in the cell, the synthesis is, in effect, irreversible. In bacteria, under favorable conditions, the rate of replication *in vivo* at 37°C is 1.4×10^3 nucleotides, or about 0.5 μ of DNA strand per second.

In the cell the polymerase displays the unusual ability to recognize simultaneously two single template strands with opposite polarity and to construct two replicas also with opposite polarity. For this reason the polymerase was considered to be an aggregate of two oppositely oriented monomers (Cavalieri and Carroll, 1968). On the other hand, the Kornberg enzyme isolated from *E. coli* (6 S; 10^5 daltons) behaves as if composed of either a single unit or a few subunits (Kornberg, 1969).

The Enzyme Involved. We begin by considering the specific mechanism of action proposed for the Kornberg enzyme. The active center of this enzyme has binding sites for (a) the template and (b) primer strands, (c) 3' and (d) 5' primer strand termini, and (e) triphosphate nucleosides (Kornberg, 1969). (Primer strands are those extended by the nascent strands in a covalent linkage.) *In vitro* the enzyme works in the 5' to 3' direction with respect to the nascent strand, hence 3' to 5' with respect to the template. A scheme was formulated as illustrated in Fig. 5 whereby the enzyme copies by complementarity both native strands in the 5' to 3' direction almost simultaneously and it does this by alternating from one strand

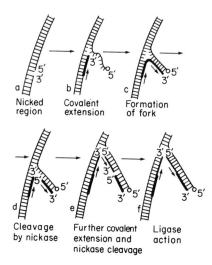

Fig. 5. Speculative model by Kornberg (1969) for unidirectional replication of a DNA helix. [Copyright (1969) by the American Association for the Advancement of Science.]

to the other at the nucleotide level. After an obligatory initial nick in the primer strand by endonuclease (another enzyme also known as nickase), the enzyme begins to elongate this strand in the 5' to 3' direction as it copies the template strand. The enzyme then switches to copy in the same direction the displaced 5' terminus of the primer strand (now acting as template) that lies ahead of the break. The nascent strand is then nicked, and the enzyme goes on again to copy the original template strand. Repetition of the cycle results in two complementary replicas: a segmented replica of the primer strand which is then sealed by ligase (still another enzyme), and a continuous replica of the template strand.

Some time ago, it was possible with these enzymes to obtain *in vitro* synthesis of biologically active φX 174 phage DNA (Goulian *et al.*, 1967). Despite earlier concern with the slow and imperfect replication of native templates (discussed by Kuempel, 1970), taken together the contributions by the Kornberg group would have seemed to dispel some of the doubt that existed in regard to the isolated enzyme being the biologically functional polymerase. Yet the possibility remained that the Kornberg enzyme could have primarily a DNA repair function (Howard-Flanders, 1968) which, incidentally, is essential to make DNA dependable as a genetic store. For a brief time it appeared as though, in fact, the versatile enzyme could be both the polymerase and repair enzyme (Kelly *et al.*, 1969). In this latter context it was shown that the enzyme could subterminally excise short runs of DNA by acting as an endonuclease in the 5' to 3' direction and then proceed to repair the T dimers produced by ultraviolet radiation. The repair sequence envisaged in the reaction *in vitro* was: (a) a nickase loosens a DNA strand at the 5' end of a faulty run; (b) the enzyme reforms the faultless strand in the 5' to 3' direction, and then (c) excises the faulty run at the 3' end; (d) the strand is sealed by ligase completing the repair. Soon after, however, it was reported that bacterial mutants with only 1% of the amount of Kornberg enzyme were able to replicate normally (de Lucia and Cairns, 1969). Hence, reopening the question, the implication of this finding was that the enzyme would, after all, be a repair editing enzyme rather than the anabolic one.

Recent work on membrane fractions of the bacterial cell has revealed an enzyme which may be distinct from the Kornberg enzyme (Knippers and Strätling, 1970). This enzyme appears to replicate DNA faithfully in the 5' to 3' direction which on present evidence is the direction *in vivo* and does so approximately at the same rate as *in vivo*, although so far for a short time. Initiation of replication is believed to depend on a break in the strand. Replication then proceeds by short fragments which later join one another; this may be true of both nascent strands (Okazaki *et al.*, 1970). The enzyme thus appears to alternate between strands as described above for the Kornberg enzyme. The precursors seem to be triphosphate nucleosides (see, however, Werner, 1971). However, the fine mechanism of replication by this enzyme remains as yet to be elucidated. A more promising but less explored replication system makes use of bacterial cells treated with toluene, whose replication fork continues to function after such treatment (Matsushita *et al.*, 1971). Hence these two approaches rekindle the hope that the functional polymerase will eventually be found. It is likely, however, that the polymerase forms part of an obligatory replication complex together with other proteins with different functions in replication, and all are associated with the cell membrane (Lark, 1969). For the purpose of analysis, the next step is to extricate the polymerase from this complex.

This discussion of DNA polymerase has gone beyond that necessary to help understand the mechanism of RNA polymerase which is our main interest, the reason being that the problem of replication remains one of outstanding biological interest.

The DNA Molecule and Its Replication. Bacterial chromosomes consist of one circular helical DNA molecule. In viral chromosomes the molecule is present as either a single strand or a helix. All single strands so far studied are covalently closed circles. The helices are either (1) linear with unique or circularly permuted sequences, and sometimes terminal redundancy; or (2) covalently closed supercoiled circles (reviews by Kleinschmidt, 1967; Thomas, 1963; Thomas and MacHattie, 1967). The first alternative means that circular genetic maps as in the case of some phages need not correspond with circular DNA.

Single-stranded DNA of a few bacterial viruses assumes a circular helical configuration as an intermediate replicative form. In this case a (minus) strand is made after infection that is complementary to the original (plus) strand. An unknown mechanism of selection determines that only plus strands are incorporated into mature progeny particles. It is clear, however, why in most organisms DNA must be double-stranded. If DNA were normally one-stranded and only double-stranded briefly during replication, then no two sister cells could receive the same genetic instructions from their parent. The replicative form of single-stranded viruses precisely avoids this eventuality.

A continuous prokaryotic genome is considered to be the basic unit of replication or "replicon." To explain unitarily the control of progressive replication from a fixed point as observed in bacteria and viruses, Jacob et al. (1963) proposed that the replicon would comprise: (1) a structural gene for the synthesis of a diffusible "initiator," and (2) a "replicator" that is acted upon by the initiator and causes DNA to replicate as the replicator moves as a single point in one direction along the chromosome. The initiator is not the polymerase itself but a species-specific molecule of unknown nature, which, at least formally, behaves as a specific nickase or an initiation factor. The replicator function is still undefined. In addition, a negative control by a replicational repressor has been proposed (Rosenberg et al., 1969). The general format of the replicon reminds one of the transcriptional unit or operon to be discussed later, except that, contrary to the operon, the replicon as originally proposed relies on positive regulation.

In bacteria the unidirectional replication of the circular replicon usually proceeds clockwise and lasts the entire cell cycle. On the other hand, the replication of certain phages whose DNA can circularize, such as λ phage, is bidirectional from a single initiation point. Returning to the bacterial replicon, this is attached to at least one point on the cell membrane, and threads through that point as replication proceeds. To visualize this dynamic configuration, imagine a wheel in motion but always in contact with the ground at some point. A model of bacterial replication was proposed by Cairns (1963) as shown in Fig. 6a, which incorporates the above-mentioned concepts of Jacob et al. (1963). According to this model, replication takes place at a fork which, under conditions of rapid growth, can become a double fork. A conjectural molecular structure or "swivel" would allow unwinding or unscrewing of the circular DNA at the terminus.

More recently, a second model was proposed by Gilbert and Dressler (1968) that tentatively intends to combine what is basically known of bacterial and viral replication while respecting the unidirectional replication from a fixed point proposed in the Cairns model. This is the "rolling circle" model (Fig. 6b) in which the plus strand, initially part of a circular helix with the minus strand, opens and becomes attached (to the cell membrane?) by its $5'$ terminus; the minus strand remains a circular template. As the plus strand peels off from its $5'$ terminus,

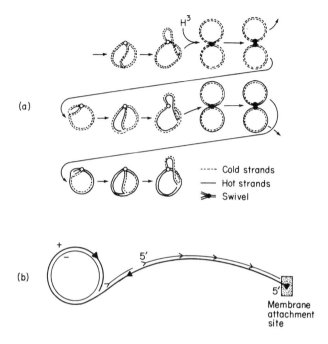

Fig. 6. Two diagrammatic models of prokaryotic chromosome replication. Model (a) represents the replication of a circular chromosome based on the assumption that each round of replication begins at the same place and proceeds in the same direction (Cairns, 1963). The model requires a swivel. The "rolling circle" model (b) represents the continuous synthesis of an (open) plus strand on a (closed) minus strand template (Gilbert and Dressler, 1968). Progeny minus strands are synthesized in short pieces using the peeling off plus strands as template. The arrows represent the direction of chain growth. The continuous plus strands are then freed by nickase, i.e., at the point of the arrow tails on the plus strand in this diagram. Further explanation of the models is given in the text.

progeny plus strands grow by addition at the open 3′ terminus using the circular minus strand as template. The parental and several successive progeny plus strands remain covalently attached end to end, resulting in a piece of DNA which is larger than the unit genome, but eventually they become free to enter mature particles. While they remain continuous, the plus strands serve as templates for the synthesis of minus strands in short Okazaki-like segments later joined together by ligase. The asymmetry of this structure obviates the need for a swivel as in the Cairns model. Time will decide on the applicability of the two models and whether or not they are compatible with each other. It could be that the models represent two different levels in a gradual approximation to the actual characteristics of prokaryotic replication, or that prokaryotic replication demands more than one model. To complicate matters, recent data have led to questioning the notion of unidirectional replication from a fixed point in *E. coli* (review by Goulian, 1971; Masters and Broda, 1971). According to these data, replication may be bidirectional as occurs in certain phages, and even may proceed at a different rate in each direction. If that is the case, both the enzymology and the molecular model of bacterial replication remain as yet to be established.

Lark (1969) has reviewed the conditions for initiation and regulation of bacterial DNA replication. The rate of elongation remains invariant and thus regulation is achieved by varying the rate of initiation at the chromosome origin and at the Okazaki segments. Concerning the various proteins involved in the replication complex, they all intervene in initiation and possibly become modified in the process, after which they are no longer exchangeable. Finally, the complex appears to be self-destroyed during chromosome segregation or cell division. This complicated machinery must be viewed realistically in conjunction with the other elements of the replicon (e.g., some of the proteins in the complex may correspond with the replicator) and also with the integration of these elements in the cell membrane. Lederberg (1966) has remarked that "organisms that have a generation time of 30 minutes cannot afford to be extravagant with DNA synthesis." They may not be extravagant, but they certainly do not lack sophistication.

Replication of Eukaryotic DNA

In higher organisms DNA replication takes place during interphase of the cell cycle. Each of the two helices resulting from one homolog chromosome is handed down to a different daughter cell. At mitosis the two homologs are partitioned between two daughter (or sister) cells, at meiosis between four. Thus in mitosis each daughter cell receives one helix from each of the two homologs, in meiosis from only one homolog. For simplicity, it was assumed here that each homolog contains only one DNA molecule. The segregation of daughter helices as described is not affected if there is more than one molecule because the segregating unit is always the chromatid. Crossing-over was disregarded.

In higher organisms there are many initiation sites per chromosome (Howard and Plaut, 1968), each chromosome then representing a multireplicon. There is a temporal replication pattern in the replicons in each chromosome, and each part of the chromosome retains the pattern after translocation. Moreover, there is an overall coordination of replication in the genome as a whole. This condition suggests a delicate regulation of initiation perhaps involving special sequences in the DNA, factors, etc., of which nothing is known. These data (from polytene chromosomes) raise the question as to whether or not each replicon corresponds with one chromosomal band (Pelling, 1969), that is, close to what may represent one functional genetic unit. Also in the meiotic lampbrush chromosome a correspondence is possible between the replicon and the indivi-

Fig. 7. A diagram of eukaryotic chromosome replication according to Taylor *et al.* (1970). The DNA from mammalian chromosomes can be dissociated into segments, about 110 μ long, by removing the proteins enzymically and raising the pH to approximately 10 in low salt concentration. These long segments have about four nicks per chain so that single chains occupying 20 to 22 μ in the native structure are released from most nuclei in actively growing cells. A variable fraction of the DNA may exist with nicks about 2 μ apart, but alternating so that the native structure is stable as large units up to about 55°C in 0.01 *M* phosphate.

The 2 μ subunits appear to be the units of replication. The template is nicked before or during replication but is closed following replication. Evidence from density labeling experiments indicates that a number of 2 μ subunits may be initiated simultaneously, probably the ten which would form the 20 to 22 μ segments. The figure shows one 2 μ subunit with initiation occurring at each end by a self-priming region (hairpin fold). There is no indication as yet as to which of the two strands is copied, the one to which the first segment appears to be covalently linked or its complementary strand. The DNA is made in segments about 0.4 μ long, numbered 0, 1, 2, 3, and 4 in the diagram, and joined as a cluster rather than sequentially. The 0 segments are assumed to be repeating sequences or sequences which have considerable similarity in large portions of the DNA complement. The mechanism of initiation of segments 1, 2, 3, and 4 is unknown, but a small oligomer nine to twelve nucleotides in length (33% AT) has been isolated from replicating nuclei and it may provide a starting sequence for the DNA polymerase at the beginning of each 0.4 μ segment with the exception of the 0 segment. The covalent link between the new 0 segment and the template must be clipped after replication. The filling of the gap in the new strands is a slow process which requires about 20 minutes to complete new 20 to 22 μ chains, but the joining of the short 0.4 μ subunits requires only a few seconds following their synthesis. The lines linking the two chains represent some kind of crosslink which appears to be present only during replication; the nature of and even the evidence for these crosslinks is still uncertain.

dual chromomere (Callan, 1970). More generally, it could be that the size of the replicon (or its higher order structure) varies according to the different amount of nuclear DNA which is characteristic of each organism.

Replication in adjacent replicons can proceed in opposite directions (Huberman, 1969; but see Lark *et al.,* 1971). A recent working model of replication proposed by Taylor *et al.* (1970) is presented in Fig. 7. This model indicates a general mechanism of replication by short segments fusing into progressively larger segments which is basically similar to the proposed bacterial mechanism by the Kornberg enzyme (Fig. 5), but considerably more intricate in detail. For example, it is not clear from this model whether the same polymerase molecule can alternate between the two DNA strands and replicate both of them in the same physical direction, as seems to be the case in the bacterium. In addition, as proposed on different grounds (Comings, 1968), it is possible that each chromosome is attached to the inner nuclear envelope, perhaps at one site per replicon.

This mode of eukaryotic DNA replication raises two extreme possibilities, namely, that each chromosome contains more than one DNA molecule, or that each chromosome contains only one DNA molecule but replication starts at many points. The last alternative would require nonterminal separation of the helix. There is no conclusive evidence to decide between these two possibilities. For example, recent work indicates that while the molecule behaves as if physically continuous, it is composed of structural subunits (each including several replicons) linked together by alkali-labile bonds which, according to Lett *et al.* (1970), may or may not be DNA. These subunits may correspond with those bounded by cross-links in the previous model (Fig. 7). It has been known for some years that the structural integrity of the amphibian lampbrush chromosome is sensitive to DNase but not to proteases (Callan, 1963), and the same is true of the circular DNP in oocyte nucleoli (Miller and Beatty, 1969b). Irrespective of the number of DNA molecules in a chromosome, it is clear that each chromosome contains many replicons.

Clearly, the final organization of DNA in these chromosomes that bears on their mode of replication still remains undecided (review by DuPraw, 1968). In contrast with bacterial DNA, eukaryotic DNA is highly supercoiled and associated with protein as well as other molecules. Thus interaction with chromosomal protein has to be considered, e.g., histone appears to participate in regulating replication (Ackermann *et al.,* 1965; Wood *et al.,* 1968). However, it is not known what happens to the proteins during DNA replication, but, since the whole chromosome is replicated, obviously their intimate association with the DNA must be affected.

The rate of eukaryotic replication is about 30 times slower than in bacteria, or 1 μ of DNA strand per minute. The same rate is found in somatic mammalian cells with widely different duration of replication (Painter and Schaefer, 1969). Hence the duration of replication is determined by the number of replicons per unit time committed to replication, that is, the overall rate of initiation. This number varies from 20 to 60% out of a total of 10^4-10^5 replicons per cell, corresponding to about 2 μ for the size of each replicon (Fig. 7). Incidentally, were it not for this number of replicons, replication by a single replicon per (human) chromosome would require 20 days instead of an actual few hours. In contrast to somatic cells, the rate of replication in (amphibian) meiotic cells is 0.2 μ of DNA strand per minute (Callan, 1970). Correspondingly, in these meiotic cells the replicative S period lasts 8 days (at 25°C) instead of 1 day for the somatic S period, suggesting a general difference in the rate of initiation.

The amount of DNA per haploid human nucleus is about 5×10^9 base pairs, or 87 cm long, divided among 23 chromosomes. Compared with the average nuclear diameter (5 μ), this length gives an idea of the degree of compaction of the DNA. In any event, the overall regulation of replication in a multichromosomal genome of this magnitude will obviously need to be more elaborate than in the single bacterial replicon.

Transcription of DNA

Synthesis of RNA from DNA or transcription is mediated by a DNA-dependent RNA polymerase (Bautz, 1967; Hurwitz and August, 1963) that uses triphosphate ribonucleosides as precursors (Chart 2). In its general characteristics such as use of H bonding to a template strand and the mechanism of chain propagation, the reaction resembles that with DNA polymerase. Two particular characteristics are that the reaction is asymmetric and Mn^{2+} can replace Mg^{2+}. In the eukaryote there is more than one type of polymerase, each responsible for the synthesis of a different RNA species; the polymerases for rRNA and dRNA require preferentially Mg and Mn, respectively. The strands of the DNA helix are believed to pull apart and only one strand is transcribed at any region, which is what makes transcription an asymmetric synthesis. The direction is from 3' to 5' in the DNA, the RNA being made from 5' to 3' in antiparallel direction. The molecular mechanism of action of the RNA polymerase is not known, although the DNA polymerase when considered to be replicating only one template strand gives some idea of what this can be. Thus the mechanism of RNA polymerase is probably more straightforward than that of DNA polymerase because it is not concerned with a symmetric synthesis. The template seldom forms a lasting part of the new molecule since a DNA-RNA helix is less favored, perhaps because of the steric configuration; artificial hybrids adopt the A con-

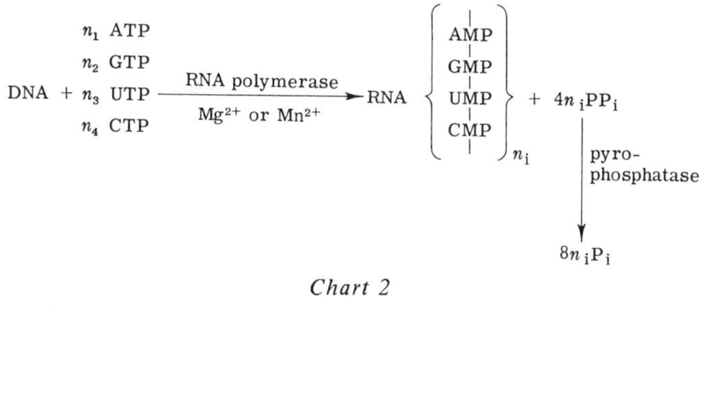

$$DNA + \begin{matrix} n_1 \text{ ATP} \\ n_2 \text{ GTP} \\ n_3 \text{ UTP} \\ n_4 \text{ CTP} \end{matrix} \xrightarrow[Mg^{2+} \text{ or } Mn^{2+}]{\text{RNA polymerase}} RNA \left\{ \begin{matrix} AMP \\ GMP \\ UMP \\ CMP \end{matrix} \right\}_{n_i} + 4n_i PP_i \xrightarrow{\text{pyrophosphatase}} 8n_i P_i$$

Chart 2

figuration that is preferred by helical RNA *in vivo* but not by DNA (L. D. Hamilton, 1968). It is thus possible that the configuration of the natural hybrid helps drive the transcript off the template, although in this process the polymerase or factors may also play a part. Hence transcription does not permanently affect the template, which, in contrast to replication, behaves conservatively. Another difference of transcription is that no breaks in the DNA are required, because RNA forms de novo and not as a primed molecule.

In bacteria the overall rate of transcription is on average about 30 times slower than for replication, or about 50 nucleotides per second at 37°C, yet the maximum rate for some RNA species may be much higher than this. The rate of eukaryotic transcription is within the same range. However, in bacteria the step time for addition of each nucleotide varies according to the preceding 3' terminal nucleotide in the chain (in msec): UMP 25, CMP 40, and GMP 12 (Manor *et al.,* 1969). The shorter time for the purine may have to do with its greater stacking ability. Thus the nature of the template could influence to some extent the rate of transcription.

In vitro, using double-stranded DNA as the template, both strands can be transcribed, which must be artifactual. The transcription product in this case reproduces the overall DNA composition (with U instead of T), and the individual chains have opposite polarities and complementary sequences potentially enabling them to form a helix. Single-stranded RNA is also transcribed *in vitro.* In general, however, enzyme behavior *in vitro* depends somewhat on single- versus double-strandedness of template. On the other hand, *in vivo* as well as *in vitro* under carefully controlled conditions, only one DNA strand is transcribed, as shown by the RNA reproducing the single-strand DNA composition. In some cases, after *in vitro* synthesis viral RNA was demonstrated to be biologically active, which is indicative of single-strandedness. In the case of single-stranded viral DNA, transcription is also asymmetric: it takes place from the minus strand neglecting the plus strand whose sequences reappear in the transcript.

Asymmetric transcription makes genetic sense in the case of helical DNA since the sequential information in one strand is not the same as in the other. As a result of complementarity the relevant sequence is that in the nontranscribed strand. It is becoming clear how a particular strand of DNA is selected for transcription, and for this the intact configuration of the template-enzyme complex in the cell is obviously important. The selection might occur at the time when the strands are believed to separate and by a mechanism related to selecting the transcription of the genome. In bacteria there is good evidence that initiation factors and certain special sequences are involved. In addition, a function of the noncoding strand could be to assist the polymerase structurally. Structural aspects of transcription are discussed in more detail in Chapter 3.

It is not clear what determines that a single strand of DNA will at one time be copied as DNA and at another time as RNA, but in bacteria initiation fac-

Nucleic Acid Synthesis

tors are again possibly involved. It is believed that this duality of DNA is made possible because it can adopt the A form typical of RNA, with which the RNA polymerase is compatible, whereas in the B form typical of DNA the DNA polymerase is compatible, and that DNA is capable of the two forms because it lacks the 2'-hydroxyl group in deoxyribose (Arnott *et al.*, 1968). Because of the lack of preference for the A form by DNA, the transition to this form could involve only a local unwinding of the helix which moves along with the point of transcription. This is most likely behavior of the helix during transcription, even though it is not definitely established.

Replication of RNA

On present evidence cell RNA molecules do not serve as templates for synthesis of their own kind. However, in the case of RNA viruses, which do not contain any significant amount of DNA, the RNA itself must be replicated and a replicase specific for the viral RNA is required. Thus, viral RNA acts as its own template, bypassing DNA as the genetic material (Erikson, 1968; Stavis and August, 1970). (See, however, footnote 2 in regard to tumor RNA viruses.) This RNA replication obeys the same strict base pairing rules for DNA replication and transcription. Yet by some still conjectural means, obviously related to the structure of the viral RNA, the viral replicase recognizes its own RNA from the cell RNA or other viral RNA that it does not reproduce. This specificity requires the qualification that three out of four moieties of the replicase are provided by the host cell (Kondo *et al.*, 1970); previous evidence had suggested that the cell contributes an initiation factor (August, 1969). In addition, the role of a 6 S RNA formed during RNA replication is not clear (Banerjee *et al.*, 1969). Peculiar modes of replication may also appear as more RNA viruses are examined.

Viral RNA is the first macromolecule whose correct self-replication was observed *in vitro*. It is now possible to follow its replication ad infinitum and study the effects of environmental pressure, e.g., leading eventually to preservation of only the sequence responsible for replication. It was found that a selected fragment equal to 17% of the genome replicates 15 times faster. Hence the replicase recognition and replication mechanism(s) do not require the entire sequence (Spiegelman, 1969). The possibilities that this *in vitro* approach opens, in general, for the study of gene function need no mention.

In a minority of viruses whose mature particles contain double-stranded RNA, at each point of the helix presumably only one of the strands is used as template for both replication and informational synthesis. A replicase for early mRNA synthesis is possibly carried in the virion (Shatkin and Sipe, 1968). This asymmetry as template of double-stranded RNA is to be expected from what is known of the vast majority of mature RNA viruses which contain single-stranded RNA. In these latter viruses the parental (plus) strand acts itself as the early viral mRNA because soon after infection it directs the synthesis of replicase using the host synthetic machinery, and is not transcribed into other mRNAs. If this strand were transcribed it would not be genetically consistent with the fact that it acts as messenger. (In certain viruses only the transcript acts as the messenger, however.) Instead a complementary (minus) strand is made that forms a RNA helix with the plus strand, analogous to the replicative form of single-stranded DNA viruses. From then on, a grossly oversimplified view is that the replicative form becomes a multistranded structure through generating a few plus strands on the minus template by means of a second replicase (Lodish, 1968a). These plus strands are late messengers or mature RNA of progeny virions into which parental RNA never enters, again as in the DNA viruses. Both plus and minus strands are made in the usual $5'$ to $3'$ direction, but in this case there may be no real distinction between a conservative and a semiconservative replication (but see Lodish, 1968b). In summary, irrespective of whether the viral RNA is single- or double-stranded, transcription is always from a single strand. The situation is entirely similar to DNA transcription, and selection of template presumably is by a similar mechanism.

GENERAL CONSIDERATIONS ON THE SYNTHESIS OF NUCLEIC ACIDS

Nucleic acid synthesis well illustrates a general principle in the formation of macromolecules with discretionary monomer sequences. This principle has to do with replicas that are oppositely shaped or complementary rather than identical to the templates used. The physical basis is that complementary but not identical molecules recognize each other accurately by means of interactions such as primarily H bonding that are individually weak but collectively strong and highly selective (Pauling and Delbrück, 1950). The principle also applies to protein synthesis indirectly (because of the adaptor tRNA), but not, for example, to polysaccharide synthesis where monotonous sequences and no templates are involved.

The peculiarities of replication and transcription demand a very special enzyme behavior. During both syntheses the new strand grows in the $5'$ to $3'$ direction, attesting to a common ancestry of mechanism. Were it not for the overwhelming interest in their products, these enzymes would have more than earned for themselves the attention of the enzymologist. They show the following properties uncommon to enzymes in general: (1) they act on and move along a template that is larger than themselves; (2) each enzyme molecule handles multiple rather than single substrate monomers and handles them sequentially; and (3) the final product is as complex as the enzyme. In effect, the product specificity has been made as dependent on the template as on the enzyme itself, which means that the enzyme must look at the template only in a general way. This behavior can only be understood if it is realistically viewed in the context of the ultimate configuration of the templates in the cell. Moreover, all elements involved in replication and transcription (but not translation), i.e., precursors, energy and information, are jointly fed into the anabolic pathway (von Ehrenstein, 1970), which could at least account for a crude coordination of the enzyme-template behavior. Also in relation to this question, the enzymes for replication and transcription (as well as translation) are assisted in selecting the template by protein factors.

It is easy to see that much more replicational and transcriptional error would occur if the various precursor nucleotides were separately handled by as many enzymes. (Kinases, by maintaining the level of precursors, are co-responsible for the fidelity of replication and transcription.) The average mistake level of introducing the wrong nucleotide during replication is considered to be as low as 10^{-9}, and there is no reason to suppose that the transcriptional error, which is more difficult to estimate, is higher. Such specificity would seem unobtainable from the energy inherent in the H bonds of single base pairs, unless the replicating enzyme somewhat assisted the template in selecting the base, which possibly is by allosteric interaction (Bautz, 1967; Lipmann, 1963). Two lines of evidence implicating the replicating enzyme directly are: (1) the mutant polymerases that influence the base sequence of the product strand (Speyer, 1965); and (2) the selection by polymerase of the correct tautomer of the base (*Cold Spring Harbor Symp. Quant. Biol.*, 1968). Conversely, there is evidence that the complementary strand makes the selection of the base more accurate (Ishihama and Hurwitz, 1969). Lastly, if the enzyme stabilizes the helix, stereochemical fit of the incoming precursor nucleotide with possible allosteric effects on the enzyme could be as important as H bonding of the nucleotide to the template strand. This brings up a fundamental question in the field of nucleic acids which is discussed by Yarus (1969): How does a (enzyme) protein recognize a nucleotide sequence?

THE GENETIC CODE

The essence of the genetic code is that the information for protein—to which it exclusively refers—is carried in units of base sequences. Genetic information is thus encoded in the internal ordering of the DNA helix which is virtually without effect on the structural or metabolic properties of the helix. It is this internal ordering of the helix which varies between organisms according to their genetic differences. The first general evidence to appear in respect of unit coding sequences was the chemically induced mutations seemingly due to a single base pair substitution (Freese, 1963) and the mutations involving changes in a minute portion of a gene resulting in a single amino acid substitution (Helinski and Yanofsky, 1962). The discussion to follow deals with the more specific evidence on unit coding sequences that has accrued in later years.

Using a linguistic analogy it is usual to refer to the bases in DNA as a four-letter language, to the universal amino acids as a twenty-letter language, and to the genetic code as the cipher or dictionary that connects both (Crick, 1967).[6] The first attempts at deciphering the code were based mainly on abstractions and developed independently of any real knowledge of the details of protein synthesis (reviewed by Crick, 1966a, 1967; Sadgopal, 1968; Speyer, 1967; Woese, 1967a,b; Ycas, 1969). Whatever ideas on the code there were to begin with, they sharpened during these exercises to the point of meaningful propositions. It was, however, not before 1966, by which time the experimental background was solid, that the genetic code became practically known (*Cold Spring Harbor Symp. Quant. Biol.*, 1966).

The *in vitro* systems used to decipher the code were bacterial cell-free systems capable of polypeptide synthesis under the direction of artificial messengers of known composition. The exception is one system [see (2) below] that detects the specific binding of aminoacyl-tRNA to ribosomes directed by triplets. In order of appearance the messengers were: (1) oligoribonucleotides of random sequence, except for homopolymers, prepared with polynucleotide phosphorylase (Nirenberg and Matthei, 1961); (2) trinucleotide diphosphates or triplets of fixed sequence (Nirenberg and Leder, 1964); and (3) oligoribonucleotides of defined, repeating sequence prepared by a combination of chemical and enzymic syntheses (Nishimura *et al.*, 1965; Khorana, 1968). With a few exceptions owing to technical difficulties, the assignments obtained with any one of the above messengers confirmed the assignments obtained with the others, gradually making certain that most were correct. Only points (2) and (3) indicated order as well as composition of the coding sequences. Observation on amino acid replacements in intact mutant cells further established certain connections between coding sequences. The mutations were in human hemoglobins (reviewed by Lehmann and Huntsman, 1966), TMV virus protein (Wittmann and Wittmann-Liebold, 1966), *E. coli* tryptophan synthetase (Yanofsky *et al.*, 1966), and phage lysozyme frameshift mutants (Terzaghi *et al.*, 1966). At present it is possible to test coding sequences directly using purified tRNAs.

---

[6] A more rigorous terminology is proposed by von Ehrenstein (1970), who distinguishes between dictionary and code. The code would pertain to the alignment of anticodons by mRNA, as it is for those that the genetic catalog codes directly. The dictionary would be at the level of the tRNA-synthetase function which translates between languages, thus representing a "code within a code." The distinction is not adopted in the text because the usually interchangeable meaning of the terms is likely, in practice, to remain with the reader.

TABLE 1

The Genetic Code [a,b]

1st	2nd	U	C	A	G	3rd
U		Phe	Ser	Tyr	Cys	U
		Phe	Ser	Tyr	Cys	C
		Leu	Ser	*Ochre*	*Umber*	A
		Leu	Ser	*Amber*	Trp	G
C		Leu	Pro	His	Arg	U
		Leu	Pro	His	Arg	C
		Leu	Pro	Gln	Arg	A
		Leu	Pro	Gln	Arg	G
A		Ile	Thr	Asn	Ser	U
		Ile	Thr	Asn	Ser	C
		Ile	Thr	Lys	Arg	A
		Met	Thr	Lys	Arg	G
G		Val	Ala	Asp	Gly	U
		Val	Ala	Asp	Gly	C
		Val	Ala	Glu	Gly	A
		Val	Ala	Glu	Gly	G

[a] The table shows the structure of the code as it is known since 1967. Most of the allocations are considered to be certain, and are mainly derived from *E. coli* or bacteriophage using the test systems described in the text. It is likely that the code in higher organisms is very similar to that shown here. Nucleotide sequence AAA stands for ApApA, i.e., first nucleotide to the left is 3', first to the right is 5'.

[b] From *Cold Spring Harbor Symp. Quant. Biol.* (1966).

The structure of the genetic code is shown in Table 1. The basic properties, not all derivable from inspection of the Table, are as follows. (1) Each coding sequence for one amino acid or codon (Crick, 1963) contains three bases, the coding ratio being three.[7] The triplet nature of the codon was shown by genetic experiments involving frameshift mutants (Crick *et al.*, 1961), coding ratio of hemoglobin (Staehclin *et al.*, 1964), virus protein (Reichmann and Clark, 1966), and direct biochemical methods (Nishimura *et al.*, 1965). Only the latter strictly proved codon size. Of the total possible 64 triplet combinations (4^3), only two (UAA and UAG) and possibly a third (UGA) remain unassigned nonsense codons used for punctuation. (2) The codons are discrete or nonoverlapping with

---

[7] Concerning the usual display of codons as in Table 1 the reader should note that (1) the codon sequences are those in mRNA; (2) the sequences in DNA complementary and antiparallel to the codons are usually represented as being analogous; and (3) the complementary anticodons in the tRNA are also antiparallel, but, when conventionally written in the 5' to 3' direction (as in the text), they are reversed relative to the codons. Antiparallelism with the codon has inspired a rather unpleasant eponym for the anticodon, "nodoc."

no intervening bases (or commas) in between. (3) The code is degenerate, meaning that each amino acid can have more than one synonym codon; for the most part this degeneracy is connected with the third codon position. Thus mainly the first two codon positions determine the amino acid. Degeneracy is manifest even within a cistron (Adams et al., 1969; Yanofsky, 1965). A particular consequence of degeneracy is that different organisms may use a certain synonym codon for one amino acid in preference to others, which can be reflected in the overall DNA composition. A more general consequence of degeneracy is that one-quarter of all possible single base substitutions may be to synonym codons (King and Jukes, 1969). In principle, because of degeneracy, a protein sequence cannot rigorously define the original DNA sequence. (4) Ambiguity, or the reading of a given codon as more than one amino acid, also occurs, but by comparison much less frequently than degeneracy. It is immediately obvious that extensive ambiguity cannot be tolerated if protein synthesis is to be accurate. However, it does commonly exist in the reading of GUG depending on its position in the mRNA, as discussed below. It might also occur at a given amino acid site in copies of the same protein (von Ehrenstein, 1966). More restricted forms of ambiguity are misreading, suppression (other than of nonsense), and the inosine wobble (discussed below). (5) The code is universal or almost so. Examples are the wide host range of viruses, the synthesis of protein in heterologous *in vitro* systems, and especially the general agreement of codon assignments derived from widely different organisms. In general, however, ambiguity restricts the universality of the code. A special case of ambiguity is the suppression of nonsense codons brought about by discrete mutations in tRNA. On the other hand, the translational machinery, in particular the species-specific synthetase-tRNA recognition function, is less universal than the genetic code.

Certain general rules or constraints are further apparent in the codon catalog. (1) For any amino acid having several synonym codons, most of these are related to one another whenever this is permitted by the base assortment in Table 1. Related codons are those connectable by a single base substitution, that is, any codon is related to nine other codons. The most similar codons are those connectable by changes in position III because they usually specify the same amino acid. Amino acids having synonym codons frequently have two or four, and in some cases six, and, except for the last cases, all synonym codons vary in position III only. (2) The most common amino acids (leucine, serine, and arginine) have six codons, while the most unusual amino acids (methionine and tryptophan) have only one—and depart the most from the code regularities. This codon plurality is reflected in the occurrence of synonym tRNA species, although not necessarily in the same numbers. (3) Related amino acids have related codons. Related amino acids means in terms of their basicity, acidity, hydrophobicity or biosynthetic pathways. For instance, amino acids with U_{II} codons (U in position II) are among the most hydrophobic; and amino acids with A_{II} or G_{II} codons often possess reactive side groups relatively important for tertiary structure or function of protein. (4) In all cases of amino acids having two codons, if one codon terminates in U, the other terminates in C. The same is true (except UGA) for codons ending in A versus G. There are still other apparent rules but these may be coincidental (Speyer, 1967). However, it is clear that taken as a whole the above rules are far from coincidental.

Two additional properties of codons are derived from the way mRNA is translated. First, a codon series is colinear with an amino acid series in protein as part of the overall colinearity between gene, mRNA, and protein. Although colinearity between DNA and protein has not been proved, it has been proved between the analogous RNA from RNA phage and protein (Adams et al., 1969). Colinearity has been proved between DNA and genetic map (Hogness, 1966), DNA and RNA (by hybridization, Hall and Spiegelman, 1961), genetic map and protein (Sarabhai et al., 1964; Yanofsky et al., 1964), and mRNA and protein (by all cell-free or genetic experiments showing direction of translation). Second, codons are translated from their 5' to 3' ends. The first translatable codon specifies the amino or N terminus of the protein, the last the carboxyl or C terminus (Bishop et al., 1960). Sometimes the first codon is not represented in the protein as discussed below.

In summary, the genetic code confronts us with a set of regularities some of which have profound implications for translation and will be encountered frequently as we proceed. Other regularities are subtle and their reason is not readily understood; for these, later sections on the evolution of the code may provide some clues.

Wobble Theory

Codon-anticodon pairing for the first two codon positions is by standard Watson-Crick base pairs. However, as previously mentioned, when one amino acid has no more than four assigned codons, that is, one full group of four codons in Table 1, the codons differ from one another in position III only. Inside each group the general pattern for any one amino acid is either (1) two top codons, (2) three top codons (rarely), or (3) all four. Hence the role of position III in relation to this pattern remained to be explained.

Crick (1966b) proposed that the anticodon "interprets" letter III of the codon according to certain base pairing rules in Table 2, which feature nonstandard base pairs in addition to the standard ones. They show that it is not possible to code uniquely for either C or A. The non-Watson-Crick base pairs are also held together by H bonds, but the separation of the ribose moieties is somewhat altered and some ambivalence in the pairing (wobble) must be allowed to accommodate this. Experimental data so far generally support the predicted wobble pairs

TABLE 2

The Wobble Hypothesis Predicting Pairing Rules for the Third Letter of Codons [a]

Anticodon	Codon
A	U
C	G
U or Ψ	A / G
G	U / C
I	U / C / A

[a] From Crick (1966b).

(Söll and RajBhandary, 1967; Speyer, 1967), except for the unique U-A pair that might perhaps be masked by a frequent A to I anticodon transition. The energetic status remains the same as in the standard pairs (Pollak and Rein, 1968). Although new wobble rules may appear, the agreement is as much as can be asked for, particularly considering that codon recognition is likely to be influenced by other factors such as modification of anticodon bases and ribosome function. Perhaps the most serious discrepancy of the wobble theory is in respect to UGA (Crick, 1968). However, the wobble is likely to stay as a further instance of the semiuniversality of the code at a more restricted level.

In brief, the wobble rules predict that a single species of tRNA can serve synonym codons only when the codons vary in position III, but at least two

TABLE 3

Codon Recognition Pattern for Two Amino Acids (a, b) Differing in the Third Codon Base [a]

Codon	Wobble pattern		
	2 + 2[b]	3 + 1[c]	All 4[d]
XYU	a {U, C} G	a {U, C} I	a {U, C} G
XYC			
XYA	b {A, G} U or Ψ	b {G} C	a {A, G} U or Ψ
XYG			

[a] The anticodon (first) wobble base is underlined.
[b] Examples from Table 1: a = Asn, b = Lys.
[c] a = Ile, b = Met.
[d] a = Thr.

species are required when all four bases occupy that position. These predictions have been substantially confirmed. On the other hand, variation in nonwobble positions I and II requires separate tRNA species. The wobble recognition pattern is schematized in Table 3. On the basis of this pattern the number of tRNAs would be 30 plus, or about one-half the number of codons in the catalog. In actuality, in higher organisms there are more tRNAs than codons.

PARTICULAR ASPECTS OF TRANSLATION

This section and the one following deal with two aspects of translation, initiation and termination, related to particular aspects of the genetic code; the next section discusses mistranslation. The more general aspects of translation are considered in Chapter 5. Thus in addition to codons for amino acid, the code contains punctuation marks for initiating and terminating translation. These functions were anticipated on general grounds because they seemed indispensable for achieving a protein with precise end-to-end sequence. Particularly the effect of polarity mutants on the translation of polycistronic mRNA strongly suggested an initiation signal (Ames and Martin, 1964). Considerable understanding of these signals has been gained since then.

Initiation of Translation

The initiation codons are AUG and GUG (Clark and Marcker, 1966, 1968; Lengyel, 1967), the first appearing to be more efficient. Both code for N-formylmethionine at the $5'$ end of mRNA. The substituted methionine has been found as the N-terminal amino acid in bacterial and viral proteins, but only as they are formed. In internal positions, AUG and GUG code for methionine and valine, respectively. There is one tRNA that reads AUG and GUG and inserts methionine, which is formylated, and this tRNA is the sole initiator. A second tRNA reads AUG only and inserts ordinary methionine into internal positions. Both are charged by the same synthetase. Formylation is at the level of methionyl-tRNA by a transformylase and obviously cospecified by the tRNA. The formyl group is then cleaved off by a deformylase at the polypeptide level. In some proteins the whole N-terminal amino acid, and even one or two subsequent ones, are removed by an aminopeptidase.

It was surprising to find that the total sequence and not just the anticodon of the two methionine tRNAs show little homology (Cory and Marcker, 1970), and a meaningful comparison of function will have to await the elucidation of their tertiary structure. However, with the initiation codons the wobble involves the first and not the usual third letter (resulting in the unusual methionine-valine ambiguity between two codon groups in Table 1 instead of one). Earlier, the apparent rule for the unusual wobble seemed to be that the base immediately after the anticodon of only the initiator tRNA is unmodified, otherwise in all other E.

coli tRNAs there is the usual wobble. Recent work on yeast initiator tRNA shows that this rule is not universally valid (RajBhandary and Kumar, 1970). A second interesting question is, what makes AUG and GUG code for different amino acids (formylmethionine, methionine, and valine) according to their position in the mRNA. Possible answers are to be found in the translation initiation and elongation factors which discriminate between tRNAs (see below and Chapter 5) and, from work on RNA phages, perhaps also in the secondary structure of messenger (Lodish, 1970).

At a physiological Mg^{2+} concentration (about 5 mM) chain initiation is thought to involve the AUG-dependent binding of formylmethionyl-tRNA to the small ribosomal subunit in the presence of GTP and a protein factor (Revel *et al.*, 1968b), together known as the initiation complex. The initiation factor determines the binding of mRNA (Herzberg *et al.*, 1969) and also excludes noninitiator tRNAs from binding.[8] A second factor controls the subsequent binding of the large ribosomal subunit to the complex. These two factors are found attached to the small subunit. However, it appears that mRNA binding to the small ribosomal subunit can also occur in the absence of tRNA, although tRNA remains essential for subsequent binding of the large subunit (Groner *et al.*, 1970). This possibility highlights the specificity for initiation of the small subunit which is further discussed in Chapter 5. One view is that the entry of initiator tRNA into the small subunit is at the aminoacyl site. Thereafter, translocation of the initiator tRNA to the peptidyl site as the first peptide bond forms on the ribosome requires hydrolysis of GTP to GDP (Sarkar and Thach, 1968). The first initiation factor concerned with the binding of mRNA may be the same one that dissociates the subunits after each complete round of translation (Subramanian *et al.*, 1968) and would then control events at both ends of translation. In this regard, it has been proposed by Mangiarotti (1969) that the distinction between initiation and termination factors is only artifactual. In any case, frequency of initiation may be regulated by the interplay between the components of the initiation complex.

The previous mechanism worked out *in vitro* seems, in general, to depend on the structure of the initiator tRNA itself, although all the participants in the initiation complex must be present. Presence of the formyl group is required (Kondo *et al.*, 1968) and this imparts a special tertiary structure to methionyl-tRNA (Stern *et al.*, 1969). By recognizing both the formyl group and the dependent tRNA structure, the initiation factors discriminate against the noninitiator synonym tRNA (Rudland *et al.*, 1969), yet recognition of the initiation codon is entirely by the tRNA and not by the factor (Rudland and Dube, 1969). Because of its analogy to an amide (peptide) bond, the N-acyla-

---

[8] The present nomenclature of factors, in particular those discussed in connection with translation in Chapter 5, is not standardized and is highly confusing. To avoid adding to this confusion, no nomenclature is used in the text and instead the factors are referred to by their function.

Translation

tion may also facilitate the translocation of initiator tRNA inside the ribosome. However, the ultimate reason for the formyl group may be in the preference by peptidyl transferase (Chapter 5) for a blocked aminoacyl-tRNA at the donor (peptidyl) site of the ribosome.

As mentioned, proper bacterial chain initiation generally requires mRNA with an initiation codon at or near the 5' end. (This is strictly not true in the case of RNA phage and perhaps the eukaryote where the mRNA contains extensive noncoding sequence, but is true in respect of the informational sequence.) This codon strictly delimits the reading in phase of subsequent codons without which, in effect, any codon sequence lacks a consistent informational meaning. It is this fact, in conjunction with the fixed direction of translation, that obviates the need for commas in mRNA. Whereas normal chain initiation depends on formylmethionyl-tRNA, at high Mg^{2+} concentration chain initiation can be by amino acids incorporated internally or toward the C end by means of noninitiator tRNAs, thus skipping altogether the translation of one or more codons. In retrospect, it has been of some practical consequence that normal initiation is overcome at high Mg^{2+} levels, otherwise the genetic code would not have been broken at the time it was. However, when normal initiation is overcome, maximum rate of subsequent translation still requires a physiological Mg^{2+} level.

N-Formylmethionyl-tRNA, discovered by Marcker and Sanger (1964), is the major if not the sole chain initiator in *E. coli*. However, although AUG and GUG are necessary for initiation it is not clear whether they are sufficient. That is, it remains possible that adjacent codons are also necessary for initiation, and this appears to be so in the case of viruses (see below). A somewhat different initiator tRNA would appear to function in the eukaryote.

Recent evidence in ascites *in vitro* system strongly suggests the presence of two methionyl-tRNAs inserting the amino acid N terminally and internally and in response to the same codons, respectively, as the corresponding bacterial tRNAs (Smith and Marcker, 1970). Initiation factors of a nature similar to the bacterial factors are involved. This would mean that eukaryotic initiation basically resembles bacterial initiation, except that apparently the eukaryotic initiator tRNA does not carry formylated methionine. However, some kind of substitution may be necessary from what is known of the structural requirements of bacterial initiation; possible alternative acyl substitutions in the eukaryote are discussed in Chapter 5. Since less eukaryotic proteins begin with methionine, this would also mean that its subsequent excision occurs more frequently than in bacteria. Eukaryotic mitochondria and chloroplasts appear to be of prokaryotic ancestry, and uniquely in their cells retain the ancestral initiator tRNA (Smith and Marcker, 1968). If existent, a narrow species specificity of initiation could have been exploited to the advantage of host-virus interactions, but this is not the case.

The formylating capacity of a cell, determined by the amount of formyl donor available, appears to be important in the regulation of translation. For example, in the case of the polycistronic histidine operon mRNA in *Salmonella typhimurium* it is possible that this capacity controls whether translation begins only at the 5' end or internally at the multiple initiation sites along the messenger. Primarily, however, the initiation codon in each of the multiple cistrons depends, for inserting initiator tRNA at the aminoacyl site, on this site being vacated by a termination codon in the preceding cistron with the attendant release of the chain made on that cistron (Grodzicker and Zipser, 1968). Any nontranslatable sequence between the termination and initiation codons could influence not only internal initiation but also whether or not the ribosomes that are translating leave the polysome. On the other hand, in RNA viruses there are nontranslatable sequences between the termination and initiation codons of successive cistrons (Steitz, 1969) that appear to be recognized by positive factors. More deterministic than the corresponding bacterial sequences, these sequences and factors specify ribosome binding and thus herald internal initiation at the appropriate codon, yet any interaction with the initiation complex and factors discussed previously remains unknown. These aspects and possible effects of the secondary structure of mRNA on translation are further discussed in Chapter 5.

Termination of Translation and Its Informational Suppression

At present, termination of translation is generally less well understood than initiation (reviews by Bretscher, 1968a; Garen, 1968). Because the mechanism governs a disassembly of components it is likely however to be somewhat less specific and simpler. The termination codons in bacteria are UAA, UAG, and UGA, trivially known as *ochre, amber,* and *umber,* respectively. All three were initially known as "nonsense" codons because they do not code for amino acid. As is now known, these codons result instead in termination of the protein at the point where they appear in the mRNA. (However, in bacteria UGA can code for cysteine and the appropriate tRNA might act to block rather than terminate translation; Caskey *et al.,* 1968.) It should be noted that the initiation codons (AUG and GUG) cannot act in termination since in an internal position they code for unblocked amino acid.

The bacterial *ochre* codon was considered to be the only or main natural termination codon chiefly because of its low level of suppression (see later). Now more direct evidence in favor of *ochre* in this role is available (Suzuki and Garen, 1969). In the eukaryote, *ochre* and *amber* seem to be used as well. In neither type of organism, however, is it clear whether or not other nontranslatable bases beyond *ochre* are required for termination (Zipser, 1969). For example, an RNA phage cistron terminates in *ochre* followed by *amber* (Nichols, 1970), and a similar situation appears to prevail in bacterial cistrons (Lu and Rich, 1971). A possibility is that the doublet ensures termination in the presence of suppressor tRNAs, the second codon acting essentially as a fail-safe stop.

While obligatory for termination, the presence of UAA plays an equally important role in the release of the polypeptide, yet it is apparently not read by any bacterial tRNA (Bretscher, 1968c). Instead it appears to be read by two protein termination factors which cause the release of the polypeptide from the last translator tRNA as the termination codon enters the aminoacyl site in the ribosome. A similar mechanism of termination is at work in the eukaryote but in this case the precise number of factors is not known. The release of polypeptide depends strictly on the ribosome and on the function of peptidyl transferase, in particular, and also seems to require an ATP split (Colombo *et al.*, 1968). One of these termination factors reads UAA and UAG, the other UAA and UGA (Scolnick *et al.*, 1968). As expected, these factors cleave N-substituted aminoacyl-tRNAs (i.e., peptidyl-tRNA) with the exception of N-formylmethionyl-tRNA (Vogel *et al.*, 1968). Termination is, therefore, an active process that does not happen by default: nonsense codons signal termination but do not terminate by themselves.

The phenomenon of genetic suppression mediated via tRNA to be discussed now is intimately related to chain termination. Certain genes known as suppressors cancel both chain termination owing to nonsense codons and misreading owing to mutant missense codons. Suppressors also exist (frequently lethal) for sense codons. This type of suppression mediated by tRNA is extragenic, that is, involves different genes (suppressed and suppressor). Because it applies to genes translatable into protein and not to structural RNA genes, it is an informational suppression. Suppressor genes act by specifying a particular tRNA that makes sense of a given nonsense codon. Not unexpectedly, the tRNA translates the nonsense with an amino acid characteristic of the suppressor gene (Capecchi and Gussin, 1965; review by Gorini, 1970). As a hypothetical example, one suppressor will cause UAA to be read as glycine, and another suppressor to be read as valine. Hence suppressor genes are the structural genes for tRNAs (Andoh and Ozaki, 1968). In some cases the suppressed protein is nearly as functional as the original. (Confusingly enough, therefore, a "suppressed" protein is one whose synthesis takes place after all.)

Because of modifications in their anticodon or tertiary structure, either one resulting in a changed codon recognition, suppressor tRNAs are thus able to read nonsense codons and essentially permit a novel or corrected recognition resulting in meaningful translation. Modification of the anticodon was first suspected from the fact that several suppressor tRNAs insert amino acids whose codons relate to the nonsense codon by a single base replacement; later the anticodon modification was found (Goodman *et al.*, 1968). Presumably these tRNAs are minor species dispensable in normal protein synthesis. In fact, their identification has been possible by the sensitive criterion of suppression. One case, however, is known that involves a major species, and in this case the suppressed bacterium needs a second (wild-type) copy of the tRNA gene to survive (Söll and Berg, 1969). It is clear that, by further backward and forward mutation (including the synthetase recognition site), suppressor tRNAs can create new codon assignments. It is also clear that a physiological termination codon such as *ochre* must be relatively immune to suppression, otherwise proteins would not terminate correctly. Moreover, the balance in favor of termination is factor determined. Generically speaking, in their function the termination factors antagonize suppressor tRNAs. Because ribosome function continues until termination, however, a screening of tRNAs by the ribosome could be an additional influence.

Mistranslation

This is the last particular aspect of translation to be considered in this chapter, and the reason for considering it at this time should become evident as we proceed. Clearly, a prerequisite for protein synthesis is that the code be faithfully read. Possible misreading can take place first at the level of aminoacylation of tRNA (Chapter 5) and later in translation during codon-anticodon recognition; only the latter will be discussed here. Mistranslation is basically an error in H bonding, but the overall higher order structure of the polysome may be ultimately involved in causing this ambiguity. Perhaps tRNA is the component most responsible for keeping the translational error down to an acceptable level, but even so, it is not solely responsible. For example, alteration in the ribosome site for tRNA could allow a wrong species to fit so well against mRNA that mistranslation ensues. A critical role of the ribosome-tRNA recognition in regard to translation was suggested previously.

In vitro conditions known to cause translational error include low temperature, high Mg^{2+} concentration, high pH, abnormal concentration of tRNA, presence of alcohol, aminoglycoside antibiotics and polycations (spermidine), base analogs in mRNA, and ribosomal mutations (Woese, 1967a,b). Interestingly, the eukaryotic ribosome mistranslates less and is less affected by the previous agents than the bacterial (Weinstein *et al.*, 1966). Errors frequently involve tRNAs for related amino acids, that is, responding to related codons. Antibiotics and mutant ribosomes also cause errors *in vivo*, as discussed in Chapter 5.

Calculation places the error of amino acid insertion in the range of 10^{-3} to 10^{-4} per codon per translation (Woese, 1967a,b; Woese *et al.*, 1966). This error is taken as the maximum permissible to allow a 70 to 95% fidelity in translation of a protein with 400 amino acids. It follows that no single translational step can have a greater error than this if it cannot be subsequently mended, e.g., when it occurs, amino acid activation error is normally corrected during acylation. Increasing the previous level of error only tenfold to 10^{-2} would make translation of most genes almost impossible. In keeping with these estimates, the error of amino acid incorporation into protein measured *in vivo* is less than 1:3000 (Loftfield, 1963), or below what is generally expected in the normal recognition of substrate by enzyme. The preceding acylation of tRNA has an even smaller error, below 10^{-4} (Loftfield *et al.*, 1963). Indeed, judging from the first-order thermal inactivation kinetics of enzymes which suggests a high structural homogeneity of enzymes, a high degree of fidelity in translation appears to obtain in the cell. (Probably most amino acid substitutions would alter the temperature sensitivity of enzymes, if not their function.)

The high level of discrimination in codon-anticodon matching which has been mentioned represents an energy differential of about 5 kcal/mole, possibly just higher than the energy gain in making a correct base pair (with two or three H bonds) in preference to an incorrect one (with one bond less). Based on this view, the fidelity of matching would not only depend on the codon-anticodon recognition per se (involving H bonding), but also on the overall tertiary structure of the molecules participating in the recognition (involving less polar interactions). Indeed, most revealing in this respect, complementary trinucleoside phosphates do not bind one another in solution. To begin with tRNA, by affecting H bonding in general, the methylated nucleotides may play an important directing or correcting role in the recognition. In this context it is claimed that undermethylated tRNA binds less well to the ribosome (Stern *et al.*, 1970) and recognizes a different codon (Capra and Peterkofsky, 1968) than the fully methylated counterpart; the validity of these claims is discussed by Starr and Sells (1969). A second example is that of the cytokinin substitution located after the anticodon of certain tRNAs which stabilizes both the codon binding (Fittler and Hall, 1966) and wobble. A third example which has been suggested is that the presence of rare nucleotides in certain anticodons may act as a structural reinforcement for pairing. Along this line, inosine present (by deamination of A) in the anticodon of yeast serine tRNA (IGA) is thought to prevent one possible mispairing (Ralph, 1968a), whereas pseudouridine present in the anticodon of yeast tyrosine tRNA (GUA) would stack better than U and reduce general mispairing (Woese *et al.*, 1966). To turn last to mRNA, it should not be forgotten that its own local structure may contribute to the precision of pairing.

In addition to the previous participants, the role of others in translation for securing fidelity must be considered. First, if enzymes or factors intervene in the codon-anticodon pairing — which is not unlikely — they would certainly play a substantial part in securing fidelity, as considered earlier for polymerases in connection with replication and transcription. Second, the same consideration applies to the ribosome, and here some evidence even suggests it. For example, (1) ribosomal mutations tend to restrict informational suppression including, significantly, missense suppression (Schlessinger and Apirion, 1969); and (2) translational error is greater in heterologous systems (such as eukaryotic ribosome and prokaryotic tRNA) than in homologous systems, although this could also, in part, reflect incompatibility with respect to translation factors. Third, it was also proposed that the ribosome recognizes a codon directly and allosterically, which might permit only the right tRNA to pair with the codon (Woese, 1970a). However, although the ribosome may assume many configurations, it remains to be seen whether these can be as numerous as is implied in this proposal.

The pattern of coding mistakes observed *in vitro* shows that when codons contain a pyrimidine in position III, this position is the most prone to error. This is not clear in the case of purines because of the lower error. It reminds one that position III absorbs most of the code degeneracy and also the wobble. The second most error-prone position is I, and the least is II. The pattern agrees well with a computer analysis indicating that substitutions in position I tend to result in an amino acid similar to the original one, and, by implication, that position II carries the hard information for the amino acid (Alff-Steinberger, 1969). This is actually seen in Table 1, and it was mentioned earlier that position II is mainly responsible for the character of the amino acid. Hence, although position III is the least defining of the amino acid, the antipode position I is not the most.

A last remark is that the greater error of translation compared with that of replication and transcription (10^{-9}) reflects the greater complexity of translation. All considered, it might be more detrimental to the cell if the order of these errors were the reverse.

BASIC NATURE AND EVOLUTION OF THE CODE

As the structure of DNA was proposed it immediately opened a meaningful vista into many functional aspects of the code such as mutation, replication, and transcription. As it were, DNA function revealed itself in the structure. By contrast, inviting though its features are, the structure of the code tells very little about the basic nature of its function for the obvious reason that the code is in relation to a different (amino acid) language. This makes translation more complex than replication or transcription, and makes its refinement probably the result of a faster evolution. However, because direct evidence is very hard to come by, to this day the fundamental nature of the code remains an intriguing question. In this and the following section, the attempt is made to gather the most cogent lines of speculation on the nature and evolution of the code. Unfortunately, to prepare the reader, it is not easy to evaluate the relative merit of these speculations and he must take his choice between them.

Any explanation of the nature of the code must conform to the constraints discussed above that include (a) intercodon and (b) interamino acid orders. Intercodon order is that which defines certain relationships ruling on the amino acid assignment of codons. Interamino acid order establishes that related amino acids are somehow grouped by codon assignments, without specifying which codons are assigned to each amino acid group. In turn, it remains to be seen whether or not these orders reflect a third and more basic (c) codon-amino acid order; this derives the codon catalog from some specific relationship between amino acids and codons. Several models summarized here have been discussed in detail by Woese (1967a,b), from whom the previous definitions of orders are taken. Recently Woese (1969) presented a reevaluation of the subject.

At the extreme, the models of code evolution to be considered are of a deterministic or stochastic type, but there are all possible intergrades in between. In general, the deterministic models presume some kind of underlying mechanism whose properties determine the code order. On the other hand, the stochastic models rely on natural selection operating on randomly established codon assignments. The main question is therefore whether a group of codons for one amino acid is related to that amino acid causally or accidentally, and similarly for each codon in the group. It should be noted that the deterministic models demand the apparent universality of the code, whereas the stochastic models, in fact, must explain it. Clearly, the stochastic models offer little hope for reconstruction (which, however, is irrelevant to their probability).

First, Sonneborn (1965) proposed a stochastic model whereby, most mutations being deleterious, selective advantage for reducing the mutation burden would be conferred on those codon assignments which were so arranged as to make any base substitution (resulting from mutation) code for the same or a related amino acid. This model ostensibly accounts for most of the code properties, but (1) it may not confer enough selective advantage to the most favorable codon assignments. In addition, in terms of further evolution, (2) the initial changes of assignment could have been catastrophic, and thus (3) the model results in either blind evolutionary alleys or several, therefore, not universal, codes.

Second, Woese (1967b) proposed a "translation error" model which is in part stochastic and in part deterministic, because a translational apparatus is taken to exist from the beginning; it also presumes some kind of codon-amino acid order. He argues that since present cells are the result of a highly evolved translation, this translation must have evolved in cells very different from present ones and the earliest translation must have been imperfect and inaccurate. He then considers that the original code catalog was random or unstructured and only groups of related amino acids rather than individual amino acids were recognized. Consequently, any gene would be using the most reactive amino acids to make not any one specific protein but rather a "statistical" family of proteins with a limited specificity. The first step to reduce translational error would have been to associate these reactive amino acids with the least error-prone codons, thus favoring both intercodon and interamino acid orders. However, as the code became structured and more amenable to precise translation, evolution of translation began. According to Woese, advantages of his model are that (a) it would eventually determine an explosive rate of evolution capable of eradicating obsolete cell types, thus enforcing universality; (b) initial evolution would not have been catastrophic; and (c) final evolution would have been canalized against mutation. Several specific predictions of the model are borne out by the codon catalog: codons differing in position III have the same or related amino acids, in position I related amino acids, and in position II (the least error-prone especially with purines) the more functional amino acids.

Third, a decidedly deterministic model advanced on several occasions is one which proposes that the codon-anticodon degeneracy is the principal cause behind the present catalog. Woese (1967b) believes that this relationship is not clear, although it could have played a part in the evolution of the code; in particular it does not explain interamino acid order. He goes on to propose a second purely deterministic model, the "codon-amino acid pairing" model (Woese, 1967b), whereby such type of pairing would have somehow determined the structure of the code. Instead, Ralph (1968b) prefers to consider an "anticodon-amino acid pairing," and envisages that the "seeing" of the amino acid by tRNA is accomplished via the intermediate adenylate. Either of the two types of pairing would be basically important, but both remain most controversial. On balance, however, some evidence stands generically in favor of them. For example, chromatography of amino acids in a pyridine solvent shows a definite correlation between codon assignment and a "polar requirement" of amino acids (Woese et al., 1966). The amino acids recognize the heterocyclic base which suggests that they also may recognize the heterocyclic purines and pyrimidines. A second correlation is that found between elution order of tRNAs in MAK chromatography and codon composition, although no explanation is known for this correlation. It is implied that even such a subtle pairing as suggested by this evidence could be significant when acting over the evolutionary scale of time. Perhaps a third promising indication stems from models of α-helical polyamino acid paired with helical polyribonucleotide in a 1:3 ratio (Lacey and Pruitt, 1969). These models permit a visualization of RNA condensing on polypeptides believed now to have preexisted RNA in a noncoded form. The models have sufficient stereospecifity to (1) predict one correct codon, (2) require the interaction of L-amino acid with D-ribose, and (3) indicate the lesser importance of position III. Clearly, if any underlying base-amino acid pairing did act during evolution of the code along with stochastic processes, it would tend to counteract these processes.

Finally, Crick (1968) has returned to a stochastic model in order to explain evolution of the code. He notes that a primitive nucleic acid system would have assimilated rather indiscriminately the first amino acids to emerge into all codons, and the subsequent amino acids into codons related to the first, thus accounting for nonrandomness. [Originally codons would have been triplets, even if they contained only two bases or had two letters read. Previously, Jukes (1965) had considered original doublets coding for 16 (4^2) amino acids, position III being a late addition.] The reason for this schedule would be to minimize, not mutation burden as Sonneborn saw it, but a generalized protein lethality as was felt by Woese. For the same reason, as soon as the code had accommodated all 20 amino acids, and proteins became too sophisticated to admit new ones, the code froze, which explains universality. The present code would then have evolved by stages toward a minimally functional one, rather than toward Woese's deterministically optimal one. Admittedly, the occult nature of the code cannot be solved by inference alone, but this tells us that the explanation may involve more than a single mechanism.

Having discussed its evolution, the final need for stability in the code is evident. Change in one major codon assignment is equivalent to a mutation spread all over the genome; in *E. coli* this event alone would amount to 2×10^4 simultaneous mutations. In actuality, it has taken an average of 7×10^6 years for two successive amino acid substitutions in hemoglobin to become established (Zuckerkandl and Pauling, 1965). Changes in initially unassigned or minor codons could have been easier, however.

EVOLUTION OF DECODING

The emphasis of the previous discussion was on the evolution of the code itself. The emphasis here is on certain basic aspects of translation, though clearly the two are interdependent (see Woese, 1967b, for further discussion).

To begin with, forerunners of modern translation are plainly not available as fossils. A coherent fossil record dates back to 5×10^8 years, marking the origin of metazoans, and it is improbable that cells have changed radically since then either in the structure or translation of the code. Thus little may remain today of ancestral translation, and, in looking at the problem, it seems hardly credible that the evolution of translation can ever be retraced. We are left therefore with the exercise of arguing evolution in reverse, provided we do not overrationalize. Again, to anticipate the outcome, no clear conclusion is possible.

A basic argument is that transcendent evolution was impossible prior to the emergence of precise translation, since this requires enzymes that are themselves the product of precise translation. In this sense, the main concern of pre-Darwinian evolution was with the translational apparatus. Both imperfect coding and translation must have prevailed at the time when cells were developing the transfer of genetic information from one generation to another. Modern Darwinian evolution could begin only after the transfer of information was efficient.

The first view proposed on evolution of decoding was for direct templating of amino acids on DNA, amounting to a direct translation, as in the totally overlapping code of Gamow (1954). It was noticed that the internucleotide distance in DNA (3.4 Å) was close to the interpeptide bond distance in protein (3.6 Å). Hence it was thought that direct templating on DNA could reflect an underlying ancestral base-amino acid pairing along the lines discussed in the previous section. Opposed to this view is another requiring an ad hoc adaptor for translation, because an adaptor of this nature dispenses with any underlying ancestral mechanism. On this basis, codon assignments could become solely determined by stochastic processes, that is, historical accidents. The prediction that the adaptor was RNA came in 1957 (Crick, 1957), although Dounce (1952) had suggested earlier that enzymes could act in this capacity. The adaptor is now of course tRNA; yet, whether it was the adaptor early in evolution is a different matter. The reason for this doubt — which questions the ad hoc appearance of the adaptor — is that tRNA could have had other functions which somehow were involved in the evolution of the code prior to it becoming the adaptor: these primitive functions could have been in relation to metabolic pathways of which certain indications persist today (Chapter 2), or to nontranslational oligopeptide syntheses which still occur (Chapter 5). In any event, the present meaning of the adaptor is that it serves to displace the recognition between oligomer and monomer (codon and amino acid) to a recognition between two polymers (tRNA and synthetase) with an attendant gain in stereochemical resolution (von Ehrenstein, 1970).

Thus, although the adaptor hypothesis has had a lasting influence in obviating the conceptual need for a direct templating mechanism, a direct mechanism later obscured by the adaptor function remains a possibility, and it is worth considering what evidence there is in its favor. First, Woese (1968b) observed with equilibrium dialysis a reciprocal polymerization catalysis between purine-rich nucleotides and basic amino acids. [Purine nucleotides are preferred to pyrimidine nucleotides in these studies because they have a greater stacking energy (T'so *et al.,* 1963).] The colinear binding of nucleotide phosphate groups to amino acid basic groups in a 1:1 ratio is reminiscent of present day DNA-histone complexes. Similar evidence was described in the previous section. However, a direct gene translation of this type, if this is what is suggested, would have been extremely ambiguous and inaccurate. Second, an indication for an autocatalytic cycle comes from the polymerization of nucleotides and amino acids under the influence of polyphosphate esters that could have abounded in the prebiotic Earth (Schramm, 1965). Third, the $2'$-ester linking of activated amino acid along the length of RNA (Beljanski and Beljanski, 1963) could also suggest the rudiments of translation. In summary, if any of these relationships intervened at all during the evolution of translation — which if far from certain is not inconceivable — it would have been prior to the appearance of the adaptor in its present form.

CHAPTER 2

Molecular Species of RNA

This chapter describes the molecular parameters and properties of individual mature species of RNA. The relative proportion in the cell and the number of molecules of each RNA species in the nucleus and cytoplasm are shown in Table 4 for a human tumor cell line, the HeLa cell. The nascent RNA species are described in Chapters 3 and 4 in conjunction with their pathways of synthesis.

MESSENGER RNA

As the name implies, messenger RNA conveys the genetic information in DNA to the translational apparatus of the cell (reviews by Lipmann, 1963; Singer and Leder, 1966). This role is unique to mRNA (Fig. 1).

Because informational function is embodied in the primary structure, it was thought until recently that mRNA (together with labile RNA) was the least interesting of all RNA species in terms of intrinsic structure. However, according to current work with RNA phage mRNA, this appears to possess some intrinsic secondary structure. In addition, during translation this mRNA appears to have some structure bestowed upon it by its product protein. Both these higher order structures, intrinsic and imposed, may also be the case with eukaryotic mRNA; in this mRNA the structurally active associated protein possibly is a spe-

TABLE 4

RNA Species of the HeLa Cell[a]

RNA species	Total cell RNA (%)	Estimated number of molecules		Estimated number of bases	Composition (AUGC)
		Nucleus	Cytoplasm		
Labile (polydisperse)	1	1×10^4	–	Max 5×10^4	28-31-21-22
45 S	1	1×10^4	–	1.3×10^4	13-17-37-33
32 S	3	4×10^4	–	6.8×10^3	14-16-37-33
28 S	53	7×10^4	5×10^6	4.9×10^3	16-16-36-32
18 S	24	6×10^4	5×10^6	1.8×10^3	21-22-30-27
5.5 S	1	6×10^4	5×10^6	140	21-23-28-28
5 S	1	6×10^5	5×10^6	120	18-22-34-26
tRNA	12	2×10^5	1×10^8	80	22-24-27-27
Small nuclear RNAs[b]	0.5	2×10^6	–	100–180	Varies
mRNA (polysomal)	3	1×10^5		Max 10^4	26-28-21-24
Cytoplasmic polydisperse (not polysomal)	<1	–	1×10^4	Max 10^4	
Total	99				

[a] Data from Darnell (1968) and Weinberg and Penman (1969).
[b] See Table 5.

cial protein to be discussed later. Nevertheless, unless controlled, excessive higher order structure would adversely affect the template role of mRNA (as also in the case of DNA). Messenger RNA is characterized by a total or almost total lack of the postsynthetic modifications that are present in most other cell RNAs such as methylation, pseudouridylation, and thiolation. Because of this relative lack of intrinsic structure and absent modification, as well as its small quantity (a few percent of cell RNA) and marked diversity, the physical properties of mRNA have been poorly characterized. This elusiveness contrasts with, or perhaps reinforces, the enthusiasm that mRNA has generated since its realization. Yet, knowledge of these properties of mRNA would be invaluable in understanding certain facets of translation.

Characterization of Messenger RNA

The analytical criteria used to characterize mRNA are based upon (1) its base composition, (2) similarity with DNA (nearest-neighbor frequency), (3) kinetics of labeling, and (4) chromatographic behavior. With some exceptions, if characterization is left at these stages, it is more a matter of faith than of fact that the RNA in question is mRNA. Under these circumstances the RNA is only appropriately described as "DNA-like." Two reasonably diagnostic criteria for prokaryotic mRNA use specific hybridization to DNA. These are (5) rate and (6) relative amount of hybridization, under usual conditions these being greater than for stable RNA. Hybridization of eukaryotic mRNA is technically difficult and its use in characterization was, until recently, questionable. Some additional evidence can be secured by means of (7) the sedimentation value of the RNA if this falls within the expected range of mRNA. However, up to this point the criteria for mRNA characterization are operational. The most stringent test for mRNA is (8) the direction of specific protein synthesis in a messenger-dependent cell-free system, a test of function unique to mRNA and not of filial relationship as in the case of points (6) and (7). Although a less stringent criterion, nonspecific protein synthesis can still be useful in conjunction with other criteria. These techniques aim, therefore, at loosely typifying total mRNA (points 1-4, and 7), or at selecting by degrees a particular mRNA with unique informational sequence (points 5, 6, and 8). (As used in the text, the term "mRNA" stands for total mRNA. Specific mRNAs are designated according to the protein they code for.) These criteria are now briefly discussed in their listed order.

The overall base composition of mRNA in higher organisms resembles that of DNA and is rich in AU (about 55%). The equivalence to DNA composition is statistical and need not apply individually to mRNAs representing, as they do, partial single-strand transcripts. In retrospect, the misconception of an obligatory equivalence to DNA composition possibly arose from the particular DNA composition of *Escherichia coli,* which is roughly an equal amount of the four bases and the same in both strands. The equality of amino (adenine, cytosine) and keto (guanine, uracil) residues has, however, been claimed for the *in vitro* transcript of *Bacillus subtilis* single DNA strands (Karkas et al., 1968). This compositional equality is known for total cell RNA, supposedly as a legacy from the compositional equality of total DNA, but it is not clear how general the suggested internal DNA strand symmetry is in this bacterium.

In general, base composition can indicate mRNA when it is sufficiently different from that of stable RNA, when synthesis of the latter RNA is suppressed, or when the mRNA in question represents a substantial part of the genome. These conditions are met in phage (Volkin and Astrachan, 1956), bacteria in a step-down culture (Spiegelman, 1962), and, ideally, in single-stranded viral DNA that yields a perfectly complementary transcript (Hurwitz et al., 1962). On the other hand, in eukaryotic cells mRNA cannot be distinguished from other rapidly labeled RNA on the basis of composition alone.

Nearest-neighbor analysis has been used in the study of *in vitro* (Weiss and Nakamoto, 1961) and *in vivo* transcription (Bautz and Heding, 1964). The technique depends on the transfer after alkaline hydrolysis of the 5'-phosphate-^{32}P in one nucleotide to the 2',3' position of the neighbor nucleotide, from which the complementarity and antiparallelism (i.e., homology) of transcript and template can be examined; it is mandatory that the ^{32}P-nucleotide be previously equilibrated with the precursor cell pool.

Rapid labeling of mRNA observed after a pulse is evidently limited to the fraction under synthesis. The criterion is therefore apt to be significant only when mRNA synthesis is predominant or exclusive, as in viral infection. In the eukaryote the kinetics of labeling may not help differentiate mRNA from precursor rRNA (Ycas and Vincent, 1960) or labile RNA (Shearer and McCarthy, 1967), which are described later. In addition, the rate of synthesis may vary with particular mRNAs or with the cell status. For example, in bacteria the rate of mRNA synthesis varies from 0.5 to 5 times the rate of stable RNA synthesis (Mueller and Bremer, 1968).

Using chromatography in a MAK column, the DNA-like RNA remains tenaciously bound to the column (Ellem, 1966) and by this means can be partly separated from other RNA species which bind less or not at all. However, in the eukaryote this DNA-like fraction includes labile RNA (Roberts and Quinlivan, 1969).

Hybridization of mRNA with purified DNA has the advantage that in principle all the coding genome can be tested. Prokaryote mRNA hybridizes readily. The first mRNA successfully hybridized was that of phage (Hall and Spiegelman, 1961). Specific hybrids form between sequences as little as 12 nucleotides long, which is close to the theoretical lower limit, indicating that, in effect, all the bases contribute to the hybridization. By suitable manipulation with mutant (partially deleted) DNA it has proved feasible to localize mRNA fragments within a cistron (Bautz and Reilly, 1966), in what amounts to a remarkable degree of purification. In the eukaryote, on the other hand, the most rapidly forming hybrids are those with the reiterated DNA sequences which appear to code mostly for labile RNA. By comparison, hybridization of mRNA with the nonreiterated DNA sequences is technically much more difficult, and has only recently been achieved. (This is discussed in Chapter 3.)

Considering the coding ratio, the molecular weight of a particular mRNA is about nine times that of the protein it codes for. For example, the mRNA for one hemoglobin chain (16,500 daltons) is about 150,000 daltons or 9 S (Chantrenne *et al.,* 1967); the mRNA for histone with 100 to 200 amino acids is about 8 S (Mueller, 1969). The average expected weights of mRNAs are in the range of 2 to 5 $\times$ 10^5 daltons; the expected S values are in the range of 6 to 30. These considerations implying monodispersity of individual mRNAs refer strictly to the informational sequence. However, particularly eukaryotic mRNAs appear, in addition, to possess extensive noncoding sequence and possibly (but not certainly) a precursor molecule, the combination of which might in principle create a certain degree of polydispersity especially (but not exclusively) at early stages. On the other hand, polycistrony is the natural condition of viral mRNA (Zinder, 1965), and also, in part, occurs in bacteria with mRNA reaching 1 to 2×10^6 daltons. In the eukaryote the breaking up of cistrons implicated in related biosynthetic pathways seems most common, and, as a rule, functional mRNA is monocistronic.

Protein synthesis *in vitro* directed by a specific mRNA (not a DNA-dependent system or chromatin) is difficult because fractionated undegraded mRNA is required, and nonspecific stimulation must be rigidly controlled. However, specific protein synthesis *in vitro* has been achieved in both prokaryote and eukaryote. An example of the latter is the synthesis of chick myosin using an homologous ribosome system, which contains endogenous mRNA (Heywood and Nwagwu, 1969). There are some indications that bacterial ribosomes respond better to viral mRNA, e.g., f2 coat protein mRNA (Engelhardt *et al.*, 1968), than to bacterial mRNA. This favoritism might be diagnostic of the host-parasite relationship, but the possibility of viral effects on the host ribosome (Chapter 5) must not be forgotten.

In summary, in the case of eukaryotic mRNA the routine characterization usually involves a compromise between rigorous and acceptable standards. When applied simultaneously, three frequently acceptable criteria are (1) stimulation of unspecific cell-free protein synthesis, (2) polysomal location, and (3) compatible sedimentation characteristics of the RNA. Firm characterization of mRNA demands special criteria as outlined above.

Some Metabolic Characteristics of Messenger RNA

Although the following two aspects are not readily distinguishable from each other, the rate of messenger decay in the cell, in part, reflects (a) the chemical instability of molecule and, in part, (b) its active enzymic degradation, the latter being probably the most important. In keeping with the rapid rate of synthesis, decay can be particularly swift in bacteria. The synthesis is more regulated, the decay more uniform. In *E. coli* it takes only 2 minutes for the induced synthesis of β-galactosidase to attain full rate, and a few minutes to cease after withdrawal of inducer (Riley *et al.*, 1960). In *B. subtilis* the rate of mRNA breakdown is exponential, and the rate of protein synthesis also decreases exponentially (Fan *et al.*, 1964). Generally, the rate of mRNA breakdown varies with the organism's rate of growth and protein formation, with a half-life of 2.5 minutes at 25°C. Similar values are found for pulse-induced specific mRNA. Thus the half-life of mRNA is somewhat related to the duration of the cell cycle. Not all bacterial messengers are unstable, however, and one example is penicillinase mRNA; sporulation mRNAs, in particular, appear to be quite stable. In contrast to the bacterium, the average half-life of mammalian mRNA is longer, e.g., 3 to 4 hours in HeLa cells (Penman et al., 1963) and at least a few days in nonnucleated reticulocytes that no longer produce any RNA. However, in any cell the half-life of individual mRNAs varies within wide limits. The stabilization of eukaryotic mRNA, as revealed in its longer functional life, appears to be in relation to the longer cell generation times and tissue differentiation (as opposed to adaptation), which are typical of the eukaryote. For this reason, the control of eukaryotic mRNA decay will probably prove to be more regulated than in bacteria. For example, it now appears that both in embryonic and mature cells mRNA is stabilized within protein-containing particles or informosomes as discussed below. The stabilization could be directed against both chemical instability and enzymic degradation.

Messenger is the RNA most susceptible to nuclease attack as a consequence of a lack of defenses, such as elaborate higher order structure and perhaps modification, which other RNAs without a template role were more free to evolve. However, during translation tracts of mRNA in contact with ribosomes are physically protected against nuclease degradation (McLaughlin et al., 1968). As suggested above, in the eukaryote the informosomal protein associated with mRNA may in part also serve to protect it against nucleases. From the previous discussion, it seems likely that bacterial mRNA becomes more rapidly accessible to enzymic degradation than eukaryotic mRNA. All of the knowledge of nucleases responsible for degradation (called by the inelegant name of "messengerases") comes from work with bacteria. A degradative bacterial enzyme is the potassium-activated phosphodiesterase RNase II, a predominant $3'$-exonuclease with strong preference for single-stranded nonmethylated RNA, which releases $5'$-mononucleotides (Nossal and Singer, 1968). Since bacterial mRNA is randomly degraded from the $5'$ end, the RNase II cannot function alone and would have to depend on the presence of an endonuclease. A more promising degradative enzyme is a $5'$-exonuclease, RNase V, which possibly is an integral part of the bacterial ribosome and depends for its function on the functional ribosome (Kuwano and Schlessinger, 1970). In particular, this nuclease activity is believed to depend on translocation rather than actual translation; it is contraarrested by cAMP. The current view is that most bacterial mRNA is degraded immediately upon translation.

Binding of mRNA to other RNA species has been reported. It is now realized that at least part of the newly synthesized RNA that appears to sediment up to 100 S is, in fact, the artifactual product of aggregation between rRNA and other RNA species, chiefly among which are labile RNA and mRNA (Hayes et al., 1966; Parish and Kirby, 1966). However, some degree of physiological association cannot be summarily dismissed. Conversely, species that show very different S values under certain sedimentation conditions can show under other conditions a uniform 16 S value (Bramwell and Harris, 1967), in this case, the sedimentation behavior describing more the molecular shape than size. From this, Harris (1968) has argued that 16 S RNA as a class could represent mRNA (see also Hadjiolov, 1967), a view which has not received general acceptance. In many cases, however, the reduction in S value can simply be due to degradation. Another serious objection to poorly controlled procedure is that factors normally present in cytoplasm can govern significant RNA interactions (Girard and Baltimore, 1966).

Looking ahead, viral RNA would appear to be the mRNA most interesting for sequence analysis (Madison, 1968). The interest is not so much in the coding sequences which, given a good guess at the codons being used could be predicted from the protein sequence, but in questions like start-and-stop translation signals and signals between cistrons, thus far investigated mostly with synthetic messengers. Progress on these questions, as well as the effects of higher order structure of viral mRNA on translation, are considered in Chapter 5.

The Informosome

As mentioned above, most eukaryotic mRNA seems not to be naked but rather associated with nonribosomal protein from its inception in the chromatin, perhaps during actual transcription, to its final destination in the polysome. Indeed, there is here an analogy with rRNA and ribosomal proteins. Messenger RNA is therefore present within particles both in the nucleus and cytoplasm. Since messenger remains in a cytoplasmic pool for about 15 minutes before entering the polysome (Penman et al., 1968), it is clear that if naked it would be readily exposed to nuclease degradation during this period. Thus, the associated protein may be related to the stabilization and transport of mRNA. Regulatory functions in translation remain another possibility for the protein. Formally, proteination is to eukaryotic mRNA what coupling of the translating ribosome with transcription is to bacterial mRNA: few if any bacterial polysomes are physically free from DNA for any length of time. Contrary to the bacterial coupling, however, the proteination of mRNA accentuates even further some of the functional consequences of the uncoupling between eukaryotic translation and transcription.

The first evidence for a particle containing mRNA was from work on fish mRNA which is produced during oogenesis and stored for use during early embryogenesis. The particle was termed "informosome" (review by Spirin, 1969). In the sea urchin embryo these free particles sediment at 15 to 65 S, that is, faster than small ribosomal subunits. Their size depends on the size of the DNA-like RNA they contain, which has a mean value of 20 S (Infante and Nemer, 1968). The free particles are difficult to distinguish by sedimentation from a combination of the RNA extracted from them and the small ribosomal subunit. Later, polydisperse DNA-like RNA complexed with 60% nonribosomal protein was found to be released from liver and L cell polysomes (Henshaw, 1968; Perry and Kelley, 1968a). However, the question of the identity of this polysome-associated complex with the free nuclear and cytoplasmic particles, on which the continuity of the protein depends, remains open. The protein complement in the nuclear and cytoplasmic particles is just beginning to be analyzed. In this respect, it was recently claimed that a nuclear and a polysomal protein associated with mRNA are identical; this protein is presumed to be a permanent component of the informosome (Schweiger and Hannig, 1970). Again, there is a serious risk of artifact in these preparations.

The difficulty with regard to characterization of nuclear particles is great. Conceivably, there are particles of many different sizes and many different types. Perhaps there is a high number of initially large particles, some transforming into the informosome by degrees and some having other functions. The technical difficulty is then to distinguish within this potential plurality. Moreover, not only mRNA but also labile RNA (most of which does not leave the nucleus) is contained in the particles. In nongrowing liver a 60 S mRNA-containing particle, but not 80 or 110 S particles that were also observed, is proposed to be the one transferred to the cytoplasm (Faiferman *et al.*, 1970). In summary, the structural and functional aspects of the informosome are in need of much further work in order to be understood. In addition, it is not known whether or not all mRNAs require informosomes.

The cytoplasmic proteins that shuttle to and from the nucleus could be considered candidates for participation in the informosome, except that they are now known to be very small (Jelinek, 1969). These cytonucleoproteins were first found in amoeba (Prescott and Goldstein, 1968) and later in other cells (Arms, 1968; Kroeger *et al.*, 1963).

Bell (1969) proposed that the informosome carries small (7 S) DNA fragments from the nucleus to the cytoplasm. From observations in embryonic muscle and adult cells, the DNA was reported to be associated with the polysome and presumed to be transcribed as mRNA in the cytoplasm. However, it remained to be explained how a myosin messenger (600,000 daltons) can originate from a smaller 7 S DNA fragment. Previously, liver microsome-associated DNA had been reported (Bond *et al.*, 1969). There are two main implications of the proposal: (a) the possibility of a coupled transcription-translation as in bacteria and (b) that selective DNA amplification and excision replace selective transcription as the primary event in the expression of the genome. According to more recent data (Fromson and Nemer, 1970), however, the association of DNA with the polysome is an artifact of extraction derived from fragmented chromatin.

EUKARYOTIC LABILE RNA

This RNA species of the eukaryote is the one surrounded with the greatest conflicting interpretation and perhaps presents the greatest challenge to understand. Although it is clear that a final function cannot as yet be ascribed to the RNA, the present section serves mainly to introduce its basic parameters and their possible functional significance. Further discussion of its function(s) is left for Chapter 3, where the companion evidence and theoretical framework being used to postulate the nature of this function are presented.

Labile RNA, also known as heterogeneous or polydisperse RNA, occurs in the nucleus of animal cells irrespective of their proliferative or differentiative status. This RNA is present in variable amounts up to about 1% of total cell RNA in some cell types. This small amount contrasts with the large amount of the RNA which is made at any time all over the chromosomes, in fact much larger than what is produced, for example, of rRNA in the nucleolus. The explanation is that the RNA shows a very rapid turnover that occurs mostly within the nucleus, with only a minor part of the RNA reaching the cytoplasm, whereas rRNA is stable and ends up by making the bulk of cell RNA. Despite the physical resemblance to mRNA to be discussed, most of the labile RNA therefore cannot be its precursor. Labile RNA is not found in the prokaryote and there is some question as to whether it appears in its typical parameters in the lower (unicellular) eukaryote, e.g., amoeba (Prescott *et al.*, 1971). It has not been adequately described in plants, but it is likely, however, to be found in them.

Labile RNA was first proposed as a separate species by Harris (1963) working on macrophages and was later demonstrated directly in several cell types (Attardi *et al.*, 1966; Penman *et al.*, 1968; Soeiro *et al.*, 1968). It is now directly observable by polyacrylamide gel electrophoresis. Although the function remains speculative, it is not (1) a template for nuclear protein synthesis or (2) the principal result of an imbalance between RNA and protein synthesis (Harris, 1964). That it has a physiological function is indicated by its synthesis in the immature duck erythrocyte (Attardi *et al.*, 1966), a cell committed to a very specific function. Harris (1964) first proposed that most of the labile RNA is degraded in the nucleus as the result of not becoming stabilized as functional messenger, the stabilization being envisaged to involve protein and perhaps the ribosome. Later, Church and McCarthy (1967b) suggested a selective transport to the cytoplasm of potential messages in nuclear labile RNA during (liver) differentiation. Since labile RNA is on the average much larger than mRNA this would require controlled scission of the molecule. Thus one view of labile RNA is that it is a source of messenger, making it a precursor rather than a separate species. If so, as the evidence stands, less than 5% can become mRNA. On the other hand, complementarity to reiterated DNA sequences is indicated by its preferential hybridization in unfractionated RNA preparations (Perry *et al.*, 1964); it is not at all clear to what extent the reiterated sequences can result in structural mRNAs. It is, however, believed that some of the labile RNA, even that which breaks down in the nucleus, derives from unique DNA sequences; as will be seen, these sequences are likely to result in mRNAs. A different concept, although by no means unanimously accepted, is that labile RNA has evolved in conjunction with the complex regulation of the eukaryotic genome. As we shall see in Chapter 3, these different functions need not be mutually exclusive.

From hybridization experiments where this RNA would have been preferentially studied, the sequences can be shown to be to a large extent tissue-specific (Paul and Gilmour, 1968), and to be replaced by new sequences during embryonic differentiation (Chapter 6). The sequences vary among different species, e.g., man and mouse (Soeiro and Darnell, 1969).

In both its physical and metabolic parameters labile RNA resembles mRNA. The RNA has little intrinsic higher order structure, although this could be imposed upon it by any associated protein. The composition is AU type, about 44% GC; like mRNA, there is no methylation. The sedimentation is heterodisperse from 10 to 100 S, the higher value surpassing 10^7 daltons or up to 50,000 bases. No doubt the range is partly due to continuous degradation of the RNA. Under the electron microscope the RNA measures up to 9 μ (Granboulan and Scherrer, 1969), which is in the order of the diameter of the nucleus and proves the extremely large but covalent RNA structure. Possibly because of degradation, there is labile RNA even smaller than 10 S (Weinberg and Penman, 1969). Therefore it overlaps all other eukaryotic RNA species from the largest to the smallest, and at its largest is much larger than any bacterial RNA species.

The reported half-life of the RNA is from 3 minutes to 1 hour, probably representing fractions of different stability. In general it is more unstable than mRNA. The actual amount in terms of nuclear RNA in mouse L cells is small, but the specific activity is the highest after a pulse of label (Shearer and McCarthy, 1967), that is, labile RNA acquires most of the label and loses it the fastest. These properties reflect, of course, the massive synthesis followed by rapid breakdown which were described earlier. The loose intrinsic structure of the RNA makes it susceptible to environmental changes, which permits its partial separation from other RNAs, e.g., by MAK chromatography. It can be extracted in particles of up to 5000 S (Penman *et al.*, 1968) under conditions suggesting a tight association with protein within "chromatin" particles. The most distinctive properties of labile RNA compared with cytoplasmic mRNA are therefore the faster kinetics of labeling, greater polydispersity and maximum size, homology with reiterated DNA sequences, and preferential nuclear-bound localization. Within nuclear labile RNA, however, none of these criteria can distinguish between the fraction that remains in the nucleus and the fraction that passes to the cytoplasm and may become mRNA (Shearer and McCarthy, 1970a).

In addition to the nuclear polydisperse RNA, several polydisperse RNAs have been found in the cytoplasm whose relationship with the former is not clear. Penman *et al.* (1968) described a polydisperse (10 to 90 S), rapidly labeled RNA present in considerable amounts (by radioactivity) in the HeLa cell cytoplasm. From its kinetics it appears not to derive from the labile nuclear RNA, thus its origin remains to be explained. A different cytoplasm polydisperse RNA (10 to

70 S) cosediments with, but is not attached to, polysomes in the HeLa cell (Penman et al., 1968) and embryonic muscle (Kabat and Rich, 1969). The relationship with the nuclear or the previous cytoplasmic polydisperse RNA is not clear. In the HeLa cell the amount of this RNA is small compared with the bulk of cytoplasmic polydisperse RNA, but very large in comparison with the polysome-associated mRNA. Clearly, the occurrence of RNA in a polysome fraction is not an absolute criterion for messenger.

Mitochondrial Labile RNA

Attardi and Attardi (1967) described in the HeLa cell a heterogeneous RNA (6-50 S) which amounts to 1% of cytoplasmic RNA and is present in polysomes that are bound to endoplasmic reticulum membranes. This RNA differs from the mRNA associated with free polysomes in having faster sedimentation, labeling and turnover rates (half-life, 30 minutes), and a higher A/U ratio. The average rate of transcription is 10-20 times that of nuclear RNA but one-tenth that of cytoplasmic rRNA. On the basis of its almost exclusive hybridization with mitochondrial DNA, the authors conclude that the RNA is of mitochondrial origin. The composition is 43% GC, compared with 45% GC for human mitochondrial DNA. Because of its base ratio, the RNA seems to be complementary to the heavy strand of mitochondrial DNA, and in its largest size to represent a significant fraction or even the whole of the strand (in which case the RNA would include also stable RNAs; see Chapter 4). They further suggest that the RNA functions as a mitochondrial mRNA in the cytoplasm (Attardi and Attardi, 1968). Although this may be the case, it is still possible that the RNA corresponds with the cytoplasmic labile RNA described in the previous section (Penman et al., 1968). In that case, since much of the latter RNA is not associated with the polysome, the messenger function would remain unclear. Undoubtedly, this RNA is being further investigated.

RIBOSOMAL RNA

Ribosomal RNA is the most prominent auxiliary RNA in the cell because of its quantity. It constitutes up to 90% of the total cellular RNA, and is found mostly in the cytoplasmic ribosome. There it serves a role in protein synthesis which is well known but little understood; on present knowledge, this is the RNA whose function most depends on permanent association with multiple (ribosomal) proteins. The various rRNA species have well-defined physical parameters and are therefore suitable molecular markers.

The RNA is relatively stable in growing bacteria (Davern and Meselson, 1960), but unstable in quiescent bacteria. In the eukaryote the RNA is also more stable in growing cells (Rake and Graham, 1962). In mammalian liver, for instance, the half-life is 2 to 5 days (Hajdiolov, 1966), which is considerably shorter than the life of the cell and perhaps of some messengers. Both major species of rRNA have an identical rate of turnover.

The RNA has considerable secondary structure as a result of approximately 60% helical content in the stem of short-looped segments. As the result of helicoidality and large size, the molecule also has a well-developed tertiary structure. Aspects of higher order structure are reviewed by Attardi and Amaldi (1970). The major species are posttranscriptionally modified. To illustrate this with methylation, in HeLa cells one out of 48 nucleosides is methylated in the small RNA (18 S) and one out of 71 in the large RNA (28 S) (Brown and Attardi, 1965). Eighty percent of the methylation is in the $2'$-oxygen of the ribose and the remainder in the bases; the situation is similar in plants. Also in bacteria, the small RNA is more methylated (1/125) than the large (1/170), but there seems to be relatively less ribose methylation than in mammals (Hayashi et al., 1966). The evolutionary trend is therefore for increased substitution of the molecules. The methylated bases occur in specific sequences and frequently in clusters (Fellner and Sanger, 1968), these clusters probably having a functional significance. (Contrast this pattern with that of chemical methylation which affects all bases depending on their reactivity rather than their position in a sequence.) Yet a particular site may not be methylated in every molecule of the RNA species (Lane and Tamaoki, 1969), the meaning of which it is premature to interpret. As would be expected, the base methylation pattern is quite specific to the animal species and varies greatly, for example, between bacteria and mammals (review by Starr and Sells, 1969). In addition, rRNA contains other substitutions which also have probably increased during evolution. For example, mammalian rRNA possesses pseudouridine to the extent of about 7% of uridine (Amaldi and Attardi, 1968). Basically, what follows is a description of the rRNA molecules, while discussion of their many interesting functional implications is reserved for later chapters. It is, however, convenient to describe these molecules with reference to some of their evolutionary parameters.

Major Species of Ribosomal RNA

In the higher eukaryote the two major rRNA species derive from separate sequences in a common precursor that passes through a series of intermediate stages as discussed later in Chapter 4. Because of the common precursor their synthesis is stoichiometric, in fact, equimolar. The biosynthetic pathway takes place within the nucleolus, therefore, during interphase. In the prokaryote the pathway is less certain, particularly in respect to a common precursor.

The large RNA species has a nominal sedimentation value of 23, 25, and 28 to 30 S in bacteria, plants, and mammals, respectively, corresponding to approximately 1.1, 1.3, and 1.7 × 10^6 daltons (5000 nucleotides for 29 S RNA). In mammals the 28 to 30 S value includes the 5.5 S piece described separately, but even without this piece the S value remains higher than in bacteria. The RNA is prone to split in two 16 S fragments, suggesting a discrete structural weakness in the molecule. The small RNA species is 16 S in bacteria and 18 S in plants and mammals (it is less compact in plants); the weights are 0.56 and 0.7 × 10^6 daltons (2000 nucleotides), respectively. Under the electron microscope, eukaryotic 28 S RNA is twice the size of 18 S and not the 2.5 times indicated by their weight ratio, which brings them in line with the weight ratio (about 2) of the bacterial equivalents (Granboulan and Scherrer, 1969). Thus bacterial rRNA is smaller than animal, and plant rRNA is intermediate. In lower algae the RNAs are bacterial-like; in dinoflagellates, plant-like (Rae, 1969).

In the eukaryote there is some sort of evolutionary trend in that, while the small RNA species has remained at 0.7 × 10^6 daltons, the large species has increased in weight: 1.3 × 10^6 daltons in plant and ciliate protozoa, 1.4 in *Drosophila*, 1.5 in *Xenopus*, and 1.7 in mammals. There are exceptions, however, e.g., amoeba is 0.9 and 1.5 (and very unstable). The emerging evolutionary picture is, therefore, that already in ciliates the complement differs from the bacterial but resembles that of the plant, and from ciliates to higher animals the large RNA species increases in size (reviewed by Loening, 1968b). The two RNA species weigh the same in all tissues of an organism.

The composition of rRNAs is GC type. Typical compositions (A/U/G/C) for the large RNA species are: *E. coli* (52% GC DNA) 25/22/32/21 (53% GC), plants 25/21/32/22 (54), and mammals 18/18/34/31 (64). For the small RNA species the compositions are: *E. coli* 24/22/32/22 (54), plants 25/24/30/21 (51), and mammals 22/22/30/26 (56). However, *Drosophila*, with 41% GC DNA resembling mammals, has RNAs of AU type, 31/27/22/20 (42) and 29/27/24/20 (44) for the large and small RNA species, respectively (the former being particularly labile). Also another insect *(Chironomus)* has an AU type RNA; its DNA is 30% GC (Daneholt and Edström, 1969). Thus, both plants and mammals, but not bacteria (or insects), have a small RNA species more AU type than the large, that is, have a greater compositional difference between the two RNA species. In addition, mammals are richer in C than plants, and therefore show greater base equality.

Considering all organisms, these data indicate that the GC content of the two RNAs has increased generally, but in the large species it has increased the most (Amaldi, 1969). By comparison, no increase is seen with GA or GU which do not represent paired bases. All bacterial microorganisms, with a widely variable GC composition in the DNA, have RNAs of rather similar GC composition; the small rDNA fraction is therefore less variable than total DNA. The similarity in GC composition for each species is also true between higher organisms with much less variable DNA. The evolutionary trend toward a higher GC composition suggests an increase in secondary structure, which is confirmed by measurement.

The two RNAs differ in sequences between them in both HeLa (Amaldi and Attardi, 1968) and bacterial cells (Aronson, 1963), indicating distinct cistrons for each of the RNAs. Sequence heterogeneity exists within each RNA species. In bacteria, according to Britten and Kohne (1969), it is even possible that each of the (few) cistrons for 16 and 23 S RNA has a different sequence; this high degree of heterogeneity, however, is uncertain. In the eukaryote, although each RNA species represents one or at most a few families of sequences (as defined later), each has a similar gross composition in all tissues of the organism (Hirsch, 1966).

Closely related bacteria show very similar sequences within each RNA species, much more so than in mRNA, but less similar sequences as the relationship and DNA composition become distant. DNA hybridization shows that within bacteria or higher organisms the rate of evolution of rRNA has been slow, since relatively similar sequences are preserved between phyla. There is therefore a certain conservatism of the molecules. However, between bacteria and higher organisms the sequences are entirely dissimilar (Loening, 1968b). As expected, the dissimilarity is to some extent a function of the number of generations intervening between bacteria and the particular present-day eukaryotic organism, i.e., more generations have elapsed from bacteria to fungi than to mammals (Bendich and McCarthy, 1970). Thus all three parameters, namely, molecular weight, sequence, and composition, have not remained completely stable over the wider range of evolution.

In *Escherichia coli* the major methylated sequences (up to about 12) occur twice in 23 S RNA, suggesting that the molecule is made of two rather similar halves (Fellner, 1969). In 16 S RNA certain methylated sequences are repeated, again suggesting duplication. Since this is also found in 5 S RNA, it may be a general property of rRNAs. Possible explanations for this modality are either the doubling and joining of an ancestral gene after a period of divergent evolution, or a convergent evolution toward a symmetrical molecule resulting from equally symmetrical functions in rRNA (Woese, 1968a).

Both RNA species are permanently associated with ribosomal proteins. From the reconstitution of ribosomes using the rRNA and ribosomal proteins from different bacterial species, it appears that a small evolutionary conserved sequence is involved in the specific interaction with the proteins (Nomura *et al.*, 1968). Cotter *et al.* (1967) concluded from physicochemical studies in yeast ribosomes that (1) the RNA conformation is similar to that in the free state with 60% base pairing; (2) the proteins are not associated with the RNA helices resulting from the pairing but may be packed into the nonhelical RNA loops and/or associated with one another in some form of quaternary structure; (3) the ribosome surface is chiefly rRNA, not protein; and, by inference, (4) the superficial rRNA is largely in the form of helices projecting outward. A linear arrangement of the various ribosomal proteins along specific binding sites on the RNA is proposed by Cox and Bonanou (1969). It is now believed that most of the ribosomal proteins bound to rRNA are internal (core) ribosomal proteins (Mizushima and Nomura, 1970); they also appear to be partly associated with RNA helices and by implication, in reference to point (4), some helices may be internal. From the models mentioned here the ribosome represents a very high order quaternary structure whose functional implications will be discussed in Chapter 5. With respect to rRNA itself, it is also apparent that it may possess an elaborate and pliable higher order structure, perhaps not essentially different from that of tRNA.

Minor Species of Ribosomal RNA

5 S RNA

This small RNA of about 120 nucleotides (41,000 daltons) occurs in all types of ribosomes, but is either smaller or altogether absent in the mitochondrial ribosome (Borst and Grivell, 1971). It is intimately associated both structurally and functionally with the large ribosomal subunit. 5 S RNA occurs in equimolar amounts with the large rRNAs and is about 1% of total cell RNA (Comb and Zehavi-Willner, 1967). At present, its structure and function are not known, but it has been implicated in subunit association (Petermann et al., 1969) and attachment to tRNA (Siddiqui and Hosokawa, 1969), and, more recently, in the mechanism of peptidyl transferase. The origin is extranucleolar.

This stable RNA molecule has an overall composition approximating that of total tRNA: in the fungus *Blastocladiella* 5 S RNA is (A/U/G/C) 19/24/32/25, and mixed tRNA is 16/26/33/26. It differs from tRNA in having only traces of or no pseudouridine or methylation, different termini, and no amino acid acceptor function. Hybridization to DNA indicates that the two sequences are different (Zehavi-Willner and Comb, 1966), as expected for distinct RNA species. The total sequence in bacteria is now directly comparable with that of individual tRNA species.

One molecule of 5 S RNA is bound to the large ribosomal subunit by Mg^{2+} bridges involving specific ribosomal proteins, and it can be removed with little loss of protein. The RNA binds more tenaciously to the eukaryotic than to the bacterial subunit (Zehavi-Willner, 1970). There is one specific binding site at the point of contact of the ribosomal subunits and another site for which tRNA competes. The bound 5 S RNA exchanges with the free RNA at high concentration. Because it had not proved possible to cleave the RNA cleanly and then restore it effectively to the subunit, until recently a strict test of dependent function was lacking. It was known, however, that at the time the RNA is being partly restored it becomes resistant to nuclease, which may mean that it enters the ribosome (Forget and Reynier, 1970). A partial functional requirement of the RNA (measured by poly U-directed incorporation of amino acid) has finally been reported after complete dismantling and reconstitution of the bacterial subunit (Nomura and Erdmann, 1970). In this regard, a certain failure of functional restoration might be expected if 5 S RNA can only enter the large ribosomal subunit during its assembly or during the period when the RNA itself is an immature precursor. However, since the same argument applies to bacterial 16 S RNA and the small ribosomal subunit, and this has been functionally reconstituted rather successfully using the mature RNA, it is not certain that the argument is valid. Hopefully, this work may soon lead to fruitful analysis of 5 S RNA function. In the meantime, the question of its universal requirement for ribosomal function must await a decision on whether or not the RNA is missing from the mitochondrial subunit.

The total sequence of 5 S RNA is known both in *E. coli* (Brownlee et al., 1968) and KB human carcinoma cells (Forget and Weissman, 1969). In the two bacterial strains investigated each contains two forms in equimolar quantities, one in common for both strains and one typical of each strain (with a single base difference), denoting separate structural genes for each congener. Comparison of the bacterial and mammalian RNAs reveals an authentic sequence similarity, that is, a partial preservation of sequence (Dayhoff, 1969). The bacterial molecule contains both terminal regions of homology and internal regions of homology, yet a lengthening process more complex than a simple doubling of the sequence is not excluded. The RNA from two mouse strains and human KB cells have identical fingerprints indicating a similarity or even identity of sequence (Williamson and Brownlee, 1969).

From hydrodynamic studies the tertiary structure of 5 S RNA appears to be rather rigid but more asymmetrical than in tRNA (Boedtker and Kelling, 1967). It has a measured helical content greater than 60%. The actual shape of the molecule is prolate (axial ratio 1:1:5) with a 31-35 Å radius of gyration (Connors and Beeman, 1970). As with tRNA, a highly ordered structure is indicated by the spontaneous reconstitution of the molecule (as measured by ribosomal subunit binding) starting from split molecules (Jordan and Monier, 1971). Based upon physical and biochemical data a number of models of secondary structure have been proposed for 5 S RNA. The first of these (not in chronological order) is that of a hairpin studded with narrow loops (Forget and Weissman, 1969). Second, from models of tRNA, Raacke (1968) tentatively proposed a cloverleaf configuration as shown in Fig. 8a. Third, from computer analysis, Lapidus and Rosen (1970) proposed the cloverleaf shown in Fig. 8b with a maximum possible base pairing (68%). Of these models, the second is worth discussing since it raises some interesting points.

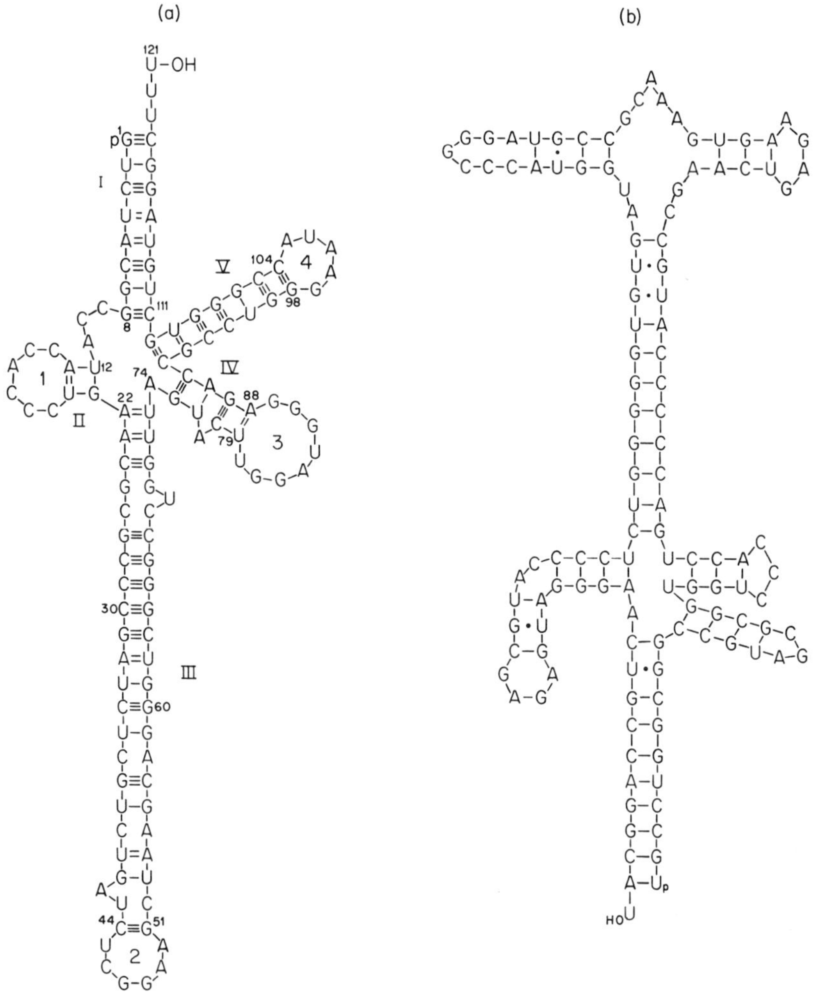

Ribosomal RNA

Fig. 8. Secondary structure of 5 S RNA. The two hypothetical models are (a) for human KB cells based on the cloverleaf model of tRNA (Raacke, 1968) and (b) for a computer-designed molecule with a maximum double-helical content (Lapidus and Rosen, 1970). To compare these structures, (b) should be viewed inverted with the free ends of the molecule pointing upward.

In the bacterium, Raacke visualizes a molecule with a helical content of 68%, the nonstandard GU pair being used extensively, while in KB cells the corresponding figure is 70% (Fig. 8a). (Note the good fit with the computer data above.) The stem is 11 nucleotides long as in all known tRNAs, equal to one complete turn of the RNA helix (Arnott *et al.*, 1967). The deployment of arms and loops is the same as in tRNA, and possibly capable of folding flexibly into various tertiary configurations. Loop I, with a common pentanucleotide in the two 5 S RNAs, is thought to bind to the site in the large ribosomal subunit. Loop II is species-specific and may hold the subunits together. The speculation is that tRNA and 5 S RNA lie side by side on the ribosome, but, whereas tRNA would just reach the small subunit and mRNA, 5 S RNA would extend some 30 Å beyond and across mRNA. Loop III contains the common UGGG sequence that by suitable folding of the structure can pair with the common GTψCG sequence in Loop IV of tRNA. The final suggestion is that 5 S RNA is directly bound to the ribosomal subunit, and in turn binds the juxtaposed tRNA through base pairing and Mg^{2+} bridges between phosphates. It should be stressed that, insofar as interaction with tRNA is concerned, this model of secondary structure of 5 S RNA largely depends on the tertiary structures of the two RNA molecules, which so far remain hypothetical. However, the model makes two testable predictions: (1) the binding of 5 S RNA to the ribosomal subunit is not species-specific, but (2) the binding between subunits is. Unfortunately, as described in Chapter 5, (2) is not true. In conclusion, the higher order structure of 5 S RNA is not known.

Interesting if not readily interpretable data are available from denaturation studies on the molecule. Bacterial 5 S RNA occurs in two forms after denaturation in the absence of Mg^{2+} (Aubert et al., 1968). Form A is chromatographically indistinguishable from the native, but does not bind the large ribosomal subunit too strongly during ribosome reconstitution. Form B has a more expanded or asymmetric conformation with a little higher sedimentation value, has lost the binding ability, and tends to aggregate under conditions where the native or A forms do not. Form B completely reverts to A by heating in the presence of Mg^{2+} or high ionic strength, but may not renature completely. Form A has as many bases as the native form, but arranged differently and less stably. Curiously, form B has less bases (Scott et al., 1968). Altogether, it would seem that form A is a conformer of the native molecule and form B is a degradation product.

As found in bacteria, 5 S RNA has a precursor with a more expanded configuration and slightly longer sequence.

5.5 S RNA

This 130 nucleotide long RNA is found only in ribosomes of higher organisms. A single molecule is H bonded apparently by a few bases to each 28 S RNA molecule, representing a virtual quaternary RNA structure. Apparently also, protein is not involved in this association, suggesting that it takes place in a relatively protein-deficient region of the RNAs, probably at the surface of the large ribosomal subunit (King and Gould, 1970). The RNA derives either from the intermediate 32 S RNA leading to 28 S RNA or slightly later in the pathway. The amount is 1% of cell RNA. The composition is (A/U/G/C) 22/22/29/27 with much less methylation than 28 S RNA, i.e., less than one methyl group per molecule (which suggests a heterogeneous molecular population). The structure and function of 5.5 S RNA, or its effect on the conformation of 28 S RNA, are not known (Pene et al., 1968). The present S value is that given by Weinberg and Penman (1969); originally it was taken to be 7 S.

Nucleolar 8 S RNA

This species of ribosomal-like RNA (58% GC) was described in Novikoff hepatoma and consists of several subspecies (Prestayko et al., 1970). The RNA is H bonded to 28 S RNA in the nucleolus (again representing a virtual quaternary structure), but not in the cytoplasm. The molar ratio is 0.5, indicating that not all molecules of each RNA species are bonded to the other; it is no longer associated with 28 S RNA as this reaches the cytoplasm. The RNA enters the rRNA pathway after the 36 S RNA stage. The origin is possibly nucleolar, but the derivation from the common rRNA precursor is not certain at this date. The unknown function of the RNA would however appear to be intranucleolar. The rate of synthesis is slow; the half-life is not known.

8 S RNA is apparently not methylated. A methylated RNA species with a sedimentation value of 7 S, called U3, seems to have the same behavior with respect to 28 S RNA as 8 S RNA. It also consists of several subspecies. However, both the ribosomal character and derivation of U3 are not clear.

Mitochondrial and Chloroplastal Ribosomal RNA

Although less is known about these RNAs, mitochondrial and chloroplastal rRNAs of the eukaryote possess certain characteristics which warrant their separate mention from the cytoplasmic rRNAs discussed thus far. In the lower eukaryote the mitochondrial RNA species are closer in size to the 16 and 23 S bacterial RNA species, i.e., ranging from 14 to 19 S and 21 to 25 S, but their GC content is lower. (A low GC RNA has an S value somewhat smaller than a high GC RNA of equal length.) In *Neurospora,* for example, the average composition of mitochondrial rRNA is 32% GC (Edelman et al., 1970) compared with an average 53% GC in bacteria. In the higher eukaryote the mitochondrial RNA species are smaller and extremely fragile, which until recently prevented their identification. The S values are 13 and 21 in *Xenopus* (Swanson and Dawid, 1970) and 13 and 17 in mouse (Montenecourt et al., 1970). The average composition in *Xenopus* and cultured human HeLa cells (Zylber et al., 1969) is 44% GC. All mitochondrial rRNA has therefore 15 - 20% less GC than the surrounding cytoplasmic rRNA. In addition, contrary to size, the composition of higher eukaryote mitochondrial rRNA is therefore closer to the bacterial. These comparisons with the bacterium are of interest because, as discussed later, there is the strong presumption that the organelle has originated directly from a prokaryotic ancestor. Yet it is also clear from the present data that there has been a subsequent evolution of mitochondrial rRNA in respect to its physical parameters. On this basis, future structural correlations between bacterial and mitochondrial RNA species may prove to be rewarding.

One such correlation refers to the presence of 5 S RNA in the mitochondrion. Three possibilities according to Lizardi and Luck (1971) (see also Borst and Grivell, 1971) are that this RNA is (a) undistinguishable by size from tRNA; (b) retained in a joint molecule with the large rRNA species (as is postulated for the nascent bacterial molecules; see Chapter 4); or (c) altogether absent.

Concerning the mitochondrial ribosome, this is 70 to 80 S in the lower eukaryote, suggesting some differences with the 70 S bacterial ribosome. The higher eukaryote mitochondrial ribosome is the smallest one so far discovered, e.g., 60 S with 32 and 43 S subunits in *Xenopus* (Swanson and Dawid, 1970), which are in accord with the small size of the rRNA.

Chloroplastal rRNA is also closer, yet not identical, to bacterial rRNA. In the alga *Euglena* the chloroplastal RNA species are 16 and 23 S and have an average 44% GC content (Scott *et al.*, 1970). The chloroplastal ribosome is 70 S as in bacteria.

TRANSFER RNA

This best characterized of RNA species was named by Allen and Schweet (1960) on the basis of the tRNA-mediated transfer of amino acid to the nascent peptide (von Ehrenstein *et al.*, 1963). This process (a) involves, first, adapting the amino acid to a molecule (tRNA) in the language of the code; second, the actual reading of the code transcript (mRNA); and third, donating the amino acid to the polypeptide. Transfer RNA intervenes in many other cell functions. It participates (b) as a cofactor for group transfers to amino acid, e.g., the amidation of glutamic acid to glutamine destined for protein synthesis (Wilcox, 1969); (c) in direct transpeptidation of amino acid to the N terminus of acceptor protein; (d) in nontranslational, enzyme-mediated oligopeptide synthesis (Chapter 5); (e) in the synthesis of aminoacylphosphatidyl glycerol in bacteria (Gould *et al.*, 1968). In addition, the RNA has been postulated to form (f) an integral part of a transcriptional repressor (Hatfield and Burns, 1970) or translational repressor (McLellan and Vogel, 1970) in repressible bacterial systems. Either of these two effects could resemble in nature (g) the inhibition by tRNA of certain eukaryotic enzymes unrelated to tRNA function (Jacobson, 1971b). Transfer RNA also seems to have other important roles in the regulation of transcription and translation which will be considered later. This functional variety makes it likely that certain tRNA species (or their synthetases) may have specialized for many functions other than the usual amino acid acceptor role. Part of the molecular degeneracy of tRNA, to be discussed, could therefore have purposes other than translation. The molecule is synthesized mostly in chromosomal loci, possibly partly in the nucleolar locus, and also in cytoplasmic organelles (mitochondria, chloroplasts). Excellent reviews on various aspects of tRNA were presented recently by Zachau (1969) and von Ehrenstein (1970).

During translation tRNA intervenes in three enzymic (covalent bond) reactions: (1) 3'-terminal turnover by cytidylyladenylyl transferase(s); (2) aminoacylation by synthetase; and (3) amino acid transfer to polypeptide by peptidyl transferase. Only (2) is specific to each amino acid and its corresponding tRNA. Several factor-dependent reactions take place during translation and are described in Chapter 5. These reactions are mediated by H bonds and hydrophobic interactions. All reactions depend, however, on the tertiary structure of the molecule.

The RNA contains about 75 ± 5 nucleotides (26,000 daltons). At the 5' end there is generally G, and at the 3' end always the labile CCA sequence required to accept amino acid. The composition of mixed tRNA from all sources, including bacteria with widely divergent DNA composition, is very roughly (A/U/G/C) 20/20/30/30. In contrast to rRNA, the base ratio by itself immediately suggests extensive pairing, hence, a highly ordered structure. As expected, however, individual sequences now known from bacteria to mammals depart significantly from the general composition given above. The actual number of tRNA species is considerably higher than the basic 30 plus predicted by the wobble hypothesis. In all, while preserving an essential similarity, these tRNA species display a marked microheterogeneity. A species is assigned preferably on the basis of adequate criteria of purity and function of the isolated molecule. Many peaks and shoulders in various chromatographic separations may only correspond to transitional states of the molecule (described below). At high resolution, however, those tRNAs for amino acids with the greatest number of codons (Table 1) also tend to show the greatest number of peaks and shoulders, which are close to the expected number of species.

The primary structure of tRNA is the most modified of all RNAs. All modifications are introduced at the polymer level (i.e., are not coded for directly) by means of a veritable enzyme cosmos that links the tRNA molecule to many pathways in intermediary metabolism. The modifications, adding up to 60 or more, are species specific, and can involve all bases and as much as one-quarter of the molecule. Among the modifications are methylation, pseudo- and dihydrouridylation, cytokinin formation (Hegelson, 1968), thiolation, and other less usual or less known modifications (reviews by Hall, 1970a,b; Söll, 1971). Methylation in mammals is to about one in ten nucleosides, and mostly in the base. A greater proportion of bases is therefore methylated than in rRNA. Methylation is both base and species specific. The methylases recognize base sequences longer than ten bases (Baguley and Staehelin, 1969) and do so directly and to a large extent irrespective of the steric accessibility of the bases (as revealed by heterologous methylases) (Svensson *et al.*, 1969). In addition, an intriguing possibility is that certain methylases are shared by tRNA and rRNA as substrates. Pseudouridylation of tRNA is in greater proportion than in rRNA, 25% of U. Thiolation is of interest because, in principle, it permits the formation of intramolecular disulfide bridges. Other modifications have a double or triple chemical character. A universal modification in tRNA is of course the terminal CCA group which is obligatory in order for translation to occur; yet this group undergoes continuous turnover, the meaning of which is not understood. The degree of total modification (except CCA) appears to be of evolutionary consequence. To begin with, phage-coded tRNA contains some ψ and thiol groups (Daniel *et al.*, 1970). Thereafter, the range of modification is minor in mycoplasma (Hayashi *et al.*, 1969) and increases progressively in bacteria, yeast, and mammals. A similar trend was described in rRNA. Probably, the structural complexity of the modifications has also increased during evolution.

The ancestral role of tRNA, suggested by its function in translation and outlined during discussion of the code (Chapter 1), is reflected in the molecule at every level of structure. Thus, present families of tRNA retain many traits of a common history, and sufficient structural conservatism is emerging perhaps to permit uncovering one day of the evolution of individual species. In contrast to the loops, there is greater conservation of the base-paired stems whose length tends to be preserved even if the individual bases are not, which must mean selective pressure in favor of complementary mutations in the stems. Other general homologies in the molecules are compiled by Dayhoff (1969). For example, for a given tRNA, e.g., serine tRNA in yeast and mammals, the change in sequence is only in one-quarter of the number of bases. Even greater similarities are preserved between more closely related organisms. Two examples of specific compositional regularities are that, when present, certain minor nucleotides in different yeast tRNAs all occur at identical positions; yeast and liver tRNA share all their six methyladenine-containing oligonucleotides (Baguley and Staehelin, 1969). In all, the molecule displays a fixed size as well as fixed distances between landmarks in its primary structure (Bayev *et al.*, 1968). Moreover, any additional base that appears is accommodated mostly in the variable length arm of the secondary structure (to be described). To give an idea of the overwhelming structural regularity, according to Levitt (1969), the universal secondary structure (cloverleaf) could occur by chance with a probability of 10^{-20}. By way of summary, the consensus is that the still unknown tertiary structure will turn out to be just as regular as the secondary structure, and, perhaps, more universal.

The secondary structure is represented in the Holley cloverleaf arrangement (reviews by Miura, 1967; Novelli, 1967) in Fig. 9. This arrangement has now been observed with electron microscopy (Ottensmeyer, 1969). It allows a maximum of base pairing (60%). Basically, the arms of the structure consist of helical

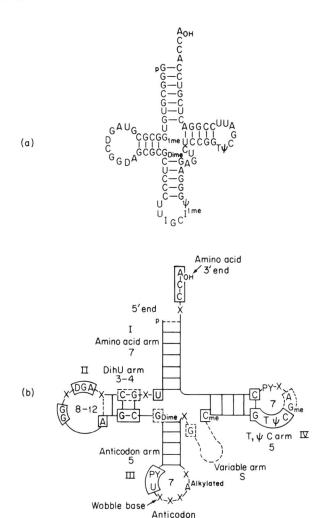

Fig. 9. Secondary structure of transfer RNA (tRNA). The models show (a) yeast alanine tRNA arranged in the cloverleaf structure (Holley *et al.*, 1965) and (b) a generalized cloverleaf structure. Base pairs are indicated by a line between two regions of the chain. Dashed lines indicate base pairs occurring in some tRNA species; X indicates a nucleotide that varies with the tRNA species. Sequences enclosed in solid boxes are common to all tRNAs except for nucleotide modifications. Dotted boxes enclose sequences common to most tRNAs. Numbers in the center of the loops and next to the arms indicate the number of nucleotides in the loop and of base pairs in the arm, respectively. Roman numerals indicate regional notation of the structure (adapted from Fuller and Hodgson, 1967, as modified by von Ehrenstein, 1970).

stems and terminal loops stabilized by the stacking of bases.[9] Arm I has a rather constant length and few if any minor bases. The extra arm S between arms III and IV is variable in length. Dihydro U occurs mostly in loop II, ψ in arms III and IV, and the common sequence GTψCG in loop IV.

The tertiary structure is compact, highly ordered, and behaves cooperatively during such treatments as denaturation, solubilization, and resistance to endonuclease and formylation (review by Cramer, 1971). The crystalline structure of mixed tRNA crystals attests to an overall regularity of the tertiary structure (Blake et al., 1970). Several proposed configurations with differently oriented arms (H, stem, Y) are reviewed by Levitt (1969), including one of his own (Fig. 10). For example, the folding may be with arms II and IV up, I and III down (Fig. 10A). A few other proposed models have a central loop (Fig. 10C); a coaxial helical stem made of arms I, III, and IV (Fig. 10D); and interloop base pairing (Fig. 10F). By this time the reader should have realized that the planar representation of the models in Fig. 10 does not permit one to visualize readily their three-dimensional appearance. However, arm I probably will be largely inside the structure, except for the protruding 3' terminus. Most of the few exposed unpaired bases are in the anticodon loop (Metz and Brown, 1969), explaining its chemical susceptibility. Of course, exposure of both the 3' terminus and anticodon is a functional necessity. As measured in yeast phenylalanine tRNA, the 3' terminus and anticodon are more than 40 Å apart (Beardsley and Cantor, 1970); it can be seen in all of Fig. 10 (except model B) that these two structural landmarks are not neighbors. There is still the question as to what conformation of tRNA is the one represented in these models, the most modern of which attempt in part to conform with recent crystallographic data. For example, in Levitt's model (Fig. 10E) the axial ratio of 4 approaches the ratio of the denatured rather than native molecule (see below). According to Cramer et al. (1970) the crystalline conformation resembles the native one, although for technical reasons this can be difficult to prove (Fresco et al., 1968). That the answer to the present question may not be a simple one derives from the real possibility that different conformers may be required by the manifold functions of the molecule, that is, the structure of tRNA may behave dynamically in regard to function. All considered, the native structure will probably be elongated, the 3' terminus and anticodon in somewhat antipolar positions, and the side arms wrapped around transversally.

[9] The cloverleaf is not simply a "secondary structure" as is usually described. First, it involves strictly secondary structure in the overall stacking of bases, but also a low order tertiary structure in the base-paired helical stems (see footnote 5). However, as pointed out by Zachau (1969), base stacking confers as much rigidity to the molecule as base pairing, and it is not very meaningful to distinguish between these orders of structure (secondary and low order tertiary) in the case of tRNA. Second, to anticipate the text, there is a high order tertiary structure consequent to folding of the cloverleaf which is usually described simply as a "tertiary structure." It is this latter highly ordered structure, governed by H bonds, electrostatic, and nonpolar forces, which gives the molecule its decidedly protein-like character.

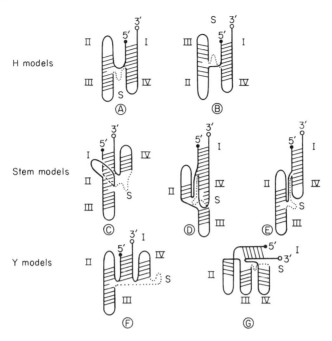

Fig. 10. Planar models of tertiary structure of transfer RNA. Model (A) by Lake and Beeman (1968); (B) Doctor *et al.* (1969); (C) Ninio *et al.* (1969); (D) Fuller *et al.* (1969); (E) Levitt (1969); (F) Cramer *et al.* (1968); and (G) Melcher (1969). Helical regions of the molecules are shown as straight lines joined by the tilted base pairs. The variable S region is dotted. For description of regional nomenclature of the structure in roman numerals, see Fig. 9 (adapted from Levitt, 1969).

At least four specific sites are required for tRNA function: (a) the 3' terminus for aminoacylation; (b) the anticodon in arm III; (c) the nonspecies specific ribosomal (or 5 S RNA) site that might involve the common GTψCG sequence in arm IV (Ofengand and Henes, 1969); and (d) the synthetase recognition site centered either on the stem of arm I and particularly in the first three base pairs (Schulman and Chambers, 1968), or in the four base pairs in the stem of arm II (Dudock *et al.*, 1971). By analogy with general enzyme structure, it was pointed out [for site (d) in particular] that a more sensitive recognition would be achieved if the relevant bases were brought together by tertiary structure (Zamecnik, 1969), which is probably what actually happens. Thus, because of the involved tertiary structure, these various sites [particularly (c) and (d)] are influenced and indeed probably shared by other regions of the sequence. However, despite constraints the tertiary structure is highly resilient, which could affect the function of the sites. This consideration applies particularly to sites (c) and (d), but, more generally, it means that different regions of the molecule do influence one another's function according to the conformation of the molecule at any time. Thus, to the extent to which this interaction occurs, it might not be permissible to speak of a discrete regional differentiation of tRNA in respect of its functions. It might also be difficult, if not impossible, to assign unique sites to all the interactants with tRNA during translation.

Proof of the existence of the anticodon is provided by the general experimental agreement with the wobble hypothesis and, formally, by one suppressor tRNA that arises from a single base mutation in the anticodon (Goodman *et al.*, 1968). Anticodon composition may in some obscure way be correlated with molecular composition of tRNA (Woese *et al.*, 1966), which could imply a causal influence of the anticodon on the evolution of the molecule. A model by Fuller and Hodgson (1967) shows that maximum base stacking in the single-stranded anticodon loop uniquely solves the structural relationship with the codon in that both come to lie in the same generic helix and all the wobble is absorbed by the anticodon. The last point was recently confirmed by oligonucleotide binding (Uhlenbeck *et al.*, 1970). In addition, the substituted base present at the 3' side of the anticodon prevents the reading of triplets overlapping the codon and stabilizes the stacking in the anticodon loop; the latter aspect is supported by physicochemical measurements by Leonard *et al.* (1969). In summary, according to von Ehrenstein (1970), because of its orderliness and the many facts it explains, the Fuller model may be essentially correct.

A general pattern of substitution at the 3' side of the anticodon is related to the first letter of the codon (Hall, 1970a). Thus, if this letter is U, the substitution is a cytokinin, and, in the case of the eukaryote, also a Y derivative; if C, there is a modified G; and if A, a modified A. No clear pattern has emerged for G.

An accepted function of the minor bases of tRNA is to influence the tertiary structure as their predominance in loops and single-stranded regions seems to indicate, e.g., in the anticodon loop just mentioned. These base substitutions may do for tRNA what proteination does for rRNA and mRNA, hence their prevalence in tRNA which lacks proteination. Single base substitutions can affect more than one tRNA function such as codon binding and aminoacylation (Thiebe and Zachau, 1968). For example, base methylation can suppress some Watson-Crick pairs and change other pairs to non-Watson-Crick. Methylation could also affect enzyme recognition sites in general [as suggested by the phage restriction enzyme (Meselson and Yuan, 1968) that recognizes a methylation pattern]. Undermethylation of tRNA is reported to cause alterations in conformation, codon response, and amino acid acceptance (review by Starr and Sells, 1969). However, owing to the difficulties of adequate physiological testing, with few exceptions, the actual function of the modifications is more assumed than known. Perhaps it may soon be possible to examine this function further using artificially produced tRNAs (which are not modified), but, unfortunately again, not in the cell. Nevertheless, the long-range effects of the modifications along the molecule such as are known are consistent with the presence of a highly cooperative tertiary structure, and seem to affect the higher order of this structure. A plausible argument for modification is that it achieves a more efficient tertiary structure than could be specified directly, in other words, the primary structure provides a framework for enzymes to achieve the final structure, among other things more resistant to nucleases. Lastly, it must not be forgotten that modification may not only represent a tooling of the molecule for translation, but also provide a means for recognition in certain regulatory pathways.

As described above the molecule contains essential elements of tertiary structure which result from interaction between the unpaired residues mainly in the loops and the rest of the molecule after folding. Partial helicoidality makes tRNA more cation-sensitive than, for example, DNA. Apart from the biologically active form, a few tRNA species can be trapped in a metastable, inactive or partly denatured state (Adams et al., 1967; Ishida and Sueoka, 1967) which is incapable of any enzymic reaction. It is thought that this transition manifests an underlying physicochemical property common to all tRNAs. The transition is a reversible first-order reaction requiring a variable activation energy as high as 60 kcal. The transition depends on withdrawal of Mg^{2+}, but the cation stabilizes both native and inactive forms. The stability of each form sometimes depends on aminoacylation (Novelli, 1967). Compared with the native form (axial ratio 1.5), the partly denatured form has a more asymmetrical conformation (axial ratio 5); the denatured form corresponds to a revolution ellipsoid of 40 by 60 Å. The denatured form is probably less base paired but without overall loss, though with possible significant deviation, of secondary structure. (Under more extreme conditions, however, it can denature completely or unfold irreversibly to a random coil.) Because of conformational slippage, the denatured form changes in codon response (see below). Full renaturation takes place (in contrast with 5 S RNA) in the presence of monovalent or a lesser concentration (5 mM) of divalent counterions, or under conditions that labilize H bonds. The rate of renaturation varies with the tRNA species. The divalent ions are the physiological ones, and they alone elicit cooperative behavior in the molecule. Thus, the molecule fluctuates between two relatively stable tertiary structures, one "right" (native) and one "wrong" (partly denatured). For those tRNA species requiring renaturation, the active conformation would seem therefore to depend on elements (one or a few Mg^{2+} atoms) other than those in the primary structure, and the counterion would provide the balance of interaction required to maintain the native structure. However, it has been challenged that Mg^{2+} is obligatory to the native form (Ishida and Sueoka, 1968), and the question of the precise element required must remain open. In general, it should be stressed that apart from these stable denatured states, any partial unfolding or change of shape of the molecule is completely reversible.

A well analyzed structural rearrangement in conformation caused by the previously described treatments is that of tryptophan tRNA of *E. coli* resolved by MAK chromatography into two interconvertible forms (Gartland *et al.,* 1969). Form I, enzymically active, binds to the ribosome in response to the appropriate codons (UGG, poly UG). Form II, enzymically inactive, does not respond to these codons but responds instead to poly AC. However, the poly AC-dependent binding is not conducive to transfer of amino acid and hence it cannot be of the normal codon-anticodon type. Moreover, form II uniquely shows other unusual binding reactions, such as (1) binding to poly AC without the ribosome being present and (2) binding to the ribosome without polymer being present. The codon response of form II is therefore abnormal as the result of a structural rearrangement exposing a different sequence as anticodon. In the cell, 90% of this tRNA species is in the active form.

From the previous observations it seems likely that allosterism, a typical protein behavior, is applicable to tRNA. For example, when acylated the tertiary structure of tRNA becomes looser (Kaji and Tanaka, 1967). The RNA can adopt three distinct conformations according to whether it is unacylated, acylated, or acylated and N-blocked (Stern *et al.,* 1969). This allosterism implies a virtual functional substructure of the molecule, such as, in fact, is underscored throughout this section. Substructure is further revealed by the substantial recovery of acceptor activity that follows reconstitution of tRNA split in halves (Philippsen *et al.,* 1968), which must reflect an inherently ordered structure such as is reminiscent of an enzyme. The same is revealed by the fact that yeast phenylalanine tRNA undergoes no less than five conformational transitions during melting (Römer *et al.,* 1970). In conclusion, from what has been said it is clear that firsthand knowledge of structural-functional relationships may come from the physicochemical study of tRNA. The recent crystallization of several species (e.g., Clark *et al.,* 1968) augurs well in this respect.

Although tRNA is not bound to protein, evidence from mammalian cell extracts indicates that the RNA is in equilibrium with soluble protein(s) as a dissociable complex sensitive to the environment (Hess *et al.,* 1961). These proteins could in part be synthetases or translation factors, but the point is that "free" tRNA is dynamically even if not stably associated with protein. Transfer RNA also appears to be loosely bound to endoplasmic reticulum membranes in the cytoplasm.

Binding of tRNA with the ribosome occurs at two sites—one for aminoacyl-tRNA in the small ribosomal subunit and one for peptidyl-tRNA in the large subunit—and requires Mg^{2+} (Rudland and Dube, 1969). Perhaps a more realistic view is that, although the sites are in separate subunits, each one shares the site with the other to some extent. Ribosomal proteins are directly or indirectly involved in the binding of tRNA, and rRNA might be involved also, depending on its arrangement with the ribosomal proteins. Binding of both aminoacyl-tRNA and peptidyl-tRNA is however not passive, for they are directly involved in the binding of the two subunits to each other (Belitsina and Spirin, 1970). In the presence of the appropriate codon, each tRNA species binds to the ribosome specifically, which represents, of course, the physical basis for the specificity of translation.

To date, it is impossible to predict the number of tRNA species in the cell (Novelli, 1967). By the latest count there are at least 80 species in mammals to correspond with about 60 codons (Yang and Novelli, 1971). Consider that leucine may have six species (Yang and Novelli, 1968a) when only three would suffice, according to wobble hypothesis. A convenient distinction within a synonym set of tRNAs for one amino acid (or isoaccepting tRNAs) is therefore between synonyms varying in the anticodon sequence (degeneracy) and synonyms varying in the sequence elsewhere than in the anticodon (redundancy). By this definition, degeneracy of tRNA corresponds clearly with degeneracy of the code, but it is equally clear that the number of codons in a set (Table 3) does not mean as many anticodons because of the wobble; however, when wobbling cannot occur (e.g., because of the tertiary structure), this would tend to increase degeneracy. Also by this definition, redundant species within a synonym set recognize the same codon. As the result of gene duplication and mutation these redundant species are not rare (Abelson et al., 1969), and in fact they alone are responsible for the total species outnumbering the codons. As can be expected of their highly ordered structure, redundant species have different thermosensitivities in regard to their functional properties—again resembling proteins. Irrespective of the two types of synonyms, particularly in the eukaryote each tRNA species is coded by several identical structural cistrons, that is, there is cistronic multiplicity (Chapter 4). Crossover between adjacent cistrons coding for the same tRNA species is detectable (Russell et al., 1970).

Transfer RNA has a precursor that is longer than the mature molecule (Bernhardt and Darnell, 1969; Altman, 1971) and is biologically inactive (Sirlin and Loening, 1968). A less compact structure of this precursor suggests also conformational changes during maturation (Burdon and Clason, 1969). Synthesis of the precursor depends partly on methylation and its structural maturation is likewise somewhat dependent. From what was previously said, it would be the functional maturation of the precursor which is the most dependent on modification.

OTHER LOW MOLECULAR WEIGHT RNAs

Chromosomal (Derepressor) RNA

Chromosomal RNA is present in the chromatin of eukaryotic and higher plant nuclei in amounts up to about 5% of the DNA, or 1-2% of total cell RNA. The amount depends on the transcriptive activity of the nuclei and the RNA is almost absent from inactive nuclei. The RNA is believed to function in derepression of the genome (Bonner et al., 1968), and on this premise it assumes a great interest. However, as explained later in Chapter 3, the RNA is conventionally rather than as a matter of proved fact referred to as "derepressor" and mainly to distinguish it from other nascent RNA in the chromosome. The RNA will be considered as a bona fide species, yet until its structure and function are proved it remains a putative species; some recent work (Heyden and Zachau, 1971) has questioned its existence.

The sequences of this small RNA species are very heterogeneous, representing a family with many subspecies each present in but a small number of copies. The sequences are partly organ specific. Consequently, by comparison to other species, the amount of DNA coding for chromosomal RNA seems to be unusually large, i.e., about 4% of total DNA, according to Bonner et al. (1968). As far as is known the species are not related to any other outside their family.

The length of the sequences is relatively homogeneous, about 50 nucleotides with an average sedimentation coefficient of 3.2 S, which in itself suggests a similar function. The composition is AU rich and the RNA is little methylated. It contains 5-25% dihydropyrimidine, possibly present in an open ring configuration and the result of posttranscriptional modification.

The kinetics of synthesis is not known, but the RNA is made at approximately the same rate as tRNA and therefore cannot be pulse labeled (Dahmus and McConnell, 1969). However, this could still mean rapid synthesis and slow modification (Bonner, 1969). For reasons that will become clear, it remains important to measure the kinetics of synthesis.

The RNA is covalently bound in the chromosome to an acid-soluble protein, and both, in turn, H bonded to aggregated histone. The binding protein consists of a single species of about 10,000 daltons which resembles histones even though it is not one of them. The RNA in the complex is ribonuclease-resistant, therefore probably base paired to DNA (or protein).

About half of the sequences of the RNA are found in the chromosome and the other half in the nuclear sap in association with protein. They are also reported to be associated with protein in the cytoplasmic supernate (Dahmus and McConnell, 1969). If verified, the cytoplasmic localization could have interesting implications.

4 to 7 S Nuclear (Nucleolar) RNA

Although the lowest range of sedimentation coefficients of these RNA species overlaps that of tRNA, they are different species (Table 5). These species, first observed in nuclei of rat liver (Hodnett and Busch, 1968; McIndoe and Munro, 1967) and HeLa cells (Weinberg and Penman, 1969), are exclusively eukaryotic

TABLE 5

Small Nuclear RNA Species of the HeLa Cell [a]

RNA species	Estimated number of molecules		Estimated number of bases	Composition (AUGC)
	Nucleoplasm	Nucleolus		
A	–	2×10^5	180	23-29-28-20
B	–	1×10^4	170	–
C	4×10^5	1×10^5	165	21-31-27-21
D	1×10^6	7×10^4	150	19-27-29-25
F	–	3×10^4	125	–
H	2×10^5	1×10^5	100	26-27-30-17

[a] Data from Weinberg and Penman (1969). Only well defined species larger than H are included in the table.

and include about 25 distinct monodisperse nonribosomal species. Their occurrence is known to be widespread in vertebrates, and to a certain extent to be species but not cell type specific. According to the species, they contain from 100 to 180 bases with an average composition (A/U/G/C) 22/29/29/21. The GC content varies between 48 and 60%. Most of them are extensively methylated.

At least one species predominates or resides exclusively in the nucleolus and the others reside in the chromosome. The nucleolar association depends on normal nucleolar function. The RNAs have a sensitivity spectrum to antibiotics similar to that of rRNA, but can be made in the absence of rRNA synthesis. Thus the derivation remains unclear but even in the case of the nucleolar species it is unlikely to be from the common rRNA precursor. The RNAs are not precursors to cytoplasmic RNA. All species appear to be stable but their half-life varies widely from one to several days according to the species. This range means that in the HeLa cell (generation time, 1 day) some of them reenter the nucleus upon mitosis. Their monodispersity, stability, and analytical reproducibility exclude the possibility that the molecules are degradation products.

Individually, the 4-7 S RNA species can amount to 0.5% of total cell RNA. In their variable numbers (maximum 10^6) they remain fewer than cell ribosomes (5×10^6), yet the most abundant of these species are the most numerous of any RNA in the nucleus (compared with an estimated $1\text{-}2 \times 10^4$ molecules each of precursor rRNA and labile RNA). Both the function and the functional interrelationship between species are not known, but they are not concerned with nuclear protein synthesis. Their relationship to the smaller chromosomal RNA is of interest, particularly in view of their nucleolar occurrence. Their dissimilar stability, numbers, and localization perhaps suggest more than a single function. An attractive possibility that comes to mind is a regulatory participation in transcription. If that is the case and these 25 species recognize DNA sequences, then it is clear that the DNA sequences must be common to many genes. Another possibility is some kind of "structural" function in transcription, perhaps still regulatory. Whatever their function, these species might well open new vistas into the expression of the eukaryotic genome.

Some species are transcribed throughout the cell cycle and others only after DNA replication (Clason and Burdon, 1969). The timing of transcription is not related to that of the major RNA species (Weinberg and Penman, 1969). Their posttranscriptional maturation, if any, is short. These species may remain loose in the nucleus, and some are possibly not associated with either protein or chromatin; this condition would require a higher order structure to prevent enzymic degradation. According to Enger and Walters (1970), however, these RNAs may all be in RNP particles. Several monodisperse, metabolically unstable molecules are extractable from the cytoplasmic supernate, whose electrophoretic properties, if not identical with those of the nuclear species, nonetheless "closely shadow" them. Speculation on the relationship between the nuclear and cytoplasmic counterparts must await proof to exclude procedural leakage from the nucleus, even though, at present, their different stability points against this artifact.

Bacterial 7 S RNA

So far, this recently found RNA species (Hindley, 1967) is perhaps the only bacterial RNA whose function and relationship to other bacterial species is not understood. No relationship with the previous eukaryotic 4-7 S RNA is implied here. This homogeneous RNA (55,000 daltons) occurs free in the cell and does not accept amino acid. The fingerprint is distinct and the composition (A/U/G/C) is 22/23/29/26, with U present at both termini. The structure of the RNA has a considerable degree of H bonding (60%) (Brownlee, 1971; Harewood and Goldstein, 1970). The kinetics of synthesis is fast and the molecule is stable. The amount in the cell must be small but is not reported.

Homopolymers

Several reports in the literature deal with the presence of different homopolymers in the cell nucleus, of which poly A is chosen for discussion here. Ascites cell nuclei contain a polynucleotide rich in A (90%) of about 300 nucleotides (8-10 S), which amounts to 1% of total cell RNA (Edmonds and Caramela, 1969). In HeLa cells, it was found that a fraction (0.9%) of labile RNA larger than 45 S includes runs of poly A and these are detectable only in the nucleus (Edmonds and Vaughan, 1970). Poly A has been described in nuclei of other vertebrates by Kato and Kurokawa (1970). However, it appears as if the homopolymer is also present up to 5% in DNA-like RNA obtained from ribosome and reticulum membrane fractions of the cytoplasm (Lim and Cannellakis, 1970), in which case the possible relationship with the nuclear counterpart has to be considered. Bacteria also have poly A, but its covalent linkage to other RNA species is not clear. The amount relative to total bacterial RNA is only one-tenth of that in the eukaryote (Edmonds and Kopp, 1970). Whereas some speculation will be offered for the role of eukaryotic poly A, none can be offered for the bacterial polymer.

The synthesis of nuclear poly A appears not to depend on DNA (Darnell *et al.,* 1971) which distinguishes the enzyme involved from RNA polymerase. Thus poly A would not be transcribed covalently with RNA but would be added after transcription. A different nuclear poly A, poly ADP ribose, is made from nicotinamide adenine dinucleotide by a DNA-dependent enzymic activity; curiously, this activity is also relatively insensitive to actinomycin (Chambon *et al.*, 1966). The role of the eukaryotic homopolymers is unknown and translation of poly A into lysine-rich histone was first mentioned as a possibility (Edmonds and Caramela, 1969), although not an extremely likely possibility. Other more realistic possibilities in relation to transport or translation of mRNA will be considered in Chapter 5.

Viruses also contain homopolymers. A quarter of total reovirus RNA consists of copies of a single strand (mainly A) of at least 15 nucleotides (Shatkind, 1969). The possibility of a linker for the ten two-stranded segments of this RNA virus (total 1.5×10^7 daltons) was mentioned. However, it seems that the polymer may be present in the infected cell but not in the virion (Krug and Gomatos, 1969), in which case the function remains open. Poly A (about 10 S) is reported to be transcribed from poly dAT clusters in vaccinia DNA virus and perhaps to be covalently bonded with the viral mRNA (Kates and Beeson, 1970). This viral poly A could then be related to the pyrimidine-rich clusters which are implicated in the initiation of transcription in certain phages and are described in the next chapter. Yet, conceivably, it could instead be related to functions similar to those which will be considered for eukaryotic poly A.

Part II

Functional Aspects of RNA

Chapters 3-5 discuss the transcriptional and translational aspects of the various RNA species with the accent on the regulation of these processes in the cell, including the participation of the RNAs themselves in the regulation. We will therefore be dealing with the expression of genetic information in the cell regulated via the function of RNAs acting in several different capacities.

CHAPTER 3

Transcription of DNA-Like RNA: Messenger RNA and Eukaryotic Labile RNA

This chapter and the one following deal with the transcription and regulation of RNA species. Ideally, by the end of these chapters, the aim would have been to establish some basic organizing principles for these processes. However, the trouble with these kinds of principles is that they are not always easy to delineate, particularly in respect to the eukaryote. For the sake of inclusiveness, therefore, the broader evidence on transcription will be presented first and, as we proceed, the emphasis placed on what appear to be certain guidelines for regulation.

Previously, in discussing the general mechanism of transcription and the parameters of RNA molecules, we looked at the molecules chiefly by themselves. In approaching them now rather in terms of their many interactions, we are likely to be confronted by two realities: first, the one created by the questions we ask about these interactions, which is connected to a second – the intrinsic reality – in measuring the validity of these questions. This chapter first discusses in general terms the structural basis of transcription and then the needs with regard to the cell that are to be fulfilled by the regulation of transcription in the prokaryote and eukaryote. Following this, the regulation of the various RNA species is treated individually for the two types of organisms. The regulation of informational RNA (mRNA) synthesis is discussed, which reveals a different strategy used in the two types of organism. Then, the possible function of eukaryotic labile RNA is examined in relationship to informational and regulatory transcription. Finally, the regulation of stable RNA synthesis is considered (see Chapter 4).

GENERAL STRUCTURAL BASIS OF TRANSCRIPTION

Prior to discussing the transcription of individual RNA species, the purpose of this section is to review the structural and molecular data that form the basis for understanding the general process of transcription at the biological level. Relatively less has been learned with the electron microscope of the physical framework for RNA synthesis in the nucleus than has been learned, in general, of protein synthesis in the cytoplasm, particularly in the eukaryote. This disparity of knowledge has several reasons (discussed below), but, because of it, the interpretation of structural observations on the nucleus must lean heavily on the molecular data.

In viruses, the folded DNA within the phage head is, perhaps, loosely associated with protein, but any special advantage for study that might be gained from this arrangement is lost by the fact that transcription occurs within the bacterial host cytoplasm. By contrast, recent important progress has been made in bacteria. Due to the facts that, first, bacterial DNA is largely (though probably not entirely) devoid of protein and not highly supercoiled, and, second, that mRNA transcription is physically linked with translation, it has been possible to identify complexes of DNA, RNA polymerase, mRNA, and ribosomes (Revel et al., 1968a; Miller et al., 1970a). This complex is illustrated in Fig. 11. Prior to these direct observations, all that was known of nascent mRNA was by way of biochemical reconstruction. Nascent bacterial rRNA, undergoing proteination as it is made, has also been identified. Thus, a fruitful analysis of the physical attributes of bacterial transcription may be forthcoming.

The ultrastructural complexity of the eukaryotic nucleus is consonant with its macromolecular complexity and a more sophisticated methodology is awaited to resolve it (Monneron and Bernhard, 1969). As far as the study of RNA synthesis is concerned, this ultrastructural complexity offers little hope of immediate success in the interphase chromosome (Wolfe, 1969). Progress has, however, been made in relation to the understanding of chromosome organization (DuPraw, 1968; Ris, 1969), although this relies considerably on observations made on metaphase chromosomes. Much of the difficulty in the field of chromosome organization lies, on the one hand, in reconciling the limited structural observations with what is expected to be the optimal configuration and, on the other hand, with the restrictions imposed by cytogenetic data. Most of the evidence on eukaryotic chromosomes points to the fact that one chromatid contains one continuous DNA helix (20 Å diameter), but this interpretation is by no means universally accepted (see Wolff, 1969). It is believed that a basic fibril of 40 Å diameter containing coiled DNA-histone and wrapped in protein to make a larger fibril of 100 Å, is actively transcribed in the euchromatic regions of the nucleus. During replication the fibril is partly denuded of protein (Moses and Coleman, 1964), and to some extent this must be true also of transcription. In the heterochromatic regions of the nucleus the 100 Å fibril is supercoiled to give an inactive fibril of 250 Å (cf., Davies and Small, 1968).

Structural Basis of Transcription

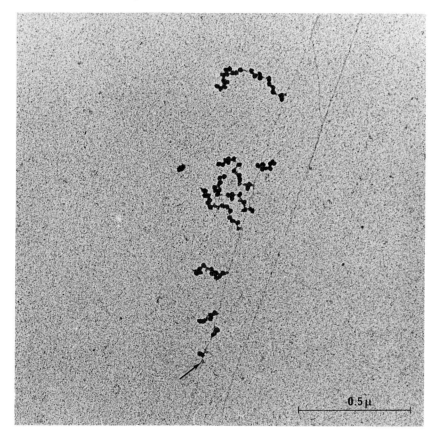

Fig. 11. Visualization of a presumptive operon in *Escherichia coli.* A strand of DNA is seen transcribed into nine fibrils of putative mRNA with ribosomes already attached. The mRNA fibrils are shortened relative to the length of the operon, which extends from just in front of the shortest fibril up to or just past the longest fibril. The direction of transcription is, therefore, from bottom to top. Lighter particles on the DNA axis at the base of the fibrils are believed to be RNA polymerase molecules; one of these is shown by the arrow probably near the promoter site. (Courtesy of Drs. O. L. Miller, B. A. Hankalo, and C. A. Thomas.)

An alternative view on these structures is that the basic 40 Å fibril is supercoiled directly to give the inactive 250 Å fibril. The coiling and supercoiling of fibrils are to a great extent due to association with protein and, presumably, mostly histone.

The intimate molecular organization of the basic fibril is also disputed. However, it is believed that at least some of the histone occupies the major groove of the DNA helix and is disposed at 60° to the helix axis (Zubay, 1964), whereas the RNA polymerase is believed to move along the minor groove of the helix. This leaves unaccounted for the sites occupied by a substantial part of the chromosomal constituents, mainly acidic (residual) proteins. While part of these proteins are probably permanent chromosomal constituents, part of them represent enzymes, transcription factors, etc., some of which may be in a continuous dynamic flux with respect to the chromosome. Because this flux is likely to apply to many nonprotein constituents as well, the chromosome in effect must be viewed as a "statistical" structure. Even the permanent proteins probably change their disposition according to their functional condition. Divalent metals are also an integral part of the chromosome which have both structural and metabolic functions. All these chromosomal constituents are known to affect markedly the physical properties of the DNA helix (see, for discussion, DuPraw, 1968), which however retains a B-like conformation. It can only be hoped that future reconstruction work on the chromosome starting from the isolated macromolecules will provide some ideas about their mutual arrangement.

Concerning the common structural basis of transcription, one possible generalization is that the chromosome has to be in a loose state to maintain RNA synthesis, as seen in euchromatic regions of the nucleus and in puffs of polytene chromosomes. The inactive chromosome remains condensed, as seen in heterochromatic regions, partly no doubt because of limited space within the nucleus. During transcription, the supercoiled DNA described earlier is thought to uncoil and stretch out in loops, as was first observed in lampbrush chromosomes (Swift, 1965). Presumably, this uncoiling of DNA is to facilitate destabilization of the helix (and perhaps the flux of molecules about it), while retaining a template continuity. Concomitant changes at the gross nuclear level are seen in the form of the swelling of chromatin that precedes transcription in nuclei transplanted from cells in which they were not transcribing (Gurdon, 1968; Harris et al., 1969). During this swelling more template becomes available as judged by the increased affinity of chromatin for acridine orange (Ringertz and Bolund, 1969). This change is considered to be preparatory rather than causative for transcription. Thus, nuclear volume becomes proportional to the amount of transcription (Stöcker, 1964). Conversely, actinomycin may prevent transcription in part by tightening DNA aggregation (Cavalieri and Nemchin, 1968). By and large, however, our picture of the fine structure of RNA synthesis in the eukaryote is biochemically inferred rather than directly observed. Nevertheless, a turn of the tide is marked by recent work on amphibian oocyte nucleoli where the complex of nascent proteinated rRNA, proteinated DNA, and RNA polymerase has been observed under the electron microscope (Miller and Beatty, 1969a) (see Chapter 4). It was found that the rRNA polymerase is larger (125 Å diameter) than the bacterial polymerase (75 Å) now described.

The following description of polymerases and other elements responsible for transcription deals first with the prokaryote and then the eukaryote. The simplest RNA polymerases thus far known are the eukaryotic mitochondrial polymerase (Küntzel and Schäfer, 1971) and the T7 phage polymerase (Chamberlin et al., 1970) which weigh 64,000 and 100,000 daltons, respectively. Contrary to all other polymerases to be described, these two polymerases consist of only one protein chain, which undoubtedly has to do with the simplicity of the genomes they transcribe. However, the question of whether or not all functions of these polymerases are discharged by the single protein has to await clarification as to whether or not they require the initiation factors that are required by the complex polymerases. The bacterial enzyme has a molecular weight of 0.4×10^6 daltons (Lubin, 1969; review by Richardson, 1969). What is known as the core enzyme consists of four chains: two α (39,000 daltons each), one β (155,000), and one β' (165,000); the enzyme is therefore $\alpha_2\beta\beta'$. This enzyme is associated with the DNA from which it can be isolated. It was first observed that at least part of the core enzyme intervenes in the synthesis of all RNA species (Yura and Igarashi, 1968); it is now believed that the complete core enzyme may be the sole bacterial vegetative enzyme. However, it was also found that associated proteins act as selective initiation factors (Burgess et al., 1969). Hence the concept developed from work on both bacteria and phages is that the core enzyme is concerned with transcription and an associated σ factor (90,000 daltons) with positively selecting the right template for transcription (Sugiura et al., 1970). The holoenzyme then becomes $\alpha_2\beta\beta'\sigma$. Lastly, it was found that a termination factor ρ complements the σ factor by ensuring that transcription is terminated at the correct point on the template (Roberts, 1969). The ultimate role of these factors (σ, ρ) is, therefore, to provide an accurate translation greater than that provided by the basic signal in the translation initiation and termination codons. The specialized DNA sequences that these factors in part seem to recognize (see below) are presumably near to, but outside, the informational sequences. In the eukaryote, the nuclear polymerase forms an integral part of chromatin (Bonner et al., 1968). It is recovered as an "aggregate enzyme," which can be subsequently solubilized. This integration is perhaps to be expected for a multimolecular chromosome, but it also means that the polymerase may interact dynamically with other chromosomal molecules. This polymerase includes more than one distinct class each responsible for transcribing different RNA species such as rRNA, mRNA, and perhaps tRNA (Pogo, 1969; Roeder and Rutter, 1970). These distinct classes of polymerase all have a similar molecular weight but a different multiple subunit structure. It is possible that these nuclear polymerases are separately controlled, but some degree of joint control, particularly if they share subunits, is possible. Nothing is known with certainty about transcription initiation and termination factors in the eukaryote, but there is little doubt that they are required for most if not all RNA species; initiation factors for rRNA will be considered later.

The asymmetry of bacterial transcription probably depends upon the intact configuration of the template-enzyme complex in the cell (review by Smellie, 1968), some structural aspects of which were considered in Chapter 1. Clearly, this complex represents a very high structural order which is essential to understand before the transcriptional mechanism can be understood. More immediately, the asymmetry also depends on the presence of the appropriate initiation factor (σ). It is proposed that transcription initiation (and possibly termination) signals are provided by clusters of 15-40 pyrimidines (especially T) in the one or two DNA strands transcribed according to the organism (Taylor *et al.*, 1968). In synthetic DNA with pyrimidines in only one strand, these clusters result in a slightly altered helix, which signifies an effect of DNA composition on structure. It is supposed that a similar type of structural singularity in the native template strand somehow signals initiation (termination) to the polymerase, presumably in the presence of σ (ρ) factor and with the participation of Mg (Wilson *et al.*, 1970). In the case of initiation, these pyrimidine-rich sequences correspond with the promoter site of the bacterial operon. [Another device proposed for termination in λ phage involves collision of polymerase molecules traveling on the two strands in opposite directions (Szybalski *et al.*, 1969).] What determines asymmetry of transcription in the eukaryotic nucleus remains to be seen, but, as mentioned, initiation factors also appear to be involved. In this regard, the promoter of a mammalian virus, SV40, is recognized by the bacterial polymerase, suggesting a certain universality of promoter sequences. An interesting possibility suggested for the eukaryote is that an enzymic conversion from C to T (Sneider and Potter, 1969) could selectively destabilize the template and thereby mark initiation. In the case of single-stranded RNA viruses, higher order structure and/or specific sequences in the template are important in RNA replication. For example, *in vitro* the replicase will not use randomly segmented Qβ phage RNA (Mills *et al.*, 1967), but segments can be selected to replicate presumably by retaining the 3' terminal structure relevant to initiation. Also in organisms with rigid helical DNA, a higher order structure of the template may play a role in transcription, but to what extent this occurs is less clear than in the case of viral RNA replication. Finally, concerning the direct participation of polymerases in transcriptional regulation of the cell, some of the data presented earlier in this section suggest that qualitative changes in the complex enzyme substructure may be significant in that regard. Contrary to older views, however, quantitative enzyme changes appear not to be significant.

Topographical considerations in bacteria require that during transcription the DNA spin about the polymerase (rather than the other way around), perhaps utilizing the energy from triphosphate nucleosides (Maaløe and Kjeldgaard, 1966). The DNA would be spun out from the "nucleus" and transcription would occur at the interface with the cytoplasm. This is supported by the visible inaccessibility of the "nucleus" to the translating ribosomes, which are known to be coupled with transcription. These considerations do not apply to replication during which the genome is thought to be threaded through a single replication point at the cell membrane (Chapter 1). They would also not apply to transcription if this took place at the cell membrane and immediately following replication (Mueller and Bremer, 1968). Oddly enough, the point remains that transcription cannot be inside the bacterial "nucleus." It shows that, although the bacterial "nucleus" and cytoplasm are in intimate physical and physiological contact (there is no nuclear envelope), to some extent they remain discrete from each other. Evidently, the introduction of the eukaryotic nucleus has reversed this spatial constraint in the ancestral transcription.

GENERAL BASIS OF REGULATION IN THE PROKARYOTE AND EUKARYOTE

The primary reason for regulating transcription is that, by turning it on and off, protein synthesis is controlled at the origin. Most of the available information concerning regulation at the level of transcription comes from bacterial genes that make mRNA. For bacterial genes making other RNA species, the views on regulation are largely inferred and derived from what is known of the genes that make mRNA. Regulation of eukaryotic transcription is less well understood because of its complexity.

To begin with, in both types of organisms regulation must enable cells to cope with fluctuation in their immediate environment, which is the external habitat for the unicellular prokaryote and the internal milieu for the multicellular eukaryote. In general, the prokaryotic cell reacts much faster metabolically to a changing environment (as well as to its endogenous requirements). In the eukaryote the regulation reflects the division of labor as represented in an enormous variety of cell types. To explain this point further, in the prokaryote all or most vegetative genes (excluding the spore forming) come into operation sooner or later during each cell cycle (because this is also the life cycle), but this is not true of many genes in the eukaryote (with separate cell and life cycles). Hence, part of the basic regulatory differences between the two organisms stems from the cellularization (and attendant neural and hormonal connections) typical of the eukaryote. As we shall see, this eukaryotic property (cell differentiation) has resulted in a strategy of regulation unlike that in the prokaryote. By contrast, the lower (unicellular) eukaryote has a certain regulatory resemblance to the prokaryote.

This changed relationship between organism and environment has created nutritional differences. In general, in contrast to the prokaryote, on becoming a partial auxotroph the higher eukaryote has lost the ability to manufacture some essential metabolites and relies for their origin on its biotic environment. As it happened, higher eukaryote evolution could not have been energetically possible without the advent of photosynthesis.

The two kinds of organisms also differ in the turnover of macromolecules. In growing bacteria the rate of macromolecular synthesis is high and the rate of degradation (except for mRNA) is comparatively low; protein is removed mainly through mRNA instability or is diluted by growth. However, resting bacteria lose considerable protein and RNA at a rate of about 5% per hour, and resynthesis balances this loss. In 6-8 hours one-third of the protein will have cycled through the amino acid pool, ensuring that enough amino acid is available for newly induced enzyme (review by Mandelstam, 1968). On the other hand, the growing mammalian cell turns over macromolecules at a rate of 1% per hour. For a cell doubling time of 1 day this contributes to one-quarter of the new protein, or to more in proportion to a longer doubling time. In the resting mammalian cell, protein and RNA are in dynamic equilibrium, that is, continuously being removed through their instability and simultaneously replaced (Mandelstam, 1960). For example, adult liver cells, although dividing about once a year, replace their protein and stable RNA and possibly much of their mRNA every few days. Thus, the growing mammalian cell resembles the resting bacterium, and the difference between the growing and resting cell is less than in the bacterium. However, although to a different degree, in both organisms an increase in protein breakdown accompanies a reduction of RNA synthesis and vice versa, representing a complementary metabolic relationship in terms of cell growth. The mechanism behind this relationship is not known, but it is conceivable that RNA (possibly rRNA) regulation is involved.

Other general differences between the prokaryote and the eukaryote are as follows. First, the genetic organization (not the genetic code) changes significantly in respect of the number, redundancy, and active proportion of genes and their mode of regulation. Second, on present knowledge the virus or factor-mediated transfer of genetic information between bacterial cells, i.e., transformation and transduction, plays no major role in the normal eukaryotic cell, although it is not possible to say that it plays none. The same is true for the direct bacterial cell-to-cell transfer of information; DNA-induced transformation has been described in *Drosophila* (Fox *et al.,* 1971) but its physiological relevance is not known. Third, bacterial sexuality involves the sequentially ordered transfer of a single (haploid) somatic chromosome in radical difference with eukaryotic sexuality featuring meiosis at the diploid level and final transmission of an haploid germinal chromosome complement. Meiotic recombination, in particular, is greatly responsible for the unprecedented tempo of eukaryotic evolution, to which we shall return.

The reason for mentioning these general differences between the prokaryote and eukaryote — leaving more cell-specific differences for later — is to justify why, beginning with the study of mRNA, it is advisable to deal with the two types of organisms separately. For mRNA more is known of regulation in the prokaryote and, consequently, this will be considered first. In dealing separately with these organisms, the possibility should be left open that certain prokaryotic aspects of the regulation of mRNA may still be found in the eukaryote, particularly the lower eukaryote. For example, polycistronic operons inducible by substrate may occur in the fungus *Neurospora* (Rines et al., 1969). The reverse possibility, that the most typical eukaryotic aspects of the regulation will be found in the prokaryote, is, to be sure, unlikely.

Regulation of mRNA synthesis will be discussed mainly in terms of specific mechanisms or of data that point to specific mechanisms, when these are not clarified. Unspecific mechanisms resulting from conditions of metabolism, growth, etc., are mentioned but not considered consistently because they border on cell physiology and, as such, lie outside the scope of this book.

PROKARYOTIC MESSENGER RNA

Specific Regulation in the Prokaryote: The Operon Model

Bacterial regulation of mRNA transcription is typically expressed during enzyme induction-repression. During induction, presentation of an exogenous nutritional metabolite (or analog) rapidly causes synthesis of mRNA at full rate, and withdrawal causes rapid cessation of synthesis. Repression is the reverse, namely, the cancellation of synthesis with presentation of exogenous or endogenous metabolite, followed by resumption of synthesis upon withdrawal. Another form of repression is the latent or prophage condition of temperate viruses when integrated as episomes within the bacterial genome.

These phenomena are collectively explained in the operon model of Jacob and Monod (1961), whose elements are illustrated in Fig. 12. (a) A regulator gene controls (b) an operator gene that opens or closes, as the case may be, the transcription of (c) the structural genes (cistrons of Benzer, 1961) to be eventually translated. The regulator gene can be separate and in a different (*trans*) strand but the operator is always adjacent (*cis*) to the structural cistrons. The regulator is transcribed into a regulator mRNA, which produces an (apo)repressor protein (generally but not always acidic) that directly combines with the operator. This protein is therefore a freely diffusible genotropic molecule (Waddington, 1962). When the system is inducible this combination repressor-operator blocks further progress of the polymerase along the template and the operon as a whole is not transcribed. Conversely, in a repressible system the repressor by itself does not block the polymerase and the operon is transcribed at full rate. The operator may itself be transcribed, although presumably it is not translated. The status of possible nontranslatable sequences between structural cistrons is also not clear.

Two new genetic elements have been added to the original operon model. The polymerase initiates transcription from (d) the promoter gene in front of the operator. The promoter is thought to contain sequences rich in pyrimidines whose possible function in initiation have already been discussed. A promoter genetically inserted in an operon will not stop ongoing transcription, that is, can itself be transcribed. In addition, for initiating transcription correctly the polymerase requires the presence of positive factors. These σ factors, (1) possibly act more by determining an allosteric transition in the polymerase that enables its binding to promoter, than by recognizing the promoter directly; (2) are possibly more concerned with the binding of polymerase than with destabilizing the DNA; (3) are required for initiation of transcription (up to the first internucleotide linkage) after which they separate from the polymerase and are therefore not required for subsequent polymerization (Bautz, 1970). Another gene in front of the promoter, (e) the initiator, somehow specifies the particular repressor for the operon (Miller *et al.*, 1968); the initiator function is most clear in operons with an unusual positive type of regulation discussed below. It has also been found that cAMP interacts with the promoter and/or the polymerase via a binding protein which might, in effect, be the active element (Zubay *et al.*, 1970; Pastan and Perlman, 1970). This interaction is supposed to result in a general stimulation of the rate of initiation of ongoing transcription, as opposed to the specific derepression of each operon by the cognate inducer. Most operons for inducible enzymes are sensitive in this way to cAMP. The cAMP effect is further explained later in connection with the phenomenon of catabolite repression.

In an inducible system a specific inducer acts as an allosteric effector (Monod *et al.*, 1963) to cancel a normally active repressor, thus permitting transcription. In a repressible system the interaction of an end-product metabolite (corepressor) is supposed to activate a normally inactive aporepressor which thus binds to the operator and suspends further *de novo* transcription. Both systems are formally and mechanistically equivalent, the repressor acting always essentially in a negative fashion and withdrawal of effector restoring either system to its original condition. The reason that a system is either inducible or repressible is supposedly adaptive and in relation to the function of the pathway concerned. Nevertheless, the aporepressor of repressible operons has not to date been identified. Instead, feedback inhibition has been hypothesized as the repressor mechanism (see later), but this has yet to be established. Work on induction of the arabinose operon has revealed a different picture, however. It has indicated that the inducer detaches repressor from the operator gene, and converts it to an activator that attaches to the initiator gene, after which transcription ensues (Englesberg *et al.*, 1969). In this system the repressor acts rather like an initiation factor. For operons of the arabinose type the regulation is therefore positive.

In summary concerning the proximal regulatory genes of the operon (Fig. 12), these must be involved in complex interactions with several kinds of molecules such as inducer, repressor, polymerase, factors, and perhaps metabolites (cAMP). The DNA sequences involved in these genes are essentially for the purpose of these interactions. Understanding of these interactions has just begun. Perhaps a pertinent suggestion here is that the interaction may depend on a distinct local higher order structure of the DNA, so far unknown. (The reader is referred to footnote 5 for clarification of higher order structure as it pertains to DNA in particular.) Far from passive, these proximal regulatory genes also play a dynamic role in modulating the cistron response, but the primary and most important regulation remains with the discontiguous regulatory gene.

Two specific, mutually exclusive affinities of the repressor in the operon model are (1) to operator and (2) to either inducer (repressor does not bind operator) or corepressor (repressor binds operator). The first affinity requires native operator DNA (review by Bretscher, 1968a). The repressor, being an allosteric protein (16,000 daltons), contains four subunits. There are five to ten molecules of repressor of the inducible lactose operon per cell, and even less of the repressor mRNA (Müller-Hill et al., 1968). Repressor mRNA of viral origin has possibly been observed in lysogenic cells. As mentioned, to date the aporepressor of a repressible operon has not been isolated. Instead, it was proposed that this type of operon has a holorepressor which is a complex of the operon's own immature enzyme product and aminoacylated tRNA (Hatfield and Burns, 1970), that is, the repression would be by end-product feedback inhibition (hence not truly negatively determined as in inducible operons). This hypothetical holorepressor remains to be substantiated.

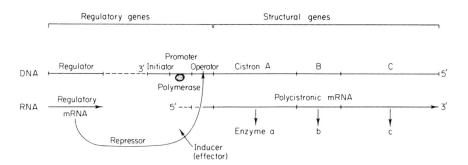

Fig. 12. The operon model of regulation of bacterial transcription represented for an inducible polycistronic system. Exogenous metabolite such as galactoside is the effector that releases the normally repressed system, hence subject to a negative type of regulation. The function of the noncoding regulatory genes is explained in the text; the function of the initiator gene is most clear or exclusive to the rare operons which are subject to a positive type of regulation. Also explained in the text are the seven types of molecules or molecular aggregates (not all represented) that participate in transcription: (1) DNA; (2) repressor protein; (3) inducer; (4) polymerase (coupled with σ initiation factor); (5) a stimulating protein (with cAMP); (6) a ρ termination factor that is most studied in bacteriophage; and perhaps (7) a regulatory ψ factor (with a guanosine derivative). The genes are not drawn to scale.

Certain genetic situations render immediately evident the regulatory nature of the operon. Mutation in the regulator or operator genes can make a system behave constitutively, that is, synthesis of mRNA escapes regulation. Regulator-constitutive synthesis can surpass in amount the fully repressed or so-called basal synthesis of the wild-type operon. A mutant promoter can abolish synthesis. A single base change in promoter or operator can destroy their respective recognition by polymerase or repressor. In other words, an operon can become constitutive because of a nonfunctional regulator or operator but must retain a functional promoter. A most dramatic proof of regulation is that if all proximal regulatory genes are genetically removed from an operon, then, provided the next operon at the 3' side is transcribed in the same direction, the structural cistrons come under regulation of the next operon. In contrast, inversion of an intact operon does not affect its regulation, even though this means moving it into the other DNA strand and reversing the direction of transcription relative to the chromosome.

A multicistronic operon explains coordinate enzyme synthesis in a biochemical pathway starting with a polycistronic mRNA (review by Epstein and Beckwith, 1968; Martin, 1969). However, an operon can be monocistronic, and most in fact are, with the same characteristics described. Thus the operon model does not set out to explain polycistronic regulation in particular. The model builds controlling elements into a system responding to the classic "one gene-one enzyme" concept, now more appropriately "one gene-one polypeptide." Since there is one separate and three adjacent regulatory genes per operon, the relative investment in control in terms of number of genes is in fact considerable, particularly in small monocistronic operons. (The relative investment is much less in terms of DNA sequence, because the adjacent regulatory genes, contrary to the separate regulator gene and the structural cistrons, occupy very short sequences.) The regulation is more of a quantitative than an all-or-none type, because, first, fully repressed operons are still transcribed, if only subliminally, leading to a few enzyme molecules per cell (basal transcription); second, derepressed operons are transcribed at full rate but the rate depends on such factors as the particular promoter and the level of cAMP.

The ultimate regulation of the regulator gene itself remains a complete mystery. In general, it is believed to be regulated constitutively, that is, not to depend directly on any other gene. Recent evidence suggests that this constitutive regulation is not by self-regulation (Morse and Yanofsky, 1969a). Regulation of the regulator gene is considered again later.

Operons of related functions under the control of a common regulator, but not necessarily adjacent, form a "regulon" (Maas and Clark, 1964). An example are the operons of the arginine pathway. A regulon could derive from an operon whose structural cistrons became disseminated, or, conversely but less likely, from single cistrons with whom replicas of regulatory genes of established operons became associated. A hypothetical arrangement of several operons whereby a gene in one operon makes repressor for a second operon, and so on, is said to constitute a "cascade" regulation (Waddington, 1962).

Regulation as discussed above is at the DNA level. In this respect, two noteworthy points can be considered. (a) The DNA is only indirectly within control by metabolite (inducer or corepressor) through repressor. (b) The polymerase itself is outside control of the operon, and the number of polymerase molecules is never limiting. A qualification concerning point (a) is the mentioned effect of cAMP (via a binding protein) on the promoter and/or the polymerase. Concerning point (b), several polymerase molecules simultaneously transcribe one operon (Kasai and Bautz, 1969); the molecules are seen to be present also in nontranscribed DNA regions (Miller *et al.*, 1970b). Additionally, in the bacterium there is no protein or RNA permanently associated with the DNA that could be involved in regulation. Bacterial polyamines, however, by neutralizing DNA phosphates, perhaps have an important regulatory role. Such a role, although so far not clear-cut, would be somewhat reminiscent of the role of eukaryotic histones.

Although the bacterial genetic system was treated here only in terms of DNA in the chromosome, in reality the system is more complicated. There are extrachromosomal genetic factors, e.g., colicinogenic and resistance transfer plasmids, also containing DNA and with a variable and complex degree of independent regulation. Eventually, these plasmids may be unified under a common function in cell surface structure (Luria, 1969; review by Richmond, 1970).

Less Specific Regulation in the Prokaryote

Certain aspects of regulation, representing a feedback from the physiological status of the cell, coexist with the specific induction-repression of enzyme synthesis via the operon system. Collectively, they are just as important in maintaining the homeostasis of the bacterial cell.

A crude general regulation of synthesis is due to the interweaving of metabolic systems, whereby depletion of nitrogen, oxygen, or carbon source all stop cell metabolism (review by Mandelstam, 1968). This lowering of basal metabolism preserves cell viability at least for a period of time.

An important factor that initially controls the inductive-repressive effects of exogenous metabolite is cell permeability. The responsible permeases are made constitutively or derive from the basal (repressed) transcription of the operon eventually responsive to the metabolite.

A pervading unspecific aspect of regulation at the transcriptional level is the so-called catabolite repression, whereby accumulated catabolite appears to generally block informational transcription. This type of repression is believed to be mediated by a reduced intracellular concentration of cAMP that would normally interact with the core polymerase and/or promoter (Pastan and Perlman,

1970). In effect, cAMP counteracts catabolite repression (Ullmann and Monod, 1968). As explained, this metabolite stimulates physiologically the ongoing rate of initiation of transcription of operons in general, while the specific derepression of each operon remains under the control of its appropriate inducer. Evidently, the single bacterial vegetative polymerase is compatible with such a general regulation by cAMP, yet this appears not to have any effect on synthesis of stable RNA. Actually, the stimulation of informational transcription is by a specific protein which binds, and thereby is activated by, cAMP. The metabolite could however also in part act on transcription indirectly via translation (Yudkin, 1969). Clearly, this kind of regulation is most important because, for example, when sufficient carbon source is available, the reduced cAMP level spares enzyme synthesis even in the presence of inducer.

In addition to the operon system and catabolite repression, regulation at the transcriptional and translational level is reinforced by the allosteric end-product (feedback) inhibition or activation of enzymes. The enzymes inhibitable by this means always belong in repressible pathways. This quite specific mechanism is not gene mediated and does not affect enzyme levels but acts at a posttranslational level to control enzyme activities. Again, the inhibition spares enzyme synthesis. In turn the end-product inhibited enzymes may in some manner affect the expression of the operons of which their own structural cistron is a part. As proposed by Hatfield and Burns (1970), these enzymes may act as part of a transcriptional repressor. This is not, however, the only possible interpretation of the enzyme effect, which is further discussed in Chapter 4 in connection with tRNA and amino acid metabolism.

General Aspects Related to Bacterial Messenger RNA

Having discussed the specific regulation of prokaryotic mRNA by means of the operon model, we now examine some of the more distant implications concerning mRNA in relation to various aspects of cell function. To be sure, because other areas of regulation are involved (the full evidence cannot be presented here) the connection with the operon model will not in all cases seem direct. In these cases, the main purpose is to call attention to some of the most important macromolecular aspects related to mRNA.

Enzyme synthesis in anabolic pathways (amino acids, purines) is often repressible by the terminal anabolite. In catabolic pathways (carbon or energy sources), enzyme synthesis is inducible by the substrate. This contrast in the regulation of enzyme synthesis is clearly built into the operon model, which ensures in both pathways the same rapidity of response typical of the bacterium. However, most essential intermediary pathways are not closely regulated and the operons in these pathways could be regulator-constitutive. More generally, the question arises as to whether or not all enzyme synthesis is regulated according to the operon model. This question touches on the problem of the ultimate control of metabolic homeostasis and cannot, of course, be fully answered.

At least for those enzymes in the same pathway requiring coordinate synthesis, it may be said that a polycistronic arrangement offers a common and efficient regulation, even though a very large operon would make regulation too rigid (as in the case of some bacteriophages). In certain operons, like the lactose operon, there is a translation polarity, that is, the molar amount of different enzymes produced decreases as a function of distance from the operator. Despite the overall decrease, the molar ratio of the enzymes remains fixed regardless of the extent of induction-repression in the operon or the presence of polar (nonsense) mutations (review by Ames and Martin, 1964). Therefore, even if its mechanism is not fully understood, the physiological translation polarity now under discussion is different from the translation polarity introduced by nonsense mutations. (With these mutations, less distal mRNA is available probably as a consequence of its rapid degradation in the absence of ribosomes translating past the nonsense codon.) In these operons with physiological polarity, the order of cistrons corresponds with the decreasing number of copies required of each enzyme (because of their different turnover number) and not with their order in the biochemical pathway. The allocation of cistrons is therefore presumably adaptive. More generally, that is, during evolution a polycistronic operon can be formed by duplication and then modification of structural cistrons, or by the coming together of related cistrons under common regulatory genes (Mc Fall and Maas, 1967); contrast this with previous views on the evolution of a regulon. Additional advantages of a polycistronic operon in relation to bacterial sexuality will be mentioned later.

The attractive features that were pointed out notwithstanding, the polycistronic operon is not a prerequisite for coordinate regulation. Enzymes not requiring coordinate synthesis can still be regulated as monocistronic operons. One possible generalization is that genes for single-purpose pathways (amino acids) tend to be associated in polycistronic operons, and those for multi-purpose pathways (aromatic compounds and purines) scattered as monocistronic operons. This generalization could suggest that the polycistronic operon format may fit particularly those enzymes which form aggregates (Ginsburg and Stadtman, 1970). It might, therefore, be that some of the intricacies of regulation will not be understood before the quinternary structure of these enzyme aggregates is elucidated (review by Bock, 1969). Concerning those monocistronic operons functionally related in a pathway, it is clear that they also must be harmoniously regulated in relation to one another. The regulon and cascade arrangement of operons suggest themselves for this purpose.

Analysis of the *E. coli* tryptophan operon during derepression (Morse *et al.*, 1968) reveals that translation is coupled with transcription. The mRNA of this operon represents five structural cistrons with a total of 6700 nucleotides

(33 S; 2×10^6 daltons). An intrinsic feature of the promoter is that it determines the periodicity of initiation of transcription by a mechanism as yet unknown but probably independent of the repressor. Only rarely is transcription of the operon initiated internally owing to the presence of a second, less efficient promoter. Under the experimental conditions used, transcription of the operon takes 6.5 minutes with a new round beginning every 2.5 minutes. At most three polymerase molecules travel at any one time along the operon. Each mRNA molecule is translated once by each of 100 ribosomes moving uniformly as a tight cluster behind the polymerase. Translation starts at full rate as the first mRNA molecule is being transcribed and ends soon after the last is transcribed. Two conclusions from the study of this operon are that (a) mRNA production is regulated first by the rate (frequency) of initiation and, second, by the rate of transcription, and both rates vary in the same direction; (b) transcription and translation are physically coupled and the time taken by each individual round of transcription and translation is the same. Recently, however, the Yanofsky group has revised their previous estimates to 15 polymerase molecules per operon and 20 ribosomes per mRNA, and have questioned the regular periodicity of initiation initially proposed (*Cold Spring Harbor Symp. Quant. Biol.*, 1970). These data can now be compared with observations on disrupted cells under the electron microscope by Miller *et al.* (1970a). These authors identified polysomes in general with a maximum of 40 ribosomes each, concluded for an irregular periodicity of initiation, and found few free polysomes, thus proving the coupling of transcription with translation. (These observations were conducted on cells under conditions of catabolite repression which could have depressed transcription-translation and affected translation independently, as well.) Considering the whole cell, the overall quantity of mRNA present per ribosome is constant and therefore independent of individual operons (Maaløe and Kjeldgaard, 1966). This relationship indicates uniform spacing of ribosomes on all mRNAs. An important rule follows: the number of ribosomes in the cell is proportional to the overall rate of protein synthesis.

Degradation of tryptophan mRNA starts somewhere at the free (5′) end and may progress in a random fashion as the cluster of ribosomes travels behind the polymerase. Under some experimental conditions, degradation starts, in fact, before transcription is completed (Morikawa and Imamoto, 1969). Whether the onset of degradation determines the completion of ribosome loading, or vice versa, is not certain, but the latter alternative is favored. The general direction of degradation is therefore compatible with the view that ribosomes protect the mRNA. Moreover, degradation is specifically stimulated by (corepressor) tryptophan, which could represent an end-product inhibition effect. Possible enzymes involved in mRNA degradation were mentioned in Chapter 2.

Of special interest is how the tryptophan operon regulates the output of the enzymes it specifies when the rate at which the cell is growing undergoes variation (Rose *et al.*, 1970). This variation in the growth rate can be up to sixfold depending on the composition of the culture medium. In the normal case of the repressible operon, the specific activity of the enzymes decreases as the growth slows down. The reduction of enzyme is believed to be the result of repression of the operon, since this would explain the lower enzyme specific activity sufficient to maintain the steady-state level of tryptophan in the slow growing cell. By contrast, when the operon becomes regulator-constitutive (not repressible), the specific activity of the enzymes remains constant as growth slows down. This observation reveals that when repression of the operon is not functional, some other regulatory mechanism ensures that the output of enzyme remains proportional to the overall rate of protein synthesis (rate of cell growth). What is this regulatory mechanism in the constitutive operon? It is known that the rate of tryptophan mRNA elongation as well as the rate of peptide elongation vary with the growth temperature, e.g., the two rates decrease by one-half when the temperature changes from 37° to 25°C. At any given temperature, however, the two rates of elongation (as well as the rate of mRNA degradation) are largely invariant with respect to the rate of cell growth. It follows that the mechanism of constitutive regulation of enzyme synthesis probably involves mainly the frequency of initiation of transcription. As described, the frequency of initiation determines the frequency of translation, hence the final output of enzyme.

The fact just mentioned that transcription and translation are coupled could imply that the product RNA inhibits the polymerase (Stent, 1966). It is now known that ribosome binding or binding of the small ribosomal subunit alone, in the presence of translation initiation factors and the appropriate codon, stimulate both transcription and release of mRNA from the template *in vitro* (Revel *et al.*, 1968a). The factors are thus what appears to couple transcription with translation. Messenger RNA synthesis *in vivo* is likewise proportional to the amount of small ribosomal subunit available (Forchhammer and Kjeldgaard, 1968). It would seem from these data that either ribosomes or proper translation initiation conditions, rather than translation per se, are able to stimulate transcription. Further evidence excluding translation comes from the RC control which will be discussed later, namely, stringent cells deprived of amino acid synthesize mRNA but not protein. The effect of stimulation could well be by avoiding polymerase inhibition as suggested above, but more subtle explanations involving the higher order structure of transcription are possible. In the normal course of events, however, translation in the cell proceeds to completion after the conditions for initiation are given. Hence the operon becomes both the unit of transcription and translation. In addition, because transcription is also partly related to replication in time and location (see later), and replication is absolutely dependent on translation (Lark, 1969), it follows that a good deal of macromolecular synthesis in the bacterium is effectively integrated.

Coupling of transcription with translation is one explanation for the phenomenon of transcription polarity, that is, the progressive reduction of transcription operator distal to a nonsense codon. It is thought that mRNA is rapidly degraded as it becomes exposed to endonuclease activity beyond the nonsense codon at which point the translating ribosomes fall off (Morse and Yanofsky, 1969b). A different explanation is that there is a premature termination of transcription near the nonsense codon (Imamoto, 1970), perhaps caused by a jamming of the transcribing polymerase by its now ribosome-free product. In agreement with either explanation in terms of (nonsense) translation polarity, transcription polarity is canceled by informational suppression or by an initiation codon next to the nonsense codon.

Even if it is obvious that the major macromolecular processes in the cell are integrated, this integration will not be the probabilistic compounding of individual processes but a function of a series of subtle links between them. Consequently, and partly as a result of experimental reductionism, we know relatively little about how macromolecular integration is controlled at the cell level. The balance between bacterial cell division and growth is maintained by an elaborate series of molecular events. The remainder of this section deals with the interrelationship between the underlying mechanisms.

There is evidence at the level of single enzyme, e.g., galactosidase, for one physiological burst of synthesis during the cell cycle, although the enzyme remains inducible throughout the cycle. For some enzymes the burst order is parallel to their gene replication order, from which it was originally thought that the burst order was due to gene dosage affecting those operons in an active state. In reality each burst is possibly quite complex and composed of increments such as would correspond with the translation rounds described for the tryptophan operon. On further examination, each burst could be due to a fluctuation in repressor concentration, dependent on the doubling of cell mass per generation counterpoised against the doubling of genes during replication (Pardee, 1965). This seemed to be so because at a particular growth rate the rate of translation of an enzyme depends solely on the number of copies of the structural cistron and on the concentration of specific repressor, and there is also no evidence of any period in the cell cycle when the activity of the cistron is totally inhibited (basal transcription disappears). It was later proposed that the burst of enzyme synthesis is caused by a cyclic variation in the concentration of corepressor. This self-generating oscillation of enzyme-corepressor would arise from the end-product feedback on transcription (Donachie and Masters, 1969). Thus, a little mRNA synthesis at the time of each gene replication, as a result of momentary derepression, would entrain gene and enzyme synthesis in the same sequential order, with enzyme synthesis lagging somewhat behind that of mRNA (Goodwin, 1966a). In reviewing the evidence, Mitchison (1969) weighs this oscillation concept against the concept of a "linear reading" of enzymes during the cell cycle. According to this concept (developed mainly for the eukaryote) transcriptions also begin in the gene order in the chromosome as in the previous gene dosage model, but for other purposes. For further discussion the reader is referred to the original article by Mitchison (1969). At the present stage of thinking it is envisaged that translation in general, although not immediately dependent on replication, shows a periodicity similar to and formally coupled with the cell cycle. Biosynthetic enzymes may not be important in the regulation of cell growth during the cycle, but other products under the same periodicity could be. In summary, according to Donachie and Masters (1969) the most obvious aspects to be regulated for during the cell cycle are DNA replication (dependent on cell mass) and cell division (dependent on replication). The point not to miss here is that these two aspects are accompanied by the doubling of everything else in the cell, the coordination of which is what we have been discussing.

Some of the previous views on transcription called for a rather tight coupling with replication, although the indications are that not all genes behave in this manner (review by Sueoka, 1967). Perhaps the discrepancy is a quantitative rather than a fundamental one, but the question is not settled. For example, bacterial pulse-labeled RNA is richer in RNA (by a factor of two) for the segment of the genome that has been replicated (Cutler and Evans, 1967). The enrichment in RNA could be due to a doubling of early replicating genes relative to late replicating ones, that is, the above mentioned dosage effect resulting from linear replication, but proof is still lacking. The explanation would, however, be at odds with the facts that transcription does not uniquely follow the replication point, for a significant portion of the genome is transcribed at all times; nor can a RNA polymerase molecule closely follow a DNA polymerase molecule which moves faster on the template; in addition, the replicated template is initially fragmented (Chapter 1). These facts therefore clearly tell us that the coupling between replication and transcription cannot be absolutely tight. In that case, formal coupling may not even always be required to benefit from a certain degree of dosage effect, one potential example being the genes for stable RNA. Because most tRNA genes in this example are not in multiple numbers, a general possibility is that the frequency of initiation in individual genes varies across the cell (replication) cycle in order to dose their product. Nevertheless, for those genes whose transcription may be somewhat coupled with replication it can be visualized that they locate to positions in the chromosome where their timely transcription most benefits the cell: most of these genes will be expected to be concerned with constitutive functions. A different physiological correlate that remains possible is that of transcription-translation with the distribution of substrates for the enzymes that are made, e.g., polysomes near the cell membrane could be producing enzymes distinct from those in the interior of the cell. This correlate would be most meaningful in the case of constitutive functions, again perhaps implying some coupling with replication.

Lastly, it is fitting in dealing with macromolecular integration to conclude by mentioning the coordination of the bacterial genome. The amount of specific mRNAs correlates with the degree of expression of each operon. It is doubtful that differential mRNA stability plays a regulatory role in this correlation (Epstein and Beckwith, 1968), in the manner that the overall rate of transcription most probably does. (This virtually means that, like translation, degradation of mRNA is also coupled to transcription.) More significantly, it is the transcription of the genome as a whole, not just operons, that appears to be coordinate. As determined by hybridization, the entire bacterial genome is simultaneously and continuously transcribed (McCarthy and Bolton, 1964). This is plain from the fact that it would take some 30 hours instead of the normal 30 minutes to transcribe the genome sequentially. However, individual segments do show characteristic changes in the extent of transcription during the cell cycle depending on the rate of growth (Cutler and Evans, 1967). These changes are determined by the frequency and duration of transcription, the point made in the preceding paragraph. In addition, Kennell and Kotoulas (1968) conclude that only part of the genome makes most mRNA, most of the genome makes extremely little, and part of the genome may possibly make none. The implication of this finding, which is supported by other evidence, is discussed in a later section.

A theme recurrently implied in this section was the organization and perhaps compartmentation of the bacterial cell cytoplasm (review by Van Iterson, 1969). The importance of this aspect for understanding metabolic integration cannot be overemphasized even if, because of reasons of scope and paucity of data, it was barely touched upon directly.

General Regulation of Bacterial Messenger RNA

We have previously discussed the specific regulation of mRNAs in terms of the operon model. We have seen that bacterial mRNA is transcribed directly into its final translatable form, except for possible promoter or operator sequences that may be transcribed, but are not translated. In addition, as a consequence of the coupling between the two processes, the rate of transcription determines the rate of translation. To recapitulate briefly, those aspects in the operon most likely to control the rate of transcription are the affinities of both the operator for repressor and promoter for polymerase. The promoter affinity is apparently able to regulate intrinsically but perhaps not very orderly the rate of initiation; it depends on σ and cAMP-binding factors. The present aim is to discuss the general regulation of mRNA as may be necessary to adjust the overall rate of protein synthesis to the rate of cell growth; this regulation involves general effects on the previous affinities mediated other than by specific inducers or corepressors. This general regulation encompasses two categories of mRNAs: the inducible-repressible and the constitutive mRNAs. In the latter, their operons are either regulator- or operator-constitutive, but retain a functional promoter, thus being subject to general regulation.

The general pattern of mRNA regulation is revealed in the changes in the overall rate of transcription dependent on the conditions of growth. In exponentially growing bacteria all classes of RNA are transcribed at about the same rate (Mueller and Bremer, 1968). However, relative to stable RNA synthesis, the overall rate of mRNA synthesis changes upward or downward depending on the rate of cell growth. With respect to constitutive mRNAs, this effect appears to depend particularly on the temperature of growth rather than on the composition of the culture medium which at any given temperature codetermines the rate of growth (Rose *et al.*, 1970). The rate of transcription is adjustable by the rate of initiation and the rate of elongation and both these rates depend on temperature (Manor *et al.*, 1969). It seems that at any given temperature the adjustment to cell growth involves mainly, but not exclusively, the rate of initiation. Growth thus affects mostly the rate of constitutive mRNA initiation. With respect to inducible mRNAs, their regulation in the operon likewise involves mostly the rate of initiation, irrespective of the dependence of this rate (and the elongation rate) on the conditions of growth. Hence regulation at large is coextensive and mainly at the level of initiation, whether by the relatively coarse constitutive mechanism (probably involving the promoter affinity) or by the fine inducible mechanism at work in the operon (involving the repressor affinity). Having made this point, we turn to examine, in some detail, certain aspects of general regulation.

According to the evidence, under certain conditions the regulation of mRNA is not tightly coordinated with that of stable RNA. An empirical observation is that the two RNA syntheses are dissociable both in step-up and step-down conditions. In stringent cells (with the wild-type allele for the RNA control locus known as RC) but not in relaxed cells (with a mutant RC allele), stable RNA will not be synthesized unless the full complement of amino acid is available. By comparison, mRNA synthesis is much less dependent on amino acid (review by Edlin and Broda, 1968). Amino acid starvation inhibits protein synthesis in both stringent and relaxed cells, so that the effect of the RC gene does not depend on protein synthesis per se. The differential effect of RC on RNA synthesis is particularly well illustrated during induction of the tryptophan operon in *E. coli* auxotrophs (by means of tryptophan deprivation), since stringent cells synthesize this mRNA but not stable RNA (Edlin *et al.*, 1968) even though the mRNA is not translated. The same is true of other stringent auxotrophic cells when a specific mRNA is transcriptionally derepressed by the lack of cognate amino acid (Venetianer, 1969). By contrast, under these conditions relaxed cells would, if anything, tend to produce more stable RNA to keep pace with the new mRNA. The situation regarding mRNA regulation, as distinct from inducibility, is clearer in constitutive cells. Irrespective of the RC status, the regulator-constitutive synthesis of tryptophan mRNA is inhibited by deprivation of the amino acid (other than tryptophan) for which the cells are auxotrophic (Stubbs and Stubbs, 1970). Similarly, synthesis of T4 phage mRNA in stringent strains responds or does not respond to amino acid deprivation according to the particular strain (Olsnes and Hauge, 1968). In conclusion, these last examples indicate that mRNA synthesis is under a generalized control by amino acid which is not directly related to the stringent control by the RC locus (as stable RNA is) yet also occurs in the relaxed cell. It was proposed that the difference in control of mRNA and stable RNA in the absence of amino acid may ultimately depend on (a) separate transcription initiation factors or (b) other initiation signals creating a different promoter response; but other mechanisms affecting initiation differentially are still possible. In addition, (c) stable RNA could also be synthesized but rapidly degraded.

There is no good evidence for any of the previous mechanisms of mRNA regulation, except that, as pointed out earlier, control of initiation must be at play. A rapport could exist between the regulation of mRNA and the status of growth reflected in some aspect of protein synthesis. It is simple to imagine that the above-mentioned generalized amino acid control affects this aspect of protein synthesis which, in turn, controls initiation. Since mRNA is derepressible in the absence of protein synthesis, this aspect cannot be protein synthesis per se (in similarity to the RC control). Perhaps initiation factors are geared somehow to such aspect of protein synthesis, in which case the control would probably affect groups rather than single operons. Clearly, the general regulation of mRNA is not understood. With regard to the regulation of rRNA, it also appears that initiation is controlled indirectly by protein synthesis (this time via the function of RC). Yet, by now, the point should have become obvious that the relationship between the regulation of mRNA and rRNA is unresolved. This relationship is essential for adjusting the overall rate of protein synthesis to the rate of cell growth.

A rather special case of mRNA regulation is that of certain amino acids (histidine, valine) for which the corepressor of the corresponding anabolic enzyme synthesis appears to include the cognate aminoacyl-tRNA (Williams and Freundlich, 1969). As mentioned, in these repressible operons the repressor appears to include both aminoacyl-tRNA and the biosynthetic enzyme made by the operon. This question is discussed in connection with tRNA in Chapter 4 but has certain implications here. Generalizing from these operons, it was suggested that several "regulatory" genes may control repression at different levels including translation by converting end-product substances into active effectors (McFall and Maas, 1967). In contrast to the operon regulator which produces a genotropic substance (repressor), these other regulatory genes would produce a plasmotropic substance (Waddington, 1962).

Viral Messenger RNA: A Survey

Without prejudice as to what the regulation may be in other viruses where it is not as well understood, the regulation of lysogenic bacteriophages is patterned on the bacterial operon, but introducing special modalities of its own. All known temperate phages are DNA phages. The mechanism that brings forth lysogeny by these phages is not known, but the mechanism is partly virus specified. Interesting as a model for regulation, in λ phage a repressor protein may stimulate transcription of its own structural gene (Heinemann and Spiegelman, 1970). This single protein ultimately controls in a negative fashion (inhibits) the transcription of all other 30-50 viral genes (Ptashne and Hopkins, 1968), probably by affecting the binding of RNA polymerase (Hayward and Green, 1969). This repressor acts then as a master switch for chromosome rather than operon repression; it also renders the bacterial cell immune to superinfection. The little viral RNA detectable during the prophage state, representing 2-4% of the viral genome, is probably the regulator mRNA responsible for making repressor. The intricate details of viral regulation now to be briefly schematized are reviewed by Szybalski *et al.* (1969).

A general postulate is that phage induction depends on the appearance of an antirepressor substance capable of neutralizing the repressor mentioned above (Oppenheim *et al.*, 1970). Thus phage induction seems to depend both on prior interference with host DNA replication and on cancellation of the repressor by certain viral genes. Early during induction it is believed that a gene (N) becomes activated and its function positively controls the phage attachment to the host cell membrane, whereby replication and late transcription can ensue (Hallick *et al.*, 1969). The parallelism with the attachment of the bacterial chromosome to the cell membrane for replication is evident. After induction, or during the lytic cycle, there follows a sequential but complex (not linear) ordered appearance of phage functions. A distinction can be made between "immediate early" through "true late" mRNAs controlling correspondingly timed viral functions (Bolle *et al.*, 1968). The informational cistrons concerned with the early and late functions are associated in three separate groups, in fact operons, within the genome. Early functional mRNA is transcribed from the parental genome, and later functional mRNA from the progeny. Late functions therefore depend on replication, but previous viral protein synthesis was required. The sequential appearance of viral function shows that several interdependent polycistronic operons can achieve a fine control of transcription as in the regulon model and with an orderly yet inflexible time control built into transcription. Some functions, even early ones, may however be partly controlled at the subsequent level of translation as will be discussed in Chapter 5. In any event, as proteins are completed there follows a sequential interaction of these proteins to assemble the various phage structures. In virulent DNA phages roughly the same general pattern obtains as in temperate phages, that is, a sequential appearance of functions and a replication dependence of late functions, except that there is no master repressor because there is no prophage state (review by Geiduschek and Haselkorn, 1969). Basically, therefore, the regulation of the two types of phages appears to be similar.

The fine control of transcription is best known in virulent phages, the T even series. This control is clear from the transcription requirements of the progressive classes of genes. The first transcribed genes or immediate early genes have no requirements and use both the host RNA polymerase and the host σ initiation factor. The next transcribed genes or delayed early genes require prior phage protein synthesis (an immediately early function). This protein synthesis results in the appearance of, first, an initial phage-specified σ factor (Travers, 1969) and, second, a modified host polymerase (Walter et al., 1968), these two being the initial elements of fine positive control. Whether any element of negative control also appears is not known. A modified host polymerase rather than a completely *de novo* made polymerase is therefore required for subsequent viral transcription. The induced modification is on the α subunit of the polymerase and involves addition of AMP to a serine residue. Other subunits of the enzyme, but not all, remain functional throughout the viral cycle (Geiduschek and Sklar, 1969). The modified polymerase is believed to interact more readily with the viral σ factor. In effect, the binding of polymerase and factor is now looser than it was between host polymerase and host factor, and this might facilitate their exchange. It is clear that the viral initiation factor subverts the enzyme to read only viral templates, adding an interesting dimension to the parasitic relationship. In addition, another protein factor specified by the phage (antitermination factor) may oppose the host ρ termination factor and permit the polymerase to continue transcribing past the immediate early cistrons into delayed early cistrons (Schmidt et al., 1970). The significance of antitermination is not clear but could be great. This is because in T4 phage, for example, the genes are disposed along the chromosome in the same order in which they are expressed; therefore, after transcription begins the subsequent control of the polymerase run-through could be a most simple and effective form of regulation. Curiously enough, antitermination may result in the formation of anti-mRNAs (because the overtranscription may run into sequences that should not be transcribed from that strand), the significance of which, if any, is even less clear (Geiduschek and Grau, 1970). Finally, late transcribing genes again require viral protein synthesis (a delayed early function) and also viral DNA replication. These viral functions result in a second viral σ factor and a second modification of host polymerase, this time in the β subunit (Bautz, 1970), both being presumably needed for late transcription. A different explanation for the replication dependence of late functions is that they require DNA modification (Bolle et al., 1968), an interesting concept which cannot be enlarged upon here. In contrast with the previous description for T even phage, in T7 phage an early made, phage-coded polymerase (rather than a factor) is used for late transcription (Chamberlin et al., 1970); contrary to the complex bacterial polymerase, this viral enzyme consists of a single protein. Conversely, in T7 phage (but not T4) late transcription is partly coupled to viral replication, perhaps implying a measure of gene dosage. Certain large DNA virions such as vaccinia carry their own RNA polymerase and other virions may carry their own DNA polymerase to be used in the next infective cycle.

Certain replicational characteristics of RNA viruses may be briefly considered here in addition to their general characteristics described in Chapter 1. Each terminus of the RNA virus may have special functions, although their nature remains uncertain (Dahlberg, 1968). The common pppG 5' terminus might be solely involved in termination of the antiparallel minus strand, or, conversely, in initiation of the mature plus strands, because in several phages (Billeter et al., 1969) there is no translation initiation codon for at least the first 150 bases (and similarly no termination codon for the last 50 bases or more). In agreement with this fact, translation does not necessarily start at the 5' terminus (Bassel, 1968; Lodish, 1968a). Whatever the function of this terminus, its general function is attested to by the possibility that it has been conserved in different phages. On the other hand, at the 3' terminus of the RNA virus a common hexanucleotide sequence (ending in A) is present subterminally in both plus and minus strands, and may be concerned with replicase recognition; terminal 3' A-free RNA remains infective, however (Kamen, 1969). A factor may be required at this terminus to initiate the minus strand which begins with G instead of the expected U (August, 1969). This kind of heterogeneity is also found at both termini of other RNA phages, e.g., Qβ (De Wachter and Fiers, 1969), suggesting a frequent replication "error" not affecting translation. Surprisingly, the 3' terminus of tobacco yellow mosaic RNA virus has a tRNA-like structure which (when freed) accepts valine just as well as valyl-tRNA does (Yot et al., 1970), the meaning of which, if any, is puzzling yet may not be a translational one; it could be an RNA cistron for a tRNA species that the virus requires. Not improbably, some of these seemingly disjoined molecular characteristics of RNA viruses may be related to translational aspects and they should become clear with increasing understanding of both replication and translation.

Several mammalian RNA viruses that direct their own cycle in the cell will be briefly mentioned. They differ from transforming viruses (not described) which integrate with the cell genome and depend more directly on cellular function for their cycle. Some large mammalian RNA viruses, such as single-stranded polio virus, seem to behave like an oversize operon with a single polycistronic mRNA. This mRNA is translated into one or a few polypeptides which are then cleaved into several viral proteins (Holland and Kiehn, 1968). Other large single-stranded RNA viruses, such as Newcastle disease, are apparently first transcribed into several small complementary mRNAs that are then separately translated. Still other mammalian viruses contain, to begin with, a number of either single- or double-stranded RNA fragments that result in several mRNAs. Supposedly, these adaptations obviate the fact that the mammal, with mainly monocistronic mRNAs, does not provide for internal translation (Jacobson and Baltimore, 1968). In any case, polio virus pays for its polycistrony with a very inflexible regulation. The regulation of these mammalian viruses is not as well understood as that of the bacterial DNA or RNA viruses, yet it is hoped that eventually they will help clarify some of the cell regulatory mechanisms.

As a closing note, this schematic section cannot convey the importance and amount of basic information available on viruses. The interested reader will find this information in several recent monographs.

Extent of Genomic DNA Transcribed in the Prokaryote

Bacterial and viral DNA sequences are not noticeably reiterated as revealed by the homogeneous rate of renaturation. Repeat sequences of more than 12 bases are relatively rare (Britten and Kohne, 1968; Thomas, 1966) and most of these probably concern the stable RNA genome. Thus in *E. coli* both the size and complexity of the genome are the same, or 4×10^6 base pairs.

Essentially all the genome of a DNA virus is transcribed. In mammalian adenovirus with helical DNA, for example, 80-100% of the sequences in one-half of the DNA are transcribed during the cycle, as followed by sequential DNA-RNA and DNA-DNA hybridization (Fuginaga *et al.*, 1968). This fact confirms that only one of the two viral strands, or the equivalent made of unique portions in each strand, is transcribed. Concerning the distribution of the active genome in a DNA virus, we can expect that much of it, apart from a regulatory portion, represents mRNA. Ribosomal RNA is not coded but many species of tRNA are (Chapter 4). A small but not necessarily negligible fraction of the genome represents internal or terminal signals for replication, transcription, and translation, and these signals may not all be transcribed or translated. The same would be true of terminally redundant sequences which are used for circularization of the genome during recombination. Functionally related genes are to some extent clustered in the chromosome and in some cases they are very much so, e.g., in temperate λ phage early and late functions and functions for DNA regulation are disposed in three operons; in T-phages the genes are arranged in the same sequence in which they are expressed. It was seen how this orderly topography of genes can be of importance for their joint regulation as a genome. In addition, there is no compelling reason to believe that all viral genes are transcribed in every cycle. For example, the lysogeny genes of temperate phages are probably not.

The situation in bacteria is possibly more complex than in viruses. In *E. coli* also approximately the equivalent of one DNA strand is transcribed into mRNA as revealed by a study of the whole genome (McCarthy and Bolton, 1964). From the minor size of the stable RNA genome (and other considerations below), this means that most of the bacterial genome is informational.

The previous observation was confirmed by separately hybridizing fragments equal to one-tenth the size of the genome under conditions where only 10% of DNA that would never be transcribed could have been missed (Cutler and Evans, 1967). However, the relative abundance of mRNAs differs enormously as the result of their differential rate of transcription (McCarthy and Bolton, 1964). As a first approximation, 1% of haploid DNA makes 20% of mRNA, and 10% makes 75%. On the average, five copies of each mRNA are present per cell generation, the actual range being from 1 to 200 copies (Cutler and Evans, 1967). It follows from these data that each cell in a growing culture cannot contain copies of all mRNAs at any given instant. Kennell and Kotoulas (1968), working on the same genome, confirm the previous disparity but go on to show by hybridization that virtually all mRNA is complementary to only 20% of the DNA and little is complementary to the remainder, in fact, most of the DNA. The implied severe limitation of the genome should be taken with caution in view of the occasional imprecision in the interpretation of hybridization data. Taken at face value, this limitation would mean that not all DNA is simultaneously transcribed, which appears to be confirmed under the electron microscope (Miller *et al.*, 1970a). A possibility exists, therefore, that growing cells may never in their cycle possess copies of certain messengers. This is undoubtedly correct in the case of spore-forming bacteria (other than *E. coli*); even if a figure cannot be given, a moderate fraction of the genome codes for the spore which, oddly enough, is only sporadically transcribed. These points question the notions that (1) bacteria do not harbor silent genes and, as in previous considerations, that (2) transcription obligatorily follows replication.

In bacteria, (a) stable RNA requires a negligible DNA fraction, (b) there is no RNA other than mRNA requiring a substantial DNA fraction and, as previously mentioned, (c) the DNA has no noticeable reiteration. In terms of mass, stable RNA vastly outnumbers informational RNA only because of the latter's instability. At any given instant, a half or more of the total transcript is mRNA (Winslow and Lazzarini, 1969a) and the total output per cell generation is more than that of stable RNA. Of these transcripts, the greater part of mRNA represents unique sequences, but part of stable RNA probably represents redundant sequences. As mentioned, possibly some of the bacterial DNA may not be transcribed at every cell cycle, which is evident for the DNA coding for the spore. It does not necessarily follow from this that in bacteria there may be noninformational RNA apart from stable RNA, though at present this possibility cannot be eliminated. For example, there could be a noninformational RNA of some kind required for regulation every few generations of the bacterium.

We can therefore state with some confidence that most of the bacterial genome codes for RNA and indeed most of this codes for mRNA. Thus it is probably true that most bacterial DNA-like RNA is translated. On present evidence, the informational sequences are distributed uniformly along the chromosome.

Apart from polycistronic operons, the structural cistrons for metabolically related proteins are scattered rather than associated, as one might perhaps have expected. The same scattering is true of the cistrons concerned with the spore, except for some minor clustering (Takahashi, 1965). One notable exception are the clustered cistrons for ribosomal proteins.

Is there any bacterial DNA that is never transcribed? Indeed, one can think of sequences that are part of initiation and termination signals for replication, transcription, and translation or of possible sequences that intervene between cistrons. One can further adduce hypothetical DNA sequences with a mechanically structural or a genetic reservoir role. However, the likelihood of these structural sequences is at present a matter of conjecture; the presence of reservoir sequences is made unlikely by the apparent lack of noninformational DNA-like RNA.

Conclusions on the General Regulation of Prokaryotic Messenger RNA

Having discussed the various facets of prokaryotic mRNA regulation, we may ask if our present understanding of the process is conceptually sufficient. The area in mind here is the general regulation of informational syntheses which, as just seen, engage most of the genome. This regulation is directed to operons in general, rather than to operons in particular, in the case of induction-repression. This regulation, therefore, involves inducible-repressible syntheses and constitutive syntheses; the latter are regulated because their operons retain functional promoters. The general regulation must be attuned to the progression of the cell cycle, that is, growth. The present task is thus to discern by which means in general these two classes of mRNA syntheses are geared to growth.

A first question concerns the primary regulation of the regulator genes themselves and applies only to inducible-repressible syntheses. These regulator genes may be (a) periodically repressed, or (b) always repressed. Possibility (a) implies that derepression is coupled to the cell cycle, for example, via some event in replication or translation which are both practically continuous during the cell cycle. This event would primarily determine when the regulator gene is to make repressor during the cycle. Possibility (b) implies that the cell cycle does not primarily control the regulator gene whose transcription remains basal (repressed), except insofar as the level of unspecific stimulants of transcription like cAMP could fluctuate rhythmically during the cell cycle; it remains an assumption, however, that basal transcription can be so stimulated. Possibility (b) is particularly attractive when it is considered that there are very few molecules of repressor per cell, and that mechanistically it is more economical than (a) in the sense that the regulator gene would function more nearly constitutively. According to this view, any control of the basal regulatory transcription would be at the level of initiation, as occurs in other known constitutive genes like those which can become constitutive in the tryptophan operon. However, the few copies of repressor made suggest also that there may be some translational control over their production, this resulting in an overall regulation with formal characteristics somewhat intermediate between those of (a) and (b) at the transcriptional level.

A second question has to do with the coordinate regulation of the informational genome which is essential if the cell is to function as such. This question concerns both inducible-repressible and constitutive syntheses, and refers to the common elements (other than repressor) that are involved in transcription such as polymerase, factors, and metabolites like cAMP. What is asked is how great can the contribution of these common elements—only a handful at best—be to the coordinate regulation of the genome? However, before starting to consider these elements, a short digression is necessary. In the case of inducible operons, coordinate regulation would ensue if these operons were to be jointly subregulated as a limited number of operon blocks or regulons all having the same operators, thus requiring only a limited number of specific repressors. At present there is no evidence that this operator-mediated type of subregulation occurs. (A conceivable possibility, but difficult to evaluate in view of the previous considerations on the constitutivity of regulator genes, is that blocks of these genes are jointly regulated: this would result in a joint operon subregulation without any limitation on the number of repressors.) Returning to the common transcriptional elements, a subregulation equivalent to the one just mentioned might be achieved if blocks of operons were to have a similar if not identical class of promoter responding to one class of σ factor. In principle, this type of promoter-mediated coordination applies to both inducible and constitutive operons. Recent evidence shows that this type of subregulation does occur in bacteriophage. Thus, about 35 promoters belonging to the immediate early genes of T5 phage all respond to the host (*E. coli*) σ factor (Bautz and Bautz, 1970). Considering the probable number of these early genes (60 to 80) and that some of them may be grouped in polycistronic operons with a single promoter, it would appear that 35 co-responding promoters can sufficiently account for the joint subregulation of the immediate early gene class. If these findings can be generalized to the later classes of genes, this time using phage-coded σ factors, then at least one aspect of the coordinate regulation of the phage genome is beginning to be understood. However, at this time it cannot be excluded that other mechanisms (viz., phage-coded polymerases, host polymerase modifications, and control of internal termination) are just as important for the regulation.

In the bacterium, where, in contrast to phage only one σ factor has been found, it remains possible that the operator-type subregulation of operons mentioned above is also at work. However, in the case of inducible-repressible operons no evidence is at present available for this type of subregulation, although its possibility makes it deserving of further investigation. Recent information of a different kind has come to light from work on bacterial sporulation. During this process many vegetative genes are turned off and many sporulation genes are turned on. This new template specificity of the RNA polymerase seems to be determined by the appearance of a modified β subunit of the enzyme, and perhaps but not shown, of a new σ factor (Losick et al., 1970). The resemblance of the coordination to the promoter-mediated coordination in phage is evident. Whether or not these findings can in the future be extended to the selective switch-on of vegetative genes, nevertheless, they suggest that the promoter-mediated coordination may also intervene in the bacterium. As noted above, this type of coordination applies in principle both to inducible and constitutive operons, but the latter, in particular, would appear to require it (assuming mostly initiation control as in the constitutive tryptophan operon).

A different aspect of promoter-mediated coordination involves cAMP, another common transcriptional element. In this case, the rudiments of a general theory of endogenous bacterial regulation is beginning to emerge (Travers et al., 1970). As mentioned, when cAMP is available a binding protein seems to stimulate informational transcription without regard to any particular transcription. Mechanically similar but in the opposite direction, it will be described in Chapter 4 that, when a tetranucleotide derivative (GT_4X) is not available, a factor appears to stimulate stable RNA synthesis also in general. Thus in both types of synthesis the effect on the polymerase and/or promoter is by a protein factor cospecified by (the presence or absence of) a nucleotide derivative. Fluctuation of these derivatives during the cell cycle could then allow a rapid and reversible synthetic response as is typical of the bacterium; indeed, this response is more flexible than the σ factor-mediated informational response of the virus or bacterium. Furthermore, considering that both mRNA and stable RNA syntheses are affected by amino acid and this seems to affect GT_4X, it cannot be excluded that the GT_4X and cAMP systems interact with each other to balance the syntheses as required during growth.

In conclusion, the individual regulation of operons is well explained in the operon model and some understanding, though by no means complete, of the basic coordination of most of the informational bacterial genome has been gained. However, it is not clear how this genome is attuned to the cycle of cell growth—which is what we set out to find. In this regard, it may well turn out to be that the interaction between the common transcriptional elements considered here is much more complex than we think it is.

With some exceptions, the inducible or repressible bacterial syntheses directed by exogenous metabolite are of the same nature as the numerous syntheses directed by endogenous metabolite. This is to say that exogenously directed syntheses do not represent either a special class or a specialized genome (a few of them may even be gratuitous to the cell), but many are vegetatively required and therefore endogenously directed in the course of cell growth. It follows that any endogenous inducer-corepressor appears in order to contraarrest repressor as dictated by the cell cycle of growth. This delicate balance is again far from understood, although entrainment to replication is a possibility mentioned earlier. Regarding constitutive syntheses, however, the situation may be different. It appears as if many of these syntheses may be obligatorily constitutive for ancestral reasons pertaining to the establishment of central metabolic pathways, etc. This is to say that part of the genome may have been retained exclusively for constitutive syntheses. If that is the case, the regulation of the bacterial cell involves two functionally differentiated genomes, namely, the inducible-repressible genome which can be mobilized endogenously or exogenously and the constitutive genome.

EUKARYOTIC REGULATION: THE PROBLEM AT THE CELLULAR LEVEL

After first pointing out the general differences in regulation between the prokaryote and eukaryote, we studied the regulation of prokaryotic mRNA. We can now examine which additional requirements at the cellular level are likely to be needed for the regulation of eukaryotic mRNA. Regulation in higher organisms must be principally viewed as the means to attain and maintain the specialization of tissue. Initially, this differentiation is achieved by activating a part of the genome that otherwise would not express itself. The selecting force is toward the harmonious rapport between specialized cells rather than the reproductive success of individual cells in bacteria. The extent of this rapport is difficult to verbalize but must be obvious to anyone who has ever looked at the structure of tissue under the microscope; it involves a very great number of specific cell-to-cell interactions. Clearly, most of these interactions cannot be directly encoded in DNA but instead must depend on second or possibly higher order information which is recognized in the form of different cell properties, themselves ultimately depending on DNA; for example, the estimated 10^{14} neuronal synapses in man greatly surpass in number the informational content of the genome, which points to epigenetic components being coresponsible for neuronal specification. A further characteristic of the eukaryote is that programmed cell death plays a part in the development and equilibrium of tissues. Even dead cells have a function, as shown by the keratinized cells of mammalian skin forming the impermeable barrier of the body.

To contrast at the extreme (a) the bacterial inductive phenomena with (b) the stable eukaryotic differentiation, the essential points according to Paul (1968) are: first, there is a relative increase of induced protein in (a) but an absolute appearance of new protein, that is, a positive activation of syntheses, in (b); second, (a) is reversible while (b) in general is not and is highly directional; third, (a) requires the continuous presence of inducer but (b) typically requires inducer (or the equivalent agent) for brief periods. By the criteria employed, these points define cell differentiation. Frequently, however, the contrasts with the prokaryote are less well defined. Nonetheless, the emphasis in the eukaryote is on the whole on stability rather than rapidity or amount of response. In their amount per cell, both induced and differentiated protein in the prokaryote and eukaryote, respectively, can be equally substantial. In the higher eukaryote, however, cell specialization brings about the production of a great number of proteins (such as keratin mentioned already) that the organism as a whole, but not the individual cells require.

In a morphological and functional sense, the eukaryote represents the phylogenetic emergence of both nucleus and cytoplasm as discrete but not separate compartments. Somehow, this compartmentation has permitted multicellularity and differentiation, although the first is not a prerequisite for the second, as shown, for example, by the unicellular lower eukaryote. The main genetic apparatus is sequestered within the nuclear envelope and part of the apparatus is further sequestered within a nuclear organelle, the nucleolus, with no envelope. The outer nuclear envelope may contribute to the cytoplasmic endoplasmic reticulum membranes, the present view being that of a dynamically (not physically) unitary system extending from the nuclear envelope, through the endoplasmic reticulum and golgi complex, to the cell membrane. Evolution of this system or vacuome was accompanied by acquisition of the chondriome, probably from a prokaryotic ancestor, as a separate event. These systems, nucleus, nucleolus, vacuome, and chondriome serve to compartmentalize the cell for the purpose of macromolecular syntheses. As explained, nuclear DNA is so intricately complexed with other chromosomal constituents that in the eukaryote DNA by itself is only a biochemical abstraction. Thus, any ideas concerning gene regulation have to account for the following chromosomal facts not encountered in bacteria: (1) a large amount of nonbasic and (2) histone protein, (3) chromosomal RNA, and (4) supercoiled DNA with a high sequence reiteration.[10] It is evident that the molecular and functional properties of this chromosomal DNA-RNP complex must be vastly different from those of the bacterial DNA chromosome. As with bacterial polyamines, the possibly important role of these substances in chromosomal function is not understood (review by Bachrach, 1970). In summary, both the chromosome and cytoplasm are enormously more structured molecularly, hence functionally, in the eukaryote than the prokaryote.

Polycistrons in the eukaryote are generally absent above the level of fungi (a lower eukaryote), although an exception is the nucleolar rRNA cistrons which of course are noninformational. Significantly, informational cistrons that are in common operons in bacteria, are in different chromosomes in yeast

---

[10] The relative composition of interphase chromatin is: DNA 1.0, histone 0.7-1.0, acidic protein 0.25-1.0, RNA 0.05-0.25 (DuPraw, 1968).

and fungi. Clustered genes in the eukaryote, e.g., the well known mouse tail genes, appear to be developmentally rather than regulatorily linked. However, functionally related genes such as the 100 or so known neurological genes of the mouse are spread over all chromosomes. A general loss of physical connections between genes is dramatically revealed also by the vastly variable chromosome numbers among mammals with the same size genome. To explain the contrast between eukaryote and bacterium, apart from a different regulation, it has been argued that a further reason why there are so many polycistronic operons in bacteria is because of the sexual transfer of the chromosome in an ordered linear fashion; the risk of disrupting a metabolic pathway during this transfer increases with the distance between cistrons. According to this view, its novel mode of sexuality would have spared the eukaryote this constraint. On the other hand, eukaryotic sexuality and particularly meiotic recombination demand a diploid genetic condition.

From the complex chromosome constitution of the eukaryote, an involved regulation of transcription can be anticipated. Moreover, eukaryotic translation is physically separated from transcription, thereby also anticipating the problem of transporting the nuclear transcripts as well as a more independent regulation of translation. The separation of transcription and translation has permitted (and probably depended on) the evolution of more complex transcripts requiring more maturation (processing) and proteination, qualitatively new classes of transcripts possibly involved in the more complex regulation, and a more complex translational apparatus (the ribosome in particular). As a whole these eukaryotic innovations must reflect the need for a more complex genetic system and a suitable synthetic machinery to express it, all as part of the adaptation to different biological characteristics than are found in bacteria.

From these preliminary remarks one would not be logically compelled to accept a priori the bacterial operon or regulon format for the regulation of eukaryotic mRNA. What then is known about eukaryotic mRNA regulation? There is no question that a mechanism of repression-derepression is at work, yet the available data form a growing collection in search of a theory—which is why none has been mentioned in this introduction.

EUKARYOTIC MESSENGER RNA: PARTICIPANTS IN ITS REGULATION

Contrary to the situation in bacteria where the regulation of informational transcription can be explained by one and probably final theory, the operon model, eukaryotic regulation has attracted several ad hoc hypotheses to explain its partial facts. Because no comprehensive theory has emerged, by and large the approach to eukaryotic regulation is still pragmatic. Two main difficulties responsible for this dearth of hard facts are, first, the inaccessibility of the material to fine genetic analysis and, second, its enormous biochemical complexity. The particular genetic analysis in mind is of the type dealing with regulatory genes which in bacteria led so successfully to the operon theory and from this to the definitive biochemical work on regulation. Paradoxically, these difficulties combined make the biochemical approach the only effective one available in the eukaryote. In this approach a special strategy is to deduce the informational function of the DNA templates from the properties of their transcripts. Because these RNAs cannot be directly analyzed as yet, their characterization leans heavily on three methods: the response to inhibitors, the sizing of molecules, and molecular hybridization. Chromosomal proteins are studied by more conventional means. There are technical difficulties which further complicate the interpretation of the data: two of the most important (which will not be pointed out each time in discussion) are the frequent use of heterologous enzyme-substrate systems and the imprecision of molecular hybridization procedures.

Thus, for the lack of established concepts about regulation, what appear to be the most relevant pieces of experimental evidence on eukaryotic mRNA will be presented first, and this will be gradually followed by conceptual appraisal of the evidence in terms of hypotheses that explain regulation. This is in diametrical contrast to bacterial regulation whose analysis in previous sections began with the operon model for the opposite reasons stated. However, much more than in the case of the prokaryote, the interpretation of the evidence often remains inferential and particularly so with respect to the causal relationships being examined. As mentioned, possibly some of the aspects to be considered are not as yet developed in the lower eukaryote (below the level of fungi) and in such cases the regulation might considerably resemble the bacterial regulation.

Heterochromatin: A State of Chromatin

We will first discuss the process of heterochromatinization as a structural correlate of transcriptional regulation. In classical cytology heterochromatin refers to chromosomes or part of them whose condensation and replication cycle is out of phase (late) with respect to euchromatin. This phasing-out is characteristic of the cell lineage (Lima-de-Faria and Jaworska, 1968). On the basis of cytogenetic and genetic inference, the functional significance of heterochromatin has long been a matter of dissent which will not be revived here (reviews by Baker, 1968; S.W. Brown, 1966). Both genetic activity and paralysis have been traditionally attributed to heterochromatin; it is now ascribed with subtle aspects of directed genetic change or paramutation (Brink et al., 1968). The term heterochromatin as used by the biochemist means chromatin lacking transcriptional activity at the time, without any further connotation (Allfrey et al. 1966); it is in this sense that it is used here. Conversely, euchromatin means chromatin active in transcription. Our main interest is in facultative heterochromatin which is a nonfunctional state of chromatin as defined. (Unless otherwise stated, heterochromatin in the text means this type of heterochromatin.) On the other hand, constitutive or permanent heterochromatin at least partly includes a different type of DNA, which makes it a different type rather than a state of chromatin. In accordance with this, contrary to facultative heterochromatin, its chromosomal localization is constant throughout cell function and development (Lee and Yunis, 1971); it occurs in chromosomal regions such as centromeres, telomeres, around nucleolus organizers, and intercalated in euchromatin. More important in the present context, constitutive heterochromatin appears to be permanently inactive in transcription, which is why it is of no immediate interest to us. Finally, as required by a complex life cycle, heterochromatin repression has a much more stable character than bacterial operon repression. However, within the concept of selective derepression of the genome to be discussed, at some stage heterochromatin must be able to revert to euchromatin, and this is supported by cytological observation.

Structurally, heterochromatin is more condensed than euchromatin. Heterochromatin is found widespread in the nucleus as clumps, often near the nucleolus and inner nuclear envelope, and occurs both in autosomes and sex chromosomes. It tends to vary in a characteristic pattern for a tissue in relation to its metabolic and differentiative status, e.g., the characteristic heterochromatic pattern of the plasma cell. Heterochromatin increases in relative amount with terminal differentiation. Thus it increases in aged cells, and, somewhat surprisingly, also in tumor cells, according to Harbers (1968). It is generally inferred that the cytological pattern of heterochromatin corresponds with the pattern of genome repression, that is, condensed clumps of chromatin represent inactive chromatin. Two examples should caution against hasty generalization, however. First, the neuronal nucleus barely has any heterochromatin, and, second, the nucleus of the smooth muscle cell has a great deal, but only while it is contracted (Kelly and Rice, 1969). Hence the cytological picture need not accurately reflect the pattern of repression, and this is substantiated biochemically.

That condensation of chromatin is associated with inhibition of transcription is certain for mitotic (Prescott and Bender, 1962) and interphase chromosomes (Frenster, 1969). Observations on subnuclear systems point in the same direction, but the evidence is not unequivocal because in practice these systems are also subchromosomal. More convincingly, the correlation between heterochromatin and gene repression has been validated statistically for sex-linked enzymes in clonal cell cultures (Klinger *et al.,* 1968). However, since occasional female sex chromatin-negative cells are as repressed as the chromatin-positive cells, there must be a more subtle factor at work, at least for those enzymes tested.

With regard to its cytological appearance, therefore, the contrast drawn between transcriptively active euchromatin and transcriptively inactive heterochromatin should be understood as in the nature of a general but not absolute distinction. As mentioned, not all euchromatin observed in the nucleus need be active and most probably not all is, or need all heterochromatin be completely inactive. In addition, if heterochromatin decondenses to replicate (Milner, 1969), one necessary consideration to exclude residual transcription would be that any heterochromatic region should not be simultaneously replicated and transcribed. This might seem acceptable on grounds of template restriction to one polymerase at a time in short regions of the DNA helix, and has been shown for the nucleus of ciliates at the cytological level (Prescott and Kimball, 1961), but still remains an unproved assumption for heterochromatin at the molecular level. With these qualifications firmly in mind, in what follows we shall regard facultative heterochromatin as manifest repression of transcription. If this heterochromatin has functions other than repression, these do not concern us here.

Histones: Not Exclusively Repressor Molecules

Since the early 1960's histones have been increasingly considered in relation to regulation of RNA synthesis (reviews by Butler *et al.*, 1968; Fambrough, 1969; Georgiev, 1969a). This relationship was first proposed by the Stedmans in 1950 on the basis of the near stoichiometry of histones to DNA in most tissues (Stedman and Stedman, 1950). Depending on the tissue, the mass ratio of total histone to DNA is from 0.7 to 1.30 (Bonner *et al.*, 1968), or below the stoichiometrical ratio (1.35) representing one basic amino acid residue to one DNA phosphate. Not all the DNA phosphate is therefore bound to histone: per unit length, and depending on the disposition of histone, there is probably even less bound DNA than these figures suggest. Since histones are present in both euchromatin and heterochromatin of most cells, the present effort is to discriminate between their function in these two states of chromatin. (One cell type which in many animals lacks histone and has the smaller and more basic protamine instead is the spermatozoon.) Before proceeding, it bears mentioning that, despite a great amount of work, the causality of histone function remains to a considerable extent interpretative. The understanding of histone interactions within the chromosome does not equal the chemical knowledge on histones.

Histones are rather small basic proteins with 100 to 200 component amino acids, that combine strongly in salt linkage (more properly, ionic bonds) through their cationic residues to the DNA phosphates. That is, histones mask DNA. As expected, these bonds are highly ion sensitive. Because of this masking, histones stabilize helical DNA against strand separation and therefore prevent transcription. *In vitro* chromatin is transcribed to a few percent of the DNA, but closer to 50% (the equivalent of one strand) after histone is removed; full unmasking necessitates removal of nonhistone proteins. By contrast, the DNA remains replicable to an appreciable extent in the presence of histone (Schwimmer and Bonner, 1965). To a degree unknown, other more complex interactions of histones, depending on H bonding and hydrophobic forces involving noncationic residues, may participate in their masking function directed to DNA and probably also to protein. As seen later, however, masking reflects a particular disposition of histone rather than a total exclusion from DNA of any other molecule.

There are only a few families of histones (about five) with a low internal heterogeneity, that is, few molecular species each. The important point is that, all told, there are not very many kinds of histones (Panyim and Chalkley, 1969). Histones are broadly divided into arginine-rich and lysine-rich fractions. The two fractions may differ in their disposition: the arginine-rich histones may run parallel to DNA; the lysine-rich histones may cross-link with DNA, and, when required, presumably mediate its condensation (Mirsky et al., 1968). Condensation means supercoiling of the DNA with a concomitant reduction in length by 35%. On this basis the lysine-rich histones could play a leading structural role in heterochromatinization, whereas, as is discussed below, the arginine-rich histones could play a more metabolic role. However, apart from the fact that histones somehow fold DNA on itself and neutralize its phosphates, their disposition is largely speculative and could well vary with the individual histone. So far, moreover, no clear-cut distinction between the effect of the various histone fractions on either stability of the chromatin or repression of the genome has been demonstrated by the selective removal of fractions in any different order. In addition to interacting with DNA, histones interact with other chromosomal proteins (including histones) but the nature of this interaction is unknown.

Coupling chromatin isolated from pea cotyledon with a mRNA-dependent bacterial cell-free system made it possible to recognize immunochemically the synthesis of cotyledon globulin. This was not possible when chromatin from other tissues lacking globulin was tested, unless proteins were first completely removed. This particular experiment requires confirmation and is mentioned here as a model to analyze the role of histone. Evidently, not only histone had to be removed in order to make globulin-improductive chromatin a producer (the implications of this will be considered later). Nevertheless, accumulated experiments of this and of the type previously described have led today to the postulation that histones have a certain regulatory role in transcription. The postulate remains credible even if, first, because of their general cross-reactivity, histones have indirect effects at the pool and enzyme levels, and, second, they also subserve an unspecific structural role in the packing of chromosome materials (review by Bonner et al., 1968). The regulatory specificity of histones is, however, not decided. Yet, for example, in the isolated thymus nucleus the arginine-rich histones inhibit RNA polymerase more than the lysine-rich histones and by a mechanism primarily not involving DNA precipitation (review in Allfrey et al., 1966). This action of histone is mainly by template inhibition rather than polymerase exclusion (Spelsberg et al., 1969), but some direct inhibition of transcription is not excluded. The same amount of polymerase binds to pure DNA as to any type of histone-containing chromatin (Bonner et al., 1968).

However, euchromatin and heterochromatin, even metaphase chromosomes, do not markedly differ in the amount or type of associated histone. Puffs in polytene chromosomes retain their histone when they become active. Much of the histone remains stably associated with DNA through several cell generations. In other words, histones are permanent chromosomal constituents. There is also a surprising if not absolute compositional uniformity of histones in different tissues, between which the differences are mainly quantitative (Panyim and Chalkley, 1969). The same holds true between organisms. Perhaps to be expected, however, differences are found during development (Chapter 6). These observations suggest, although they do not prove, that regulation of RNA synthesis is mainly by changes in the structure of the DNA-histone complex rather than by the presence or absence of a particular histone.

Efforts were directed at detecting these changes in the DNA-histone complex. Resting cells have, in general, little histone synthesis. In addition, when cells become activated, histone synthesis does not precede RNA synthesis. By contrast, preceding activation of RNA synthesis there is an increased turnover (independent of synthesis) of the acetyl, methyl, and phosphoryl groups present as postsynthetic modifications in histones. For example, in the phytohemoagglutinin-activated lymphocyte, preferential acetylation of the arginine-rich histones in euchromatin precedes RNA synthesis (Vidali et al., 1968). The acetyl group is thereafter preserved in histone for the length of the active period. Upon subsequent repression, further acetylation ceases and the acetyl group is lost. Apparently, as would be expected, at least part of the histone acetylation occurs in the chromosome, although to what extent in the active loci themselves is debatable (Clever and Ellgaard, 1970). (Curiously, heterochromatin is unable to acetylate histone seemingly not because acetylase is lacking but because it cannot convert the precursor acetate into acetyl-CoA, which is the immediate donor.) The role of this acetylation is currently viewed as that of weakening the binding of histone to DNA and thus altering the properties of the complex, presumably by neutralizing the amino acid cationic residues. Acetylation is therefore assumed to be in the manner of a preparatory step to making the template more available for transcription. As shown with acridine orange, the template becomes increasingly available *pari passu* with the increasing degree of acetylation (Ringertz and Bolund, 1969). Euchromatin DNA whose associated histone is most acetylated is in effect more accessible since it melts at a lower temperature than heterochromatin DNA, indicating a partial physiological denaturation in euchromatin.

Methylation of histone during liver regeneration is not obviously correlated with RNA synthesis as acetylation is. The same might not be true of phosphorylation of lysine-rich histones, in particular, for this precedes RNA synthesis as does acetylation. Although histone phosphorylation occurs predominantly in euchromatin, it is less in degree than the phosphorylation of nonhistone protein. The biological meaning of methylation and phosphorylation of histone is therefore unclear. It is certain, however, that both can occur as multiple modifications simultaneously with acetylation. Perhaps then these modifications also serve to modify the DNA-histone complex. An additional modification is the SH-SS transition in histone (Stocken and Ord, 1966); whether it plays a regulatory role is again unclear, but it has clear structural implications in chromosome condensation, as was pointed out for lysine-rich histones in general.

Returning to acetylation, in the thymus arginine-rich histone this modification occurs exclusively in the free ϵ-amino group of lysine in residue 16 (De Lange et al., 1969). By contrast, N-terminal acetylserine does not turn over and is only formed during histone synthesis. The turnover of the acetyl group is therefore residue specific. The complete sequence of this histone reveals that the N-terminal portion of the molecule is highly positively charged, e.g., four basic residues occur immediately next to the lysine in residue 16. This portion is therefore almost certainly involved in binding the DNA phosphates. The C-terminal portion of the molecule is negatively charged and contains the total hydrophobic residues in the molecule; it is also more α-helical. This portion is therefore capable of a more typical protein interaction with other chromosomal protein or DNA, but with the latter not involving the phosphates. Whether the functionally polarized structure of this histone applies to other histones is not known.

As mentioned, there is a striking compositional similarity between plant and animal histones, particularly the arginine-rich type just described, revealing that they are primitive proteins that have remained immutable since their appearance in the eukaryotic nucleus. This is also reflected in a rather poor antigenicity. The overall conservatism within and between organisms attests not only to an essential and economical function of histones but also to a restricted functional specificity (which could, however, be enhanced by modification). It also tells us that the whole of the molecules, including the C-terminal portion, is functional and therefore its total steric relationship is of interest. Although there is some predilection for certain DNA base compositions, no specific interaction of histones with any particular isolated DNA fraction has been found, which is not really strange considering that histones bind strongly to the DNA phosphates in particular. Different histones seem to be randomly scattered throughout the genome, although the possibility of some fine orderly pattern is certainly not excluded. It was mentioned that histones probably do not cover the DNA completely. There is evidence from reconstruction experiments that certain histones bind preferentially but not exclusively to the highly redundant DNA sequences (Georgiev, 1969a). Under the electron microscope, histone I molecules locate to every 400 Å along purified DNA (Bonner, 1969), a spacing equal to about eight lengths of the histone molecule; the meaning of this regularity is not clear. As to the mean number of histone molecules on the DNA, it was calculated that there are about ten molecules of each main histone species per average informational cistron (Georgiev, 1969a). This calculation does not take into account the interaction of histones with other proteins which is bound to affect their disposition on the DNA. To date, this pattern of histones on the DNA raises two unanswered possibilities: first, that some of the histones may after all represent some kind of regulatory signal in the chromatin; second, that the polymerase somehow has to make its way through histones as they loosen with the onset of derepression. This last point signifies that perhaps much of the histone associated with each cistron may be near the initiation site of transcription, for it is there that repression would make the most sense.

In terms of histone-gene relationships the weight of implication seems therefore to be that histones at large are not gene specific and are indeed shared, even though not all histones or the effects of their various modifications may be absolutely unspecific. Contrary to a first impression that this combined implication of unspecificity and excess molar multiplicity runs counter to the basic function postulated for histone, Goodwin (1966b) pointed out that it could be just what is needed. He argues, using a computer analogy, that this lack of specificity and molar multiplicity may be required to achieve simultaneously the finite number of cell types in the eukaryote and the stabilization of its embryonic decision-making machinery. Irrespective of the detailed merits of this argument, the main point to derive from it is that rather unspecific histone may be a regulatory necessity of genetic complexity. Even if difficult to prove, this is an important concept.

In conclusion, histones emerge from this discussion for the most part as generalized repressors of the genome. Since it is unlikely that any permutation or combination of histones could satisfy a high specificity of transcriptional repression, other more specific molecules must be involved. Several special modifications of histones are, however, thought to be instrumental in derepression and perhaps with a not negligible specificity. Conversely, since histones are not removed from euchromatin on derepression it cannot be excluded that they in fact participate in derepression together with the following elements: only in this event should histones be considered to act as true regulators rather than repressors. It also cannot be excluded that the few exceptional tissue-specific histones are for the purpose of a nonregulative, permanent repression.

Thus far, in discussing heterochromatin and histones we implied that decondensation of chromatin is an obligatory step in derepression. We shall next discuss certain other molecules, acidic proteins and chromosomal RNAs, believed to be proximally involved in the mechanism of derepression, and then proceed to more distal effectors in regard to the genome. This has been the actual progression of experimental strategy in order to circumvent the conceptual deadlock brought about by the very function of histone in regulation. Clearly, however, derepression is a functional continuum and treating its elements separately is purely a practical convenience. It will be apparent as we proceed that the attention concerning derepression has centered on the processes that take place at the general level of activation of transcription rather than on those which occur at the level of the mRNA molecule in particular. This situation prevails because it is not yet possible to relate these two aspects to each other with any degree of confidence (as is possible in the case of bacterial mRNA), although they obviously must be intimately related; as in bacteria, however, activation must involve foremost the initiation of transcription. At this point we may assume that the most immediate derepressive functions are preliminary to informational transcription and at least partly consist of regulatory transcription. Ultimately, by definition, derepression means the release of informational transcription. At some early stage of either type of transcription there is presumably proteination of the transcript, whose functional relationship to the mechanism of transcription remains equally to be understood. The hypothetical processes at the level of the mRNA molecule are discussed later in the light of additional information.

Stable Regulation of Messenger RNA

A first aspect to consider is that only part of eukaryotic DNA is available for transcription. The evidence is that in the thymus nucleus 80% of DNA is extractable before affecting transcription, which ceases, however, after all DNA is extracted (Allfrey *et al.*, 1966). Similarly, in several types of tissue 90% of DNA is not transcribed as shown by hybridization and, as expected, what is transcribed is partly specific to each tissue (Paul and Gilmour, 1968). Since, as a first approximation, the genome can be taken to be equally represented in all cells, it is apparent from the previous figures that in the eukaryotic cell the procedure used is to selectively activate some genes rather than to selectively repress most of them. This procedure is directly observable in differentiating embryonic cells (Chapter 6). The task at hand is therefore to discuss the mechanism for selective transcription of part of the DNA, even though at some subsequent stage selective repression may be equally necessary. In practice, however, only derepression is considered in the discussion, the logical but still unproved assumption being that, mechanistically speaking, repression is the reverse of derepression.

Selective derepression converts a common genome into many facultatively functional genomes in one organism, in principle, as many of these genomes as there are cell types. Basically, however, this should not be taken to apply only to differentiating or differentiated cells, because a dividing undifferentiated cell also must use a different part of its genome at different times in the cell cycle. In any cell, therefore, both endogenous and exogenous signals for derepression have to be considered. Again, in practice one refers mostly to the latter.

The regulation under immediate discussion is of a stable character, although its permanence varies with the particular cell or cell status. This stable regulation is to be distinguished from the more adaptive type of regulation to be discussed at the close of this chapter and which involves mechanisms from different areas of regulation.

Acidic (Nonhistone) Proteins

A basic if oversimplified assumption was that heterochromatin as manifest repressed genome is causally connected with the repressive function of histone. We may now consider the role of certain polyanionic ligands in chromatin, a complex of acidic proteins, which appear to affect the behavior of histone in the opposite direction, that is, derepression. It is necessary to stress our ignorance concerning the function of this complex of acidic proteins (review by Hnilica, 1967; Stellwagen and Cole, 1969), except that on the whole they behave quite differently from, and frequently antagonistically to, histone. By weight, acidic proteins amount to from 0.1 to 1.0 of the DNA depending on the tissue (Bonner *et al.,* 1968), and in some tissues the acidic protein content appears to vary inversely with histone content. They are more difficult to remove from the chromosome and from this property they are also known as residual proteins. Yet, contrary to histones, some of them would seem to be able to move in and out of the chromosome. Some, but not all, acidic proteins are small proteins, 2.7 S, and may form physiological aggregates (Marushige *et al.,* 1968). Their acidity derives from their amino acid composition and content of phosphorylated amino acid (about 1% phosphorus), much higher than histone.

Within the small fraction of acidic proteins studied so far, there are about 40 major species, as revealed by gel electrophoresis. Part of them but by no means all vary with the particular cell type and animal species (Teng *et al.,* 1971). According to Elgin and Bonner (1970), the number and heterogeneity of the proteins may be less than the previous estimate, and it is clear that technical difficulties stand in the way of deciding this question. In several tissues these proteins have been shown to increase greatly in amount with cell differentiation, during puff activation in polytene chromosomes, and in metaphase chromosomes (where presumably their role is mainly structural). Invariably, template-active chromatins possess more acidic protein (roughly fourfold more) and without this protein they lose their activity. These correlations point very neatly to a preferential association of acidic protein with euchromatin. Evidently also, acidic protein does not inhibit RNA polymerase. It is important to note that single species of acidic protein are preferentially made during hormonal stimulation and the newly made species vary according to the target tissue (Teng et al., 1971).

Many acidic proteins are in addition metabolically active, that is, the molecules turn over, in particular, their phosphorus. The phosphorylation pattern depends on the tissue, protein, and metabolic rate, in other words, is very specific. As in the case of histone acetylation, an increase in phosphorylation (mainly to serine) precedes activation of RNA synthesis, for example, by hormone. Upon activation the pattern changes drastically according to the proteins; in addition, phosphorylation precedes their own synthesis. In contrast with histone acetylation, the increased phosphoryl groups are continuously released from the activated chromatin, so that the role of the negative charge density in the acidic proteins (a general charge neutralization) is essentially dynamic. Similar to histone modification, phosphorylation is assumed to modify the interaction of the proteins with DNA, thus contributing to their functional specificity.

As mentioned, histone masks DNA for transcription unspecifically. Chromatin reconstruction experiments have led to the postulate that the organ-specific unmasking of transcription depends on acidic proteins (Paul and Gilmour, 1968). However, only transcription of redundant DNA was tested in these experiments. In any case, as in the reciprocal experiments with histone, based on this evidence the exclusive assignment of unmasking to acidic proteins is not unequivocal. For example, the proteins tested could be contaminated with chromosomal RNA which is similarly postulated to have an unmasking function. The observation was made that dehistoned chromatin has less transcriptional activity than pure DNA, and that DNA combined with only acidic proteins has about the same activity as native chromatin (which is also less than the activity of pure DNA). Taken together, these observations strongly suggest that the acidic proteins, apart from specifically unmasking certain transcriptions, also repress others. It is possible, therefore, that the protein in part interacts with certain histones to unmask DNA and in part interacts with other histones to mask DNA. Mechanistically, the function of acidic proteins revealed by these experiments involves, on the one hand, DNA sequence recognition, and, on the other hand, structural effects on other chromosomal proteins. Significantly, in the first respect, some of the acidic proteins specifically anneal with homologous DNA, and, when annealed, promote transcription (Teng et al., 1971); which protein functions are involved (see below) is however not known.

It now seems pretty clear that an interpretation compatible with all the evidence presented would be that this unmasking complex of acidic proteins includes proteins of such diverse function as RNA polymerases, binding proteins, structural chromosome proteins, enzymes for chromosome derepression (e.g., acetylase) or metabolism, hormone receptor proteins, perhaps transcription factors, and even specific repressor or derepressor proteins (see, for example, Mondal et al., 1970); many counterparts to bacterial and viral proteins in control of transcription are therefore found in the complex. To say the least, it would be a mistake to think of a unitary role for such a complex of proteins. As expected from these functions, at least some of the proteins should be in a dynamic state of flux between the chromosome and its nuclear environment, or, more generally, between the nucleus and cytoplasm. From cytochemical evidence (Sirlin and Knight, 1958), a steady-state concentration of the bulk of acidic protein is maintained in all chromosomes of the very active insect salivary gland.

It is not known which acidic proteins interact specifically with their own histone, DNA, or chromosomal RNA, yet some degree of specific interaction depending on the function of each protein is expected and is, in fact, suggested by the evidence. Evidently, solving this manifold interaction (even less known than for histone) would help understand the function of each protein. At present, therefore, the total role of acidic protein is difficult to state precisely, but from the several types of proteins involved, their response to hormone, the probable unmasking of DNA, and the binding to DNA, this role will probably be a multiple one and, in some cases, show regulatory specificity. The current effort is to authenticate these regulatory effectors of the genome. In summary, broadly speaking, a positive regulatory role in derepression is assumed for some of the proteins in the complex, though this role need not be exclusively theirs. The evidence also suggests that some of these proteins are coresponsible for repression, but definite functional assignments must await the characterization of individual acidic proteins in different tissues. However, everything considered and especially the relatively limited number of these polyanions, it remains that as a class by themselves they can hardly supply all the required regulatory specificity of chromatin.

Chromosomal (Derepressor) RNA

As proposed by Frenster (1966), the difficulty just pointed out with regard to sufficiency of regulatory specificity would be partly obviated if locus-specific RNAs intervened in derepressing chromatin. This postulate envisages that the RNAs would displace histone from DNA by combining with histone nonselectively while selectively base pairing with the nonsense strand of the now destabilized DNA; the sense strand would then be capable of transcription. That is, the RNA would unmask DNA positively. A prediction of this particular scheme is that derepressor RNA and mRNA are complementary, and this complementarity is seen as a possible basis for the cell-to-cell transmission of the derepressive system (review by Frenster, 1969).

From this point on the reader should be aware that all RNA-DNA recognitions to be discussed involve either a partial DNA denaturation followed by a duplex RNA-DNA strand formation as in the preceding scheme, or as an alternative considered below, the formation of a triplex RNA-DNA helix without denaturation. The general possibility also exists that upon recognition the oncoming RNA becomes a primer for the resulting transcriptions, but for lack of evidence this possibility will not be mentioned further.

Heterogeneous chromosomal RNA that appears to fit the characteristics of a derepressor was described in Chapter 2. As shown by hybridization, chromatin requires the RNA to produce its specific transcript *in vitro* (Bekhor et al., 1969b), a finding that must be viewed in conjunction with the parallel finding for acidic protein. The observation is that the RNA specifies the correct transcription of reconstructed chromatin without stimulating the overall transcription. The most plausible inference from this observation is that as the result of the function of the RNA some specific transcriptions are favored at the expense of others. The postulated derepressor role for the RNA rests so far on this inference and the remainder of the evidence is circumstantial. It has been proposed by Heyden and Zachau (1971) (but see Bonner, 1971) that most of the RNA represents degraded tRNA extracted from chromatin. It is at some risk, therefore, that the RNA is provisionally accepted here as a bona fide species. In any case, it has yet to be proved that it functions derepressively and not mainly structurally, pending which proof the derepressor role remains putative. Still, the RNA may interact derepressively in conjunction with certain acidic proteins, or it could more directly control the repression by histone. As pointed out also for acidic proteins, none of the present evidence excludes the possibility that the RNA in part corepresses together with histone.

New chromosomal RNA appears very soon in response to hormone and accompanies in amount the subsequent increase of template activity of chromatin (Teng and Hamilton, 1969). The RNA appears to exchange between the chromosome and its intranuclear environment. Of regulatory interest, there are preliminary indications for the possibility that chromosomal RNA shuttles to and from the cytoplasm, but this awaits confirmation.

A nonhistone protein binds chromosomal RNA in a complex which is linked to histone, and, in turn, the RNA is possibly base paired with DNA (Bonner *et al.*, 1968). The pairing characteristics of the RNA are revealing. First, the RNA hybridizes only with homologous DNA, indicating that it can read sequences. Second, the RNA uniquely hybridizes in a triplex with native DNA, as this is one way expected to destabilize the template (Bekhor *et al.*, 1969a). It is presumed that the RNA can form this unique triplex because of its extensive base modification (Chapter 2). The question arises as to how closely related to its own structural DNA sequences are the sequences that the RNA is supposed to recognize. It may be pertinent to this question that the RNA anneals preferably to reiterated sequences scattered in the DNA (Sivolap and Bonner, 1971), for this behavior suggests that the RNA may not only recognize its own sequences of origin but also related reiterated sequences. As will be seen, this possibility has considerable regulatory implications. Because of the promise of this family of RNAs there is a great need first to validate and characterize them conclusively, then to define the molecular mechanism of their function, and finally to find out whether they associate exclusively with euchromatin as is expected from their putative function.

In summary, we have examined a series of hypotheses concerning derepression which in their barest form read: (a) histone normally binds and masks DNA and (b) the binding in this complex is contingent to the degree of histone modification; (c) complexes of acidic proteins and (d) chromosomal RNAs intervene positively in derepression by an essentially unknown mechanism, but supposed to be accompanied by (e) modification of these proteins. All three classes of molecules (histone, acidic protein, and chromosomal RNA) have been shown to recognize DNA sequences. Being most unlikely that all these molecules function passively, it becomes almost axiomatic that some of them will participate in regulation. Thus, according to these views, regulation of derepression would combine, at a coarse level, certain preparatory changes in histones with, at a fine level, the more discriminating action of activated nonhistone proteins and RNAs. The derepressive system appears therefore not to be specified by any single molecular process but instead to be cospecified by many coordinated processes. Thus, although we do not have an explanation for this coordination, it implies something important concerning the nature of the system. The immediate result of these molecular operations is also not clear except that eventually they allow the polymerase to transcribe its template; whether, in addition, the polymerase itself plays an active role in these operations (prior to transcription) is not known. In the final consideration it is possible that the series just described corresponds with a hierarchy of regulatory equilibria still to be defined and, most relevant, not necessarily in the temporal order given here. Equally relevant is the implication above that probably no single repressive or derepressive function in this series is the exclusive attribute of any particular class of molecule, but, instead, these functions represent different stages of interaction within a multimolecular complex involving all chromosomal components. Certain interactions in the complex would, for example, be more typical of heterochromatin than euchromatin, and vice versa. The present view places the final specificity of derepression in a continuum of function within the chromosome, yet finally in relation to specific effectors acting externally on the chromosome—a view which underscores the simple fact that the exceedingly high complexity of the chromosome must be for a purpose. Unhappily, serious analytical difficulties will very likely delay the unravelling of this complexity.

Hormones

It would seem questionable that in a eukaryotic organism the above counterions, protein and RNA, can by themselves be entirely responsible for controlling the genome. One basic consideration which will be elaborated later is that each cell probably carries in its genetic make-up the program for its own programming and thus each cell largely depends upon its own control circuitry, which includes the function of these counterions. Indeed, this self-organization of the cell is ultimately responsible for specifying what, when, and how much of it is to be made—which constitutes a fundamental difference with nonliving systems. In many circumstances, however, if the organism is to function as such, the genome in almost any cell must be accessible to further control by the organism. As we know it, an important means of organismal coordination of cell function is by specific signals, namely hormones, whose function is therefore to act on cells other than their own cell of origin. In the usual case the target cell responds to the hormonal signal with an alteration of its spectrum of protein, which, depending on the hormone, can be either permanent (growth) or transient (secretion). Generally then, this response leads to growth or a change in functional activity but, occasionally, directly to cell death (Drews, 1969). Evidence has accumulated showing that a general characteristic of hormone-dependent tissues is that their mRNA synthesis is hormone sensitive (reviews by Carriere, 1969; Paul, 1968; Tata, 1968; for plants, Key, 1969). This is not to say that hormone acts only on mRNA. For example, in the immature uterus, estradiol markedly stimulates stable RNA synthesis (Billing *et al.,* 1969a). This response is not surprising considering that the translation apparatus equally needs to be built up in the growth-stimulated cell.

Practically nothing is known about the crucial mechanism of hormonal action on mRNA synthesis. Earlier, this was believed to be rather generally on polymerase or template activities. Today, it is felt that certain hormones more specifically direct their effects toward regulatory genes that regulate informational genes in charge of mRNA synthesis, but without excluding that other hormones act directly on informational genes. Such effect on genes could be at the level of DNA or DNP, of which either has some evidence in its favor, yet no incontrovertible proof. On the other hand, as implied above, it is now clear that the activation of RNA synthesis is paralleled by a sequential activation of distinct Mg- and Mn-dependent polymerases, respectively, implicated in rRNA synthesis in the nucleolus and DNA-like RNA synthesis in the chromatin (Pogo, 1969; Roeder and Rutter, 1970). Thus the assumption is that the primary site of hormonal action is somewhere in transcription, but it remains imprecise whether this site involves the gene itself or its associated chromosomal constituents, the polymerase, or any combination of them. In short, the action of hormone on genes could be direct and/or indirect. In this respect, in most instances, it is difficult to exclude an effect on transcription secondary to a primary effect elsewhere, more about which will be said later. In addition, there is a cascade of other well-known hormonal effects of a kind which does not concern us here; for example, effects on precursor pools can lead to overestimation of RNA synthesis from incorporation data (Billing *et al.,* 1969b). These last remarks are not meant to disparage the previous assumption regarding the genetic action of hormone, but to qualify it.

It appears that there are other RNA species such as labile RNA (T. H. Hamilton, 1968; Kidson and Kirby, 1965) and chromosomal RNA (Teng and Hamilton, 1969) that respond to hormone earlier than mRNA (within minutes). The increase of labile RNA is in excess of what might be expected for a single mRNA species (O'Malley et al., 1969), and the sequences of this RNA appearing in different organs in response to the same hormone are specific to each organ (Church and McCarthy, 1970). The actual amounts of labile and chromosomal RNAs are probably too small to be reflected in a correspondingly early increase of the Mn-dependent polymerase activity. Curiously, in cold-blooded animals such as amphibians the early production of these DNA-like RNAs prior to the appearance of rRNA is greater than in warm-blooded animals (Wittliff and Keeney, 1969). Further, in keeping with transcriptional derepression in response to hormone, total acidic proteins increase in the chromosome and single species of these proteins make their appearance (Teng et al., 1971). Regarding specific mRNA itself, the time of activation of synthesis after hormonal stimulation cannot be unambiguously assigned. An indirect estimate is however possible from the fact that the first enzymic or cytological manifestations of hormonal stimulation are observed generally within a few hours, but, in some cases, within the hour (De Angelo and Gorski, 1970). Despite the present emphasis on synthesis, it should be noted that in certain systems hormones also stimulate the posttranscriptional processing of mRNA, an aspect which will be considered later.

A proposal by Tata (1968) attempts to unify the mechanism for multiple hormonal activation of RNA species. He proposed for growth hormones, in particular, that their effect is to produce endoplasmic reticulum, rRNA, and specific mRNA simultaneously, and that these elements then move together to the cytoplasm where they function as a translation unit. A basic observation by Tata is that, under conditions of marked hormonal stimulation, endoplasmic reticulum membranes undergo turnover of their constituents at the same rate as ribosomes. He notes that different hormones have additive effects on tissue and each hormone draws from a quota of rRNA, although this RNA itself need not be specific to the hormone. A discussion of the general applicability of this interesting proposal is deferred to Chapter 5. It needs mention, however, that the effect of growth hormone on stable RNA could be less direct than on mRNA.

Independent of any particular scheme of hormonal action, a general view is that hormone stimulates in tissue what the tissue has been previously programmed for. This is to say, the cell responds permissively, and, to this extent, the mechanism of hormone is noninstructional. Several general facts suggest a programmed response of the cell: (a) hormones have pleiotropic effects on many enzymes in a target cell and (b) on different enzymes in different tissues; (c) a cell can respond similarly or differently to different hormones, but (d) does not respond to its own hormone; (e) during evolution the same enzymic activity has come under more multivalent hormonal control. The mediators discussed below do play an important role in conditioning the cell response, however. A necessary qualification to this noninstructionality of hormone would be that the increase of acidic protein could be caused by direct hormonal stimulation of translation.

If that protein were to function in a regulatory manner and specifically on the cistrons ostensibly responding to the hormone, then the hormone could act, though indirectly, by instructing transcription, that is, overruling a preexistent program. A similar but weaker argument may be made for chromosomal RNA, namely, that the RNA could be indirectly controlled by hormone via translation. Without further belaboring these possibilities, they should serve as a reminder that there are potential hormonal mechanisms for regulating transcription other than those commonly accepted. Whatever the mechanism, it appears that replication and cell division, the S phase in particular (Griffin and Ber, 1968), are required for certain hormones in order to affect transcription (Lockwood *et al.,* 1967). This might indicate that as the DNA or chromosome undergoes replication, it becomes more accessible to hormone.

Hormone Mediators. So far, hormones have been considered as a class which, to say the least, is a gross oversimplification. Certain mediators, such as the following, permit us to separate the mechanism of hormonal action into different categories.

Receptor proteins. It was shown in the endometrial cell that estrogen is first bound to a cytoplasmic 9 S receptor protein that appears to become 5 S as it enters the nucleus (Shyamala and Gorski, 1969). It is not clear whether it is the same protein in the nucleus and cytoplasm (after undergoing an allosteric transition) or whether different subunits are involved in the two compartments. Vitamin D metabolites are similarly transported to the nucleus (probably the nuclear envelope) in intestinal mucosa, where they exert a hormonal-like effect (Haussler and Norman, 1969) which could be less direct than that of typical hormone. Hormone receptors also occur in plants (Matthysse and Phillips, 1969). In general, they are long-lived proteins (Sarff and Gorski, 1969). The suggestion was made that degeneracy in the receptor protein could govern the access of one steroid hormone to different target organs (McGuire and Lisk, 1968). Within each target organ, therefore, the diversity of receptor proteins would be commensurate with the number of hormones that have access to it.

Accompanied by receptor protein, intact hormone finally reaches the chromatin within minutes after administration. There it is believed to bind stereospecifically and noncovalently to promote transcription (T. H. Hamilton, 1968). In particular, the steroid hormone forms H bonds with GC-rich sequences in one DNA strand (Cohen *et al.,* 1969), perhaps involving specific gene recognition. The mechanism of action of the receptor protein itself is unknown. In part the protein binds directly to the DNA (Matthysse and Abrams, 1970), as expected if it is to deliver hormone there. However, although these data point fairly clearly to the presence in chromatin of specific receptor sites, it remains unclear which (hormone or protein) binds first or foremost to the DNA. An earlier proposal was that the receptor protein acts as a transcriptional repressor in the absence of hormone (Talwar *et al.,* 1964) since, for example, estrogen does not stimulate isolated chromatin in the absence of receptor protein. Somewhat different, it was also suggested that in the presence of hormone the receptor protein sequesters a repressor (Jensen et al., 1969). On present evidence, contrary to these two proposals, the protein behaves more in an auxiliary manner than as codeterminant of hormonal specificity. Its recycling role with respect to steroid hormones seems to be to select and facilitate their continuous access to the chromatin—or to be selected and transferred to this place by the hormones.

Cyclic AMP. A different line of evidence concerning hormone mediators is that protein hormones (and biogenic amines) on arriving at the cell elicit a change in the amount of cyclic 3′,5′-AMP (or of cGMP) and this in turn acts as a second signal which mediates, hence precedes, the hormonal effect (review by Robison et al., 1968). Certain hormones are themselves released by cAMP, whereupon cAMP functions as a third signal. This function as hormone mediator is by no means the only one of this versatile metabolite, which, for example, also functions in translation. The enzyme which directs the conversion of ATP to cAMP, adenylcyclase, is present at the cell membrane where many hormonal effects are well known to impinge. In the cell this conversion is irreversible because the released PP_i is rapidly removed by pyrophosphatase. In itself, cAMP is stable except to a specific phosphodiesterase responsible for its breakdown. Moreover, cAMP is not consumed in its function as a signal. Hence the limits of cAMP in the cell are carefully controlled at both the level of synthesis and degradation. As envisaged here, hormone intervenes mainly at the level of synthesis, but while most hormones stimulate the synthesis, a few hormones inhibit it. We shall consider only the effects of stimulation, but, in practice, the effects of inhibition will just be the reverse.

A model for the mechanism mediated by cAMP is provided by the physiological response of the fat cell. A single catalytic unit of adenylcyclase is coupled at the cell membrane with selectivity sites for a set of lipolytic hormones, to which the unit responds nonadditively with the release of cAMP (Bär and Hechter, 1969). These sites determine the specific cell response to each hormone, but the unit sets a limit to the overall response. The model can be generalized to other protein hormones by assuming that they operate on comparable sets of adenylcyclase units and selectivity sites.

According to this model, cAMP is what acts as the signal within the cell where hormone never enters. This raises the question of the limited specificity, if any at all, in respect to the genome of which by itself the metabolite is capable. Related to this question, Kerkof and Tata (1969) in their studies with thyrotrophic hormone concluded that cAMP mimics the purely metabolic hormonal functions (permeability, enzyme adaptation), without reproducing the more specific functions on syntheses of RNA and endoplasmic reticulum. Clearly, the last word on cAMP, including its site of action, has yet to be stated.

In conclusion, although they share a mediative role with respect to hormone, receptor protein and cAMP would seem to act dissimilarly. In the case of cAMP the (protein) hormone apparently need not come in contact with chromatin whereas the (steroid) hormone bound to receptor protein does come in contact with chromatin. Without wishing to imply an absolute rule, this dissimilarity suggests that, with regard to transcriptional effects, it might be possible to draw a distinction between a pathway for hormones acting via a receptor protein on chromatin, and another pathway for hormones acting via cAMP more generally on cell metabolism. (That growth hormone, which is protein, seems not to act via cAMP would certainly indicate that this is not an absolute rule.) The latter metabolic pathway obviates the improbability that cAMP by itself may act specifically on the chromatin, while permitting one to envisage how some of its effects eventually reach the chromatin. For example, cAMP is known to affect the phosphorylation of histone (Langan, 1969) and perhaps also acidic protein, in which manner it can influence transcription. Indeed, prior to these effects, serial mobilization of protein kinases and phosphorylases seems to be one of the main functions of cAMP (Corbin and Krebs, 1969). According to this view, these enzymic pathways (in conjunction with the membrane selectivity sites) would determine which proteins are to be phosphorylated, so that cAMP activates a specificity already present. One may thus speak of the cAMP hormonal pathway acting on the metabolic or enzymic program sustained by the organization of the cell, as much as one can speak of the receptor protein pathway acting on the genetic program carried in the chromosome. As a reminder that the bifurcation of hormonal pathways may not be as simple as outlined here, steroid hormones also appear to stimulate cAMP (Szego and Davis, 1967), and not all protein kinases seem to depend on cAMP. Some relationship between the two pathways is therefore not excluded and perhaps is likely. By contrast to the eukaryotic cAMP pathway, in the prokaryote the metabolite has an effect on the polymerase and/or promoter which it stimulates for general transcription (without, as far as is known, phosphorylation being involved), and the metabolite is served by a protein directly responsible for its effect. Were there no similar special protein to be found in the eukaryote, it could well be because, in this case, the metabolite is acting on intermediary enzymes rather than close to the DNA, in other words, these enzymes are the special proteins.

Ions

Data by Kroeger (1968) indicate that the titer of steroid hormone affects the Na/K balance in the nucleus, possibly as the result of inducing changes in membrane permeability. In the case of protein hormones the cation balance is affected via cAMP (Rasmussen, 1970). A basic observation by Kroeger is that, when the cation balance is shifted experimentally, the *in vitro* puffing pattern of polytene chromosomes resembles that produced by moulting hormone (ecdyson). This suggests that the cation balance has different effects on different transcriptions. Kroeger proposes that the cation balance acts by affecting the binding of histone to DNA in a locus-specific fashion, that is, more specific than histone modification. Yet it could also affect the function of other chromosomal polyanions or even have less specific effects, which he did not consider. Thus, although the cation effect on the chromosome is clear and may be regarded as an influence from the immediate chromosomal environment, the regulatory significance of this effect in respect to transcription is not as clear. In particular, even if it is true that ions and hormone act in relation to one another, there is no real basis for assuming that the specificity of transcriptional regulation rests causally with the ions rather than the hormone. In addition, ions may influence transcription not only in relation to hormone but probably also in relation to many cellular physiological events. This whole area of regulation deserves further exploration.

Irrespective of any possible regulatory function, ions are essential constituents of the internal cell milieu. Cells within epithelia are freely ion permeable through their tight junctional complexes and the resulting ionic communication may be important for coordinating cell behavior. One thus wonders how much the ionic impermeability of cancer cells (Jamakosmanović and Loewenstein, 1968) can be responsible for their ominous breakdown in cell-to-cell coordination. In this connection, atypical derepression of cancer cells is manifest in the production of extraneous hormones (Ross, 1968). It should be pointed out, however, that the communication between normal cells mediated by junctional complexes need not involve only ions but also more complex signals, namely, the supracellular signals to be discussed in Chapter 6.

Factors

Finally, the possibility exists that cytoplasmic factors are capable of influencing transcription according to the metabolic status of the cell. As a possible general example, chromatin activity may be affected by the diurnal cycle of the polysome which is under control by diet (Steinhart, 1971). This kind of influence could either be to facilitate or to control transcription more directly. Factors need not only be protein but could represent products of metabolism or differentiation or even indirect products of hormones. Most factors would act positively either by themselves or in concert with other signals that could affect them or even add their own specificity. Conceivably, as the behavior of acidic proteins might suggest, some of these factors can, at times, be an integral part of the chromosome. These factors are seen here as being equivalent to transcriptional factors (see below), but with a wider range of specificities over many aspects of chromosomal function.

An example of a transcriptional factor could be the reactivation of transcription in the inert avian erythrocyte nucleus made into an heterokaryon with a HeLa cell. This reactivation involves rRNA and DNA-like RNA. It points to a signal from cytoplasm (Harris, 1968) which is heat-stable but lost on freezing and thawing (Thompson and McCarthy, 1968). The nature of the signal has not been clarified, except that evidently it is not species specific or hormonal in the usual sense. Factors, this time for initiation in particular, are also indicated (Roeder *et al.*, 1970) by the lack of preference by the amphibian nucleoplasmic polymerase for total DNA as opposed to rDNA; and, conversely, by the lack of preference by the nucleolar polymerase for rDNA. Many other good examples of factors are to be found in embryonic development.

To sum up the final part of this discussion, we have considered a series of molecules thought to mobilize the transcriptional machinery and bring about accurate and selective transcription. These molecules are hormones and their mediators, possibly factors, and, under certain circumstances, perhaps ions. Whatever the mechanism of action of these molecules—and this remains very vaguely defined by comparison with the bacterial operon—the final effect is to evoke a specific transcriptional response from chromatin.

The Format of the Stable Regulation of Messenger RNA: The Concept of Transcriptional Program

Important as recent progress has been, it is no understatement that, as yet, we do not understand how informational transcription is regulated in the eukaryote. As a first step we can ask whether or not the data presented fit in with the format of the bacterial operon. It is evident from the outset that regulatory genes must exist with their effects going back to informational cistrons. For example, strong regulatory genes of this type are known in maize (McClintock, 1967). It is useful therefore to compare the regulation of the two types of organisms by focusing on the classes of molecules that participate in the regulation. It is necessary, however, to bear in mind that in the eukaryote the functional assignment of molecules is not unequivocal, and, what is more important, that no single function may be exclusively assigned to any class of molecules. With these reservations, it appears that repressor histone is not specific to any DNA, but certain acidic proteins may repress and these may be specific to DNA. Likewise, it appears that putative derepressors such as other acidic proteins and chromosomal RNAs are specific to DNA. As it stands, this set of specificities differs from that in the bacterial operon: here the repressor protein is specific for operator DNA but the derepressor (inducer) is not. Moreover, the repressor is the principal macromolecule that represses and derepresses the operon. From these differences and from the general structure of the system, one is entitled to conclude at present that the eukaryotic regulation does not conform with the operon format. This format, it should be recalled, was designed by Nature for cells with their DNA not ensconced within a chromosome of high molecular complexity. As the main genetic apparatus of the eukaryote, this complex chromosome apparently necessitated the predominantly positive type of regulation that has evolved, in contrast with the predominantly negative type of regulation prevalent in the prokaryote. Contrasted with the bacterial repressor, were the positive regulatory function of the eukaryotic molecules to be gene specific, it would also be instructional.

To date the information in the eukaryote is consistent with an intrinsic, stable repression seated in the heterochromatin and presumably geared to a long-term masking of the genome. Stable regulation is the keynote in the eukaryote. Signals or effectors such as hormone, embryonic inducer, or even smaller endogenous molecules seem to act on the DNA via the chromatin, but the template specificity seems to reside with the chromatin. The overall impression one gains is that the expression of the genome is self-programmed and can respond with precision to this constellation of noninstructional signals, only because of the interposition between genome and signals of a complex apparatus, the chromosome. This concept of intrinsic interactive complexity poses new topological questions for the chromosome; to the measure of the validity of the concept, these questions await molecular formulation in the future.

A concept of regulatory program implies, first, that the constituent steps of the program are reached progressively and are affected by previous steps, an aspect that comes out clearly in the study of the cell cycle and development; second, that regulation itself is programmed. Hence programming implies an organization of the eukaryotic chromosome commensurate with its higher genetic function. As an orientative hypothesis, a program may be envisaged as a preordained series of regulatory transcriptions leading to informational transcription and necessitating a series of molecular rearrangements in the chromosome related to these transcriptions, some of which depend on external effectors.
The organization of the chromosome should correspond with the progressive activation of genetic circuits. This hypothesis is further elaborated later but a molecular definition of program must await recognition of its postulated elements.

Regardless of the final format, there is at present a great need for material clarification of the function of the eukaryotic elements known to intervene in the transcriptional regulation. To date no dependable molecular models have been built representing the interaction of these elements. With this superficial knowledge one may not exclude the possibility that entirely new interactions of major functional significance remain to be discovered. For all we know, therefore, this area of regulation could be the one whose conceptual understanding is the most wanting. A few of the essential questions that need answering up to this point in the discussion are: which are the effective derepressors; how are they controlled; what is their range of specificity; and what is the nature of their interaction with repressors, that is, are there any molecules that exclusively repress and others that exclusively derepress; what is the dynamic molecular organization of the chromosome in relation to these elements; what are the cytoplasmic factors reflecting the cell functional status that feed back on the chromosome to induce or maintain differentiation; is there in the nuclear space a functional topology in regard to informational chromatin?

Molecular Redundancy of Eukaryotic DNA

Current estimates place the number of informational cistrons in bacteria at 4×10^3. This number corresponds to both a genome size and genome complexity equivalent to 4×10^6 base pairs. A similar estimate in mammals with a 1000-fold greater genome complexity (Wetmur and Davidson, 1968) is not available, but, based on the complexity and assuming the same cistron size as in bacteria, the mammalian genome (3×10^9 base pairs) should have 3×10^6 informational cistrons. This number raises the obvious question as to whether the increase in mammalian DNA can be proportional to the number of informational cistrons. An indirect answer comes by way of the certainly disproportionate amount of DNA in certain amphibians and primitive fishes, e.g., *Amphiuma* has 10^{11} base pairs and the lungfish one-half this number. It is not very likely that these creatures would require more informational DNA than a mammal, and in fact other members of these classes have no more DNA than the mammal. Moreover, as postulated by Crow and Kimura (1970), the mutational load imposes an upper limit on the number of informational genes that the organism can carry. The possible inference is that the amount of mammalian DNA is disproportionate to the number of informational cistrons by a factor that is unknown. What is the disposition and function of this possible "excess" DNA in the mammal?

The physicochemical approach to the genome provides an answer to this question. Before examining this answer, it should be pointed out that it is important for a different reason. In *E. coli* somewhere around one-fourth of the genome has been accounted for either by genetic or molecular analysis. In the mammal we are aware on various grounds of perhaps a few thousand genes but, because of the serious analytical difficulties in this material, only a handful of them have been examined thoroughly at the genetic or molecular level. At best this means that in the order of 1 in 1000 genes have been "accounted for" and one can legitimately wonder if the mammalian genome will ever be sufficiently understood on this piecemeal basis. The advantage of the physicochemical analysis is precisely that it looks at the global genome, in particular, the structure of sequences. Clearly, however, for the mammal the gap between this genome level and the level of individual cistrons which can be approached by genetic and molecular analysis will some day have to be filled. Direct data on the structure of sequences are essential if one is to make a comparison with the structure derived from physicochemical analysis.

Although the evidence is still fragmentary, work on DNA reassociation has revealed that the eukaryotic genome from protozoa onward, except perhaps fungi (but this is not clear), is highly redundant in a molecular sense (Britten and Kohne, 1968, 1969). The kinetics of reassociation (in solution) is related to the sequence complexity defined in Chapter 1.[11] Moreover, the thermal stability of the resulting hybrid is a function of the degree of homology between sequences. These kinetic and thermal stability data combined indicate several classes of repetitious DNA sequences apparently similar in all tissues of any eukaryote. Considering all eukaryotes, the proportion of repetitious sequences is very roughly proportional to genome size. Taking as an example the most studied genome of the mouse, this is made of: (a) 70% of unique sequences (about 1 copy); (b) 20% of related but not identical sequences with 10^2 to 10^5 copies each which are arranged in families; and (c) 10% of almost identical sequences in 10^6 copies. These three categories should give an idea of the gradation of repetitious sequences which will become clearer as we proceed; for convenience these types of sequences are referred to as unique (a), redundant (b), and satellite (c). The kinetics of reassociation is slowest and most accurate in (a), fastest in (c), and intermediate but variable in (b). Considering mammals in general, (a) and (b) vary in their proportion but are always present; (c) may or may not be recognizable but to a variable extent and under different guises it is probably also always present. The following description applies to the eukaryote, in general, with emphasis on the mammal. Let us begin with (c) which is a reasonably well understood type of sequence.

The 10^6 copy mouse DNA can be centrifuged apart from main DNA (42% GC) as a light AT-rich satellite (34% GC). Separation as a satellite is possible because of the partial clustering of the copies and their rather uniformly dis-

---

[11] A parameter Cot is defined by Britten and Kohne (1968) to study the genome complexity based on the second-order kinetics of the renaturation or hybridization reaction carried out in solution. Cot means concentration of nucleic acid multiplied by time of reaction (moles nucleotides × seconds per liter) as defined for a standard temperature and ionic strength; it is convertible to an equivalent Cot for other ionic strengths. Cot is approximately equal to: optical density units (at 260 mμ) × hours × 0.5. The complexity of a genome is proportional to the Cot value at half-slope of the renaturation reaction. Thus, a small unique genome like the bacterial and the unique sequences of the much larger eukaryotic genome half-react at values several hundredfold apart. In combination with chemical analysis, this kinetic procedure enables one to measure the proportion of unique to redundant sequences in eukaryotic DNA. As will be seen, the procedure is also used to study the RNA that forms hybrid with each type of sequence. The distinction between unique and redundant DNA sequences is somewhat arbitrary, in the sense that a greater procedural stringency may lead to an ostensible higher proportion of unique sequences derived from the least redundant sequences. However, these operational difficulties do not detract in any significant way from the real structural complexity of eukaryotic DNA. A useful evaluation of the kinetic procedure in the light of these difficulties is written by McCarthy and Church (1970).

tinct base composition. The two strands of satellite can also be isolated because of their different composition (see below). From reassociation kinetics studies, satellite DNA was first thought to consist of a single sequence of about 300 bases which was repeated end-to-end 200 times in blocks of 5×10^7 daltons (Walker, 1968). If evenly spread there could have been 300 of these blocks, together larger than a bacterial chromosome, in each mouse chromosome. From partial sequence analysis it is now proposed that the satellite has derived from a basic sequence 8-13 bases long which has undergone a linear multiplication greater than that stated above and in the process has accumulated an enormous number of mutations (Southern, 1970). (If valid this reassessment reveals the risk inherent in the interpretation of kinetic analyses.) Using either estimate, it is likely that the present day sequence of the mouse satellite is both highly repetitive and molecularly redundant.

Mouse satellite sequences are physically joined with main DNA in the chromosome and some in fact are included in main DNA. Satellite DNA is the most firmly bound to the chromosome. Its subsistence in cell strains in culture for 30 years first suggested that the satellite is functional. Fractionation of metaphase chromosome into several classes by size indicates that the satellite is uniformly present (Maio and Schildkraut, 1969). Previously, isolated mouse nucleoli were shown to contain an associated 10% of total DNA but as much as 30% of the total satellite. This enrichment in satellite suggested that there was a topographical relationship *in vivo,* perhaps its presence in the nucleolus-associated heterochromatin. In fractionated interphase chromatin the satellite is associated with autosomal constitutive heterochromatin (Yasmineh and Yunis, 1969), which includes the nucleolus-associated heterochromatin. All these data now fit nicely together. Using cytological hybridization, Jones (1970) has found the satellite DNA in metaphase chromosomes to occur mainly at the centromeres, and in the interphase nucleus to be present both in regions located around the nucleolus and scattered in the nucleus. The centromere and probably these other regions with satellite represent constitutive heterochromatin. Preliminary evidence indicates that the satellite replicates preferentially late in normal cells (Odartchenko and Pavillard, 1970) and early in cells infected with polyoma virus (Smith, 1970); these results require confirmation. Hence, at least in the mouse, but probably in all mammals as well, constitutive heterochromatin in general can be chemically defined as that chromatin in part including this most reiterated DNA. This type of DNA may soon be recognized by staining procedure (Gagné *et al.,* 1971) and rapid progress can be expected in this important field.

The two strands of mouse satellite DNA differ in composition, A-rich and T-rich, and contain many of all the oligonucleotide fragments possible from their composition. The sequences are periodically reversed between strands and have regions of intrastrand homology resulting in a number of complementary loops.

The sequences are species specific, in fact more so than those of main DNA. The divergence of satellite sequences is therefore much less within than between species. In rodents the differences in the composition of sequences do not follow a taxonomic order, suggesting a recent evolution (Flamm et al., 1969). The composition of satellite sequences in mammals is not, however, restricted to the low GC type in the mouse. Thus the composition and amount of complexity (frequency distribution) of sequences are likely to result in a variety of satellites with different buoyant densities and different relative sizes; this has been observed within and between species. Certain rodents possess more than one type of satellite, yet they all seem to predominate in constitutive heterochromatin (Lee and Yunis, 1971). In certain mammals, as man, the amount of satellite is small (1%); in other mammals the satellite does not separate from main DNA, although there are reasons to believe that the sequences are present. In some species, therefore, the amount is difficult to determine molecularly. In conclusion, the characteristics of satellite DNA are frequently typical of each organism.

A conclusion from the previous paragraph is that, considering different genomes, the most redundant sequences are present in a gradation from an extremely repetitious and redundant form as in the mouse satellite to other more common forms with less repetition and redundancy in most other species (for lack of a better term still called satellites). This may mean that satellite DNA is an extreme form of the typical redundant DNA to be described below. This gradation may have important functional implications. First, satellites tend to predominate in constitutive heterochromatin, whereas typical redundant sequences need not. This says that not all redundant DNA need be in heterochromatin. Yet whether all heterochromatin contains this type of DNA is not known. Second, although transcribable *in vitro,* mouse satellite is apparently not transcribed in the cell (Walker *et al.,* 1969), where, by contrast, many redundant sequences are transcribed. Third, and related to the previous point, satellite DNA forms part of the mouse centromere and this suggests a mechanical function, presumably similar in other species, whereas, as we shall see, structural functions (in a coding sense) are envisaged for redundant sequences. Hence the overall picture appears to range from, at one extreme, exceedingly repetitive and redundant sequences which do not code for RNA and are part of constitutive heterochromatin merging into, at the other extreme, less redundant sequences which code and are interspersed with the remainder of the genome. An example in the middle of this range is provided by *Drosophila* where it is possible to locate several satellites in the salivary chromosomes using cytological hybridization (Hennig *et al.,* 1970). Sibling species in this genus contain different numbers of distinct species-specific satellites (judging by their relative proportion and buoyant densities) which, in fact, represent most of the redundant sequences in their genomes, approximately 10% of DNA. These satellites differ markedly in their pattern of chromosomal localization but they are all present mostly in heterochromatic regions, including the chromocenters (which contain the centromeres). By definition of sibling species, each of the satellites remains distinct in the viable interspecies hybrids. The relationship between these less extreme satellites and the more typical redundant sequences will be clarified as we now discuss the latter sequences.

Typical redundant DNA (10^2 to 10^5 copies) is present in every eukaryotic species examined in the form of a few families with up to hundreds of thousands of sequences each. This DNA is uniformly distributed in the genome and all sequences show a considerable degree of physical interspersion with total DNA. Among chromosomes, however, the X chromosome seems to carry more than a normal share of redundant sequences (Gall *et al.,* 1971), the explanation of which is not obvious. Within a family the degree of homology ranges from a few sequences perfectly identical to one another to a few sequences with only about two-thirds homology, yet the sequences are largely species specific. By this operational definition of a family—the only definition available—the sequence divergence within a family is less than the divergence between families whether of the same or a different organism. The reassociation product of these redundant sequences therefore cannot be absolutely specific or perfectly thermostable. (A minority of sequences have sufficient intrastrand homology to fold back on themselves with the stability of native DNA.) The most related sequences possibly stretch for a rather uniform few hundred nucleotides in all eukaryotes. These families thus represent different components of typical redundant DNA. Between related species, the complexity (frequency distribution) and composition of sequences in each family are likely to be heterogenous, resulting in distinctive patterns of sequences. In their evolution the largest families of sequences (highest frequency) have diverged the least and show the greatest homology between member sequences (highest thermal stability). These families are presumably the youngest and are not shared even by closely related species; because of greater divergence, the opposite in all these respects is true to a point of the oldest families. To give an idea of the evolutionary divergence of this DNA, Walker (1968), studying rodents, measured 15% homology within one taxonomic family but none between families. During evolution the ratio of redundant to unique sequences appears to have increased from barely marginal in bacteria to about one in protozoans, four in insects, and twenty in mammals (Bonner, 1969). All the evidence presented so far, and more to be presented below, indicates that most of the redundant DNA is actively evolving and, in particular, much of it is transcribed. How many sequences within each family are transcribed, and how this relates to their age, is not known. From these data the fact is that this DNA has a functional purpose.

A divergent trend between families of redundant sequences was said to be noticeable between related taxonomic species. To Britten and Kohne (1969) this suggests a slow evolutionary drift of the families with the occasional abrupt (saltatory) emergence of a new family. According to these authors, a family would arise from (a) the manifold replication of an initial sequence, itself of unprecise origin, or (b) a series of events by which shorter sequences, somehow related to one another but essentially also of unprecise origin, are replicated to an approximately equal extent. Both (a) and (b) have a general model in the phenomena of oocyte rDNA amplification and DNA puffing of polytene chromosomes. To discuss (a) here, each saltation resulting in a new family of sequences must occur more frequently than the average period that it takes to lose a family by divergence, or about 10^8 years. One consequence of these saltations is that over evolutionary time a considerable amount of DNA would be added to the genome, or 20% of the vertebrate genome in 10^8 years. These occasional saltations emphasize a genetic reservoir role for redundant DNA. A more formalistic explanation of the origin of this DNA resorts to the degeneracy of the genetic code (Walker, 1969). According to this explanation, a set of proteins using only one of the available codons for each amino acid would be organized into a DNA having a 400-fold faster reassociation rate than the same set of proteins using all available codons randomly: the first DNA could reassociate as redundant and the second as unique. A corollary of this argument is that unique DNA is that part of the genome for which there has been either no selective advantage or selective pressure to use specific codons. The species range of unique DNA (20-80% of total DNA) is compatible with this view. The implication is therefore that the transition from unique to redundant DNA has speciation value. Since this kind of selection acts at the level of protein, it follows that part at least of unique DNA must be translated; which it is. However, with respect now to redundant DNA, the biological reality of this interesting argument is less clear, for it is not known to what extent this DNA is translated. Finally, addressing themselves to the extreme satellite DNA in particular, Walker *et al.* (1969) consider improbable a saltatory origin like one of the two origins postulated for redundant families. These authors propose instead a differential preservation and replication of sequences because they are needed for a certain function in the chromosome they already have; in the mouse this can now be equated with a centromeric function, but the problem that remains is how the sequences have spread to all chromosomes. (A similar preservation has certainly occurred in the case of rDNA, another highly repetitive, if less redundant, DNA.) Such proposal for the evolution of satellite sequences differs from (a) above for redundant sequences essentially in that this time a function for the original sequence is required. A different origin of redundant and satellite sequences as envisaged in these proposals could therefore have to do with the different genetic function of each type of sequence. These functions are now discussed beginning with the redundant sequences.

The view was expressed that not all typical redundant sequences may be translated or even can be expected to be meaningfully translated, which is certainly unlikely for the most recent evolutionary sequences. Moreover, not only that not all redundant sequences are translated but also that many are not even transcribed is suspected from the fact that *Amphiuma*, with about 50 times the amount of DNA in the mammal, has a much greater proportion of redundant DNA. As mentioned already, there can be little doubt that, whatever else it does, this DNA serves to generate new genetic function. What is now being implied is that it may be easier for an organism to experiment with novel genetic function by testing the evolving information in DNA at the RNA rather than the protein level, since more rapid change (whether selected for or not) is feasible at the RNA level. Thus a non-Darwinian type of eukaryotic evolution may operate in accordance with this "principle of economy of transcriptional testing." If redundant DNA functions in regulation at the RNA level as is further suggested below, the RNA would necessarily be tested without protein being involved at first. A different possibility, namely, that this DNA is used to amplify informational transcription, can be dismissed outright inasmuch as the majority of redundant sequences are imperfect copies of one another. This imperfection of copy is less true of noninformational DNA such as rDNA whose redundant sequences are repeated more accurately.

Another view which has been voiced for some time is that the genetic self-organization of the sophisticated eukaryotic genome demands a good deal of regulatory as opposed to informational genome. Given that the basic biochemical capabilities are similar, this substantial commitment of the eukaryote to regulation could well be an essential difference with the prokaryote. Evidently, this commitment would go a long way to explain the "excess" eukaryotic DNA. The point being made is that this regulatory genome may essentially be represented in the redundant genome. In relation to this, for example, Goodwin (1966b) has stressed the need for redundant DNA in order to achieve effective repression by histone. By now the analogy between certain of these views on redundant DNA and those mentioned for labile RNA in Chapter 2 will not have escaped notice. One of the possible functions listed for this RNA which is coded by the redundant genome is in a regulatory capacity. One may then conclude at this point that there is a good case for associating the molecular redundancy of DNA with genetic regulation, particularly the production of noninformational RNA. One cannot exclude, however, that some RNA may not be discrete but part of another (e.g., mRNA), or that the regulation may involve no transcript.

In contrast to these dynamic functional possibilities contemplated for typically redundant DNA, the extremely redundant (satellite) DNA may discharge rather more mechanical functions in the chromosome such as linking or buttressing, folding, and interaction with protein. An obvious mechanical meaning is attached to the relationship between satellite DNA and constitutive heterochromatin, viz., the mouse centromere. According to Walker (1971), the function of the centromere satellite is to enable the chromosome to withstand the rigors of meiosis. Other mechanical functions to consider for different chromosomal regions are: (1) to prevent exchange in evolutionary stable regions of the chromosome, or (2), conversely, to stabilize the rare exchanges that are advantageous; examples of (2) could be the centromere and the B chromosomes to be mentioned later. Moreover, because of the well-known "stickiness" of constitutive heterochromatin, a discretionary location of this satellite-rich chromatin could (3) bring together certain chromosomal regions in the nuclear space, or (4) anchor these regions against the nuclear envelope, either contributing to the spatial organization of chromatin. One possible example of (3) will be discussed later in connection with 5 S DNA abutting on the nucleolus-associated chromatin; another example of both (3) and (4) could be the "bouquet" formation in meiotic chromosomes. Lastly, any effects of satellites on meiosis could (5) reinforce the reproductive barriers that exist between organisms. (Any mechanical function may involve the Y chromosome less than others, since satellites are rarely found in this chromosome, especially of the more extreme type.) It follows that if satellites of different species share a mechanical function, this cannot be sequence dependent, for satellites are highly heterogeneous between species. In that case the structure of satellite sequences would only tell us something about their past history but little that concerns their actual function. The general concept which appears to emerge is that the more repetitive the redundant DNA is, the less likely it will have a coding function and the more likely it will have a mechanical function (with the notable exception of rDNA). Contrary to a coding function, however, a mechanical function is probably carried out more by the DNP complex formed with the DNA than by the DNA itself.

Finally it remains to consider unique DNA. This DNA has a relatively low degree of interspecies homology, the lower the homology the greater the phylogenetic distance. It is uniformly distributed in all genomes of which it generally forms about one-half. Little can be said of this DNA since in the present context its unique sequence structure does not lend itself to special comment. However, in contrast to redundant DNA, possibly including regulatory sequences, unique DNA definitely includes informational sequences. Judging from the DNA capable of forming hybrid with putative cytoplasmic mRNAs, these informational sequences are distributed to all four size classes of HeLa cell chromosomes (Huberman and Attardi, 1967), which is in accordance with the known uniform distribution of unique DNA. Hence the concept to be elaborated in the following sections is that the facultative genome consists of regulatory (mostly redundant) sequences and informational (mostly unique) sequences. On this basis, part at least of the typically redundant and unique DNAs should occur in euchromatin or its facultative heterochromatic form.

To conclude with an answer to the initial question regarding the disposition and function of the "excess" eukaryotic DNA, this section reveals that we have some knowledge about its gross disposition but only gross conjectures regarding its function. The intention in this section was to introduce the various possibilities which exist for this function. Pursuing the matter, the following section deals with how much of the DNA may actually be in excess of informational requirements. As far as genetic function is concerned, looking ahead into the future, what now appears to need most study is, on the one hand, the relationship between the organization of sequences in redundant and unique DNAs, and, on the other hand, the organization of informational cistrons in respect of their regulatory cistrons. As must be clear, these two closely interrelated aspects again refer to the general problem of the genetic organization of the chromosome.

Extent of Genome Transcribed in the Higher Eukaryote

How much of DNA in the higher eukaryote makes RNA? Prior to dealing with this question, a brief comment on the DNA-RNA hybridization technique employed in this field of work is pertinent. To measure efficiently the hybridization of DNA-like RNA as opposed to stable RNA, the first involving less relative amounts of RNA but a larger and more differentiated fraction of the genome, high RNA/DNA input ratios are necessary (Kennell and Kotoulas, 1968). To measure the hybridization of unique above redundant transcripts, even a higher input ratio and also a higher DNA concentration (higher Cot values) are required, such that until recently in practice only redundant transcripts could be hybridized (Britten and Kohne, 1968, 1969). It is a common misconception regarding unique genes to believe that if they are sufficiently amplified during transcription, the RNA hybrid will readily be observed (Melli and Bishop, 1969). Nevertheless, the higher the input ratio the more noticeable become the less frequent transcripts.[12] Evidently, these fastidious requirements stem largely from the great complexity of higher eukaryote DNA. It follows from these considerations

---

[12] Higher Cot values are feasible by hybridizing the nucleic acids in solution than when one of them (DNA) is immobilized on a filter or some other physical support. The filter technique, the one used initially, assays only redundant sequences. Moreover, when used to study sequence homology between DNA and RNA this technique serves to reveal differences rather than to prove absolute homology between sequences (Shearer and McCarthy, 1970b). This additional limitation is because the technique detects differences between but not within families of redundant sequences, being impossible to ascribe a given transcript to any sequence of a family. Nonstringent conditions of hybridization will permit one transcript to react with the DNA sequences coding for all the transcripts in the family. Stringent conditions will limit the reaction to the most closely related sequences in the family. These remarks mean that families of redundant sequences thus far are definable only operationally by their hybridization behavior. It follows in general that it is not possible by this reaction to deduce the true complexity of these sequences from that of their transcript, which only reveals a minimal shared complexity. Regarding these questions the reader is referred to the critique of McCarthy and Church (1970). Recently transcripts derived from unique DNA sequences were specifically hybridized in solution at very high Cot values (Davidson and Hough, 1971). In this case, under stringent conditions of hybridization, the complexity of the unique sequence is deducible from that of its transcript.

that present data on RNA hybrids are representative not of the genome as a whole but mostly of the redundant genome. Therefore, only a partial assessment of the genome will be possible here.

Returning to our question, we need first to establish the fraction of redundant DNA that is transcribed. By hybridization studies, (a) stable RNA requires a fraction of this DNA less than 1% of the genome, (b) chromosomal RNA requires from 2 to 4% of the genome depending on cell type. (c) Nuclear labile RNA hybridizes to about 4% of the genome, for example, in mouse L cells (Shearer and McCarthy, 1967) and HeLa cells (Soiero, 1968). Assuming these values to be representative, the estimate for transcribed redundant DNA is therefore of the order of 10% of total DNA. In actual fact, the previous values resulting in this estimate are probably maximal ones because of nonstringent conditions of annealing and the indeterminacy inherent in this procedure; in particular, the values for chromosomal RNA seem to be too high. Be this as it may, the estimate agrees well with the proportion of total DNA found to be transcribed in HeLa cells (Scherrer and Marcaud, 1968), several mammalian organs (Paul and Gilmour, 1968), and isolated thymus nuclei (Allfrey et al., 1966); it also agrees with the proportion of total DNA in the polytene chromosome which is represented in the transcript of puffs described later. Being that a substantial part of the transcript comes from redundant DNA, these biological data suggest that the molecular estimate of 10% of DNA transcribed as redundant DNA is basically correct, but, if anything, exaggerated. For simplicity, let us now consider that on the average the redundant mammalian DNA including the extreme satellites is 50% of total DNA (Britten and Kohne, 1968). On this basis, redundant DNA in an average mammalian cell is less than one-quarter active. The most hybridizable of these redundant transcripts are in labile RNA, which is partly specific to tissue (e.g., Paul and Gilmour, 1968). Using present techniques, however, the individual DNA sequences responsible for this RNA cannot be identified beyond their occurrence within a family. It follows that it is impossible to compare individual sequences transcribed in different tissues, that is, the extent to which these active sequences differ between tissues. Yet it is important to note that the biological data as they stand clearly establish a selective derepression of the genome which mobilizes a different spectrum of genes in different cells. It should be noted that mRNA was not included in this account of redundant DNA.

To establish the fraction of unique DNA that is transcribed presents serious difficulties. The first evidence was by Gelderman et al. (1968, 1971) who studied the terminal mouse embryo. They found that a minimum 12% of unique DNA (8% of total DNA) forms hybrid with RNA. This represents a complexity in the RNA equivalent to a minimum of about 3×10^8 DNA base pairs. The authors extrapolated their data to conclude that up to 36% of unique DNA (25% of total DNA) in their material is transcribed. Davidson and Hough (1971) were able to hybridize unique amphibian DNA with RNA made in the oocyte lampbrush chromosomes. Because of the subsaturation conditions attainable, they could only estimate that a minimum 1.2% of unique DNA was represented in the mature oocyte RNA. The equivalent complexity is 2×10^7 DNA base pairs. Brown and Church (1971) report that 10% of the adult mouse unique genome is transcribed in the brain and less than 2% in various viscera. In the brain this represents a complexity of 2.5×10^8 DNA base pairs. Indirect estimates by competitive hybridization at low Cot values of the DNA represented in mRNA are 1% in mouse L cells (Shearer and McCarthy, 1967) and 5-10% in HeLa cells (Soeiro and Darnell, 1970), the range no doubt partly reflecting the analytical conditions. In general, it is assumed here (from arguments given in Chapter 7) that part of the unique transcript includes informational RNAs. That these in turn include mRNAs with ultimate regulatory functions is possible in view of the high complexity of the transcript.

As explained, the complexity of unique DNA is deducible from its transcript. The transcript complexity in the mouse embryo, mouse brain, and amphibian oocyte therefore represents 3×10^5, 2.5×10^5, and 2×10^4 diverse cistron equivalents each of 10^3 base pairs, respectively. These figures are about 75 and 5 times the total genomic information in the *E. coli* chromosome. A comparison can be made in the oocyte with the transcript complexity derived from redundant DNA, of which about 6% is transcriptively active (Chapter 6). The complexity of this type of DNA cannot be directly derived from its transcript. Assuming, however, that the several redundant DNA components of the oocyte are proportionally represented, the complexity would be commensurate with 300 cistron equivalents each of 10^3 base pairs. This is about 100-fold less than the unique transcript complexity (Davidson and Hough, 1971). Most of the information in the oocyte stems therefore from unique DNA. Concerning the number of copies per active cistron in the two types of oocyte DNA, the situation is reversed. Each unique cistron can be represented by 3×10^4 transcripts, and their sum total amounts to 0.01% of cell RNA. Because the total redundant transcript has a lower informational content, yet in terms of mass is 100 times the amount of unique RNA, the number of copies per active redundant cistron of the oocyte must be enormously greater. Similar comparisons in the mouse embryo are less secure. According to Gelderman *et al.* (1968) in this embryo the percentage of heterogenous RNA that hybridizes with the 20% of redundant DNA present in the genome could equal the percentage of redundant DNA, that is, is less than the percentage of this RNA (70%) that hybridizes with the 70% of unique DNA. This suggests that there is both a lower complexity and copy frequency represented in the redundant mouse embryo transcript. In conclusion, the two systems—one the complex oocyte and the other the total ensemble of differentiating embryonic cells— indicate that most genetic information derives from unique DNA. The same holds true for the adult mouse brain, but in this case no comparison with the redundant transcript is available. There is also the suggestion that the relative numerical representation (copy frequency) of redundant transcripts in a differentiating cell pool (mouse embryo) may be lower than in the case of the oocyte. Pending data in other developmental systems, this suggestion cannot be elaborated further.

With regard to the total genome, the conclusion from the data above is that a fraction, at most, of the genome of any cell type is involved in RNA synthesis, the amount depending upon the extent to which unique DNA is transcribed. To illustrate this point in the mammalian genome, if only 1% of unique DNA is transcribed, then 13% of the total genome is transcribed (based on the data for the unique murine visceral genome and the previous estimate that one-quarter of the redundant genome is transcribed, and again assuming that these two genomes are each one-half of the total genome). The next obvious question is what might be the absolute fraction of genome transcribed in the organism, that is, all cell types included. This question is not tractable at present. All that can be said is that, first, the overlap between cell types in the one-quarter redundant genome transcribed is not known, and, second, in the whole mouse embryo up to 36% of the unique genome may be transcribed; in this embryo the absolute fraction transcribed is therefore only about 30%. A personal impression is that this absolute fraction will indeed be much less than the total genome. In addition, if the genome may not all be transcribed, it is possible that not all of that which is transcribed is translated. This may be particularly true of the redundant genome which, although it contains at least one unique sequence per each of the few families of sequences, consists mainly of redundant sequences and these cannot all be expected to be translated. It might also be true of the unique genome that not all of it is transcribed, or what is transcribed is not all translated. This possibility becomes particularly likely when it is recalled that related genera vary significantly in the absolute size of the unique genome.

In summary, this section continued to explore from a different angle the functional significance of the components of the eukaryotic genome which were first outlined in the preceding section. The genetic hypotheses to be described in the following section will contribute further to a conceptual definition of these functions.

Hypotheses in the Eukaryote: Models of Genetic Organization and Regulation

The purpose of this section is to present several formal genetic hypotheses proposed separately from the working hypotheses previously considered in connection with the biochemical studies on the regulation of informational transcription. Modern hypotheses concerning genetic units in the eukaryote fall into two categories, according to the prevalent ideas about the structure of DNA sequences at the time. Roughly, these hypotheses span the knowledge derived cytologically from exceptionally large chromosomes to the knowledge derived molecularly from DNA. The hypotheses in the first of these categories are constructed around the question of "excess" DNA relative to genetic requirements, and the more recent ones, in the second category, around the question of DNA redundancy. In the present context the difference between "excess" and redundant DNA is that the first concept admits precise copy of sequences (amplification) and the second, by definition, means imprecise or familial reiteration of sequences. It will be seen that this difference implies a profoundly different functional potential. Furthermore, the first category of hypotheses only examine the organization of cistrons into units, while the second category elaborates both on the organization and regulation of cistrons. Conceptually, the linking of redundancy with regulation in the second category of hypotheses is most important.

The first two hypotheses to consider, in which the aspect of "excess" DNA is foremost, are those that Beermann (1967) and Callan (1967) proposed for the individual bands and chrommomeres of polytene and lampbrush chromosomes, respectively. These cytological elements are known on cytogenetic evidence to represent, at most, a few genes. Beermann argues that puffing of bands is not required for transcription proper as much as it is required for packing and disposing of the transcript; for analogy, replication does not require puffing. [The argument need not be generally applicable, however (Berendes, 1968).] Another of Beermann's arguments is that a point mutation is known to alter the character of a puff as a whole. Part of the "excess" DNA would therefore act as a template or "master," but most of it would intervene in the packing of transcript. This hypothesis was presented in general terms and no detailed mechanism was offered.

Callan, addressing himself to the meiotic lampbrush chromosome in amphibians, proposes a terminal "master" sequence only in which recombinational and mutational events take place and are preserved. This sequence is followed by repeated "slave" sequences made congruent with it by a matching process once in a life cycle. This well thought out hypothesis explains neatly how mutation and recombination become established and are expressed, for by the time meiosis occurs all slave sequences are identical to the master. It is proposed that, as they are being matched, the slaves form the typical lampbrush loops and only they do then become transcriptively active. This hypothesis depends on the polarized displacement of the chromosomal axis along the length of the loop, which was previously disputed and is now claimed to have been established (Snow and Callan, 1969); the universality of this phenomenon remains an assumption, however. For this reason, independent evidence bearing on the hypothesis must be considered. Evidence on the cyclization of amphibian and fish DNA was recently put forward in support of an identical seriation as in Callan's hypothesis (Thomas et al., 1970). However, lacking the data on renaturation kinetics and thermal stability, it is equally possible that these observations reveal the extensive and highly molecularly redundant fraction known to occur in these DNAs. This possibility was later verified in *Drosophila* (Hennig et al., 1970) when it was shown that the cyclization involves preferentially the highly redundant fractions of the DNA. The important point, as explained below, is that these redundant fractions do not correspond with the slaves in Callan's hypothesis. In addition, directly against the master-slave hypothesis is the fact that by far most of the genetic information derived from the same material that Callan studied, the amphibian meoitic nucleus, is present in unique transcripts (Davidson and Hough, 1971); this is contrary to what is predicted from identical slave transcription.

To circumvent certain genetic restrictions, namely, that during recombination genes must be single copy and contiguous, Whitehouse (1967) extended Callan's hypothesis. He proposes that the slaves separate from the master during intrachromatid exchange and adopt the form of closed rings. Eventually, these episomes would reattach to the master. However, although commonly observed in prokaryotes, oocyte nucleoli, mitochondria, etc., circular DNA has been reported only once in eukaryotic chromosomes (Hotta and Bassel, 1965). Obviously, it requires substantiation.

Edström (1968) further modified Callan's hypothesis by submitting that the first-order slave in each generation was in fact the master of the previous generation. By this model, the series of slaves retain a continuous record of allelomorph sequences in the population and this would explain several features of genetic adaptation. The principle of enforced congruency with the master was abandoned.

The first three of these hypotheses propose an amplified transcription in terms of a generalized amplification of informational sequences, that is, exact copy of sequences. Callan's concept of a master gene set aside genetically might have proved significant in the event that eukaryotic mRNA started off from a polycistronic repeat, since, in principle, owing to the single master, this concept would allow for an informational polycistrony consistent with requirements for unique genetic specification. This concept therefore requires identical linear amplification along the DNA. Is there any evidence for this in informational sequences? A massive linear amplification is not in the least revealed by the structure of DNA sequences (Britten and Kohne, 1969). Linear amplification has been demonstrated for the histone cistrons (Kedes and Birnstiel, 1971), but is was not demonstrated that the sequences are all identical. Of other sequences that are transcribed, yet whose informational status is uncertain, the redundant sequences appearing in all genomes as families are related to one another but are not identical. Hence these sequences coding for DNA-like RNA do not qualify for slaves in Callan's hypothesis, even though part of the transcript of the lampbrush chromosome loop comes from these redundant sequences. Even assuming that each of these families of sequences could produce several related proteins, in the mouse, for example, the total redundant fraction of the genome could not code for more than a few hundred proteins (Walker and Hennig, 1970); it is necessary to invoke the unique fraction for the remainder, in fact most of the proteins. Similarly in *Drosophila* about 10% of DNA is redundant, but the bands, supposed to include most genes, contain 90% of DNA (Hennig *et al.*, 1970), so that again mostly unique DNA seems to be informational. In conclusion, there is no structural basis for a generalized amplification of informational sequences on the scale suggested in these three hypotheses, even though amplification has been demonstrated in one case. Lastly, Edström's fourth hypothesis does introduce the notion of nonidentical reiteration (as the result of cumulative mutations) but for an adaptive, not regulatory, purpose. In general, except for Beermann's, these hypotheses are principally devoted to the meiotic chromosome and attempt to explain the organization of cistrons but not their regulation. Obviously these hypotheses, particularly Callan's, also have implications for the somatic chromosome, but this does not concern us here.

The second category of genetic hypotheses now to be considered is aimed at the functional meaning of redundant DNA. In this category the hypothesis by Britten and Davidson (1969) emphasizes the complex circuitry of eukaryotic regulation and the need for a large fraction of regulatory DNA. In format this hypothesis transposes the parlance of the 1950's on gene circuitry into the contemporary one of molecular circuitry. In essence, redundant DNA is the link for regulation of unique informational cistrons as shown in Fig. 13. The format

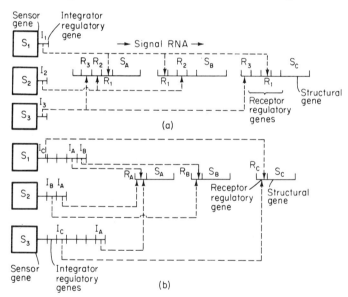

Fig. 13. Model of integrative regulation in the eukaryote according to Britten and Davidson (1969). The integrative models depend on molecular redundancy of DNA among the receptor regulatory genes (a) or among the integrator regulatory genes (b). After the sensor genes have been stimulated to initiate transcription of their integrator regulatory genes, the resulting RNA (broken line) acts as a signal to receptor regulatory genes. These genes then initiate transcription of the structural cistrons (S_A, S_B, S_C), each representing a unique DNA sequence; only one structural cistron corresponds to each set of receptor genes. At a more complex level of regulation, the integrative system is envisioned to combine both models and result in overlapping gene assemblies whose joint activation depends on the stimulation of several sensor genes.

is as follows: (1) sensor genes, reacting directly or indirectly but allosterically to effectors such as hormones (or their binding proteins), embryonic inducers, and other endogenous signals, are responsible for integrating the function of discontiguous informational cistrons; (2) redundant regulatory sequences are contiguous to both the sensor genes and informational cistrons, the postulated signal in between them being RNA. This signal, hypothesized to be represented in labile RNA, would elicit informational transcription, which prior to the arrival of this signal is intrinsically repressed. The result is (3) a pleiotropism of interaction between one sensor and several informational cistrons (Fig. 13a) or many sensors and one informational cistron (Fig. 13b), in line with the known patterns of response to hormone. At a more advanced stage of the model, sensors become themselves sensitive to products of regulatory or even informational cistrons, thus instituting a cascade regulation. This format, the authors feel, is the minimum required for the coordinate regulation of the genome that they have reviewed extensively. A period of intense coordinate regulatory activity would then precede the appearance of each mRNA. By being capable of "figuring out" the target of external signals (hormone), at some point in the model the sensor genes become essential elements of the genetic program. The sensor gene is in fact the primary and most important regulatory gene, even though the remainder of the regulatory genes retain a dynamic role (apart from their connecting role in the regulation) which, as will be seen later, is essentially that of modulating the informational response to the primary (sensor) signal. The sensor gene is therefore capable of distributing a single signal to many cistrons by using circuits which are part of a preestablished program. Yet the program cannot respond to the signal before it has matured to a certain stage in each particular cell. In brief, an obligatory part of the genetic program precedes the facultative informational response. The central point in this hypothesis remains however that molecular redundancy is linked with regulation and the pleiotropic potential for function inherent in redundancy is exploited fully. Conceptually, the point negates previous hypotheses in that it does away with strict cistronic amplification as in Callan's hypothesis. Instead, much of redundant DNA is envisioned as being used in circuits for regulatory purposes. The sharing of regulatory circuits provided for by the model is an economical necessity of the eukaryote.

For the sake of simplicity, Britten and Davidson chose first to consider regulatory RNA by itself, but did not exclude its possible translation into regulatory protein. Their first choice is adhered to in this discussion not the least because, if not unquestionable, it is more economical and might well be a correct one. The choice of regulatory protein is taken by Georgiev (1969b) in his own hypothesis explained in Fig. 14. This hypothesis, patterned on the operon model, offers little regulatory potential additional to that in the inflexible bacterial system. As in this system, the regulatory connection between different cistronic assemblies is entirely via regulatory protein, this presumably recognizing similar

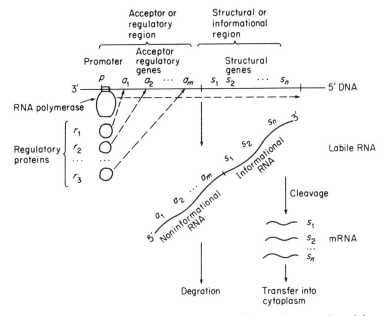

Fig. 14. Precursor model of eukaryotic messenger RNA specification adapted from Georgiev (1969b). The diagram represents a length of DNA which contains a regulatory portion and an informational portion corresponding to several structural cistrons. The regulatory portion involves redundant DNA sequences, the informational portion represents unique DNA sequences. The postulated function of the regulatory portion and the regulatory proteins acting on it were discussed in connection with genetic hypotheses on eukaryotic regulation. The length of DNA is transcribed into one labile RNA molecule. The regulatory portion of the RNA becomes effete and is degraded within the nucleus, whereas, after scission from one another, part or all of the informational sequences move to the cytoplasm.

receptors in the assemblies. However, if less elaborate, this hypothesis bears substantial resemblance to that of Britten and Davidson in predicating the regulatory use of redundancy. As will become clear later, the question of the nature of the regulatory molecules has a considerable theoretical importance.

In both hypotheses the mechanism of selective derepression discussed earlier would take place at the level of regulatory genes contiguous to the informational cistrons, even though the signals to which this mechanism will be responding in the two hypotheses are different. In addition, in Britten and Davidson's hypothesis, derepression would also occur at the level of the discontiguous sensor genes activated selectively by hormone or any other equivalent signal. (In the case of hormone, according to previous discussion, this probably means steroid hormone which can enter the nucleus. Protein hormones whose effects are mediated via cAMP could act indirectly on any of the genes in the model. Since it would be unrealistic to speculate on these latter effects at this time, protein hormones will not be considered further.) These general remarks may seem reasonable enough, but any specific comments are fraught with uncertainty. For example, histones being widely distributed on the DNA, would be associated with both regulatory and informational cistrons, yet it remains possible that the histones associated with each type of cistron are different or differently arranged in respect to other chromosomal components. Both models require a colligative response of the histones present along the several contiguous regulatory genes upon a single derepressive hit on any of these genes, but again, judging from the general distribution of histone, this response must then pervade the entire informational region. Therefore, to be realistic again, derepression should be regarded at present as a response of the total chromosomal apparatus in the entire region concerned, leaving for the future the question of the distribution of the specific components of the apparatus to regulatory and informational regions. It follows that it is premature to attempt to relate these regulatory models with the biochemical data for the chromosomal apparatus, which are poorly understood.

In conclusion, contrary to the previous hypotheses, the hypotheses by Britten and Davidson (1969) and Georgiev (1969b) incorporate present concepts of the molecular biology of DNA, and, equally important, attempt to explore with these concepts the fundamental questions of intragenic and intergenic regulation. Britten and Davidson's hypothesis is the first to reach back to the origin and regulation of the regulatory signal itself, which they propose is RNA. To prove (or improve upon) this theoretical model will take a serious experimental effort, which, in view of the difficulties of genetic analysis, means also a considerable amount of time.

Cytological Evidence Bearing on Eukaryotic Messenger RNA

Before moving into the evidence for direct visualization of gene activity in two types of chromosomes, lampbrush and polytene, we shall consider some evidence on aspects of gene repression rather than activity in interphase chromatin.

Interphase Chromatin

Cessation of mRNA (and other RNA) synthesis is shown by the physiological or, in the case of mammals, physical enucleation of the mature erythrocyte. What other changes can be seen in the chromosome that, if less dramatic, are perhaps more instructive in regard to specific patterns of repression? Involving remarkable interactions, in coccid insects some of the chromosomes become heterochromatic and inactive in what amounts to a facultative heterochromatinization on a massive scale. Sometimes, as in *Ascaris,* the inactive chromosomes are regularly eliminated from the complement in the somatic but not germ cell lineages, which is called chromatin diminution. In some insects the total genetic information derived from one parent is lost in this way (S. W. Brown, 1969). Some of these chromosomal patterns depend on interaction with the cytoplasm where the nuclei come to lie. These examples therefore show that while chromosome behavior is intrinsically programmed, in many cases the behavior is affected or even overruled by the cytoplasm. In higher mammals, according to Lyon (1968) there is a permanent somatic (not germinal) facultative inactivation of one at random of the two female X chromosomes during early development. Studies on aneuploids have shown that the number of heterochromatic X chromosomes depends on a balance with autosomes. Certain genetic elements in the mouse X chromosome that control this balance are now known (Cattanach *et al.,* 1970). By contrast, in primitive mammals, such as the kangaroo, the inactivation is not random but restricted to the X chromosome of paternal origin (Cooper, 1971). The single X chromosome inactivation represents a quantitative (not qualitative) equalization of sexes in respect to X chromosome-linked genes, that is, it is simply a sex dosage compensation. According to Grüneberg (1969), however, what formally results in the one X chromosomal level of function is a variable but complementary inactivation of both X chromosomes; under this model complete inactivation of one X chromosome occurs only in a small proportion of nuclei. Irrespective of model, yet more difficult to align with the Lyon's model, there are subtleties concerning the mammalian X chromosome that remain to be explained. In man, for example, a missing X chromosome (XO) causes the severe Turner syndrome. The mammalian Y chromosome is largely constitutively heterochromatic but in part it is also genetically active. Diametrically contrary to mammals, sex dosage compensation in insects is not by facultative inactivation of one female X chromosome but by superactivation of the single male X chromosome (Lakhotia and Mukherjee, 1969), therefore, the reverse of repression which we are discussing.

The changes in facultative heterochromatin just mentioned at the gross level of chromosomes or even sets of chromosomes are also known to occur in small finite chromosome regions. One simple way of looking at this vast range of facultative heterochromatinization of the chromosome is that it spans from the classical, cytologically visible form to the modern form defined operationally by molecular criteria. From the previous examples the visible heterochromatin appears to involve a significant segment of the functional genome. Part of this heterochromatin (as well as the constitutive type discussed below) is, however, believed to involve chromosome regions relatively poor in genes. Operationally defined heterochromatin (repressed chromatin) at times probably does not involve a large number of genes in any region, but, in all likelihood, many times these genes are intimately related functionally. One may thus speak of a gross and fine level of control represented in facultative heterochromatin but only in a very loose quantitative sense, since it is not established that this heterochromatin is a single functional state or responds to the same molecular mechanisms across its entire range. Repressed chromatin, as pointed out earlier, may represent a disposition of chromosomal proteins which is characteristic of the repressed state. In view of these dilemmas it is hardly surprising that in the past heterochromatin could not be defined by cytogenetic criteria alone, which is even truer of constitutive heterochromatin. We shall return to the fine level of molecular control when we discuss lampbrush and polytene chromosomes.

In the case of constitutive heterochromatin of the interphase nucleus, its structural and molecular characterization seems as distant as the functional one. Ultrastructurally, it is indistinct from other heterochromatin. Yet a few characteristics, even if not unequivocal, are beginning to emerge. First, the centromeres include extremely redundant (satellite) DNA as seen in the mouse. This extreme type of DNA is not present in all animals, but other less highly redundant DNA takes its place. Although not exclusive to it, redundant DNA in general appears therefore to be a characteristic of constitutive heterochromatin. Of course, the greater the redundancy the clearer transcriptive inactivity of this heterochromatin becomes. In fact, given the abundance of highly redundant DNA with its possible effects on the disposition of protein, it is rather surprising that no major ultrastructural difference has been found. Second, heterochromatin associated with the nucleolus, especially the chromatin internal to the nucleolus, contains rDNA. However, in germinal cells such as some spermatocytes and oocytes this nucleolus-associated chromatin contains a preponderance of redundant DNA other than rDNA; this other DNA might be related to synaptonemallike complexes present in this particular heterochromatin (Chapter 6). Contrary to rDNA, its transcriptive capabilities are not known. Third, and similar to facultative heterochromatin, constitutive heterochromatin shows a different cycle of condensation (late replicating) than euchromatin. By itself, this indicates an effect of chromosomal protein on DNA replication. It will be interesting in the future to see how all these characteristics fall in place with one another and with the functional meaning of constitutive heterochromatin, provided this is only one type, as was assumed here for simplicity. The functions considered earlier for satellites suggest the possible range of function.

One particular functional aspect of heterochromatin in the interphase nucleus that eludes interpretation is the quantitative variegated effects (known as position effects) on the phenotype that result from relocating euchromatin to a position near constitutive heterochromatin (review by Baker, 1968). A conceivable explanation, especially when this relocation is to an intercalary rather than a polarized position with respect to heterochromatin, could be that the observed phenotypic effects are due to a pattern of reversibility or instability in the repressed-derepressed state of euchromatin related to the proximity of heterochromatin (Sirlin, 1968). These effects show that constitutive heterochromatin, although transcriptively inactive, can influence genetic expression, although the nature of this influence is unknown. A paradox suggesting a distinction between euchromatin and heterochromatin on a minute scale, is the unexpectedly high transcriptive activity of the heterochromatin associated with the nucleolus mentioned above. That is to say, this condensed chromatin may include uncondensed regions (facultative heterochromatin?) that are, in fact, the active regions. For example, the large classical heterochromatic bodies near the nucleolus of certain insect oocytes contain 1% of their DNA as rDNA (Lima-de-Faria et al., 1969), which could be the only one transcribed in the bodies. The interesting question is whether these active regions could be under some sort of position effect (as defined above) from the surrounding heterochromatin in the bodies. These examples point rather clearly to a consideration, by this time perhaps intuitively evident, that the distinction between euchromatin and heterochromatin can be contingent on the level of observation—cytological versus molecular. They also show that any distinction between euchromatin and heterochromatin is essentially at the level of chromatin rather than any type of DNA it contains.

In addition, parts of euchromatin can lack typical genetic function yet influence this function, as pointed out for constitutive heterochromatin. Certain plants and animals possess supernumerary chromosomes, called B chromosomes, which are not essential to the normal phenotype or reproductive performance and are not homologous to any chromosome in the normal complement. The function of these B chromosomes is not known. They contain no known genetic loci yet affect the function of other loci; they can be deleterious to the individual carrying them yet are maintained in a balance within the population, presumably because of their beneficial effects on recombination. Such a deleterious effect of supernumerary DNA will later be pointed out to be general at any evolutionary level. It was observed that the presence of B chromosomes in plants increases the amount of histone and decreases the amount of RNA and acidic proteins per nucleus (Kirk and Jones, 1970), which may bear some relation to their adverse effects on the individual. In insects, the B chromosomes are heterochromatic and include a substantial fraction of all the redundant DNA (Gibson and Hewitt, 1970). General effects of B chromosome-type may, of course, occur in other parts of chromatin yet not be visualized as easily as in these distinct chromosomes. In this regard, it should be recalled that there is no general proof of the transcriptive ability of all euchromatin. Information on this point would not only depend on the level at which euchromatin and heterochromatin can be defined from each other as indicated above, but also on distinguishing at the ultrastructural level between transcriptive and replicative changes in the chromatin. This latter distinction might present a formidable problem in view of the multireplicon organization of the chromosome.

Ultrastructural studies have so far not contributed much to the understand-

ing of nuclear syntheses. Ultrastructurally, the interphase nucleus generally reveals the presence of various orders of chromatin fibrils (described earlier) and RNP granules (Ris, 1969), some of which possibly correspond with those observed in polytene chromosomes. With a few exceptions, the functional significance of these structures is not understood. This rather unimpressive contribution of ultrastructure to nuclear function will soon almost certainly be made good by the current use of molecular and structural methods in combination.
In contrast, the interphase nucleus has provided most of the information on all biochemical aspects of chromatin discussed in this book. It is also anticipated (Monneron and Bernhard, 1969) that the nuclear envelope will become an important area of research into the mechanisms that control macromolecular fluxes both into and from the nucleus.

Lampbrush Chromosomes

The lampbrush chromosome is especially modified for a high RNA synthetic activity. Since it is a germinal chromosome, any modification must be compatible with its subsequent reproductive career. These chromosomes are present during the diplotene stage of meiotic prophase in the oocyte of probably all animals, but have been most thoroughly studied in the amphibian oocyte (Callan, 1963). They are also present during the pachytene stage in the spermatocyte; in this cell of *Drosophila* the Y chromosome, in particular, behaves as a typical lampbrush chromosome (Hess and Meyer, 1968). In all, the phylogenetic ancientness speaks for a basic functional modification of this type of chromosome.

The lampbrush chromosome is formed of two homologs, each of which, in turn, consists of two intimately associated chromatids; because each chromatid remains dissociable from its sister, they all can proceed with their germinal career (yet only one will reach the zygote). A seemingly continuous DNA backbone in each chromatid—with the 1C haploid amount of helical DNA—is stretched out at intervals as loops. Thus each homolog exhibits numerous perfectly paired loops, each pair being attached to one of approximately 4000 chromomeres; this arrangement gives the chromosome its descriptive name and contributes to its usefulness to the biologist. A calculated 5% of the total DNA is included in the loops (Davidson, 1968). Many loops are individually recognizable by their RNP products in the form of granules and masses of material. At the cytological level, loops show a typical Mendelian segregation for single functional genetic units.

A structure apparently related to the presence of lampbrushlike chromosomes is the synaptonemal complex (Moses, 1968) which occurs during the zygo-diplotene stage of meiosis in both sexes of organisms that show chiasmata. The complex forms between two homolog chromosomes and is believed to pull them together for chromomere-to-chromomere pairing. Intimate molecular pairing then occurs as a separate process (Comings and Okada, 1970).

Many lampbrush chromosome loops are active in syntheses, although the exact proportion is not known. Most of the RNA present in the loops is a product of the loops (Gall and Callan, 1962) but the associated protein is probably a cytoplasmic product. About 3% of DNA in the oocyte lampbrush chromosome contributes RNA that functions as mRNA later during early embryogenesis (Davidson et al., 1966) and this RNA has received considerable biochemical attention (Chapter 6). Thus at least part of the RNA product of this chromosome is not for use by the oocyte itself. Hamster testis lampbrush chromosomes produce a heterogeneous DNA-like RNA (50% AU) that is both rapidly labeled and degraded (Muramatsu et al., 1968). In *Drosophila* spermatocytes, 5-10% of the Y chromosome DNA involved in loop formation anneals with a stable DNA-like RNA believed to decode fertility factors (Hennig, 1968). These data suggest great genetic activity in the chromosome loops.

While all the evidence supports the notion of a vigorous genetic function in the lampbrush chromosome loops, it also questions their informational status. It has been known that each chromomere contains "excess" DNA relative to the amount expected to code for an average size protein. It was partly from knowledge of this kind that the notion of slave genes originated. A single loop (average length 50 μ) could include 50 cistrons with 3000 nucleotides each; a somewhat higher figure is obtained in *Xenopus* after dividing the total length of DNA by the number of loops. A possibility to consider, therefore, is that the loops include active families of redundant DNA along with informational DNA. There is some evidence, although indirect, for this dual character of loop DNA. First, concerning redundant DNA, this generates most of the polydisperse RNA (by mass) in the oocyte nucleus, and it is known that most of this RNA is made in the loops. This is now confirmed by electron microscopy which permits visualization of RNA molecules several tens of microns long as they form in the loops (Miller et al., 1970b). Moreover, the amount of DNA (5%) physically estimated to be present in the loops does not disagree with the amount of redundant DNA (6%) transcribed in the oocyte measured biochemically. Second, concerning unique DNA, the informational content measured in the transcript (2×10^4 unique cistronic equivalents) indicates that, for a total of 4000 chromomeres, there are on the average five unique cistrons active per loop. (This is a minimum estimate for it refers to the mature oocyte transcript which need not represent all chromomeres.) If the assumption is made that the loops are stationary with respect to the chromosomal axis, which may be true but is not certain, one could then say that each loop is informationally complex. In support of this, the same order of complexity will be indicated later for the polytene chromosome chromomere. In conclusion, as it stands, the biochemical evidence is that the loop is a site of intense transcriptive activity in both redundant and unique DNA. This conclusion has to be viewed in relation to the biological evidence indicating that the loop is a single functional genetic unit, although not necessarily a single gene. The important implication here is that the eukaryotic genetic unit may consist of more than one gene, for example, regulatory and informational genes, all functionally related. This concept was touched upon in the discussion of modern hypotheses on genetic organization (Fig. 13) and will be further considered in relation to the derivation of the mRNA molecule.

In summary, lampbrush (and polytene) chromosomes show the reverse of heterochromatinization exemplified earlier with the interphase nucleus in that they provide visual examples of activated genes. Several of the hypotheses addressed to the cistronic organization of the eukaryote which were presented originated in these chromosomes. Indeed, the lampbrush chromosome provided the first model of basic organization of the chromosome, i.e., each chromatid containing a continuous DNA helix (with the 1C amount of DNA) stretched out as loops at intervals. This model is now favored for chromosomes in general. For example, a transient loop arrangement of the same nature as in the lampbrush chromosome was proposed in interphase (Swift, 1965) and metaphase chromosomes (Stubblefield and Wray, 1971). In this respect, it is as well to remember that the typical lampbrush chromosome (and the puffs somewhat resembling them in the polytene chromosome) is also a transient structure.

Polytene Chromosomes

The polytene chromosome is another type of chromosome modified for a high RNA synthetic activity. Because it is a somatic and nondividing chromosome (contrary to the lampbrush chromosome) polyteny does not have to be compatible with reproduction and is in fact a terminal modification. Although these chromosomes occur in plants (Nagl, 1969), they are typical of insects, the best documented of which are in larval salivary glands (review by Beermann, 1967). The salivary gland cells are huge and nondividing. The polytene chromosomes therefore represent an adaptation to a large synthetic demand, but, unlike the lampbrush chromosome, the synthesis is for the use of the gland cell itself. This demand is best illustrated in insect pupal footpad cells. These cells grow to a tremendous size in a matter of days, continuously secreting cuticle as the chromosomes increase their polyteny (Whitten, 1965). In salivary glands the DNA replicates in up to 13 steps from 2 to 13,000C (Daneholt and Edström, 1967) and all strands remain together, which is why this kind of chromosome cannot have a reproductive future. The degree of polyteny increases above normal in larvae with a prolonged lifespan (Rodman, 1967), suggesting an hormonal control of polytenization. Curiously, protozoan or viral infection of the salivary gland also increases the polyteny severalfold (Pavan and da Cunha, 1969), but the mechanism in this case is unknown. By contrast, in insect ovarian nurse cells as the chromosomes polytenize in concert they become disperse (Jacob and Sirlin, 1959), in what is known as endoreduplication (in these chromosomes part of the RNA produced may be for the oocyte). Because of its high strand multiplicity and permanent alignment, the typical polytene chromosome is a functional interphase chromosome but with all the appearance of a gigantic prophase chromosome. This is what makes it a convenient object for study.

The polytene chromosomes contain in the order of 3000 transverse bands that include most of the total DNA in a very supercoiled condition. Interband DNA is probably continuous with band DNA as part of the same strand but is much less supercoiled, therefore less prominent. Cytogenetically, each band corresponds with a single genetic locus but in cases possibly with more than one (Beermann, 1967). Each band shows Mendelian segregation, polymorphism in populations, and is suppressed by point mutations. Thus a band probably represents a single functional genetic unit, the potential genetic complexity of which was discussed for the lampbrush chromosome. The average chromomere (1C DNA) in a band contains on the order of 10^5 base pairs, which could give a transcript of 3.3×10^7 daltons sufficient to code for 50 proteins of average size. Considering only the unique genome (85% in *Drosophila*), a recent estimate is ten cistrons of 3000 bases each per chromomere. About 10% of the bands make RNA. These bands appear somewhat engrossed in the chromosome as what are called (in order of increasing size) bulbs, puffs, and Balbiani rings. In these actively transcribing regions the individual strands may adopt a configuration similar to the lampbrush chromosome loop.

Balbiani rings, the largest puffs, are few yet they make as much as one-fourth of the total RNA made by the chromosomes, and almost as much as all nucleoli combined. Many puffs common to all tissues presumably discharge a general function. About 5% of the puffs, the interesting ones from the viewpoint of selective derepression, occur, however, in patterns characteristic of each tissue and its developmental stage and probably have specific functions (review by Kroeger, 1968). For example, certain puffs in the salivary gland are concerned with puparium secretion (Doyle and Laufer, 1969) and in the prothoracic gland with ecdyson production (Abd-el-Wahab and Sirlin, 1959); in nurse cells the polytene X chromosome may discharge sex-specific functions (Schultz, 1965), as the Y chromosome probably does in the spermatocyte under the appearance of a lampbrush chromosome. Different puffs respond characteristically, either positively or negatively, to different ecdyson levels (Clever, 1968). However, differences in the timing of the puffing response at different loci within a target tissue suggest an additional intrinsic, nonhormonal regulation (Bultmann and Clever, 1969). This difference in timing may indicate a regulatory program in control of the puffing pattern. This kind of experiment and many other classical experiments on these chromosomes provide a convincing demonstration of hormonal action at the chromosomal level, even though it is not possible to conclude definitely from these experiments alone that hormone is acting directly at that level. In a few instances, a correlation between the appearance of a certain puff and a particular granule, vacuole, secretion (e.g., Baudisch and Panitz, 1968), or individual protein units (Grossbach, 1969) has been established. Needless to say, one assumes that these are mRNA-directed products (the informational status of a puff will be considered below). Yet the first phenotypic effects of hormonally induced transcription are sometimes not very immediate (Clever, 1969), again as expected if regulatory transcription were to precede informational transcription. Indeed, as a whole, this evidence strongly suggests that puff activity represents regulatory and informational transcription, even though the correspondence between the various stages in the puffing pattern and these two types of transcription remains to be established. Clearly, however, if it were to become feasible to locate specific mRNA by cytological hybridization, these and the lampbrush chromosomes could serve this purpose.

Cytochemically, polytene chromosomes have been intensively studied. Of interest is the association in the puff of preformed protein with nascent RNA and their joint passage to cytoplasm (Stevens and Swift, 1966). If protein synthesis is prevented, RNA is not released from the puff (Clever, 1968). In *Drosophila* puffing is necessary in order for transcription to occur, and this in turn to maintain the puffed condition (Berendes, 1968). A different view on puffing is that it may be more the consequence than the cause of transcription, that is, puffing would be more related to proteination of the transcript than to transcription. These observations would therefore pertain to genesis of the informosome, yet it would be unlikely that they did not also reflect some aspects of transcription. On a per chromosome basis, a remarkably uniform turnover pattern is found in acidic proteins (Sirlin and Knight, 1958), which, in retrospect, is what was to be expected from the continuous flux of this complex of proteins. The amount of acidic protein, but not of histone, increases in any locus during puffing.

What are the biochemical data on puffing? First, the 10% of DNA present in active puffs invites comparison with the 10% of redundant DNA that is transcribed in the average cell mentioned earlier. Second, the composition of puff RNA as revealed by microphoresis differs from total DNA in a manner which suggests the expected single strand transcription (Edström and Beermann, 1962). The RNA that is formed is polydisperse (20-90 S, Edström and Daneholt, 1967), and is therefore probably partly derived from redundant DNA. In single Balbiani rings the polydispersity is less but still high (40 S peak value), yet the average transcript is shorter than would be predicted for the 1C amount of DNA present. Whereas the C/U ratio in the RNAs made in the Balbiani ring and the chromosome are different, they remain the same for all sizes of molecules; the average molecular size decreases, however, with time (Daneholt, 1970). These observations may be related to rapid breakdown of the RNA or redundancy of template, or both. In general, the observations indicate the presence of labile RNA in the puffs. This is further supported by selective elimination from the chromosome, by means of chemical inhibition (Sirlin and Loening, 1968), of RNA that binds tenaciously to MAK column and which, in part, represents labile RNA (Roberts and Quinlivan, 1969). Thus the overall evidence is that both labile RNA and mRNA (described earlier) are made in puffs, but unfortunately all the evidence does not come from the same puff.

What is the informational status of puff transcription? First, considering the unique genome, it was mentioned that the number of cistrons per average chromomere is estimated at ten. Second, in contrast with the lampbrush chromosome, there is some information at the level of the protein product. Grossbach (1969) found that, taken together, three Balbiani rings of the salivary gland code for five polypeptide units each about 10^6 daltons. These units may represent the main polypeptides secreted by the gland. This lack of a simple relationship such as one Balbiani ring-one polypeptide is common (Wobus *et al.*, 1970). Unfortunately, not knowing the possible covalent substructure of the polypeptide units, these data cannot be transformed into informational complexity in the DNA. However, as concluded earlier on biological evidence, it is likely that both informational and regulatory transcriptions coexist in a puff. The parallel biochemical evidence suggests the presence of both informational and redundant (regulatory?) DNA. Thus, even if a Balbiani ring represents a single functional genetic unit as is not uncommonly assumed, the status with respect to the unitary "one cistron-one polypeptide" concept remains undecided. What the informational status may be in the case of smaller puffs which constitute the vast majority has not been examined. The present conclusions on the informational status of the large puffs should be viewed in relation to the earlier conclusion on the status of the ubiquitous lampbrush chromosome loop. The possible complexity of the functional genetic unit was mentioned at that time. The greatest value of the polytene chromosome in the study of this unit might be to permit the allocation of regulatory and informational genes to their well-known band-interband pattern—a statement which suggests the functional importance of the interband, so far almost a genetic nonentity.

Certain important cytological observations on polytene chromosomes cannot as yet be fully interpreted molecularly. During its formation a puff frequently engulfs several adjacent inert bands, perhaps as a passive mechanical phenomenon. Other puffs, however, grow unidirectionally within their own locus and their transcription is exclusively from the starting point of growth (Berendes, 1968), suggesting a serial activation within the locus. Exceptionally, as in the case of an individual heterozygous for one particular puff, the puff in the wild-type homolog spreads to the other homolog only if the two are intimately paired (Kroeger, 1968), indicating derepression in a trans fashion yet dependent on intimate pairing. Ecdyson at high titer induces sequentially two sets of puffs, only after which can moulting occur. Puffing of the second set, but not the first, depends on previous protein synthesis (Clever, 1964). The relevant protein is believed to be specified in the first set of puffs, but whether or not it then acts directly on the second set is not clear. At variance with these findings, it was shown that the temporal pattern of puffing is independent of the physical removal of other portions of the same or different chromosomes (Kroeger, 1968). In this case the apparent self-programming of puffs is believed by Kroeger to be in response to the nuclear electrolyte balance, but the sufficiency of this explanation was criticized earlier in this chapter. The basic mechanism in the control of gene dosage compensation is not known. Judging from the amount of transcription in aneuploid bands, the degree of dosage compensation may vary between bands (Korge, 1970). Hopefully, these observations suggest that it might be feasible in polytene chromosomes to eventually gather information of the kind obtained in microorganisms by means of molecular genetic dissection.

Certain observations on polytene chromosomes suggest a role of cytoplasm vis-a-vis the nuclear environment in controlling chromosomal function, although the interpretation of these observations is far from unequivocal. Chironomid salivary gland chromosomes explanted into fresh cytoplasm show a specific pattern of puff alteration which usually is of a regressive nature (Kroeger, 1968). Similar *Drosophila* chromosomes present in isolated nuclei cannot be hormonally induced to form puffs which are normally inducible in the cell (Berendes and Boyd, 1969). This observation may suggest the absence of a cytoplasmic receptor protein, and thus provide circumstantial evidence against a direct hormonal action on the chromosome. In addition, bacterial RNA polymerase given to isolated nuclei increases only the activity of the loci already active, possibly suggesting that lack of enzyme in the cell is not limiting the inactive loci. This material could therefore also possibly serve to study the function of cytoplasmic factors in maintaining the differentiated state of the cell.

Cytologically demonstrable "excess" DNA in polytene chromosomes may be attributable to amplification, that is, faithful copying of sequences (themselves unique or redundant). The presence of a few puffs showing such a disproportionate DNA synthesis, the so-called DNA puffs as opposed to the regular RNA puffs, has been known in sciarids since 1952 (Pavan and Breuer, 1952). Studying interspecific chironomid hybrids, Keyl (1965) presented evidence for a geometric increment of DNA (2:4:8:16) between about 30 pairs of homologous heterozygote loci. The increments occur in the germ cell line, possibly as the result of tandem duplication following misreplication, and remain stable for the life of the individual. Similarly, the mechanism of sudden DNA increase in DNA puffs of somatic (salivary gland) cells (Crouse and Keyl, 1968) is also by tandem duplication; in particular, meiotic unequal crossover is excluded. This excess replication of puff DNA is intimately related to the general polytenization of the chromosome. The DNA thus increased remains integrated with the chromosome, except in some puffs at a terminal stage of disintegration; in other words, the degree of polyteny is not increased locally. These exceptional DNA puffs produce polydisperse RNA much as the usual RNA puffs do (Lara and Hollander, 1967). As speculation, the function of the "excess" DNA could be with the amplification of template for DNA-like RNA synthesis. [By extension of

EUKARYOTIC LABILE RNA

Large amounts of this highly heterogeneous RNA are made in the eukaryotic chromosome, in fact a much larger amount than that of rRNA which is made in the nucleolus. Contrary to rRNA, however, labile RNA turns over very rapidly inside the nucleus, but some of the RNA appears to be more stable (or be stabilized) and moves to the cytoplasm. The base composition of the nucleus-restricted RNA differs from that of the RNA that moves to the cytoplasm. The highest molecular weight range of this RNA in the nucleus, up to 100 S, surpasses that of any cytoplasmic RNA species, with the exception perhaps of cytoplasmic polydisperse RNA of unknown origin and function (Chapter 2). Because the function of labile RNA is largely undecided, we will first consider observations that illustrate the behavior of the RNA relative to other RNA species, and then discuss the functional implication of this behavior in relation to informational and regulatory transcription. This discussion expands that in preceding sections concerning the function of the various components of the genome.

Biological and Molecular Aspects of Labile RNA Synthesis

Synthesis of labile RNA is at first very pronounced in the early amphibian embryo (Gurdon, 1968), the mature avian erythrocyte nucleus as it begins to be reactivated within a heterokaryon (Harris *et al.*, 1969), and the resting lymphocyte before phytohemoagglutinin stimulation (Cooper, 1969). Rather generally, as synthesis of rRNA in these cells increases, the synthesis of labile RNA decreases relative to the rRNA synthesis. This sequence is the same during hormonal stimulation. Similarly, the immature avian erythrocyte (Attardi *et al.*, 1966), the mammalian primary spermatocyte (Muramatsu *et al.*, 1968), and the early growing pollen tube (Mascarenhas and Bell, 1970) all seem to make mostly or exclusively labile RNA. Superficially at least, the two RNA species behave, therefore, as if they were inversely regulated. Also, resting cells or starved cells in culture (Holoubek and Crocker, 1968) have a preferential synthesis of DNA-like RNA (mostly labile RNA); the same is true of both plant cells in culture (Bellamy, 1966; Rogers *et al.*, 1970) and heat-shocked HeLa cells (Warocquier and Scherrer, 1969). From what was said before, these cells are expected to make little rRNA. They show that labile RNA tends to predominate by itself in resting cells, much as in the metabolically stepped-down bacterial cell producing mostly mRNA. However, whether there is any causal relationship between the regulation of labile and ribosomal RNAs, as might perhaps be suggested by these data, is impossible to tell. The functional significance of the observations is, therefore, not clear.

Tumor liver cells have a genome less repressed than normal liver cells and a greater variety of polydisperse (labile) RNA, as shown by competitive hybridization. Accordingly, the CCD pattern of labile RNA is altered in tumor cells (Kidson and Kirby, 1965). In regard to this derepression, tumor cells resemble regenerating liver cells (Church et al., 1969). Also a greater variety of sequences present in nuclear RNA are recovered from tumor than from normal cytoplasm, which Church et al. (1969) interpret as suggesting a more efficient transfer of tumor labile RNA into functional mRNA. The least efficient transfer to cytoplasmic mRNA is found in highly differentiated cells such as the maturing avian erythrocyte. It follows that labile RNA constitutes a greater proportion of the nuclear RNA in differentiated (nondividing) cells than in cells undergoing differentiation (dividing). In addition, in the uninephrectomized mouse kidney, polydisperse RNA is converted metabolically into smaller RNA molecules faster than in control cells, thereby heralding a compensatory hypertrophy (Willems et al., 1969a). The faster conversion of RNA in the reactivated kidney possibly preludes a greater transfer to cytoplasm (as in reactivated liver), but this aspect was not pursued. In general, the transport of labile RNA to cytoplasm is considered to be a highly selective process. Were any of the nucleus-restricted labile RNA to escape from the nucleus in mitosis, then the evidence is that this RNA is as unstable as it was in the nucleus.

Irrespective of the functional interpretation, this selective transport of RNA to the cytoplasm implies, indeed depends on, the existence of a posttranscriptional level of control over the immediate products resulting from the primary transcriptional regulation. Selection means that some informational transcripts which are made are not used. For some reason, which presumably is flexibility in the regulation, the eukaryotic cell thus resorts to an ostensible wasteful two-step regulation in lieu of a one-step rigorous transcriptional regulation. It follows that the posttranscriptional regulation shares equal responsibility with the preceding regulation in respect to making genetic information available to the cell, and the specificity of the mechanisms involved must be equal to this responsibility. Unfortunately, understanding of these elaborate posttranscriptional mechanisms, including a concurrent maturation and selection of mRNA, has not yet begun. Of these two processes, the maturation of mRNA has an operational model in the maturation of precursor rRNA, although this model becomes very sketchy when transposed from the monotonous rRNA to the highly heterogeneous population of mRNAs. Selection of mRNA, however, has no known mechanism or even plausible model: it remains a "black box" with thus far a single postulated functional element, the informosome. Recent evidence, to be considered later, suggests that certain special sequences in mRNA may collaborate in the selection.

Despite uncertainties, in quiescent cells the relative preponderance of labile RNA synthesis over rRNA synthesis, in part, may reflect the fact that these cells have a full rRNA complement and therefore require (almost) no additional synthesis. However, although the continuing labile RNA synthesis remains to be explained, in the stimulated cells exemplified above certain aspects of the RNA reveal that its behavior is far from passive. Under these stimulatory conditions, there is (a) less intranuclear turnover of labile RNA concurrently with (b) a more noticeable exit to cytoplasm, and (c) new sequences of labile RNA appear indicating that new genes are transcribed (Church and McCarthy, 1970). It is commonly known to most workers in this field that labile RNA molecules can (at least some of them) be divided into smaller molecules with different degrees of stability. Fractions across the entire molecular weight range of labile RNA, including the largest but possibly excluding the smallest weights, appear to consist of both unstable molecules confined to the nucleus and stable molecules that pass to the cytoplasm (Shearer and McCarthy, 1970a). On such basis, the largest of the stable nuclear molecules reach the cytoplasm only in a much smaller size—since this is what is found in the cytoplasm. Of interest in this respect is the claim that an endonuclease present in nuclear 30 S RNP particles (perhaps informosomal in nature) specifically cleaves 50-80 S labile RNA into more stable molecules, first about 30 S and later 4-16 S (Niessing and Sekeris, 1970).

Labile RNA is coded mostly in redundant DNA sequences, but some of it, including that which breaks down in the nucleus, is coded in unique sequences. In common with other DNA sequences, the redundant sequences are repressed by histone and derepressed when histone is sequestered (Georgiev, 1969a), which indicates that the regulation of these sequences falls within the molecular mechanisms applicable to derepression in general. Four times more genome codes for nuclear labile RNA than for cytoplasmic mRNA produced in cultured mouse L cells, i.e., 4 and 1%, respectively (Shearer and McCarthy, 1967), as measured under relatively nonstringent conditions of hybridization. Whether or not this ratio remains the same in other cells will probably depend on their function. For example, the immature avian erythrocyte produces labile RNA from 0.6% of its genome but produces no mRNA at all (Attardi et al., 1970).

Different cells or tissues vary in the spectrum of labile RNA sequences they possess, which is best seen in early developing embryos. The sequences differ also in different cells in response to the same hormone (Church and McCarthy, 1970). The sequences of labile RNA tend to be more similar among different tissues of an organism than their unstable cytoplasmic RNA, presumably including mRNA (Sullivan, 1968). In agreement with this observation, the sequences within the redundant families of DNA transcribing for the nucleus-restricted labile RNA are informationally less diverse than the sequences whose transcripts appear in the cytoplasm (Shearer and McCarthy, 1970b). Assuming that the latter are unique transcripts, this situation resembles that in the amphibian oocyte (Davidson and Hough, 1971) where the informational content is found to be greater in the transcript of unique than of redundant sequences. Labile RNA sequences differ also between species, e.g., mouse and man (Soeiro and Darnell, 1969). Indeed, the sequences transcribing for the nucleus-restricted RNA have diverged between related species more than the sequences transcribing for cytoplasmic RNA (Shearer and McCarthy 1970b). An evolutionary divergence of the nucleus-restricted transcript parallels therefore the divergence and replenishment of redundant DNA itself, much of which obviously cannot be informational. In conclusion, from these data the sequences represented in nuclear RNA are less diverse in a species but more diverse between species than the sequences represented in cytoplasmic RNA. This suggests that uniquely nuclear RNA has a more recent evolutionary origin than cytoplasmic RNA. The informational content represented in the nucleus-restricted RNA is less than in RNA that moves to the cytoplasm.

Depending on the extraction procedure, labile RNA is recovered either bound to chromosomes (Arion *et al.*, 1967) or large nuclear supernate particles (Penman *et al.*, 1968). As shown by autoradiography, rapidly labeled RNA is formed all over interphase chromatin (Harris *et al.*, 1969). Formation of labile RNA in puffs, in general, and single Balbiani rings, in particular, was documented in the previous section on polytene chromosomes; an average 40 S RNA is produced in the Balbiani ring. In conclusion, within the serious limitations of the evidence, labile RNA synthesis is topographically indistinguishable in the nucleus from mRNA synthesis.

Function of Labile RNA: Precursor to Messenger RNA or Regulatory RNA?

By now it should be amply clear that there is a need to discuss the relationship between labile RNA and mRNA. The first and trivial possibility is that these two RNAs merely coexist in the same nuclear RNA fractions. The second and obviously interesting possibility that concerns us here is that the RNAs are functionally related. Two hypotheses that consider such a functional relationship are currently available.

The first hypothesis is that *labile RNA is a precursor of mRNA* which is derived after a series of controlled enzymic scissions (Georgiev, 1969b; Scherrer and Marcaud, 1968; Soeiro and Darnell, 1970). This derivation from a larger precursor molecule is found in all major stable RNAs of the eukaryote. The biochemical evidence (such as it stands) in favor of labile RNA being a precursor of mRNA was presented in Chapter 2 and the previous section. Given the occasional imprecision of hybridization data, it can be seen that (a) the extent of genome involved in each RNA and (b) the competition between the two RNAs for sites in DNA are quantitatively not incompatible with a partial precursorship, nor are other aspects of labile RNA such as the much larger size, faster kinetics of synthesis, and higher proportion of poly A runs (Chapter 2). It needs to be stressed, however, that this interpretation of the data, plausible as it may seem, is speculative and not unequivocal; the precursor-product relationship between the RNAs remains to be proved. Nevertheless, this interpretation offers a conceptual basis for looking at the events taking place at the level of the nascent mRNA molecule as opposed to those at the level of the surrounding mechanism of derepression. It may be recalled that the evidence on derepression did not deal with nascent mRNA as such. For this reason alone, the precursor hypothesis deserves attention. An alternative within the framework of this hypothesis is that labile RNA is a template for mRNA synthesis (Baltimore, 1971); this possibility, implying an RNA-primed RNA synthesis, is conceivable, but cannot be considered further for lack of sufficient evidence.

Some of the difficulty inherent in establishing a precursor-product relationship between the RNAs can be illustrated as follows. In the previous section, according to Shearer and McCarthy (1970a), the competition by cytoplasmic mRNA against large nuclear labile RNA molecules for sites in DNA was taken to imply a precursor-product relationship. According to Soeiro and Darnell (1970), also, the similar hybridization kinetics of DNA to nuclear labile RNA and polysomal mRNA in HeLa cells does not exclude such a relationship in respect to the large labile RNA molecules. Yet this interpretation of the observations is not necessarily the only one. Codon restriction per se can create a certain redundancy of sequences (Walker, 1969). Thus if the use of codons were to be tightly restricted and affected all DNA-like RNA, the RNAs could show a similar hybridization behavior, particularly under nonstringent hybridization conditions (low Cot values), even though they may not be filially related. Since the degree of codon restriction in the eukaryote is not known and the stringency attained in the hybridization is likely to be suboptimal, the previous argument should be given serious consideration. A different source of error in the interpretation of hybridization data is the possible contamination of mRNA with cytoplasmic polydisperse RNA (Chapter 2) whose sequence structure and relationship to nuclear labile RNA is uncertain. What this critique is saying more generally is that the hybrids studied may well be irrelevant to the precursorship question in hand until such time when hybridization conditions for authenticated (purified) informational sequences are unequivocally met. In conclusion, the present biochemical evidence in favor of a precursorship of mRNA in labile RNA merits favorable consideration but is not as yet indisputable. An argument in particular which ostensibly supports this precursorship needs a word of caution. This refers to the fact that, whereas in the eukaryotic nucleus viral mRNA is derived from a large precursor covalently bonded to nuclear RNA (Wagner and Roizman, 1969), this may not be used as a fair argument in support of the same mechanism for cell mRNA without in principle begging the question. Only after the mechanism was established for cell mRNA, would the viral evidence be relevant supporting evidence. An instance of a faulty kind of extrapolation would be the polycistronic translation typical of certain mammalian viruses, which surely does not apply to cell mRNA.

The precursor hypothesis envisions that large molecules of labile RNA are divided into fragments, of which some are retained as different functional mRNAs and others perhaps as regulatory RNA. The rest of the RNA, in fact the majority of it according to the evidence, is discarded within the nucleus. Of logical necessity, the relevant evidence offered in support of the hypothesis must therefore deal with labile RNA molecules much larger than the average cytoplasmic mRNA. A subjective rationale behind the hypothesis is that it is more economical to transcribe large blocks of DNA and then select fragments of the transcript than to regulate independently a large number of informational cistrons. Thus this model offers a greater flexibility of regulation than the polycistronic bacterial operon without sacrificing physically coordinate regulation. It should be noted that the model requires a virtual polycistrony of mRNAs but only at a prefunctional level. According to most proponents of the model, mRNA could derive from either redundant or unique sequences which are known to be interspersed with one another in the genome (to the extent that most unique sequences are interrupted at least every 4000 bases by redundant sequences). That is, although labile RNA comprises many redundant sequences (as shown by hybridization), it still contains a few unique sequences, and a sequence of either type could become functional mRNA. According to the model by Georgiev (1969b), however, redundant sequences function exclusively in regulation and unique sequences are the sole source of structural mRNAs (Fig. 14). In this model, labile RNA molecules represent a covalent transcript of both regulatory (redundant) sequences and informational (unique) sequences in about equal proportion. Of these, only some unique sequences would reach the cytoplasm, redundant sequences with no further use remaining in the nucleus. On genetic and structural grounds, most mRNA sequences are expected to occur once in the genome. As must be clear, redundant sequences may not fulfill this requirement as well as unique sequences. Thus in this respect Georgiev's model is attractive.

It is considered in the precursor hypothesis that proteination of the transcript during or just after transcription has a dual discriminating role in specifying the enzymic cleavage of the precursor and stabilizing that part which is to be preserved as a functional mRNA. This processing of the molecule has a clear model in eukaryotic rRNA. It is tempting to equate the incipient proteination of the transcript with formation of the informosome, much the same way as nascent rRNA is immediately sequestered inside a particle. Cleavage by a free endonuclease could either be specific or random depending on the enzyme specificity vis-a-vis the distribution of substrate sequences in the RNA. Speaking subjectively, cleavage cospecified by proteins (present in the informosome) could perhaps secure a greater specificity of cleavage, for which there may be a precedent again in the rRNA pathway. A second level of selection of prospective messages is proposed to be at the nuclear envelope. Since it is uneconomical at this time to postulate a multiple mechanism, this last proposal is best regarded as a topographical refinement of the informosomal mechanism. Having mentioned economy, one may wonder how much of it that was made by coordinating transcription can be left after all this posttranscriptional specification. Nevertheless, hormones could control specification. An interesting concept is therefore that, by analogy with a proposal for rRNA (Tata, 1968), each hormone draws a quota of mRNA from the precursor (Kenney et al., 1968). However, touching as it does on the basic mechanism of hormonal action which is poorly understood, this concept cannot be meaningfully discussed here.

Recent work (Sidebottom and Harris, 1969) has shown that, when the avian erythrocyte nucleus is present in a heterokaryon with a HeLa cell, it first makes labile RNA but appears not to begin to produce detectable antigen, therefore functional mRNA, until its nucleolus begins to produce rRNA. Similarly, irradiation of the HeLa cell nucleolus prevents nuclear RNA from migrating to the cytoplasm. One interpretation of these findings, if not the only possible one, is that newly made rRNA somehow facilitates the appearance and perhaps transit of functional mRNA. The switch from predominant labile RNA synthesis to rRNA synthesis previously described for other cells could then correspond with this facilitation. A corollary of this interpretation is that in cells such as the immature erythrocyte and spermatocyte, where little rRNA is being made, also less functional mRNA is made; this is supported by the lack of mRNA synthesis in the immature erythrocyte. These cells therefore possess a continuous (if diminished) labile RNA synthesis whose function remains to be explained. According to the precursor model, this continuous synthesis is a gratuitous one in the sense of representing precursor that fails to become functional mRNA.

Different from the bacterial operon, the envisaged specification of mRNA in the precursor model is entirely posttranscriptional. This, however, has to be viewed in connection with the pretranscriptional specification (selective derepression) on which the posttranscriptional specification is mechanistically dependent. Theoretically, this opens two possible avenues of action for any agent with effects on mRNA, the transcriptional and subsequent posttranscriptional specification.

The second hypothesis regarding the function of labile RNA contends that *labile RNA is regulatory RNA* (Britten and Davidson, 1969). This hypothesis was already presented in the section on genetic hypotheses in the eukaryote and only those aspects touching directly on the derivation of mRNA need to be considered here. In particular, according to this hypothesis, labile RNA is the product of regulatory genes (integrator genes) which acts as the transcriptional signal between them and other regulatory genes (receptor genes) contiguous to the informational cistrons (Fig. 13). The model envisages informational sequences discontiguous to one another but contiguous to several regulatory sequences, for which there is some indirect support from the numerical relationship between unique and redundant DNA sequences described previously. It tacitly denies the derivation of mRNA from large precursor molecules. The model depends absolutely on redundant DNA being part of both types of regulatory genes and, of course, on RNA being the signal between them. Since labile RNA is transcribed mostly from redundant sequences and has mainly a regulatory function in the model, by implication most mRNA is assigned to unique sequences, which is reassuring on genetic grounds as mentioned earlier. The model does not however exclude that some of the regulatory transcript is unique, particularly in view of the now known high level of informational complexity of this transcript. The question really becomes whether or not this unique regulatory RNA requires translation, for if it does it remains informational.

This model proposes very clearly a separate regulation and transcription of individual informational cistrons and, in addition, coordination between them which does not depend on physical contact. It should be stressed that the monocistronic transcription which is implied is hypothetical, for what is established is that eukaryotic mRNA is monocistronic during translation which does not necessarily exclude polycistrony during transcription. However, it is important to note that the proposed individual specification of mRNA in the model inevitably renders a polycistronic specification, as in the previous model, unlikely; in this model only a scission between the proximal regulatory genes and each mRNA is required. An argument in favor of monocistronic specification is therefore that it obviates the truly staggering number of specific nucleases which would be required under a polycistronic specification to separate the mRNAs eventually, unless of course there is some other mechanism (so far unknown) to achieve this purpose. On balance, the concept of monocistronic specification seems to be the most acceptable. A similar argument can be made why control of the initiation rather than termination of transcription must provide the primary basis for specification, to say nothing of its far superior possibilities for coordination. Some terminal control must exist, but the observed selective derepression of the nucleus (the rRNA cistrons particularly) establishes a primary control of initiation.

Without prejudice to the validity of the regulatory model, some of its implications are as follows. First, on this model practically one-half of the mammalian genome can be regulatory (redundant DNA), the other half informational (unique DNA). Second, some labile RNA molecules are 50,000 bases long, from which one interprets that the regulatory transcript can be 15 times larger than the informational one. These oversize molecules, although not as critical to this model as they are to the precursor model, require explanation. The authors

(Britten and Davidson, 1969) favor the possibility that the regulatory (integrator) sequences discontiguous to the informational cistrons are longer than the contiguous regulatory sequences: 100 integrator sequences of 500 base pairs each could produce an RNA signal of 50,000 bases. The impression is that the actual signal could be considerably shorter (a minority of molecules), very dynamic (fast turnover), and not much proteinated. Third, and related to the previous point, a functional kinship needs to be considered between labile RNA and chromosomal RNA presumed to act in a regulatory manner like labile RNA itself but more proximal to the informational cistrons. Chromosomal RNA binds to redundant DNA sequences which a priori may mean that it controls unique sequences (Sivolap and Bonner, 1971). Little is known of the kinetics of chromosomal RNA synthesis, but the RNA is known to increase rapidly in chromatin within 15 minutes after hormone administration (Teng and Hamilton, 1969), and appears therefore to fall within the kinetic range of labile RNA. Yet little or nothing constructive can be said concerning the relationship between these RNAs because of their enormous difference in mean size (chromosomal RNA has approximately 50 bases). If a precursor-product relationship exists, then the 1000-fold reduction in size is to be compared with a reduction at most by half in all other known RNAs derived from their precursors. Three possibilities for labile RNA would be that (a) part of the sequences are supernumerary; (b) by extension, the sequences are thrown into a selective pool (as Britten and Davidson suggest); and (c) different fractional lengths of DNA are transcribed but only part of the transcript serves as precursor. However, since any precursor-product relationship between labile and chromosomal RNAs is purely speculative, it may be after all that each functions in regulation separately from the other and by a mechanism still to be explained. In this connection it should be recalled that the function of the small molecular weight RNAs of the nucleus (Chapter 2) is not known and could be related to this mechanism. Finally, the continuous synthesis of labile RNA in cells making little or no mRNA takes a related yet different explanation to that in the precursor hypothesis. In the present hypothesis, these cells would be gratuitously making regulatory RNA, as if for some reason the integrator genes continue to fire.

According to the regulatory hypothesis the specification of mRNA is entirely transcriptional. This does not however exclude a certain degree of posttranscriptional specification, that is, in the form of a mRNA precursor as in the previous hypothesis. The restriction in the present hypothesis is that the precursor must not contain more than one informational sequence, which, as was mentioned, could mean a significant biological advantage. Because in reality the two hypotheses are not incompatible and because there is a certain amount of evidence for a mRNA precursor, it is therefore worth exploring a compromise between the hypotheses. Certain labile RNA molecules might contain the transcripts of both a single informational sequence and several adjacent regulatory sequences, of which the regulatory transcripts break down eventually (after discharging possible functions considered later). This example amounts, for the precursor hypothesis, to curtailing to one the number of mRNAs in the precursor, and, for the regulatory hypothesis, to transcribing the contiguous regulatory genes covalently with mRNA. (Transcription of these genes was not made explicit in the hypothesis.) By this compromise the two hypotheses come formally close to each other in regard to the mRNA precursorship without altering any basic functional predicate intended by their authors. Even so, the hypotheses remain irreducibly different in regard to the nature of the oversize labile RNA. In the precursor hypothesis this represents a composite regulatory and informational transcript, but only a regulatory (integrator) transcript in the regulatory hypothesis. In principle, this difference should permit a direct test between the hypotheses. For example, evidence that showed that all oversize labile RNA molecules include mRNA would disprove the regulatory hypothesis at least in its current form. (An alternative form would be that some of the oversize transcript includes mRNA, say, for regulatory protein.) As discussed, the biochemical evidence in this direction is disputable and certainly not referable to every possible molecule in that size range. Being that conclusive biochemical evidence is technically difficult to obtain, it would seem at present that a decision between the hypotheses may have to wait a long time, unless other evidence of a nature not easy to predict appears in the meantime.

In conclusion, the hypotheses by Georgiev (1969b) and Britten and Davidson (1969) presented in this section take into consideration a large amount of regulatory DNA which each in its own way attempts to define, although evidently this has not yet been achieved. In the regulation of bacterial informational transcription, the unit of transcription and translation is one and the same (except perhaps for the contiguous regulatory genes) and constitutes the functional genetic unit. As taken from these two hypotheses, in the eukaryote the unit of transcription is not known and the number of sequences transcribed is greater than the number translated. Contrasted with the bacterial system, not only the mechanics of the informational pathway is different but also the functional genetic unit remains to be defined. Because much of the disparity between what is transcribed and translated in the eukaryote reflects the presence of redundant DNA, these hypotheses submit that this DNA functions in the regulation of the pathway. Thus a different pathway and type of DNA have led to postulate a regulation entirely different from the bacterial. One should note, however, that the only molecular evidence so far in support of the concept that redundant DNA functions mainly in regulation is that this DNA (a) codes for labile RNA, and (b) chromosomal RNA binds to it. This evidence is entirely inferential since it rests on the deductive postulate that these RNAs themselves function in regulation. The reciprocal concept, namely, that unique DNA is mainly informational, was until recently indirect and based on genetic and structural evidence. It is certain in the mouse, however, that redundant DNA can code for only about 100 proteins or a little more if several related proteins are considered for each family of redundant sequences (Walker and Hennig, 1970), hence the remainder and in fact most of the protein must be coded in unique DNA. Direct evidence (discussed in Chapter 7) now supports the concept, but it is also clear that not only unique DNA is informational. The importance and heuristic value of these two allied concepts should earn them full attention, yet, it cannot be emphasized enough, that regulatory DNA requires direct evidence in order to be finally acceptable.

CONCLUSIONS ON THE REGULATION OF EUKARYOTIC MESSENGER RNA

Taking stock of our present views on eukaryotic regulation as illustrated by the hypotheses of Georgiev (1969b) and Britten and Davidson (1969) just presented, we may now state what the main differences seem to be with bacterial regulation. In doing this it is most important to remember that in the eukaryotic system we are dealing with hypotheses and not with hard facts as in the bacterial system, for in no case have the progressive steps of any eukaryotic mRNA synthesis been worked out molecularly. Nevertheless, these hypotheses are all we have to explain their coordination.

In the first place, in the model of Britten and Davidson the regulatory genes not contiguous to the informational cistrons form a much more sophisticated ensemble than in the operon model. The reason for this is that these regulatory genes (sensor genes) are assumed to respond to external effectors (hormone) which to our knowledge the bacterial regulator gene does not. Another difference with the bacterial regulator gene is that, in terms of the organization of the system, the sensor genes are intimately associated with other regulatory genes whose sequences are molecularly redundant with respect to separate regulatory genes contiguous to the informational cistrons (Fig. 13). As envisaged in the model, this regulatory ensemble converts the low informational content of the hormonal signal into a highly informational and flexible RNA signal between genes; the informational genes may or may not share regulatory ensembles. Evidently, this model system is very different from that of the bacterial regulator gene which stands by itself and signals the regulatory genes (not related in sequence) contiguous to the informational genes by means of repressor protein (Fig. 12); except within polycistronic operons, these genes do not generally share regulatory ensembles. This greater overall interconnectedness of the eukaryotic system is presumably dictated by the dispersal and large number of information cistrons in each regulatory circuit, and also provides each cistron with the ability to respond pleiotropically or multivalently to the hormonal signal, as the case may be. (Pleiotropy means that one signal elicits several enzymic activities; in multivalency several signals all elicit the same activity.) A given pattern of circuits is activated in each tissue according to physiological function, but still permitting a certain modulation of this function according to changing requirements; for example, the skin of different parts of the body functions according to a basic "skin genetic repertoire" but its expression is influenced by the local anatomical and environmental factors. In summary, at the level of the organism, though not at that of the individual circuits which are relatively simple, a complex circuitry is believed to be essential to explain the integration of genetic function. A substantial amount of DNA would be required to regulate informational DNA in the eukaryote because of the integrative complexity of its bodily functions, compared with the simpler growth cycle of the prokaryote within one cell cycle. The stereotyped response of the prokaryote has become the integrated behavior of the eukaryote.

In the second place we have to consider that the proposed eukaryotic regulation by means of RNAs and acidic proteins would be in a positive rather than negative fashion as it is in most bacterial operons. The positive regulation would result from the nature of the informational circuits described above. These molecules (RNAs and acidic proteins) are candidates for derepression (a) because they are heterogenous and thus contain a vast potential for specificity and (b) can recognize DNA sequences; yet the evidence supporting their derepressive function remains essentially inconclusive. However, regulation by RNA is suggested by the biochemical evidence (such as it is) on chromosomal RNA and is postulated in Britten and Davidson's hypothesis. Regulation by acidic proteins is suggested by the biochemical evidence on the stable regulation of mRNA and is postulated in Georgiev's hypothesis. The nature and extent of possible interaction between RNA and protein during derepression, and for that matter between them and histone, is, on the other hand, not known. If any regulatory mechanisms such as these involving RNA or protein were to be substantiated, they would represent an eukaryotic novelty—which is why the study of these putative regulatory molecules is so interesting. Thus direct regulation by RNA (without an intervening protein) would be a novelty, and the same would be true of a generally positive regulation by protein (as mentioned, few operons have repressors that act positively). A difference between them is that RNA might serve a very flexible regulation because of the occurrence of DNA redundancy, whereas a more locus-specific regulation is expected of certain proteins. Their joint action could therefore achieve a very sensitive regulation, especially for the initiation of transcription—the primary control of protein synthesis that has not been clarified in the eukaryote.

An intriguing question is the actual function of molecular redundancy as such in the regulation, the answer to which is not final. A considerable degree of sequence heterogeneity is present in both labile and chromosomal RNAs, but somewhat attenuated by the imperfect reiteration of their sequences, which for technical reasons remains an aspect scarcely explored. Perhaps there is an advantage in a limited, as opposed to a precise, matching between these RNA and DNA sequences during their mutual recognition in a redundant regulatory system. Britten and Davidson (1969) propose that it is this imprecise recognition which permits a more flexible reshuffling of regulatory DNA sequences within the genome, since all that would be needed for RNA sequences to recognize them is that they belong in the same family of redundant sequences. Contrast this with the inflexible recognition of DNA by a regulatory protein. If these views are correct, redundant regulation would be an evolutionary requisite of the eukaryotic system.

A consequence of the eukaryotic system may be the lack of an immediate informational response, compared with the immediate response in the prokaryote; the eukaryotic informational response appears to be preceded by a much more immediate regulatory response. It was said that the response must somehow engage a complex series of macromolecular interactions within the chromosomal apparatus. Conceptually, this sharing of the specificity through interaction with other chromosomal macromolecules is nowhere more needed than in the case of putative derepressor proteins. Were these proteins not to interact with other chromosomal macromolecules but to act specifically and directly on the DNA, such a system would necessitate an infinite regression with respect to any antecedent system that in turn conferred specificity on the proteins. Thus, the suggestion of functional degeneracy inherent in the concept of a chromosome with a great interactive complexity is economical in that it probably reduces not only the number required of these derepressor proteins, but, especially, also the nonimmediacy of a regulation which is partly dependent at every moment on their synthesis. In the model by Britten and Davidson (1969) the specificity of regulation at the genomic level can regress no further than the sensor genes, but they prefer to consider regulatory RNA rather than protein. However, that regulatory protein remains an equally distinct possibility is implied here and is well recognized by Georgiev (1969b) in his model; as Britten and Davidson point out, their model does not exclude this possibility. (What they did consider is the binding of hormone to the sensor gene mediated by protein.) In addition, a greater immediacy of the regulation by protein would result if some derepressor protein were to be produced by the integrator genes, rather than these genes behaving solely in a noninformational capacity as in their model.

The positive eukaryotic regulation makes it likely that most informational repression will be of an unspecific nature, that is, mediated by histone. This generalized repression by histone, although probably not exclusively by histone, is an eukaryotic novelty. As a rule, the regulation makes it also unlikely that repression will be mediated specifically by RNA or nonhistone protein. However, specific repressors of this type remain a possibility and the best candidates are certain acidic proteins which appear to act together with histone; the final repressor assignment depends on first showing that each of these acidic proteins is capable of reading DNA sequences. The occurrence of a specific repressor protein for the rRNA cistrons is irrelevant to the present argument because of the exceptional characteristics of these cistrons such as their multiplicity and transcription by a distinct polymerase. Finally, but important, the positivity of regulation does not exclude a more specific regulation, as opposed to unspecific repression, by histone. This possibility is part of the concept of a chromosomal apparatus with interactive specificities.

The previous considerations represent the coming together of two different lines of approach to the problem of eukaryotic regulation. The first is the experimental approach to selective gene derepression put in motion—if one is to mention a landmark—by the findings on histones in the early 1950's. The second is the more theoretical approach prompted by the implications of the discovery of molecular redundancy of DNA in the late 1960's. However, in all justice, the conceptual framework of some of those working on eukaryotic RNAs was quite advanced by the time of this discovery. It is plain to see now that a good deal of work remains to be done to bring these complementary approaches closer together. Yet this unification seems most worthwhile pursuing at present not only on the grounds of its promise, but also, quite simply, because there is no better alternative available. To bring to a close our previous considerations, it might still be argued that a genetic system based on regulatory and informational genes such as in the models proposed for the eukaryote is not fundamentally different from the bacterial operon; or, that the regulons implied in these models were indeed first conceived for bacteria. Uncritically, the first and only serious argument is true, but it does nothing except reaffirm the ancestral blueprint of the eukaryotic system. It does not, in particular, recognize the enormously greater complexity of regulation involved in the successive stages of genetic programming required by the eukaryote, nor, perhaps, does it anticipate how much subtler than we can think this regulation may turn out to be.

Most of the discussion so far has been concerned with the exogenous regulation of transcription which is the one usually thought of when considering selective derepression. We must now turn our attention to the endogenous regulation of transcription. For a fact, at least as many obligatory as facultative cell functions depend on transcription. Good examples of the former are to be found in cells reproducing in synthetic media or in the early nonplacental embryo (both of which are hormone free). These obligatory cell functions are endogenously regulated and, qualitatively speaking, their transcription must be as selectively controlled as that for facultative functions depending on exogenous regulation. Likewise, obligatory functions must mobilize a very large number of metabolic sequences requiring a genetic program as integrated as that

Conclusions on Regulation 183

described earlier for facultative functions. For simplicity, the view will be taken here that facultative function is essentially gratuitous to the cell (but certainly not to the organism). From this it follows that the two classes of function partly exclude one another at the genomic level, which is to say, use a different part of the genome. Of some importance, however, is that certain facultative functions probably represent hypertrophied obligatory functions and in this case the same part of the genome functions both ways. Although the mechanism of regulation of the strictly obligatory genome could be simpler, it would still obey the general pattern of eukaryotic regulation inferred on the whole from facultative function–this indeed becomes a necessity if part of the obligatory genome is also to function facultatively. In the opposite direction, it is not unreasonable to consider that there may be a possible distinction between the signals for the two genomes, as is done below. Again, however, if the two genomes are to overlap in part as suggested, the final effects of the signals on the derepressive mechanism must be basically similar if not interchangeable.

A last point–a related one was made for the prokaryote–bears on whether the regulation of several hundred thousand eukaryotic mRNAs is understood in such a way which, even if general, seems at least conceptually adequate. The most basic consideration of this question offered earlier, at present, narrows down to the very hypothetical regulatory function of certain RNA species and the elements that in turn govern them. As explained then, during facultative derepression these putative regulatory RNAs are believed to be activated in response to an exogenous hormonal signal impinging on the genome's regulatory program, and thereafter to act as a signal for informational transcriptions. (It is immaterial to this consideration whether or not these regulatory RNAs are translated eventually into regulatory proteins. However, to the extent that regulatory proteins are produced from other RNAs which in themselves are not regulatory, these proteins would be considered together with the present RNAs.) Obligatory transcriptions certainly must present a different strategy, but in their case there is a total lack of evidence. Presumably, for these transcriptions the genomic program may be read directly (basally) into regulatory RNAs as in the previous case, or signaled by some internal metabolic or macromolecular event possibly geared to the cell cycle. Commenting on these events, it is not known (a) whether or not obligatory transcription is geared to multireplication in the chromosome during the S period, or (b) to protein synthesis in the cytoplasm across the entire cycle; (c) how it responds via the cell membrane to the external physicochemical and cellular environment (e.g., cell contact inhibits transcription); and (d) to what extent it requires nonhormonal signals in the systemic circulation. With regard to obligatory transcription the question is, therefore, what the specific molecular signals or factors involved are, what their number is, and how many of them can we account for. At a later stage the question might become whether there is any mechanism of obligatory regulation to obviate an otherwise large number of signals. An example would be a mechanism which bypasses most regulatory genes to act almost directly on informational cistrons and still maintain their basic coordination. Perhaps significantly, this obligatory bypass would begin to resemble by degrees a bacterial operonic system, except for its positivity. Let us ask whether there are also factors involved that, like the bacterial and viral σ factors, allow subregulation of blocks of cistrons? Is there any obligatory subregulation manifest in the massive forms of heterochromatinization discussed earlier? The answer to these questions rests with the future.

ADAPTIVE REGULATION OF EUKARYOTIC MESSENGER RNA: POSTTRANSCRIPTIONAL MODULATION?

Previous sections dealt with the regulation of stable repression typical of the eukaryote. It is accepted today that enzyme adaptation, superficially resembling that found in bacteria, also occurs in the eukaryote (Pitot, 1967). Although this adaptive regulation is a form of posttranscriptional regulation, thus an area of regulation different from stable repression, it will be instructive to delineate here which are the different consequences of these two forms of regulation and, in passing, also the differences between bacterial and eukaryotic adaptation. The major difference with stable repression is that adaptation brings about a large increase in enzyme activity, but not the appearance of new activity. Most of the evidence concerning adaptation comes from liver or cultured cells, that is, cells capable of a bacteriallike regenerative response. Most of this evidence deals with typically eukaryotic unstable enzymes, but there is also evidence for nonenzyme protein being involved in this form of regulation. As in the case of hormonal action during derepression, the adaptive phase is mainly during the S period of the cell cycle (Martin *et al.*, 1969) or some time before.

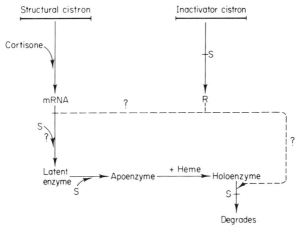

Fig. 15. Model for adaptive regulation proposed for tryptophan pyrrolase (adapted from Paul, 1968). The model is explained in the text. R represents the hypothetical inactivating protein; S, the substrate, is tryptophan.

A typical example of adaptive regulation is that of liver tryptophan pyrrolase (Paul, 1968), an enzyme which serves to degrade the amino acid and whose regulation is seen in Fig. 15. A certain inactivating gene is assumed to begin making a hypothetical protein via a short-lived mRNA (postulated from the "actinomycin superinduction" effect) as soon as the pyrrolase has risen to a certain level. This protein is short-lived (as its mRNA) and prevents either any further translation or stabilization of enzyme. Presumably under control of its own regulatory genes, a structural cistron for the pyrrolase is assumed to produce a long-lived mRNA. The substrate (tryptophan) directly enhances first translation of this mRNA into latent enzyme and then its conversion into apoenzyme; this becomes holoenzyme in the presence of heme. At the same time, the substrate represses both the inactivating gene and the effect of its protein on the enzyme. However, the substrate may also in part stabilize the enzyme directly (Griffin and Cox, 1967). Indeed, in view of the general instability of eukaryotic enzymes, a more general stabilization by substrate is conceivable. According to this model, the enzyme substrate acts at several posttranscriptional levels but not on the informational transcription itself (in contrast to bacterial substrates which act as inducers or corepressors of informational transcription). Instead, in this model informational transcription is stimulated by hormone (cortisone).

The evidence for adaptive regulation in other systems indicates that hormone may affect either transcription or translation, and sometimes both. One of these systems involves liver tyrosine aminotransferase, whose activity is elicited by adrenal steroids (not by substrate). It is proposed (Tomkins et al., 1969) that in this system hormone acts exclusively to cancel a translation inactivator function, although it is left imprecise at which level between the pre- and posttranslational the function intervenes (Fig. 16). Formally, therefore, this system shares a reversible inactivator function with the system schematized above. In the present system, both the structural and inactivator mRNAs would be obligatorily made only during the adaptive period of the cell cycle. Only the structural mRNA can be translated during both the adaptive and nonadaptive period; in the adaptive period it can be translated to a greater extent when steroid is present. It should be noticed, according to this model, that the mechanism of hormonal action is diametrically opposed to that in stable regulation where, instead of canceling inactivation, it specifies transcription. The mechanism of hormonal action remains noninstructional, however. Yet in this adaptive system steroids may possibly influence, in parallel, aspects of translation other than the inactivator function. Concerning the functional product of the inactivating gene, indirect evidence that this might be a protein comes from the effects of steroid analogs which suggest an allosteric receptor in the inactivation system, hence probably protein. Nevertheless, it cannot be excluded that the inactivation is directly via an RNA molecule. Finally, in plants in particular most of the evidence for adaptive regulation implicates activation rather than translation of enzyme (review by Filner et al., 1969). In any event, all systems known for this form of regulation are considerably more complex than the bacterial systems they resemble. Not only does this regulation not seem to act on the chromosome as it does in the bacterium, but, conversely, so far no inactivator gene has been discovered in the bacterium.

The form of regulation described in this section perhaps in general represents a modulatory regulation as contrasted to the more stable regulation manifest during repression-derepression of the genome (Clever, 1968). It needs no further emphasis that these two forms of regulation are mechanistically entirely different. As a possibility to keep in mind, modulatory regulation may serve to bypass stable regulation, and may apply in a restricted way to certain metabolic pathways. Formally, this suggestion resembles the one made earlier to the effect that certain mechanisms of obligatory transcriptional regulation may bypass regulatory genes. The great physiological importance of modulatory regulation in the eukaryote may not be apparent from this cursory discussion: as a remedial measure the reader is encouraged to consult the original literature.

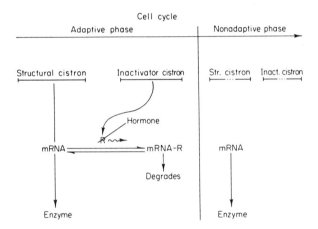

Fig. 16. Generalized theory of adaptive regulation in mammalian cells according to Tompkins *et al.* (1969). The structural cistron codes for an adaptive enzyme, while the inactivating cistron codes for an inactivator of the enzyme. During the adaptive phase of the cell cycle the structural cistron is transcribed and the resulting long-lived mRNA can be translated into enzyme. The inactivating cistron is likewise transcribed and its RNA product is perhaps translated to produce protein R. This protein combines reversibly with the structural mRNA to produce an inactive complex (MR) which leads to partial degradation of this mRNA. The protein nature of R and the existence of the MR complex are hypothetical, and all that the model implies is that an inactivating cistron somehow partly inactivates the structural function at some unspecified level. Whatever its nature, R itself is short-lived as indicated by the undulating arrow leading away from it. Inducer hormone is indicated to inactivate R by an unknown mechanism, thus increasing synthesis of adaptive enzyme. During the nonadaptive phase of the cell cycle neither the structural nor the inactivating cistron is transcribed, but the structural mRNA made previously can be translated. [Copyright (1969) by the American Association for the Advancement of Science.]

CHAPTER 4

Transcription of Stable RNA: Ribosomal and Transfer RNAs

Relatively little is known of the regulation of stable RNA species at the cistron level in any organism, and, for this reason, they will be considered for the prokaryote and eukaryote more or less in parallel. The lack of understanding of the regulation is not due to a lack of interest, but, quite the contrary, because the regulatory mechanisms involved have proved so far to be less definable than, for instance, in prokaryotic mRNA. These stable RNA species function as common auxiliaries in translation, and, because of this, they constitute the greatest bulk of cell RNA; from this fact alone their importance needs no explanation. Also obvious from this fact is that their regulation must be closely attuned to the pattern of cell growth. In contrast, this is not always the case with mRNA which represents a great variety of discretionary information. Consequently, these RNAs, in particular rRNA, show the nearest to a modular regulation in all RNA species. Certain tRNA species, on the other hand, may be discretely regulated. The structural cistrons of these RNAs represent a minute fraction of the genome, even though, particularly in the case of the eukaryote, most cistrons are present in multiple numbers. Because of this multiplicity and the slight heterogeneity between sequences, these cistrons belong in the redundant genome. Both stable RNAs have considerable higher order structure and are transcribed as a precursor which is larger than the mature molecule. This seemingly wasteful precursor is probably indispensable in ensuring (via modification) the correct final structure; this predicts that only the transcript of the complete cistron can become functional. Since these species are metabolically not completely stable, their generic name "stable" is conventional and has a historical basis, namely, to contrast them to bacterial mRNA. Especially in the eukaryote, however, these RNAs may well, after all, qualify as relatively stable, but in respect of their higher order structure and not half-life, which was certainly not the original meaning of the conventional name.

MAJOR RIBOSOMAL RNAS

Based on present evidence, the rRNA cistrons (including 5 S RNA) are the most numerous and perfectly repeated cistrons for any RNA species in the genome of both the prokaryote and eukaryote. This numerical repetition of cistrons is related to the fact that rRNA is the single most abundant RNA species in the cell. In ascertaining this fact for the eukaryote in Table 4, the reader should be aware that although tRNA involves more molecules per cell, it represents many species, hence many different cistrons. Because of this abundance as well as size, rRNA constitutes the bulk of cell RNA, or about 80% by mass. Most likely each of the two major rRNAs constitutes a single species, although some degree of sequence heterogeneity (probably of little functional significance) does occur. Taken together these points permit two generalizations to be made, which are closely interrelated with each other. The first generalization is that the large cistronic multiplicity is mainly for the purpose of satisfying the great demand for the RNA, for example, a typical mammalian cell produces from 10 to 100 ribosomes per second. The second generalization is that the molecular redundancy which is defined by these points is almost incidental rather than obligatory to the function of rRNA, unlike, by contrast, what may be the case with other RNA species. This introduction already revealed that more is known of the cistronic DNA in the case of this RNA than of mRNA, which in turn makes possible further direct analysis of the DNA.

Cistrons in the Eukaryote

In eukaryotic cells the genetic locus for rRNA is the nucleolus organizer which will be discussed in detail in Chapter 6 in connection with the organelle. This organizer locus was the first one isolated in any organism and so far one of the best analyzed at the molecular level. We will now consider the individual cistrons which form this multicistronic locus.

The following description of the rRNA cistrons refers to the amphibian somatic cell, but similar data are emerging for *Drosophila* (Quagliarotti and Ritossa,

1968), rat (Steele, 1968), human HeLa cells (Attardi et al., 1965b), and some plants (Tewari and Wildman, 1968). In *Xenopus,* according to Birnstiel et al. (1969), approximately 800 copies of each 18 and 28 S rRNA cistrons are present in one organizer locus per haploid genome. Brown and Weber (1968b) report 450 copies, but the previous estimate will be adhered to noncommittally here. Each cistron cluster weighs 8.8×10^9 daltons of DNA, or more than a bacterial genome. Because of the clustering of cistrons and their high GC content, it is possible to isolate them centrifugally as a satellite banding at 1.724 g cm^{-3} (main DNA bands at 1.700). Considering two DNA strands, the satellite comprises 0.2% of the total nuclear DNA (2C = 6 pg). When sheared the satellite separates into 18 and 28 S rDNAs that band at 1.713 and 1.723 g cm^{-3} (53 and 63% GC), respectively. Only about 40% of the isolated satellite is, however, complementary to rRNA as shown by hybridization, the implications of which we shall see later. The complexity of the sequences for each rRNA species is close to that of a single family (Birnstiel et al., 1969), yet this homogeneity does not exclude a residual degree of microheterogeneity to be described below.

A polycistronic precursor rRNA molecule (nominally 45 S; 2.5×10^6 daltons) gives rise to one mature molecule of each 18 (0.7×10^6) and 28 S rRNAs (1.52×10^6). The precursor is therefore 0.3×10^6 daltons or 14% heavier than the combined mature species (2.2×10^6). The portion of the precursor to reach maturity is 64% GC and the discarded portion is 85% GC. Assuming single-strand transcription, it follows that the cistronic DNA is equivalent to 5×10^6 daltons. As mentioned, hybridization studies show that 60% of the satellite (equivalent to 6.6×10^6 daltons of DNA per cistron) makes no rRNA. Thus, each organizer locus in the chromosome is composed of 800 cistronic repeats ($800 \times 5 \times 10^6$ daltons of DNA) and a comparable amount ($800 \times 6.6 \times 10^6$ daltons) of intercalated DNA which is nonribosomal. The composition of this latter DNA is 85% GC. The 18 and 28 S rDNAs are arranged as a linear series of singly alternating cistrons all physically integrated within the satellite, which, in turn, is integrated with the remainder of chromosomal DNA. The position of the intercalated DNA in relation to the rDNAs is now known from work described below, but its function is not. The description of the rRNA cistrons just given for *Xenopus* will vary in detail in other organisms; the HeLa cell cistrons will be described later.

From the thermal instability of mammalian DNA-RNA hybrids, it was proposed that there is a certain degree of intercistronic sequence divergence within the DNA for each rRNA species. Indeed, some degree of sequence heterogeneity must be tolerable if there is to be the high cistronic multiplicity and occasional (germinal) amplification of this DNA. However, because not all cistrons may be transcribed, it is not clear whether all the divergence is reproduced in the RNA. Even if transcribed, the sequence divergence still need not mean a functional RNA difference. This is because the RNA molecule is large to begin with and its function is probably not concentrated in any small portion, which would make the function very sensitive to divergent changes in that portion, but instead is distributed over most of the molecule (contrast this with the small active center of an enzyme). In any case, the divergence must have arisen by mutation after the original tandem duplications that resulted in the present-day loci. On the other hand, the cistronic multiplicity per se is likely to buffer to some extent any possible biological effect of mutational or transcriptional error, the first including sequence divergence. In this respect, as will be seen, pathological effects on the phenotype arise only from gross elimination of cistrons. Because of the slight sequence divergence and cistronic multiplicity, rDNA therefore qualifies for inclusion in redundant DNA (Moore and McCarthy, 1968).

In general, the rDNA of somatic cells ranges from 0.05 to 0.5% of total DNA from the lower to higher eukaryote. An exceptionally high value is found in yeast, a lower eukaryote, where the rDNA (2.4%) occupies the equivalent of one out of 18 chromosomes. Correspondingly, yeast has an extremely high total RNA to DNA ratio (Schweizer et al., 1969). On the other hand, the range of somatic multiplicity for the haploid genome in animals (and yeast above) is generally on the order of 10^2 cistrons, but appears to reach 10^3 in animals with a large genome and 10^4 in certain plants. For example, the lungfish with a genome 20 times larger than the mammal contains 5000 rRNA cistrons (Pedersen, 1971) which is however less than 20 times the number of cistrons in the mammal. Still, since a genome need not be used for informational purposes in proportion to its "excess" size (as is obvious in the comparison of lungfish and mammal), the reason for the increased multiplicity of auxiliary cistrons remains unclear. Because of the uncertainties involved in the determinations, it is impossible to decide at present whether there is any taxonomical correlation within the above range of multiplicities, except perhaps in plants versus most animals. What appears to be decided is that over the animal evolutionary scale the cistronic multiplicity of rDNA has increased but not in proportion to the size of the genome, that is, the relative multiplicity has decreased. When the increase in both multiplicity and genome size is plotted logarithmically (Birnstiel et al., 1971), they are roughly parallel.

Contrary to somatic cell rDNA which is chromosomal, the oocyte presents an additional, extrachromosomal rDNA. This additional rDNA is clearly required to support an equally extraordinary rate of rRNA synthesis. For example, in the *Xenopus* oocyte this rate is equivalent to that of the same weight of liver containing 200,000 cells. The oocyte is among the few cell types with a regular selective amplification of rDNA; it is not amplified in most developing or mature somatic cells irrespective of their rate of rRNA synthesis (Ritossa et al., 1966a). The oocyte therefore presents an augmentation of the already multiple somatic rDNA, with a final germinal multiplicity surpassing a million cistrons. In contrast to the usual two chromosomally attached nucleoli of the somatic cell, the *Xenopus* oocyte contains about 1500 free nucleoli; they first appear during pachytene as discrete but naked nucleolus organizers in proximity to the original nucleolus, although rDNA amplification actually begins in the oogonia (Gall, 1969). The organizers then move toward the inner nuclear envelope where they become attached and form definitive nucleoli (Gall, 1968). These remain active past the lampbrush chromosome stage, but the extra rDNA never functions or replicates again after meiosis. Total oocyte rDNA is about 25 pg, compared with 12 pg of nonamplified (4C) nuclear DNA (Brown and Dawid, 1968; Evans and Birnstiel, 1968). Since the 4C rDNA value is 0.024 pg, the resulting amplification by a factor of 1000 is in the order expected from the nucleolar numbers. In reality, the data indicate an average 0.017 pg per nucleolus (25 pg/1500 nucleoli). The difference with the 1C value (0.006 pg) per somatic nucleolus suggests that in the oocyte the organizer is not copied as a unit in each of the supernumerary nucleoli. Direct evidence for this disunity is that there are on the average about four DNA circles inside each *Xenopus* oocyte nucleolus (see below); the single DNA circle present in urodele oocyte nucleoli is of variable length (Miller, 1966). The buoyant density of oocyte rDNA (1.729 g cm^{-3}) differs from the somatic counterpart (1.723). This is probably due to the former rDNA containing only traces of methylcytidine whereas the latter contains 5%, which would tend to reduce its buoyant density (Brown and Dawid, 1969). This compositional difference may account for the fact that the subsisting amplified rDNA is not transcribed after meiosis.

Miller and Beatty (1969b) visualized under the electron microscope the transcribing rRNA cistrons of *Xenopus* oocyte nucleoli (Fig. 17). What is seen is a continuous circular DNP core with actively transcribing units, each irradiating nascent rRNA molecules in the form of RNP fibrils. Inactive stretches are regularly interspersed between the active units. Each unit contains a very constant and evenly spaced 100 attached fibrils disposed in a gradation of length, the longest fibrils representing the maturest rRNA molecules. Since these molecules are shorter than the cistronic length (see later), this means that they are foreshortened by coiling, probably as the result of their being proteinated as they are transcribed. In any event, the orientation of the gradient of fibril length is the same in all active units of rDNA. There are from a few to a thousand units per nucleolar core in rough proportion to its size, which measures from 35 μ to 5 mm in circumference. This again points to a rather unpredictable amplification of rDNA in each nucleolus. Hence about 100 polymerase molecules are transcribing each unit simultaneously and coordinately. These polymerase molecules can be seen one at the base of each RNP fibril in contact with the DNA core. Interestingly, similar molecules are seen equally spaced along many of the inactive stretches, suggesting, if indeed they are polymerase, that in these stretches the enzyme is repressed for transcribing

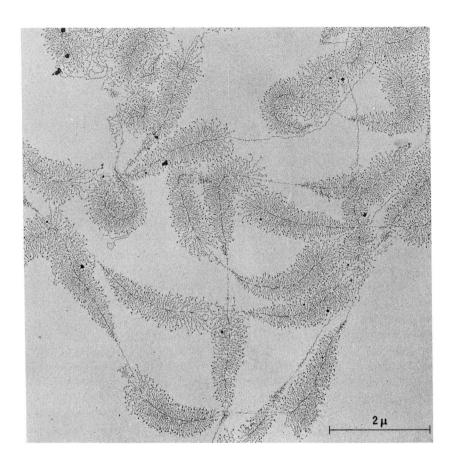

Fig. 17. The transcribing ribosomal RNA cistrons of the amphibian oocyte. Several multiply transcribing cistrons are seen along the length of the core DNP fibril, as explained in detail in the text. (By courtesy of Drs. O. L. Miller and B. R. Beatty.)

but itself is not limiting. The total number of rRNA polymerase molecules in one oocyte nucleus is about 6×10^7. (For comparison, the polymerase transcribing all RNA species in the bacterium numbers 2000 molecules; Bremer and Yuan, 1968.) The approximately 2.5 μ measured length of the active unit agrees nicely with the length expected from the rRNA precursor of 2.5×10^6 daltons. The unit core therefore represents the cistronic rDNA including the 18 and 28 S RNA cistrons. On this basis, the coiled fibrils represent the nascent precursor rRNA molecule. The fibrils would contain the future 18 S RNA, already rounded as a RNP particle, at their tip; the future already proteinated 28 S RNA, somewhere in their proximal portion; and the proteinated RNA portions of the precursor to be discarded, somewhere in their length. The inactive stretches confirm the existence of intercalated DNA deduced from molecular data, and, of course, that this DNA is not transcribed. Moreover, the length of these stretches ranges from one-third to ten times the length of the active units, suggesting that (a) there are some inactive cistrons, and (b) the intercalated DNA is probably arranged in between joint 18 and 28 S rRNA cistrons. In addition, there are elements attached to the nucleolar DNA cores such as membranes and complex filaments whose function in connection with transcription and processing of rRNA is not understood. Lastly, all visible granules in the nucleolar cortices surrounding the DNA cores are attached to long fibers, indicating that at this level the rRNA product is still not a free granule or fiber.

Amplification of rDNA in the oocyte seems to be by a primary replica that detaches from the organizer and then undergoes a series of subsequent replications (MacGregor, 1968). This "escape" of the primary replica would bear some superficial resemblance to a prophage excision, but it remains a major macromolecular event as yet unexplained. According to Crippa and Tocchini-Valentini (1971) the possibility exists that the amplification of rDNA is from rRNA by an RNA-primed DNA polymerase, similar to that recently found in oncogenic RNA viruses. This mechanism would obviate the excision of an organizer replica. The overall control of oocyte rDNA replication appears to be uniquely sensitive to the final amount of rDNA present per germinal vesicle, rather than to the size or number of replicas present in the nucleoli or the number of rRNA cistrons present in the chromosome. This is clearly shown by the heterozygote *Xenopus* mutant whose oocyte, even though it contains only half of the chromosomal rDNA, contains the full amount of amplified rDNA (and number of nucleoli) characteristic of the normal oocyte.

Cistrons in the Prokaryote

Bacillus megaterium contains about 40 sequences each of 16 and 23 S rRNA which are complementary to one DNA strand to the extent of 0.14 and 0.18% of the genome, respectively (Yankofsky and Spiegelman, 1963). In *B. subtilis* six to eight genes for each RNA species are clustered near the chromosome origin and another two to four genes near the chromosome end (Smith *et al.*, 1968); the rDNA is equal to a total of 0.38% of the genome (Oishi and Sueoka, 1965). A similar bimodal distribution of genes was found in *E. coli* (Cutler and Evans, 1967) representing a cistronic multiplicity similar to that in *B. subtilis* and about 0.5% of homologous genome (Attardi *et al.*, 1965a). Contrary to this conclusion, recent data for this species indicate that the genes occur in a single cluster (Spadari and Ritossa, 1970; Yu *et al.*, 1970); however, these data have been questioned (Birnbaum and Kaplan, 1971) and the question remains to be settled. Interestingly, *Mycoplasma*, with one-fourth the genome size of *E. coli*, is reported to have only one cistron for each RNA species (Ryan and Morowitz, 1969). Thus the prokaryote has a higher proportion of genome involved in rRNA than the higher eukaryote, but the cistronic multiplicity is at least one order of magnitude less.

Concerning the physical relationship between the 16 and 23 S RNA cistrons, one of each (jointly 1.66×10^6 daltons) is contained in a single-stranded DNA fragment of 2 to 4×10^6 daltons (Colli *et al.*, 1971), suggesting that the cistrons are in close proximity. The probable arrangement is in tandem, i.e., 16-23 . . . 16-23. The nonribosomal DNA intercalated between these rDNA tandems in *Proteus mirabilis* is at least ten times the size of the tandems themselves and has an average DNA composition (Purdom *et al.*, 1970). This physical arrangement of cistrons was directly confirmed by electron microscopy (Miller *et al.*, 1970b), in particular, that the two cistrons are extremely close. However, the central question of whether the two RNA species are transcribed separately or in the form of a common precursor remains unsolved. The bacterial rRNA cistrons have been virtually purified (Kohne, 1968).

Pathways of Synthesis in the Eukaryote

Similar pathways of rRNA synthesis are best known between higher eukaryotes, but probably similar ones obtain in the lower eukaryotes (yeast, Taber and Vincent, 1969) and plants (Loening *et al.*, 1969). The common rRNA precursor is formed in the nucleolus. This precursor then gives rise to the only defined pathway of high molecular weight RNA synthesis in the organelle, of which nothing in the end is retained by the organelle.

The ribosomal pathway in the HeLa cell is represented in Fig. 18. The following description of this pathway is based on the data of Weinberg and Penman (1970), and provides a general model for animal cells. The first stage of the pathway is represented by the precursor 45 S RNA (70% GC). Its transcription takes approximately 2.5 minutes, during which most of the RNA methylation and pseudouridylation (Jeanteur *et al.*, 1968) is achieved. As it is transcribed, the precursor is incorporated into a precursor 80 S RNP particle, thus becoming associated with ribosomal protein from the start. (The 5 S rRNA of extranucleolar origin joins the pathway at this stage and will remain associated with that part of the precursor giving rise to 28 S RNA within the large ribosomal subparticle.) It is not known whether the 80 S particle already includes all ribosomal proteins to be acquired within the nucleolus, although the particle acquires more proteins after it leaves the nucleolus. The precursor stage lasts about 20 minutes, and involves 1% of total cell RNA. The pathway that follows is non-

Major Ribosomal RNAs

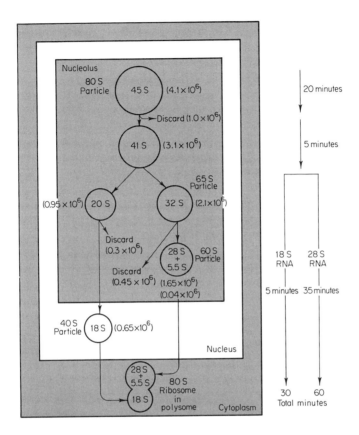

Fig. 18. Pathway of ribosomal RNA synthesis in the HeLa cell. Ribonucleoprotein particles are illustrated with their approximate sedimentation values. The S values inside the particles refer to the RNA molecules they contain and whose molecular weights in daltons are indicated in brackets. Discard pieces of RNA are explained in the text. The pathway begins with a precursor nucleolar particle and ends with ribosomal subunits in the cytoplasm. The extranucleolar 5 S rRNA (not drawn) enters the pathway at the precursor particle stage and finally appears in the large ribosomal subunit. The scale at the right shows the approximate time taken to traverse the various stages. The presence of the large ribosomal subunit inside the nucleolus is to signify that it is delayed within the organelle. Data from Weinberg and Penman (1970).

conservative with respect to 45 S RNA in that half of the molecule does not reach maturity, but it is fully conservative with respect to its methylated portions. A second stage is represented by a 41 S RNA which lasts 2 minutes. To attain this stage an endonuclease cleaves off a first discard piece that is very GC rich (80%) and not methylated. The next stage involves a second cleavage resulting in a short-lived 20 S RNA and a longer-lived 32 S RNA (70% GC) that accumulates to 3% of cell RNA. (Two possibly abnormal or unusual molecules, 24 and 36 S RNAs, may derive from spurious cleavages in place of the previous one, yet these molecules may still subsequently give rise to normal mature 18 and 28 S rRNAs, respectively.) The 20 S RNA is secondarily methylated by a further 2.5% (Zimmerman, 1968). It then converts to mature 18 S RNA (a 68% conservative conversion) with the production of a second discard piece. Being originally in the 5' portion of the RNA precursor, 18 S RNA was transcribed first; it can therefore be produced from an incomplete precursor (Siev *et al.,* 1969) which is impossible for 32 S RNA originally in the 3' portion and transcribed last. The 18 S RNA (56% GC) exits immediately from the nucleolus as an RNP particle, the small ribosomal subunit, and appears in the cytoplasm within 30 minutes. In the other half of the pathway, the transition from 32 S RNA to mature 28 S RNA plus 5.5 S RNA (80% conservative) takes place with the elimination of a third discard piece. [The last two discard pieces may be contiguous in 41 S RNA, and have a GC content identical and sequences partly similar with the first discard piece. All three discard pieces may derive from highly redundant rDNA sequences (Attardi and Amaldi, 1970).] The 28 and 5.5 S RNAs (65 and 56% GC) remain H bonded. This complex is contained in another RNP particle, the large ribosomal subunit, which is delayed in a nucleolar pool and emerges in the cytoplasm within 60 minutes. Thus, the precursor RNA molecule finally forms one molecule of each 18 and 28 S RNAs and its methylated portion is quantitatively retained in these RNAs.

The previous pathway described for the HeLa cell varies in detail in other cells. For example, according to AB and Malt (1970), the sequence in mouse kidney is 45→36→32→28 S RNA, from which there would be no 41 S RNA, but a normal 36 S RNA, in contrast to the HeLa cell. According to Ringborg *et al.* (1970), in the larval chironomid salivary gland the RNAs reach the stages equivalent to 20 and 32 S RNAs in the nucleolus, but thereafter complete their maturation in the chromosome. If not artifactual, the generality of this proposed drastic departure from the pathway remains unknown.

The maturation of the mammalian 80 S precursor RNP particle along the pathway involves a sequential loss of both RNA and protein, with a predominant loss of protein (Liau and Perry, 1969). The ribosomal subunits finally come together as they join the polysome in the cytoplasm, after having temporarily acquired a proteinaceous material in the cytoplasm that lowers their buoyant density (Perry, 1967). Hence the pathway predicts that there will be few complete and functional nuclear or nucleolar ribosomes. In accordance with this, it is a common observation that nucleolar particles are generally smaller and more irregular than the mature cytoplasmic ribosome. Indeed, the consensus (review by Monneron and Bernhard, 1969) is that there are few functional nuclear ribosomes. However, a few would have to occur if there is to be some protein synthesis in the nucleus (Gallwitz and Mueller, 1969a).

The function, if any, of the nonmethylated RNA sequences which are rapidly discarded from the pathway is at present a matter of conjecture, but some (mostly negative) statements seem to be justified. A larger amount of discarded RNA (44%) is characteristic of mammals compared with *Xenopus* and plants (12%), and this, in particular, makes the mammalian precursor considerably larger. The precursor is 4.1×10^6 daltons in mammals and 2.5×10^6 daltons in *Xenopus*, whereas the combined mature rRNAs differ only from 2.3×10^6 to 2.2×10^6 daltons, respectively, that is, by 0.1×10^6 daltons. However, in the amphibian rDNA there is a considerable fraction of intercalated (nontranscribed) DNA as shown by molecular and ultrastructural observation. Hence the difference in the mammal may be more in the length transcribed than in the actual length of DNA in the cistrons, that is, there may be less intercalated DNA in the mammal. In this respect, it is suggestive that ^{32}P-incorporation into rRNA precursor reveals a higher GC content in hepatoma than in liver (Muramatsu and Busch, 1967), which could mean that more of the GC-rich intercalated DNA is transcribed. In itself, the fact that there is less discarded RNA in *Xenopus* and plants argues against the possibility that this RNA is a mRNA for ribosomal proteins or processing enzymes. Moreover, in HeLa cells the discarded RNA is unlikely, on the basis of composition, to be capable of dictating a typical ribosomal protein (Jeanteur *et al.*, 1968). Judging from its variable size, it is also improbable that this RNA serves to direct the assembly of the rather similar ribosomes in all these organisms (Loening *et al.*, 1969).

It is also unlikely that nascent rRNA itself acts as a mRNA for ribosomal protein as is proposed in bacteria. Definitive evidence against this would be that, as is probably the case (Terao *et al.*, 1968), the protein originates in cytoplasmic polysomes, hence beyond the reach of nascent rRNA. Other evidence (Hallberg and Brown, 1969) shows that ribosomal protein is synthesized coordinately with rRNA but not in a template-product relationship. General aspects of ribosome formation, particularly with respect to ribosomal protein, are discussed by Darnell (1968) and Maden (1968).

Pathways of Synthesis in the Prokaryote

On present evidence, unlike the eukaryote, the prokaryote may have no common rRNA precursor (Adesnik and Levinthal, 1969). The 16 and 23 S RNAs are believed to be encoded in different operons which appear to be monocistronic (Bleyman *et al.*, 1969). A common precursor, however, remains a possibility. Even direct visualization of the transcribing cistrons (Miller *et al.*, 1970b) cannot exclude that a common transcript is produced and cleaves before it is completed (which still would differ from the eukaryote). This question of a common versus two individual precursors for the rRNAs is one of great comparative interest. Later in the pathway, at the eosome and neosome stages of ribosomal subunit formation, individual precursors are present which appear to have a 5-20% larger molecular weight (17 and 24 S) than the mature molecules (Hecht and Woese, 1968). This, of course, implies subsequent cleavage of these precursors.

A tentative pathway for the large ribosomal subunit in *E. coli* was proposed by Osawa *et al.* (1969) as follows: (a) the nascent 30 S particle contains newly formed 23 S RNA, which is 60% methylated, and three protein components; (b) an intermediate 40 S particle results after addition of another nine proteins; (c) the mature 50 S particle results after methylation of 23 S RNA is completed and a further seven proteins and 5 S rRNA are inserted. According to this scheme, in contrast with the eukaryote, 5 S RNA would be a late addition to the pathway. It was later found, however, that the 40 S particle already contains fully methylated 23 and 5 S RNAs (Forget and Varricchio, 1970). The duration of the pathway takes about 10 minutes in a cell whose doubling time is 60 minutes. Compared with the HeLa cell, the duration relative to the cell cycle is longer therefore by a factor of four.

In conclusion, the pathway of rRNA synthesis is simpler and faster in the prokaryote than in the eukaryote and certain species such as 5.5 S RNA are altogether absent. It is not difficult to rationalize why the bacterium does not require the equivalent of a nucleolar organelle for this pathway as does the eukaryote, but the reason for this lack of a nucleolus will become more objective as we compare the regulation of their pathways.

Regulation of Synthesis in the Eukaryote

Stoichiometrical synthesis of the two rRNA species in the eukaryote is the direct result of a common precursor and an equal number of genes. However, in contrast with the molecular organization of the eukaryotic rRNA cistrons previously described, the molecular basis of their regulation is not as well established. This situation is reversed in the prokaryote where knowledge of the organization of cistrons may be less impressive than the forthcoming knowledge of their regulation, perhaps better understood than in the eukaryote. Nevertheless, as far as one can tell, the rRNA molecule is not functionally differentiated, from which it follows that, under all cell circumstances of both the prokaryote and eukaryote, rRNA regulation has essentially a quantitative character. Eukaryotic regulation will be discussed at three different levels which, in order of treatment, are the cell, cistron, and pathway levels.

Cell Regulation

Considered at the cellular level, the rate of rRNA synthesis in the eukaryote is a direct function of the rate of protein synthesis, that is, the number of functional polysomes. This means that ribosomes and/or ribosomal subunits do not accumulate beyond a certain level determined by the rate of protein synthesis, but that new ribosomes have to be made when the rate increases. Thus it is possible that some type of negative feedback originating from the cytoplasm, presumably elicited by the number of free ribosomes or ribosomal subunits, reaches the nucleus and eventually represses rRNA synthesis (Hamilton *et al.*, 1968; Steele, 1966). For example, it is known that the polysome pool is indirectly controlled by dietary amino acid, which acts by promoting polysome stability (Chapter 5). It is suggestive that tryptophan, with the most profound effects of all amino acids on polysome stability, seems also to affect rRNA polymerase the most (Veselý and Cihák, 1970). Perhaps related to this effect, liver nucleolar RNA (mostly rRNA) is particularly sensitive to dietary amino acid (Stenram, 1962); starvation reduces the amount of nucleolar RNA which can then be restored to normal after a protein meal. These changes are accompanied by similar changes in the number of cytoplasmic polysomes. Generalizing from these observations, one can envisage a constant adjustment of rRNA synthesis to meet the translational demands of the cell, itself related to the level of amino acid.

Certain observations are in line with the last point made. Synthesis of rRNA proceeds normally in the haploid amphibian embryo whose cells are one-half the size but twice the number in the diploid embryo. Synthesis in the normal embryo remains proportional to cell numbers in regions with vastly different cell size, due mainly to different amounts of yolky material (Brown, 1967). These relationships between the amount of rRNA and yolk-free cytoplasm support the view that a correlation exists between rRNA synthesis and the protein synthetic capacity of the cell. Also significant in this respect, several molecules directly or indirectly implicated in protein synthesis are all regulated coordinately with rRNA, i.e., 5 S RNA, ribosomal proteins, certain histones, and perhaps polyamines.

The primary effect of hormones on rRNA synthesis may be at the level of protein synthesis. As seen in the previous chapter, certain hormones activate rRNA synthesis as part of their general stimulation of RNA syntheses. During stimulation the amount of rDNA remains constant indicating that hormone affects transcription but not multiplicity of template. However, as far as rRNA is concerned, it is possible that the stimulation does not depend solely or even directly on hormone. For example, the nucleolus of the inert avian erythrocyte nucleus is reactivated to make rRNA when the nucleus is made into an heterokaryon with a HeLa cell (Sidebottom and Harris, 1969). In this case the activation is certainly not hormonal in the accepted sense, but due presumably to other factors in the HeLa cell cytoplasm (whose relationship to the HeLa cell nucleus is immaterial here). Also in the organism part of the effect of hormones on rRNA synthesis could conceivably be indirect at the cytoplasmic level, particularly in the case of protein hormones which act on translation via cAMP. In the case of other hormones, particularly steroids, it is equally unclear whether their stimulation of rRNA synthesis is direct, or indirect via the polymerase (Liao *et al.*, 1965), or some cytoplasmic factor. This uncertainty indicates that caution be used in the interpretation that plant growth factors stimulate rRNA synthesis directly in the nucleolus (Zwar and Brown, 1968). Whatever the mechanism, however, the marked effect of certain hormones on rRNA metabolism is not in question. What is questioned here is rather the specificity and directness of the effect, particularly its relationship to primary effects on translation.

Preliminary data suggest a control of rRNA synthesis at the cell level involving amino acid, as already mentioned above. In HeLa cells withdrawal of one amino acid somewhat reduces stable RNA synthesis as the result of deacylation of the cognate tRNA (Smulson, 1970). In ascites cells, which withstand amino acid starvation better than HeLa cells, this reduction of synthesis is greater (Franze-Fernández and Pogo, 1971). In this effect, protein synthesis inhibition or depletion of ribosomal protein is not primarily involved. Cycloheximide, although inhibiting protein synthesis, spares some aminoacyl-tRNA and thereby partly redresses the amino acid deprivation effect. The two major rRNA species are differentially affected according to the particular amino acid which is missing and this leads to different nonequimolar ratios of ribosomal subunits (Maden, 1969). These observations suggest that the mechanism of control involving amino acid may resemble that for bacterial rRNA (known as the RC function), except that a guanosine tetraphosphate derivative which seems to play an important regulatory role for bacterial rRNA was sought but not found for the HeLa cell. Further evidence in this cell, to be mentioned later, indicates that the level of amino acid affects both the transcription and maturation of rRNA precursor. Pending further work, it would appear that the response to amino acid in respect of rRNA synthesis in the eukaryote may be fairly stringent although not quite as stringent as in the prokaryote, and have a wider range of effects.

In summary, the data presented so far in the eukaryote seem to suggest that rRNA synthesis is attuned to the functional status of the cell (whether it is dividing or differentiating), which is reflected in its level of translational activity. Even though, as will be seen, eukaryotic translation is vastly more self-regulated than prokaryotic translation, it is thus not without an effect on stable RNA.

Cistron Regulation

Is it possible to visualize regulatory mechanisms for rRNA synthesis at the cistron level? Originally, a specific low molecular weight inhibitor of the synthesis was proposed in embryonic amphibian cells to be present in amounts inversely proportional to the rate of synthesis (Wada *et al.,* 1968). The validity of this proposal has since been questioned (Chapter 6). However, the presence of a cytoplasmic inhibitor is suggested in the case of the amphibian cleaving egg which has no rRNA synthesis, although it is known to contain the relevant polymerase. Moreover, nuclei actively producing the RNA cease to do so when transplanted into the egg and, correspondingly, lose their typical nucleoli (Gurdon, 1968).

A repressor has now possibly been isolated from the amphibian oocyte (Crippa, 1970). Except for its protein nature, this putative repressor has essentially the characteristics claimed by Wada. The protein (a) specifically inhibits rRNA synthesis; (b) its activity is absent from the oocyte undergoing active rRNA synthesis; and (c) *in vitro* it binds only to native rDNA. The protein is of acidic nature and therefore not histone. An attractive postulate is that the mechanism involving this repressor is one for rRNA synthesis in general and not just peculiar to the oocyte. This is suggested by the above mentioned cessation of rRNA synthesis in nuclei transplanted into cleaving eggs. A tentative model of regulation would be that the repressor is specified by a regulatory gene and acts negatively on operator-like genes in control of the structural rRNA cistrons. Point (b) means that the repressor or its activity is absent at certain periods. In this regard, one might further speculate that the mechanism to remove repressor is to have its regulatory gene, or the assembly of repressor itself, controlled at the level of translation. The attractiveness of this model is that it brings the repressor function in line with the translational correlates which were brought up in the previous discussion. Alternatively, some factor also controlled at the translational level may determine whether the repressor is active or inactive.

The previous model of repressor function is literally borrowed from the format of the bacterial operon. This format is objectionable and was argued against in the case of eukaryotic mRNA, but there might be some justification for it in the case of rRNA. Several characteristics of the eukaryotic rRNA cistrons set them apart from informational cistrons: (a) in a numerical sense these cistrons are among the most accurately repetitious (together with the histone cistrons); (b) they are read by a distinct polymerase which does not read other cistrons (Roeder and Rutter, 1970), except perhaps 5 S RNA cistrons; (c) they are transcribed as a polycistronic precursor (a possible difference here with bacterial polycistronic transcription is that the polymerase is not released after transcription but runs through into the adjacent cistron); (d) the rDNA seems to be more denuded of protein and/or less supercoiled than other DNA in the chromosome (Hsu *et al.*, 1967); (e) in relation to (d), the rRNA polymerase requires a lower ionic strength than the mRNA polymerase for *in vitro* transcription. One possibility is, therefore, that the repressor mechanism may be exclusive to the rRNA cistrons, or more generally to all similarly multiple stable RNA cistrons, which may soon be found out experimentally. Lastly, a unitary regulation of overly multiple cistrons such as can be provided by a single repressor makes plain good sense. It is conceivable that, with up to a million cistrons involved in rRNA synthesis as in some oocytes, regulation via a common repressor becomes a necessity.

The presence of transcription initiation factors for rRNA synthesis is suggested by some preliminary evidence. During *in vitro* transcription of native *Xenopus* rDNA, the *E. coli* core RNA polymerase preferably transcribes the correct strand, and mainly cistronic rather than intercalated DNA is transcribed as *in vivo* (Reeder and Brown, 1970). Since the bacterial σ factor does not affect the specificity of this transcription, it would appear that a corresponding intrinsic (positive) factor is responsible for the specificity. However, the σ factor by itself stimulates rRNA synthesis within the amphibian oocyte (Crippa and Tocchini-Valentini, 1970), again suggesting reinforcement of an intrinsic factor for transcription. The fact that the isolated amphibian nucleolar polymerase does not show any obvious predilection for its own rDNA as opposed to total DNA (Roeder et al., 1970) further suggests a factor. The discovery of a transcriptional repressor will certainly expedite these studies of rRNA synthesis at the cistron level. It may even prove possible to examine regulation of synthesis *in vitro*.

By way of summary, the present incertitude concerning rRNA regulation lies, first, in the functional interrelationship between such elements as a possible initiation factor, the putative repressor, and a special polymerase for rRNA synthesis; and, second, in how the dependence of the synthesis on amino acid, or more generally translation, is in turn related to these elements. For comparison, in the bacterium the indirect but stringent effects of amino acid, or more generally translation, on a putative initiation factor are under study, but no special polymerase or repressor is known. Despite the distinctly metabolic character of the eukaryotic regulation, there is no reason why it should not also be partly integrated into hormone-dependent genetic circuits of the type previously considered for mRNA. If so, contrary to mRNA, it is clear that rRNA regulation remains of a quantitative type.

Pathway Regulation

As may be gathered from inspection of Fig. 18, the pathway of the rRNA molecules offers ample opportunities for physiological regulation. To recapitulate, this formative pathway involves several aspects of a precursor molecule, namely, transcription, proteination, posttranscriptional modification, cleavage, and elimination of some of its portions. All aspects after transcription represent the maturation or processing of the molecule. Association with ribosomal proteins and modification begin on the nascent chain and continue on the completed chains, during which processes the higher order structure of the mature molecules is established. What role, if any, the portions eliminated from the precursor have in these processes is not known. We shall now examine some of the evidence for points of subregulation along the pathway, that is, after the primary switch-on of transcription by withdrawal of the repressor. For convenience, the evidence will be grouped empirically under various headings but in some sort of logical order.

First, regulation of the pathway involves both the rates of transcription and maturation of the precursor. (The former will be further considered quantitatively later.) Frequently, both rates are reduced but that of maturation is reduced the most, attesting to a more relaxed control on the process of transcription than maturation under some circumstances. For example, inhibition of methylation (Vaughan et al., 1967) and of protein synthesis by puromycin or cycloheximide, as well as amino acid deprivation (Maden et al., 1969), somewhat differently reduce the synthesis of precursor, but all block its maturation at the 32 S RNA stage. As mentioned, however, these effects on synthesis can depend on the experimental system being studied. Eventually, complete arrest of maturation cuts off transcription, presumably by some sort of feedback mechanism. Sometimes, the rate of transcription remains normal but the rate of maturation is decreased (Warocquier and Scherrer, 1969) or increased (AB and Malt, 1970). Transcription is stimulated yet maturation is blocked by thioacetamide (Kleinfeld, 1966). Exceptionally, as in old plant root cells, only transcription is inhibited (Rogers et al., 1970). Thus, as the molecular details of the pathway discussed earlier, its regulation varies primarily between organisms, but possibly also between cell types in an organism, depending on their functional fate. In addition, the duration of the pathway in any cell type can be regulated according to its functional fate. The point to note is that within wide physiological limits all possible rates of transcription, maturation, and decay combine to vary the rate of net accumulation of rRNA.

Second, transcription and maturation of rRNA depend on ongoing protein synthesis, indicating that some protein (other than ribosomal protein) is continuously necessary in order for these processes to occur. However, these processes, as well as the maturation of ribosomal subparticles, ultimately depend on the presence of a pool of ribosomal proteins (Craig and Perry, 1970), whose depletion becomes a limiting factor. [These ribosomal proteins comprise (Warner, 1966): 60% of structural proteins whose appearance on the rRNA depends on normal maturation; 20% of exchangeable but firmly attached proteins; and 20% of loosely exchangeable protein.]

Third, in comparison to the bacterium where lack of amino acid affects transcription stringently, in the eukaryote lack of amino acid may also affect transcription rather stringently, but, if anything, affects maturation more stringently, the latter even after the branching of the two subpathways (Maden, 1969). As mentioned, this behavior does not primarily involve a depletion of ribosomal protein. These observations suggest again that the translational control of the pathway may not be entirely by similar mechanisms in the two types of organisms.

Fourth, methylation occurs mostly during transcription of the precursor but does not regulate the process, as shown by the uninterrupted transcription in the absence of methylation. Methylation can even be considerably postponed without seriously affecting maturation; ultimately, however, it is obligatory for maturation and function. (This is also the case in tRNA, where the dependence of functional maturation on modification has been more thoroughly analyzed.) It has been suggested that methylation stabilizes rRNA during maturation and perhaps specifies its late cleavage and/or proteination. Most of the initial methylation is on the ribose moiety, but, in respect of the previous suggestion, it may be significant that most of the residual methylation during maturation of 18 S RNA is in the bases. The dependence of the pathway on the availability of 5 S RNA has not been studied.

Fifth, it has been proposed that the amount of 32 S RNA that is present regulates both the rates of transcription and maturation (Willems et al., 1969b). However, from what we have seen already, it probably regulates transcription less stringently. It appears as if the 32 S RNA stage also acts as a physiological regulatory gate which may be used by the resting cell, probably in relation to the occurrence of a 28 S RNA pool (Papaconstantinou and Julku, 1968). This 32 S RNA stage may in fact act as a constant level gate; with cycloheximide and puromycin, for example, a reduced rate of transcription is compensated by a longer half-life of the precursor. However, it remains to be found what, in turn, controls this gate and using what mechanism. In contrast to 32 S RNA, the level of the 18 S RNA stage in the other half of the pathway is even difficult to maintain; in the resting cell (Cooper, 1970) or in the presence of the previous drugs this RNA is preferentially degraded. This condition is directly shown by the occurrence of a physiological pool of 28 S RNA, since this can only be maintained if there is a corresponding deficit in the other half of the pathway.

Lastly, there is the claim that in insects (Ringborg et al., 1970) only the subterminal stages of maturation (equivalent to 20 and 32 S RNAs in the mammal) may occur in the nucleolus, the final stages occurring in the chromosome. In the lower eukaryote it is claimed that part of the ribosomal subunit maturation may occur in the cytoplasm (Warner, 1971). If these two claims are authentic, which remains to be seen, they could open new vistas into the regulation of the pathway. It was mentioned that many aspects of this regulation may be profitably studied in the future under nucleus-free conditions.

In conclusion, the main regulation of the ribosomal pathway is at the levels of the 45 S precursor RNA and 32 S RNA stages. Both stages depend on the continuous synthesis of nonribosomal protein and the presence of a pool of ribosomal protein, while the 32 S RNA stage is also dependent on posttranscriptional modification. Any regulatory signal such as hormone may thus act at the level of one or the other stage, although evidently by affecting different processes at each stage.

Dosage Relationships in Regulation

A wealth of genetic data give us some insight into the administration of the multiple rRNA cistrons for the purpose of synthesis in the living cell. It was known that maize microspores (McClintock, 1934), chironomid hybrids (Beermann, 1960), a homozygous Mexican axolotl mutant (Humphrey, 1961), and the heterozygous *Xenopus* mutant (Brown, 1967) were viable even though they all possessed less than the normal amount of nucleolar material, as the result of deletions affecting the nucleolus organizer. Later, the *Xenopus* mutant was shown to contain half the number of rRNA cistrons (Birnstiel *et al.*, 1969), but the normal amount of rRNA per cell. The same is probably true of the axolotl mutant (Miller and Brown, 1969). A more revealing analysis of these aspects carried out in *Drosophila* will now be described.

Dosage effects are found in the bobbed locus of *Drosophila* that corresponds with the region of the nucleolus organizer in the X and Y sex chromosomes. Isoalleles of different phenotypic intensity are formed by unequal crossover in this region. It was established that these hypomorphic alleles, resulting in an underdeveloped imago, are represented at the molecular level by a proportional deletion of rRNA cistrons. This confirmed that there is no necessary correlation between the presence of a nucleolus organizer and the amount of rDNA it contains, that is, for various reasons the organizer does not always contain the same number of rRNA cistrons. The mutant bobbed phenotype is manifest within a range of rDNA from 0.04 to 0.135% of total DNA (wild-type rDNA: 0.27% of DNA by hybridization), being lethal below but normal above this range (Quagliarotti and Ritossa, 1968). According to Mohan and Ritossa (1970), in the inviable range of rDNA an insufficient amount of rRNA is produced. Within the range of rDNA which results in a viable bobbed phenotype, a normal amount of rRNA is made but at a lower rate and therefore over a longer developmental time. Exceptions are some organs which remain deficient in rRNA and these are the organs responsible for the bobbed phenotype. Thus a dosage regulation toward the normal amount of rRNA takes place above onehalf of the normal diploid amount of rDNA. A negative feedback on rRNA transcription must then operate formally beginning at the XO level, although in actuality in the normal animal both sex chromosomes contribute to such level. This formalism is also indicated by the total amount of rRNA (Schultz and Travaglini, 1965) and total RNA/DNA ratio (Kiefer, 1968) being normal in the karyotypes XX, XY, XO, and XXY, which all contain at least one of the two complements of ribosomal cistrons in one complete segment. However, genotype bb/bb, which contains incomplete cistronic complements in either of two segments, is not dosage regulated and is rRNA deficient. Considering that the bb/bb genotype can at times contain as much total rDNA as XO (Ritossa and Scala, 1969), an effect from the arrangement of rDNA on transcription is suggested by these data. In addition, deletions in the Y chromosome can invalidate a functional amount of rDNA, suggesting, as in the case of bb/bb, the importance of the normal arrangement (Ritossa, 1968). What particular genetic aspects are deranged either in bb/bb or inactive Y remains unclear. In conclusion, the general rule in animals is that a correctly arranged half-amount of rDNA is the minimum necessary for a normal rRNA synthesis. As shown by the potential for rRNA synthesis in the *Drosophila* Y chromosome, this rule cannot be explained by a mechanism for dose compensation in the hemizygotic sex to safeguard against the eventuality of sex-linked rDNA. It follows that the rule has a subtler physiological meaning, and what this may be we shall return to later. Meantime, other aspects of dose regulation require our attention.

The rRNA locus constitutes an outstanding example of adaptive equilibria at the molecular level. Bobbed stocks revert to wild-type slowly but readily. Initially, this was supposed to be by unequal somatic crossover within the rDNA locus, followed by upward selection toward the normal rDNA value (Ritossa et al., 1966c). Within a given stock both sexes reach the same equilibrium rDNA value, except that if the males in a particular stock attain a lesser amount of rDNA, the females compensate and restore the balance. It is of interest that the increase of rDNA occurs also when the flies carry one rather than two sex chromosomes with a normal rDNA value (Tartof, 1971); since this increase is achieved within a single generation, it obviously cannot be due to meiotic recombination. In the opposite direction, stocks with excessive rDNA can be genetically constructed (the flies have a normal phenotype and amount of rRNA) which revert downward to the normal rDNA value. Later, an alternative proposed for the bobbed revertants was that the rDNA is suddenly amplified somatically or, as the process was termed, magnified toward the normal rDNA value. The amount of magnified rDNA depends upon the particular stock. Although transmitted as usual through the chromosome, magnified rDNA remains loosely integrated in the chromosome (Ritossa and Scala, 1969); when confronted with the wild-type locus (in a male) the magnified rDNA, if of recent origin, once more reverts to the bobbed value. In accordance with this, it was then found that magnified rDNA is not inherited in an additive manner. Because of its loose integration, magnified rDNA probably represents a lateral regional amplification of the chromosome. The phenomenon of magnification does not therefore rectify the missing rDNA permanently, yet the partial solution is better than none, as is clearly shown by its wide adaptive exploitation. Prior to its discovery, the only authenticated somatically amplified DNA was that found in DNA puffs of certain polytene chromosomes (Chapter 3). However, puff DNA is not rDNA, and, because it is fully integrated in the chromosome, it represents a linear amplification. In conclusion, the adaptive equilibration via magnification toward the normal amount of rDNA is indicative of a cell regulated replication of rDNA. This somatic regulation of rDNA resembles the germinal regulation of rDNA amplification in the oocyte, another lateral and transient amplification. The regulatory mechanism underlying these lateral amplifications is not known.

The rDNA polymorphism just described in *Drosophila* is also found in natural amphibian populations (Miller and Brown, 1969). Several populations of *Bufo* differ in the number and size of nucleoli and correspondingly in the amount of rDNA, which in some individuals can surpass the amount in the wild-type. Since the total amount of rRNA present is apparently normal, the overall similarity with *Drosophila* is great. The possibility that such polymorphism occurs in mammals including man should therefore not be overlooked. If, as is probable, the dosage relationship rule remains as that found in the insect and amphibian, it is predictable that dwarfism in man may be caused not only by the well-documented hormonal deficiencies but by rDNA deficiency as well. On the basis of animal findings, the pathological condition in man should result from a massive deficiency of rRNA cistrons. However, even if the prediction is correct in principle, a natural defense, as it were, is that a viable individual with only one-half of the cistrons is less likely to be born in man with up to possibly five haploid nucleolus organizers than in the insect or amphibian with just one. A pilot-scale search for the human deficiency was unsuccessful (Bergman *et al.*, 1972).

Plants like animals seem able to dose compensate for a diminished amount of rDNA (McClintock, 1934). Plants are exceptional in that they appear to respond positively in their rRNA synthesis to the presence of additional rDNA. After such addition, wheat produces more nucleoli and total rRNA per cell (Longwell and Svihla, 1960). Each addition of a half nucleolus organizer in maize increases the nucleolar RNA (mainly rRNA) by 10% (Lin, 1955). This response of the bisexual plant differs from that of *Drosophila* and *Bufo* which do not respond to extra rDNA. The reason for the difference remains unexplained.

In contrast to somatic cells, there is a germinal overproduction of rRNA, which is best exemplified in the amphibian oocyte. This overproduction is based on the overly amplified oocyte rDNA, clearly intended to generate an amount of rRNA that a somatic rDNA complement could match given only an interminable time (Perkowska *et al.*, 1968). Similar to somatic rDNA magnification, the amplification of rDNA is sensitively controlled in the oocyte; the same total amplification is achieved whether the female carries a normal or half-complement of somatic rDNA. However, for all we know, half of the germinal rDNA may suffice for RNA synthesis in the normal female, and there is no compelling reason to believe that the dosage relationship rule for rRNA synthesis applicable to somatic cells does not apply in the oocyte as well. Looking at the question from another angle, Brown (1967) has argued that extrachromosomal ("episomal") rDNA permits the oocyte to escape rRNA synthesis inhibition by the excess of accumulated ribosomes (60%), in whose presence any somatic rRNA synthesis would be completely inhibited. In contrast to the maintenance function of this synthesis in somatic cells, he views rRNA synthesis in the oocyte as a differentiated function disproportionate to the actual requirements for general (nonribosomal) protein synthesis in the oocyte itself but related instead to the need for storing ribosomes for the future embryo. In agreement with serving a storage purpose, oocyte rRNA is, in effect, stable. These early views might have suggested that the regulation of oocyte rRNA was not of the usual somatic kind. At present, the discovery of a putative repressor of the synthesis obviates any such need to consider a different kind, hence basic mechanism, of regulation in oocyte and somatic cells. As expected from this new viewpoint, the repressor activity is absent in the oocyte most active in rRNA synthesis. It follows that the phenomena of oocyte rDNA amplification and excessive accumulation of ribosomes, instead of rRNA regulation, are what now require explanation.

Quantitative Aspects of Transcription

The actual rate of rRNA transcription depends upon both the rates of initiation and chain propagation. In turn, the overall or integral rate of transcription representing the total rRNA output depends upon both these rates as well as the total number of cistrons being transcribed. These quantitative aspects of transcription will now be considered. The subsequent highly resilient regulation of the pathway, the processing of rRNA, was considered qualitatively in a previous section. Both the rates of processing and decay are as important as the preceding output rate in determining the final rRNA accumulation, but these two quantitative aspects cannot be considered in this section.

Initiation of transcription is controlled by a master on-or-off switch which presumably is the putative repressor acting on an operator-like gene. This mechanical analogy implies that the repressor controls initiation on an all-or-none basis, from which it follows that the fine regulation of transcription must depend also upon other aspects. What regulates the rate or frequency of initiation after the repressor is removed from the operator? The answer is not known, but it could be the inherent binding properties of the promoter-rRNA polymerase complex, perhaps affected by any possible initiation factor. The rate of initiation determines the rate at which new rounds of transcription will occur. Formally, provided this rate is uniform in the cell, it is immaterial whether there is one promoter site per cistronic unit, or, as the high actinomycin sensitivity (Perry and Kelley, 1970) seems to suggest, one promoter site serving a block of cistrons. The latter arrangement is supported by the uninterrupted disposition of polymerase molecules along rDNA, including the nontranscribed intercalated DNA, observed under the electron microscope (Fig. 17). In these very active cistrons of the amphibian oocyte a new polymerase molecule attaches to the rDNA after each period during which the preceding polymerase has traveled a length equivalent to but a few times its own diameter, or about 100 nucleotides, along the rDNA. According to data in HeLa cells, this regular periodicity of binding should result in a new initiation of transcription every second or so. As a consequence of this periodicity, all active rRNA cistrons in the oocyte are simultaneously transcribed by about 100 polymerase molecules each. Other active cell types most probably require a similar simultaneous transcription of their cistrons in order to maintain the full complement of rRNA. However, the rate of initiation and the resulting number of simultaneous transcriptions of each cistron may well vary with the cell type and the cell status.

A second conclusion from the electron microscope studies is that the same number of polymerase molecules are associated with the cistrons that are not transcribed as with those that are. Based on this conclusion and the actinomycin data mentioned above, Perry and Kelley (1970) proposed the following model for rRNA transcription. The polymerase binds to the template at a promoter site (one site per block of cistrons) and runs unproductively along rDNA until it reaches an operator site (presumably one site per cistron) where absence or presence of repressor specifies whether or not, respectively, the cistron is transcribed. (This model may be applicable to stable RNAs in general, but not, for example, to DNA-like RNA whose low actinomycin sensitivity, on a per cistron basis suggests a single promoter per cistron. In particular contrast with rRNA, this different promoter condition is compatible with the non-modular regulation of DNA-like RNA.)

Compared with the rate of initiation, although involved in the assembly of the total rRNA chain, the rate of propagation in an active cell is likely to play a secondary role in the regulation of transcription. The rate of chain propogation of the 45 S RNA precursor in the HeLa cell is about 80 nucleotides per second (12,000 nucleotides in about 2.5 minutes). The rate of propagation in this actively dividing cell (generation time: 24 hours) is of an order similar to that in bacterial rRNA. As seen above, in the amphibian oocyte all active cistrons appear to be transcribed at approximately the same rate, that is, the rates of initiation and propagation are coordinated. This probably means that at this high rate of transcription both rates are maximized and the rate of initiation approaches that of propagation. (It is clear that initiation cannot be more frequent than the period it takes the polymerase to clear a few of its own diameter's worth of template.) The rate of propagation assumes a more important role in cells with less overall transcription, for example, in the contact-inhibited fibroblast whose transcription is reduced two- to fourfold (Emerson, 1971). In this cell the reduction appears to be principally in the rate of propagation, although the rate of initiation may also decrease. This reduction therefore implicates the mechanism of rRNA polymerase as a further regulatory possibility under certain cell circumstances. An additional possibility is that what changes is not only the previous rates but also the number of cistrons being transcribed. Perhaps the real situation in the less active cell is somewhere between these two possibilities.

The question that remains now to consider is the number of cistrons that are transcribed at different functional steady states of a cell. One would suspect that this aspect of dosage regulation depends on an equilibrium between the binding constant of repressor and its level at any time in the cell. On present evidence, the cistronic multiplicity of rRNA is the same in all somatic cells of an organism. As seen in the previous section, the overall rate of transcription in these cells depends on the amount of rDNA present in the cell only when this amount is less than half the diploid complement, but becomes independent of the amount of rDNA from the point when this reaches the half-mark to beyond the diploid amount. This is the dosage relationship rule for rRNA which says that, given the proper arrangement, half the number of cistrons suffice for a normal overall rate of transcription. The rule also says that the actual rate of transcription can be stretched by no more than a factor of two, although this does not necessarily mean that the normal animal stretches it that far and uses only half the number of cistrons; it happens, however, in certain cells that some of their several nucleolus organizers are seldom used (discussed in Chapter 6). Moreover, the selective equilibrium pressure is to maintain the diploid amount of rDNA in the cell. Since it does not seem reasonable to assume that double the amount of sufficient rDNA would be necessary just to balance mechanical or mutational loss of cistrons, the observed equilibrium may have an adaptive or physiological value and quite possibly both. A suggestion of adaptive value is the average diploid equilibrium attained in *Drosophila* populations where one sex remains below the diploid value and the other sex above it. Of a physiological value is the behavior of the amphibian oocyte where there is normal amplification of rDNA irrespective of the somatic amount of rDNA that the female carries.

The considerations above preclude a firm distinction among several alternatives concerning the number of cistrons used in the normal animal. These are: (a) a finite, discrete fraction of rDNA transcribed at a fixed rate; (b) a constant size but physically changing fraction of rDNA transcribed at a fixed rate; and (c) a variable size fraction transcribed at a compensated rate. The first two strategies could be managed entirely, for example, by controlling the level of repressor. The third would require additional control of initiation, presumably by initiation factors rather than number of polymerase molecules. A changing but constant fraction of transcribed rDNA (b) is the simplest explanation that fits the data but, at present, lacks an obvious biological rationale. A variable compensation (c) can be meaningful under certain physiological circumstances, particularly since the polymerase mechanism intervenes in the regulation, and is amenable to testing. Of interest in this regard, bobbed imagos in *Drosophila* attain a normal amount of rRNA after a protracted developmental period. To a different degree, the two more dynamic strategies (b and c) are equivalent to a qualitative subregulation of transcription within the cell. Whatever the strategy, it is probable, however, that the quantitative details will vary with the species and its own peculiar dose regulation. The differences in the known rDNA complements suggest that the dosage is variable. Indeed, because of this variation in the regulation or complement or both, it is impossible at present to estimate the modal cistronic requirement in any eukaryote.

This discussion reveals a substantial reservoir of rRNA cistrons over the maximum required under the greatest demand for synthesis. The basic dosage rule establishes that half the cistrons are expendable without any phenotypic effects on the animal, which, incidentally, is in contrast to the bacterium whose cistrons are fully engaged at rapid growth. According to different calculations, from about one-tenth (Attardi and Amaldi, 1970) to one-half (Birnstiel et al., 1971) of the cistrons present in the HeLa cell would suffice for rRNA production if they were used at full capacity as they are in the amphibian oocyte. Direct observation in the oocyte shows that a substantial proportion of cistrons are not utilized in this cell whose cistrons are expressly amplified for the purpose of accumulating ribosomes. In the metabolically active liver cell the cistrons are used at but a small fraction of their capacity (Quincey and Wilson, 1969). This surplus of eukaryotic cistrons remains an unexplained puzzle. If, as one could argue, it were related to the flexible characteristics of the regulation, then the dosage rule obscures what such relationship might be.

In conclusion, considering the regulation of the rRNA pathway as a whole, from cistron to transcription to maturation, there is clearly a great wealth of regulatory possibilities, but a relative dearth of knowledge concerning the extent to which they are implemented in the organism.

Regulation of Synthesis in the Prokaryote

Although on present evidence the two major rRNA species of the prokaryote may or may not have a common precursor, their synthesis is coordinated. It would seem logical to think that the clustering of cistrons is at least implicated in this coordination, for such clustering is one obvious parallel with the corresponding eukaryotic system. Moreover, this clustering in *B. subtilis* includes the cistrons for 16 and 23 S RNAs, 5 S RNA, and tRNA, as well as the cistrons for ribosomal proteins near the chromosome origin. This ensemble of genes behaves conservatively during prokaryotic evolution suggesting a stable genomic disposition for the control, as a whole, of the translational apparatus.

The amount of information on the pathway of rRNA synthesis is rather limited compared with the eukaryote. Nevertheless, by analogy with the eukaryote, it is possible to visualize how some of the steps in the pathway may have a potential regulatory value. Soon after transcription both RNA species are proteinated and methylated and their maturation begins. It was suggested that some ribosomal proteins could themselves act as methylases during assembly of the ribosomal subparticles (Dubin and Günalp, 1967). Little is known of the relative timing or regulatory significance for maturation of the proteination and methylation, except that, if there is no common precursor, these processes must be directed to each rRNA species individually. Ribosomal RNA accumulates in the presence of chloramphenicol or when RC relaxed mutants (see below) are starved of amino acid. Since protein synthesis ceases under these conditions, and the pool of ribosomal protein is not large, it would appear that the total amount of ribosomal protein present is not very critical in order for transcription to occur; perhaps the presence only of core proteins, the first to combine with rRNA, is what is critical (Chapter 5). As in the eukaryote, however, methylation does not regulate transcription but instead stabilizes the maturing product, as evinced by the delay of part of the methylation. Incorporation of 5 S RNA into the large ribosomal subparticle occurs at the 40 S stage and therefore cannot be involved in the initial regulation. Lastly, rRNA synthesis and the immediately following subunit assembly may take place at the cell membrane (Haywood, 1971), perhaps providing some physical support to the assembly. The maturation of rRNA and ribosome has been reviewed by Starr and Sells (1969).

During the cell cycle two main bursts of rRNA synthesis occur corresponding to the replication of the two cistron clusters in the *E. coli* chromosome (Cutler and Evans, 1967). However, although it has been argued that the early replicating cluster might benefit from a dosage effect (Smith *et al.,* 1968), particularly in view of the multiple replication forks at rapid cell growth, no such persisting effect is found. The main emphasis is therefore the continuity of rRNA synthesis during the cell cycle rather than a replication-dosage dependence. We shall see that there is a nonchromosomal control of the synthesis which would tend to counteract any dosage effect.

The rRNA cistrons are simultaneously transcribed by many polymerase molecules (Winslow and Lazzarini, 1969a). In fact, at rapid cell growth the rDNA is solidly loaded with enzyme (Manor *et al.,* 1969) and thus presumably transcribed at maximum rate, which means that no cistronic DNA is expendable. A combined total of about 70 enzyme molecules in each pair of cistrons, bearing nascent 16 and 23 S RNA chains, are visualized using the electron microscope (Miller *et al.,* 1970a). This maximum utilization of cistrons during rapid growth obviates any need for a kinetic model to explain their regulation, contrasted with the eukaryote where a model is required because of their sporadic utilization. A model for the prokaryotic cistrons utilized submaximally during slow growth is still outstanding (see, however, Maaløe and Kjeldgaard, 1966). From what follows it appears probable that the principal rate-determining step is initiation, as it is in the eukaryote. Regardless of this, the model would differ from that in the eukaryote, since, whether or not each cistron tandem is transcribed as a unit, the tandems are separate from one another; the regulation aims therefore either at individual cistrons or individual tandems. In *E. coli* at 30°C, reinitiation takes place at each putative promoter site every 1 to 2 seconds. Because of this frequency of initiation the relative yield of the minute portion of the genome represented by rDNA is very high, indeed almost as high as the remainder of the genome. In contrast, the rate of rRNA elongation is not vastly different from that of mRNA (about 55 nucleotides per second). Compared with the fully derepressed tryptophan operon, initiation of rRNA is 100 times more frequent. Thus, within a factor of two, there are as many polymerase molecules working on rDNA as on the total informational genome. During each cell generation these enzymes produce a complement of about 2×10^4 molecules of each rRNA species per genome, or the equivalent of five to ten ribosomes per second. (The total output of mRNA molecules per cell generation is somewhat more, but, because of instability, they do not accumulate as does rRNA.) By comparison, this output rate is about one order of magnitude less than in a mammalian cell. Since the prokaryotic ribosomes seem to be more efficiently utilized, the lower output rate may bear a direct relation to the fact mentioned earlier that the relative duration of the pathway is four times longer. Surprisingly, in view of their very different size, 16 and 23 S RNAs are reported to take the same time to complete, about 2 minutes at 37°C (review by Geiduschek and Haselkorn, 1969). This similar duration of transcription might indicate some sort of active constraint on transcription of the molecules in the interest of preserving their necessary stoichiometry. However, the similar duration has also been taken to indicate that 23 S RNA may be assembled starting from two half-precursors (Attardi and Amaldi, 1970). This would be in line with structural evidence compatible with the RNA being made of two identical halves (Chapter 2) but finds no support in the electron microscopic observations referred to above. The occurrence of assembly by halves in 23 S RNA is therefore not established.

In the bacterial cell the ratio of rRNA to DNA is proportional to the rate of growth (Maaløe and Kjeldgaard, 1966). This cell stores relatively fewer ribosomes than the eukaryotic cell. This means that rRNA synthesis is a function of the rate of protein synthesis (Osawa, 1968), and its rate must increase roughly in proportion to mRNA synthesis. The matched synthetic behavior of these functionally codependent RNAs is clearly to produce a balanced translational apparatus in response to demands for protein synthesis. Regulation of rRNA must therefore be as dynamic as that of mRNA. It follows also that the regulation is not geared to the cell cycle of growth as much as it is to the instantateous demands for protein synthesis. These general points suggest a narrow coordination between rRNA and protein syntheses, which will be presently examined.

Before proceeding, a word about the mechanism of rRNA regulation is necessary. Briefly, a positive initiation factor (ψ) has been reported which apparently allows the core RNA polymerase (with its own σ factor) to attach to a putative promoter gene in front (3' side) of the structural rRNA cistrons in order to transcribe them (Travers et al., 1970). At present, these appear to be all the basic elements by which regulation is operated. Based on the general evidence, no special polymerase and no specific (negative) repressor for these cistrons has been identified. For comparison, in the eukaryote both these two regulatory elements, but not unquestionably an initiation factor, have been recognized. The discussion that follows aims at clarifying how the interaction between these basic elements may be controlled.

To examine the coordination between rRNA and protein syntheses we must begin by considering that, despite the general points made above, in actual fact rRNA synthesis turns out not to be tightly coordinated with that of mRNA (Edlin and Broda, 1968). First, as far as is known, the unspecific regulation of mRNA such as catabolite repression does not affect rRNA. Second, the two RNA species have different relative rates of synthesis during shift-up or -down conditions, to wit, in the first condition mRNA is derepressed whereas rRNA is repressed initially. Third, it is known that rRNA but not mRNA synthesis is controlled by the RC locus. This locus provides the clearest insight into rRNA regulation and will now be discussed in some detail. In stringent (wild-type RC), but not in relaxed (mutant) cells, withdrawal of amino acid results in severe rRNA synthesis inhibition. Protein synthesis is inhibited in both types of cell, so that protein synthesis per se is not the operative factor. Transfer RNA remains partly acylated during amino acid starvation. Upon restoration, amino acid must first be acylated to tRNA for rRNA synthesis to resume. This need led to the early belief that the control of rRNA synthesis was mediated by tRNA acting directly at the level of transcription, which now appears unlikely. Although not unquestioned (Edlin and Stent, 1969), evidence was also presented for a limiting phosphorylation of nucleosides to produce the immediate RNA precursors under stringent conditions (Cashel and Gallant, 1969). This substrate limitation was interpreted to explain rRNA synthesis inhibition, though obviously not to explain the concurrent mRNA synthesis. The inhibitor of phosphorylation was postulated to be a guanosine tetraphosphate derivative, GT_4X, that reaches levels nearly equimolar with GTP; it results from a reaction that idles during amino acid starvation of stringent cells. This idling reaction was proposed to be a step in translation during which GTP is consumed, possibly translocation. The sequence envisaged was a decrease in the level of acylated tRNA during starvation which would first make translocation idle; the RC gene would then permit the formation of GT_4X from the unconsumed GTP. In addition, the formation of GT_4X depends on nascent RNA (Wong and Nazar, 1970), whose significance in the present context is not understood but could be an indirect one.

The previous tentative scheme attempts to relate rRNA regulation through the RC gene to events in the polysome which, for some time, have been suspected of being their final meeting ground. For example, the previously mentioned coordination of total rRNA synthesis with the rate of protein synthesis, in the lack of accumulation of free ribosomes, suggested a regulation at the polysome level. In accordance with this scheme, it is found that polysome integrity is preserved during amino acid deprivation in some (but not all) relaxed strains, all of which maintain stable RNA synthesis under this condition. On the other hand, mRNA synthesis largely escapes RC control in stringent strains, and it is not clear why it should if, according to the aforementioned proposal, there is a common substrate limitation. Thus it looks more likely that, instead of limiting substrate, GT_4X acts directly on stable rRNA transcription.

The action of GT_4X on stable RNA transcription was explained as follows. A hypothesis to unify all previously established facts is that in order for this transcription to occur there must be an uninterrupted run-through of ribosomes along mRNA irrespective of whether or not actual protein synthesis takes place. All elements required for protein synthesis such as mRNA and aminoacyl-tRNA must engage, otherwise the ribosome idles, GTP is not consumed, and GT_4X formation ensues. These conditions are apparently met in relaxed strains deprived of amino acid or under chloramphenicol treatment, for in both cases stable RNA is produced but little protein is. Under these two conditions tRNA remains partly acylated and it has been shown that the ribosome continues to run but less or not productively, respectively, along mRNA (Curgo et al., 1969). A prediction was therefore that the ribosome should idle more in deprived stringent than deprived relaxed strains. However, contradicting this prediction, it was recently found (Brunschede and Bremer, 1971) that the stringent ribosome indeed idles under amino acid deprivation but no more than the relaxed ribosome. Hence the hypothesis at present must read that a function of the wild-type RC locus is to control some essential aspect of the cell, or the ribosome in particular, which makes the production of GT_4X sensitive to the level of protein synthesis. Contrary to the earlier explanation based on the run-through of ribosomes, what this essential aspect is now remains unexplained. Specifically with respect to rRNA, the target of GT_4X is proposed to be the ψ initiation factor (Travers et al., 1970) mentioned earlier. According to this proposal, GT_4X acts as a ligand to inhibit the binding of ψ to promoter, and thus prevents the polymerase from transcribing the structural rRNA cistrons. In summary, the present model is that in the wild-type a decreasing supply of amino acid slows down protein synthesis, which in turn is accompanied by an increased level of GT_4X. This results in the lowering and eventually cessation of the binding of ψ, with a corresponding decrease and cessation of rRNA synthesis. Conversely, reversing the process, any physiological stimulation of protein synthesis (increased amino acid supply) immediately brings about stimulation of rRNA synthesis. The model therefore accounts for the basic coordination between these two syntheses. It places amino acid in a central controlling role with respect to both rRNA and protein syntheses, although clearly the mechanism of control is quite different for each synthesis. However attractive the causality in this model seems to be, it still remains to be proved. Pending this, a more cautious and general view is considered below.

The foregoing discussion did not mean to imply that there is universal agreement concerning the mechanism by which the RC locus controls rRNA, or that only rRNA is controlled by this locus. First, because all RNAs are restricted under stringency, although to a differing degree, the lack of coordination between mRNA and stable RNA syntheses is not straightforward. During amino acid starvation both the average rates of total synthesis and chain elongation of RNA are reduced by about ninefold, this magnitude indicating that both mRNA and stable RNA are involved (Winslow and Lazzarini, 1969b). During uracil deprivation of an auxotrophic cell, synthesis of the two types of RNA continues but decreases approximately coordinately (the relative rates of synthesis were not measured in these experiments). Of interest is the possible indication that stable RNA does not accumulate, apparently not because it is not made, but because it would be simultaneously made and degraded. The response to deprivation could therefore be one of destabilization rather than lack of synthesis (Lazzarini et al., 1969). However, it is now clear that under conditions of stringency the cessation of stable RNA synthesis (and reduction of mRNA synthesis) is the main effect. Second, and more suggestive, it is beginning to look as if the RC locus controls many diverse metabolites with inevitable widespread effects on metabolism in general. Irr and Gallant (1969) list the following as being somehow affected in level by this locus: energy metabolism, ATP, all four ribonucleoside triphosphates, lipids, and, of course, GT_4X. Ribosomal protein (as described below) and certain "protected" tRNAs (Chapter 5) are also affected, and so is RNA phage replication (Friesen, 1969). It may therefore be that under conditions of stringency different metabolic aspects end up by having specific effects on stable RNA. Since for rRNA the number of these effectors is probably small, the interest is in which particular metabolic pathway is the one most likely to be involved. Essentially, what was said above is that this pathway concerns GT_4X.

As far as is known, stable RNA synthesis is identical in relaxed and stringent strains under all normal conditions of growth (Neidhardt, 1963). In other words, the RC locus has no apparent effect when amino acid is available. This does not necessarily mean that the control of rRNA synthesis during the cell cycle of growth is independent of the locus, but it would mean that only the "top of the iceberg" of the RC function is what shows during stringency. At present, notwithstanding the note of caution in the previous paragraph, regulation of rRNA synthesis mediated via the ψ factor-GT_4X model is a most attractive possibility, not the least because experiments can be designed to test it. For example, the model explains well the constitutive homeostasis of synthesis in the wild-type in terms of the level of GT_4X being inversely related to the rate of protein synthesis during the cell cycle; in the extreme case, absence of amino acid, this rate becomes nil and rRNA synthesis ceases. Conversely, the model explains the basic coordination between mRNA, protein, and rRNA syntheses under different metabolic conditions. Some of the kinetic data and the multiple effects of the RC locus described above would have to be explained, but, as they stand, they do not interfere with the proposed regulation according to the ψ factor-GT_4X model. This model, nevertheless, has to be substantiated.

An interesting derivation of this model is that, according to Travers et al. (1970), the ψ factor may correspond with the host moiety of the $Q\beta$ replicase described in Chapter 1. The implication is that, by competing for ψ in order to assemble its replicase, the virus also manages to shut off host rRNA synthesis.

Attesting to its ubiquitous effects, the RC locus also controls the synthesis of ribosomal protein during amino acid starvation (Goodman et al., 1969). However, perhaps because of a pre-existent pool or other reasons given before, during starvation the decreased protein is not initially limiting for rRNA synthesis. This as well as other evidence makes ribosomal protein an unlikely mediator for the control of rRNA synthesis by the locus. Yet, as pointed out above, an effect of the wild-type RC locus on ribosomal function could still be at the root of rRNA regulation.

It remains necessary now to consider the opposite possibility to that in the previous paragraph, namely, that rRNA mediates informationally the synthesis of ribosomal protein. Ribosomal proteins are made on relatively stable templates. The pool of these proteins is relatively small, 5% of the total cell protein (Sells and Davis, 1970). As expected, their synthesis appears to be tightly coupled with their assembly into the ribosomal subunits (Beatty and Wong, 1970). A proposal was put forward that nascent rRNA functions as mRNA for ribosomal protein on the basis that, using nascent rRNA as template in a cell-free system, the product resembles ribosomal protein by electrophoretic analysis (Muto, 1968). Nascent 16 S RNA is twice as active as template in this system as nascent 23 S RNA, and six times more active than both mature species. Compatible with a mRNA function, nascent 16 S RNA is the least methylated of all these RNAs. Other evidence suggests that nascent rRNA has a secondary structure compatible with this function in the immature subribosomal particle. Nevertheless, the mRNA role requires substantiation, a difficulty being whether the ribosomes in the cell-free system are really free of endogenous mRNA. The central difficulty is, however, in the living cell. On present evidence, nascent rRNA is not associated with the polysome, but instead is bound to ribosomal protein, under which circumstances it is not easy to visualize a mRNA role. If the role, after all, were to be substantiated, it should be noted that the proximity of both rRNA and ribosomal protein cistrons could mean their partial equivalence. However, this could not apply to all ribosomal proteins since several of their cistrons are not linked to the rRNA cistrons (Smith et al., 1971). How many proteins of the small ribosomal subunit can be coded for in 16 S RNA? Assuming a single RNA species, at most about one-sixth of the proteins can be coded for. A template-product relationship as in this proposal would then mean a single control for the synthesis of rRNA and at most a few ribosomal proteins. Granted that any protein can be vital to the cell, the fact remains that there is a marked molar stoichiometry between the production of rRNA and that of most, not just a few, ribosomal proteins. For instance, both rRNA and proteins increase coordinately during a shift-up condition (Harvey, 1970). Clearly, such overall stoichiometry cannot derive from rRNA serving as a partial template but from a general coregulation whose nature remains to be established. What is known is that the primary coordination occurs at the level of rRNA synthesis and synthesis (rather than translation) of ribosomal protein mRNAs (Harney and Nakada, 1970); it is conceivable that several of these mRNAs derive from a common operon. As with other bacterial protein, it is the increased amount of these RNAs, i.e., the number of polysomes at work, which mainly controls production of ribosomal protein.

DNA, Ribosomal RNA, and Informational Aspects in Mitochondria and Chloroplasts

These cell organelles possess characteristics that almost certainly indicate their derivation by symbiotic capture of a prokaryote-like ancestor (review by S. Nass, 1969).[13] Not coincidentally, the two organelles are concerned with energy production. Since the bacterial enzymes for energy production are located at or near the cell membrane, a possible inference in the case of the eukaryotic cell is that the presence of organelles with internal membrane systems compensates for the lower cell surface to volume ratio in respect to energy production. However, that these organelles have acquired or evolved other functions not concerned with energy production also remains a possibility.

Mitochondria and chloroplasts contain DNA. Animal mitochondria contain up to six circular DNA molecules each about 10^7 daltons (5 μ circumference), which on physicochemical parameters appear to be all identical. The mitochondrial genome in somatic cells is therefore small compared with about 10^{12} daltons of DNA in the haploid nuclear genome; all mouse fibroblast (L cell) mitochondrial DNA is 0.15% of the nuclear DNA. In oocytes with large amounts of cytoplasm, however, the amount of mitochondrial DNA can by far surpass that of nuclear DNA. The kinetic complexity of animal mitochondrial DNA indicates that for the most part it represents a unique sequence with a complexity of about 10^7 daltons (review by Rabinowitz and Swift, 1970), compared with a complexity of 2×10^9 daltons in the unique bacterial genome. Uniqueness of sequence is also true of mitochondrial DNA in higher plants (Wells and Birnstiel, 1969). In lower eukaryotes such as *Neurospora* and yeast the mitochondrial DNA is about 5×10^7 daltons (25 μ) and for the most part probably represents a unique sequence, also less complex than the bacterial; this DNA is not always circular. Chloroplastal DNA in higher plants is not circular. Each of the many chloroplasts in the plant cell contains 2×10^9 daltons of DNA, the equivalent of a bacterial genome, with fractions of kinetic complexity equal to 3×10^6 (76%) and 1.2×10^8 daltons (24%). Because of the discrep-

---

[13] Several lines of evidence behind this view are in regard to certain properties of mitochondria: (a) the frequent circularity of the DNA; (b) the nakedness of the DNA that, in particular, lacks histone (Tewari and Wildman, 1969); (c) the bacterial-like size of the ribosome and rRNA; (d) the presence of bacterial-like initiator tRNA; (e) the closer functional fit of tRNA and synthetase with the bacterial than with the cytoplasmic counterparts; (f) the exchangeability of translation factors exclusively with bacteria (Grandi and Küntzel, 1970); (g) the exclusive sensitivity in the cell to antibiotics which affect only bacterial constituents such as ribosomal protein (chloramphenicol) and RNA polymerase (rifampicin). Taken together it is plainly impossible that these similarities with bacteria are coincidental. As far as one can tell, and implicit in points (d) and (e), the genetic code in use by the organelle is the universal one.

ancy between the analytical and kinetic complexity, it was concluded that chloroplastal DNA is extensively reiterated in sequence (Wells and Birnstiel, 1969). For example, the 1.2×10^8 daltons sequence (whose informational content is larger than in mitochondrial DNA) must be copied four times in the genome. Extensive reiteration has later also been found in green algae with a single chloroplast per cell. Reiteration of DNA sequence is therefore typical of the chloroplast irrespective of whether there is one or many in the cell.

The organelles display a considerable functional and genetic autonomy (reviews by Küntzel, 1969; M. Nass, 1969; Rabinowitz and Swift, 1970). Genetic evidence supporting a partial functional autonomy is that the organelles include at least some proteins encoded in their own DNA as well as proteins encoded in nuclear DNA. Hence the organelles appear not to be self-sufficient in respect to their macromolecular endowment and much depend on the cytoplasm for the provenance of molecules crucial to their metabolic function, e.g., respiratory enzymes and chlorophyll. At first view, this equilibrated endosymbiotic relationship rationalizes the simplification of the mitochondrial genome compared with the much larger bacterial genome as was described above. In addition, biochemical evidence indicates that most mitochondrial ribosomal proteins are formed on cytoplasmic polysomes, even though there is no evidence that these proteins function both in the mitochondrial and cytoplasmic ribosome. Most membrane proteins of the chloroplast are made in the cytoplasm (Hoober, 1970). Most of these cytoplasmic syntheses destined for the organelles probably depend on templates made in the nucleus. According to the genetic evidence, however, some ribosomal proteins of the organelles could be made in the cytoplasm using templates of organelle origin (Davey et al., 1969). Possible evidence for the cytoplasmic occurrence of mitochondrial mRNA, from a transcript originally representing a substantial fraction of the genome, was presented in Chapter 2. This would leave other organelle proteins, presumably membrane proteins and perhaps certain enzymes, to be made in the organelle itself, which is known to be capable of protein synthesis. This possibility presupposes the presence of functional mRNA inside the organelle, so far not demonstrated. A further possibility, namely, that cytoplasmic proteins are made in the organelle using templates of nuclear origin, is by no means inconceivable but so far is also not documented. In summary, the genetic interrelationship between organelle and cytoplasmic protein is far from being completely elucidated.

Contrary to the limited capacity for protein synthesis, DNA and rRNA (and, as discussed later, most tRNA) are made in the organelles. As mentioned, however, the presence of 5 S RNA is still uncertain. The unique characteristics of the major species of mitochondrial and chloroplastal rRNA compared with cytoplasmic rRNA, whereby they resemble more prokaryotic rRNA, were described in Chapter 2. Yet it is clear that in regard to its physical parameters organelle rRNA has evolved of its own. Nevertheless, the general implication is that the organelle system for nucleic acid synthesis has the characteristics of a prokaryotic system. [This statement would have to be qualified if, as proposed by Attardi and Attardi (1968), all mitochondrial rRNA species derive from a whole strand transcript, i.e., a joint precursor, but bacteria have separate precursors.] Actually, the mitochondrial RNA polymerase now isolated (Küntzel and Schäfer, 1971) is simpler than the prokaryotic counterpart; it consists of only one protein chain which suggests a very simple mechanism of action. In the remainder of this section we shall first discuss organelle rRNA and then evaluate the informational content of the mitochondrion.

To begin with the lower eukaryote, total mitochondrial rRNA from *Tetrahymena* hybridizes with 6.8% of mitochondrial DNA (Suyama, 1967). In *Neurospora* the two mitochondrial rRNA species are complementary to 2.8 and 6.1% of mitochondrial DNA, respectively. The multiplicity of these RNA cistrons is at least four per DNA molecule (Wood and Luck, 1969). Compared with the minute proportion of nuclear rDNA representing cytoplasmic rRNA, the huge proportion of rDNA (9%) in the mitochondrial genome reflects, of course, the smallness of this genome.

Data on mitochondrial rRNA of the higher eukaryote are similar to those in the lower eukaryote. Both mitochondrial 12 and 18 S rRNAs of rat liver hybridize with mitochondrial DNA, whereas cytoplasmic rRNA does not (Aaij and Borst, 1970). In line with this, according to Dawid and Wolstenholme (1968), no hybrid of mitochondrial DNA with nuclear DNA is detectable in *Xenopus*. Each mitochondrial rRNA species of the HeLa cell is represented by a single cistron per DNA molecule (Aloni and Attardi, 1971). The inhibition characteristics of these species are different from those of nuclear or other cytoplasmic RNAs, from which it is concluded that their synthesis is by a different mechanism (Zylber *et al.*, 1969). It was already seen that the mitochondrial polymerase is much simpler than the nuclear one. As described in Chapter 2, the small mitochondrial rRNA of the higher eukaryote has its counterpart in the small mitochondrial ribosome. Indeed, this is the smallest ribosome known so far, and is called a "miniribosome," yet its capability for protein synthesis has been demonstrated. There appears to be also a correlation between the simplified mitochondrial genome of the higher eukaryote and its minimal ribosome.

Chloroplastal rRNA forms hybrid with 0.5% of chloroplastal DNA, and this DNA does not hybridize with cytoplasmic rRNA (Tewari and Wildman, 1968). According to these authors, nuclear DNA forms hybrid with both chloroplastal and cytoplasmic rRNA (0.1 and 0.25%), but the cistrons for the two RNAs appear to be different. On a per cell basis, this would result in the information for chloroplastal rRNA being three times greater in the nucleus than chloroplast. It is probable, however, that nuclear DNA reacted with the two rRNAs because of nonstringent conditions of hybridization. The general question suggested by M. Nass (1969) of an ancestral blueprint of the organelle genome represented in the nuclear genome is therefore not answerable at present. As in the lower eukaryote mitochondrion, there are about four cistrons for each rRNA species (Wittmann, 1970) per chloroplastal genome. Of interest now is to locate these cistrons in the highly reiterated chloroplastal genome.

A final question to be examined concerns the informational content of the mitochondrion. In the lower eukaryote the mitochondrial genome with 75,000 DNA base pairs would just suffice to encode the components of the translational apparatus, assuming, as is reasonable, that they are all of bacterial size (Attardi and Attardi, 1969). Stated differently, allowing for a fourfold multiplicity of rDNA and a set of tDNAs (a total of 22,000 base pairs), this leaves room in this genome for 100 polypeptides of molecular weight about 20,000. In the higher eukaryote the mitochondrial genome with 15,000 base pairs can encode 30 polypeptides of this size, if nothing else has to be encoded. In actuality, since one-third to one-half of this genome is possibly devoted to DNA replication and stable RNA synthesis, the coding for general polypeptide would be proportionally less. This has led many to consider that the higher eukaryote mitochondrion approaches an irreducible minimum of information. Seen from this standpoint, certain observations in the lower eukaryote are particularly illustrative. Yeast mitochondrial DNA (20% GC) can be reduced to 4% GC by certain mutations that abolish respiratory function but not replication of the cell (subsequently it becomes anaerobic) or even the organelle with such abnormal DNA (Mehrotra and Mahler, 1968). The mutant DNA molecule appears to be normal in size (about 20,000 base pairs in the strain studied) but, on the basis of its GC content, has only about one-fourth of the normal informational content, or the equivalent of eight polypeptides. To explain the preservation of this organelle which is no longer functional in respiration, it was proposed that some of these polypeptides are essential for some other general cell function such as membrane formation (Attardi and Attardi, 1969). (Of course, it is not excluded that the polypeptides missing in the mutant have functions other than this, e.g., energy production, in the normal organelle.) In this respect, it is known that isolated mitochondria incorporate amino acid into proteins of their inner membrane. Several membrane proteins of the cell tend to resemble one another (Rabinowitz and Swift, 1970) and some are possibly of low molecular weight (Laico et al., 1970). Based on evidence of this kind, it was further proposed by the Attardi's that an important nonenergetic role of the mitochondrion may be related to the synthesis of these cellular membrane proteins. A possibility suggested was that the proteins are made in the cytoplasm using mitochondrial mRNAs. Returning to the higher eukaryote mitochondrion, its irreducible information could preferentially include these cellular proteins. On the other hand, most of the information required to assemble any translational apparatus in the higher eukaryotic cell must come from the nucleus.

In conclusion, the organelles represent separate self-replicating genetic systems in the cytoplasm, yet have only a partial autonomy due to a considerable functional integration with the nuclear genetic system. Thus it would be surprising that the organelles have been preserved during evolution were they not to provide some essential cell function; this function is probably energy production, but could be more than that. At this point it can only be noted that the informational content in the mitochondrial genome is greater in the lower than higher eukaryote, but this information is insufficient to account for more than the synthesis of DNA, stable RNA, and some, but not many, proteins. Of these proteins, some are involved in maintaining respiratory function as well as the function of the prokaryotic synthetic systems of the organelle. Others, according to the proposal above, may serve the cell in some more general capacity.

5 S RIBOSOMAL RNA

The general points made in the introduction to the major rRNAs apply to 5 S RNA. In the number of molecules per cell, although not in their total mass, 5 S RNA is as abundant as the major rRNAs.

Cistrons in the Eukaryote

Taken together, the evidence to follow indicates that most, if not all, 5 S RNA is of nonnucleolar origin. In the amphibian, (a) there is no enrichment of 5 S DNA corresponding to that of rDNA in the oocyte. Within the range of sensitivity of the hybridization method, (b) the enucleolate *Xenopus* mutant contains the full complement of 5 S DNA (Brown and Weber, 1968a), even though this DNA is not transcribed in the absence of rDNA. In density equilibrium centrifugation, (c) the 5 S DNA forms a satellite distinct from the rDNA satellite. Contrary to this satellite, it bands to the light side of main DNA, indicating that 5 S RNA is not transcribed as part of the rRNA precursor. (Because of the inconspicuousness of the 5 S DNA satellite, this can only be detected by the radioactive 5 S RNA-DNA hybrid separated after centrifugation.) However, since the hybridization technique might not permit one to exclude that there is less than one 5 S RNA cistron per every ten 18 and 28 S RNA cistrons in the rDNA satellite, the presence of some 5 S RNA cistrons in the nucleolus cannot be completely eliminated.

The haploid *Xenopus* genome contains about 25,000 5 S RNA genes disposed in at least 0.05% of the total DNA (Brown and Weber, 1968a). This fraction of DNA is of the same order or less than for the major rRNA species. The ratio of 5 S DNA to rDNA is therefore constant in all tissues of this organism, except the oocyte only because of its amplified rDNA. The 5 S RNA cistrons in *Xenopus* are to some extent clustered as shown by the formation of a DNA satellite and the density shift of the DNA-RNA hybrid. The lightness of the satellite contrasts with the high GC content of the RNA and makes it possible that the cistrons are interspersed with short DNA segments of very low buoyant density. The actual distribution of these cistrons in the amphibian chromosomes is not known. As seen by cytological hybridization, in *Drosophila* chromosomes many or most cistrons are located in a few closely spaced loci in one of the autosomes (Wimber and Steffensen, 1970). In the hamster, the cistrons may come to reside in the nucleolus-associated chromatin (Amaldi and Buongiorno-Nardelli, 1971), the implications of which we shall return to later. The high cistronic multiplicity of the amphibian is, however, not typical of all animals; the haploid genome has about 200 cistrons in *Drosophila* (Tartof and Perry, 1970) and 2000 cistrons in Hela cells (Hatlen *et al.*, 1969). One may justly wonder, therefore, if the high cistronic multiplicity of the amphibian is functionally related to its exceptionally high amplification of germinal rDNA, another aspect to which we shall return. In any event, in all organisms so far studied the cistronic multiplicity of 5 S RNA is variable but consistently above that of the major rRNA species.

Cistrons in the Prokaryote

In *B. subtilis* there are three or four 5 S RNA genes disposed in two blocks as determined by chromosome density transfer (Smith *et al.*, 1968). Although the majority of the 5 S RNA genes are at the chromosome origin intermingled with the rRNA and tRNA genes, the linkage is closest to 23 S RNA (Colli *et al.*, 1970). The remainder of 5 S RNA genes are at the chromosome terminus. In all, they occupy 0.005% of the total DNA (Morell *et al.*, 1967). The cistronic multiplicity is of the same order as in rRNA. The multiplicity of 5 S RNA is therefore greater than one order of magnitude below that found in the eukaryote.

Pathways of Synthesis

In the eukaryote most 5 S RNA originates outside the nucleolus, hence the pathway of synthesis must be separate from that of rRNA. However, the 5 S RNA is already present in the precursor 80 S ribosomal particle (Fig. 18) in a 1:1 molar ratio to the rRNA precursor. Thereafter, it follows the pathway along with 28 S RNA to emerge in the large ribosomal subunit. No 5 S RNA precursor larger than the mature molecule has been described in the eukaryote, although it possibly will be. In that case, the presence of di- and triphosphate nucleosides at the 5' end of HeLa cell 5 S RNA (Hatlen *et al.*, 1969) indicates that the extra sequence will not be in the precursor at this end as it is in bacteria. This finding also excludes the possibility that the RNA could start off as a polycistronic precursor then split into several mature molecules.

In the prokaryote it is not yet decided whether the pathways of 5 S RNA and major rRNAs are totally separate. In *E. coli* there is an immediate 5 S RNA precursor with a more expanded configuration, and a slightly longer sequence (one to three bases) at the 5' end, than the mature molecule (Scott *et al.*, 1968). A derivation of this immediate precursor from an even longer initial macromolecular precursor is kinetically excluded, according to Adesnik and Levinthal (1969). However, in *B. subtilis* the possibility exists that the 5 S RNA precursor is transcribed covalently with the 23 S RNA precursor (Hecht *et al.*, 1968). Thus, using transcriptional mapping in this species, 5 S DNA appears to be part of the monocistronic 23 S RNA operon (Bleyman *et al.*, 1969), although the relative positions of the two cistronic DNAs is not indicated. This disposition of 5 and 23 S DNA would agree with the linkage relationship in the *B. subtilis* chromosome, but not with the greater number of 23 S RNA cistrons that are present. Therefore, there could still be an initial macromolecular precursor other than a joint one with 23 S RNA (see Doolittle and Pace, 1970, for *E. coli*). Obviously, this question remains unsolved; it is of comparative interest since the bacterial tRNA precursor contains a considerable excess of sequence. A 5 S RNA molecule joins the large ribosomal subunit relatively late during subunit maturation, yet it remains possible that the 5 S RNA precursor itself is what joins the subunit and matures thereafter (Jordan *et al.*, 1970). This could account for some of the difficulty so far encountered in reconstituting the functional bacterial subunit.

Regulation of Synthesis

The 5 S RNA is synthesized in equimolar proportion with the major rRNA species. If it is considered that in the eukaryote the 5 S RNA cistrons are separate from the major rRNA cistrons, then this coordination of synthesis is remarkable. This means that these RNAs respond to a common regulatory mechanism. An attractive possibility that would go a long way to explain the common regulation is that the 5 S RNA cistrons are serviced by the same polymerase and repressor as the major rRNA cistrons, perhaps all responding to the same translational signals considered earlier. Similar to their model for rDNA, the partial clustering of 5 S DNA and the actinomycin sensitivity of the same order as rDNA (on a per cistron basis), has led Perry and Kelley (1970) to propose that the polymerase runs uninterruptedly through the cluster transcribing each cistron independently.

Despite their coordination there is a measure of independent regulation of 5 S RNA with respect to rRNA. For example, a nuclear pool of 5 S RNA is maintained through which new molecules must pass before entering the precursor ribosomal particle. The pool amounts to 25% of 5 S RNA in the HeLa cell (Knight and Darnell, 1967), which by far exceeds the nuclear to cytoplasmic proportion of any other rRNA species. That the synthesis of 5 S RNA is not contingent upon previous or concurrent rRNA synthesis or maturation is shown by: (a) the synthesis of 5 S RNA but not rRNA both in the early *Xenopus* oocyte (Mairy and Denis, 1971) and mitotic HeLa cell (Penman *et al.*, 1970); (b) the continuous 5 S RNA synthesis (for hours) in L cells in the presence of doses of actinomycin that prevent rRNA synthesis (Perry and Kelley, 1968b). Together with the lack of 5 S RNA synthesis in the enucleolate *Xenopus* mutant, these observations mean that the synthesis is dependent on the presence of the rDNA cistrons but not on their function. At present there is no simple explanation for this intriguing physical dependence, which is also shown by ribosomal protein synthesis (Chapter 6). In summary, the two RNA syntheses are delicately coordinated yet not coupled with each other.

The experiments by Perry and Kelley (1968b) further revealed that, in the absence of rRNA synthesis, 5 S RNA undergoes intranuclear degradation at the rate of 3.5% per hour. In the few cells still dividing, some 5 S RNA escapes into the cytoplasm, undoubtedly because there is no rRNA synthesis. Moreover, the RNA cannot be readily utilized for precursor ribosomal particle formation when rRNA synthesis is resumed after deinhibition. Thus, in addition to the general coordination between the two syntheses, a certain phasing and interdependence must exist. Unlike the major rRNA species, 5 S RNA is not methylated or pseudouridylated. Other minor posttranscriptional modifications are still possible, but they would seem to play a much lesser role in the maturation of 5 S RNA than they do in the case of the major species.

In the *Xenopus* oocyte there is no 5 S DNA amplification comparable to the 1500-fold one for rDNA (Brown and Dawid, 1968). However, keeping pace with the abundant rRNA synthesis in the oocyte is no serious problem to the enormous number of 5 S RNA cistrons present somatically in this animal. The reason for this enormous number of cistrons may very well be in order to permit their transcription in pace with that of the amplified germinal rDNA. This strategy would be because, mechanistically, the disperse 5 S RNA cistrons are difficult or uneconomical to amplify, yet their enormous number is obviously not cumbersome to the cell. Conversely, the rRNA cistrons would be expressly amplified during oogenesis because the required multiplicity is too much to carry permanently in the genome. As a consequence of the disparate multiplicity, the equimolar ratio between the RNAs in somatic cells of this animal can only be maintained by a lower relative rate of transcription of 5 S DNA, which is thus used at much less than full capacity. Different possibilities for the regulation of an excess number of cistrons undergoing such submaximal utilization were already discussed in connection with rRNA regulation, which the reader can consult. In other animals the differential utilization of somatic 5 S DNA and rDNA would of course decrease as the number of cistrons become more commensurate with one another, even though it will remain somewhat higher for 5 S DNA. Such a more even utilization of cistrons would prevail, therefore, in animals which require less germinal amplification of rDNA.

Unlike the eukaryote, in the prokaryote the possibility was mentioned that 5 S RNA is synthesized in the same molecule as 23 S RNA. A discussion of prokaryotic regulation must be deferred until such time when the pathway is better understood.

TRANSFER RNA

Considered as a class, tRNA is numerically the most abundant RNA molecule in the cell, as shown for the eukaryote in Tables 4 and 5 (pp. 52 and 81). However, in keeping with their proportion within the class, the cistronic multiplicity for any individual tRNA species is much less than for rRNA. In fact, in the prokaryote the multiplicity is almost nil. On the other hand, even in the prokaryote, several redundant species (sequence differing other than in the anticodon) or degenerate species (differing in the anticodon) of tRNA for any one amino acid are generally not hard to find. The combined figure of isoaccepting species can be as high as six or thereabouts, particularly in the eukaryote, but usually is less. This diversification, in contrast to rRNA, is because of the functionally differentiated nature of tRNA molecules which again is probably more pronounced in the eukaryote. This means that, as a class, tRNA is more molecularly redundant than rRNA. Moreover, a functional differentiation immediately suggests that tRNA is not only quantitatively regulated, as is rRNA, but is also qualitatively regulated. This tells us to expect differences in regulatory mechanism between the two RNAs. However, compared with rRNA, much more is known about the small tRNA molecule itself than about the arrangement of the cistronic DNA or its regulation.

Cistrons in the Eukaryote

In a lower eukaryote, the yeast *Saccharomyces*, about 0.07% of the DNA, or the equivalent of a total 350 cistrons per haploid genome, hybridizes with tRNA (Schweizer et al., 1969). One reason for this rather high value of tDNA (relative to the higher eukaryote discussed below) is perhaps that yeast rDNA is equally excessive, and a balance is maintained between the two cistronic DNAs to optimize the ratio of stable RNA production. Assuming a total of 60 species of tRNA (Yang and Novelli, 1968a), 350 cistrons in yeast represent a multiplicity of six cistrons per tRNA species per haploid genome. In *Droso-*

phila 0.015% of the genome, or the equivalent of 800 cistrons per haploid genome, hybridizes with tRNA (Ritossa *et al.,* 1966b). For each of 60 species of tRNA this number of cistrons represents 0.00025% of the haploid genome, or a multiplicity of 13 cistrons per species. In *Xenopus* the tDNA separates on density equilibrium centrifugation as a distinct satellite riding the bulk of the DNA, but displaced toward the heavy (GC-rich) side; it amounts to 0.016% of total DNA (Brown and Weber, 1968a). The amount of satellite tDNA that forms hybrid with tRNA represents 1000 cistrons per haploid genome, or a multiplicity of 17 cistrons per each 60 tRNA species. Higher multiplicities have been found in some but not all mammals, e.g., a total of about 1300 and 5000 cistrons per haploid genome in human HeLa cells (Attardi and Amaldi, 1970) and rat cells (Quincey and Wilson, 1969), respectively. Despite the fact that all these estimates are only moderately precise, a certain evolutionary trend is discernible. Thus, compared with the insect, the multiplicity in the amphibian and mammal is much less high than the 50-fold difference in the size of the genome. Since this applies also to rDNA, the conclusion is that during animal evolution the cistronic multiplicity of stable RNA has increased proportionately less than the size of the genome, whereas, on the contrary, the amount of cistronic DNA has relatively decreased.

In *Xenopus*, from the relatively low average density of satellite tDNA compared with the high GC content of tRNA, and the minor density shift of their hybrid, the indications are that part of the total tDNA is clustered but part is not (Brown and Weber, 1968a). If all tDNA with its GC content were clustered, it should make as heavy a satellite as rDNA. It is fairly obvious that if a portion of the cistrons are apart as the indication seems to be, it is because they represent distinct species. These data cannot, however, reveal the approximate size of the cistron clusters, much less the arrangement of cistrons within the clusters. In this respect, on logic grounds alone, Ritossa *et al.* (1966b) previously had favored the idea that each tRNA species is represented by a continuous repeating series of cistrons, which means a cluster. They hypothesized that the "minute" genetic loci of *Drosophila*, numbering about 60 loci and located in all chromosomes, represent the clustered tRNA cistrons. Of interest for later discussion, all the mutations in these loci that have been found represent chromosomal deletions and cause a hypomorphic phenotype. In the chironomid salivary gland, newly made tRNA is detected by sedimentation analysis in all chromosomes in proportion to their total DNA, suggesting a uniform cistron distribution (Edström and Daneholt, 1967) not in disagreement with the "minute" clusters proposed in *Drosophila*. Recently, in its central aspect, this proposal has been substantiated by cytological hybridization (Steffensen and Wimber, 1970). Likewise, data on actinomycin sensitivity (Perry and Kelley, 1970) suggest a certain degree of clustering of cistrons in the mammal also. It should be stressed, however, that the physical data on the amphibian tDNA satellite clearly indicate that some cistrons are not clustered.

While it is established that the majority of the tRNA cistrons are chromosomal, as first indicated by Perry (1962), there remains the question of whether or not some of them are interspersed with rDNA in the nucleolus (review by Sirlin, 1968). Briefly, the relevant evidence is as follows. First, the nucleolus was shown to contain a pool of tRNA in the starfish oocyte (Vincent et al., 1966) and an aquatic fungus (Comb and Katz, 1964), in both of which the organelle is bulky and easily isolated. In contrast, somatic cells where the nucleolus is small and laborious to isolate, were frequently reported not to contain this pool. One difficulty in these early studies involving cell fractionation therefore might have been procedural loss of the unbound small molecule. In any case, some of this early work suggested the presence of nucleolar tRNA, but this did not mean that the cistrons were nucleolar. By this time, using molecular hybridization on genetically constructed *Drosophila* strains carrying different numbers of the nucleolus organizer, Ritossa et al. (1966b) were able to show that the amount of hybridized tRNA does not correspond with the dosage of the organizer and, therefore, most tRNA cistrons were located in the chromosomes elsewhere than the nucleolus. Careful inspection of the hybridization saturation curves in the different strains studied reveals that there is a minor but consistent dosage relationship with the organizer, particularly when considering that the input tRNA fraction has a 5 S RNA contaminant (this RNA being of chromosomal origin, it obscures the dosage of tRNA with the organizer). Hence this classical work does not exclude that some of the tDNA is present in or near the rDNA region. Similarly, in the amphibian somatic cell the sensitivity of the hybridization technique cannot exclude that there is less than one tRNA cistron present per every ten 18 and 28 S RNA cistrons (Brown and Weber, 1968a). That is, considering as these authors do that there are 450 rRNA cistrons per haploid genome, there could be less than 45 but as many as 34 tRNA cistrons present in the rDNA region, which is equivalent to the average complement for two tRNA species. Since this possible number is 40 times greater than is to be expected on the basis of a uniform genomic distribution of cistrons, i.e., 1.1 nucleolar tRNA cistrons per haploid genome (0.11% of two-stranded rDNA $\times$ 1000 tRNA cistrons/100% of DNA), these negative quantitative data in the amphibian are less significant than the qualitatively positive data that follow.

Biochemical evidence was presented for the occurrence of newly made tRNA in the nucleolus of the chironomid salivary gland when this is subjected to differential chemical inhibition of RNA synthesis. Under these conditions all cell RNA synthesis is arrested, except for the synthesis of low molecular weight RNA (including tRNA) and rRNA precursor in the nucleolus; this biochemical evidence was monitored *in situ* by autoradiography (Sirlin and Loening, 1968).

The considerable amount of nucleolar tRNA radioactivity observed is not easily reconciled with an extranucleolar origin. More probably, this radioactivity represents an increased transcription of nucleolar cistrons, or a greater degree of labeling of the transcript, or its slower exit from the nucleolus, due to the chemical treatment. In untreated salivary glands of another chironomid, the relative amount of newly made 4 S RNA detected in microdissected nucleoli is likewise against a chromosomal origin, although, of course, this system also reveals chromosomal tRNA synthesis (Eghyazi et al., 1969). At 45 minutes of tracer (cytidine and uridine) incorporation, 3.5% of the total nucleolar radioactivity is in 4 S RNA. Compared with the chromosome, the much faster nucleolar kinetics of 4 S RNA synthesis suggests a larger contribution than is apparent from the radioactivity. Thus the overall evidence in the chironomid probably indicates a nucleolar synthesis of tRNA which appears to be disproportionate to the expected number of nucleolar cistrons. This expected number, according to data in *Drosophila* (Quagliarotti and Ritossa, 1968) and assuming a uniform genomic distribution of the cistrons for 60 tRNA species, is less than 8.6 nucleolar tRNA cistrons per diploid genome (0.54% of two-stranded rDNA × 1600 diploid tRNA cistrons/100% of DNA), which is equivalent to about 0.3 of the average diploid complement for one tRNA species. (Diploid values are used for calculation because each organizer locus resulting in a nucleolus represents the paired loci of two salivary chromosomes.) The nucleolus-made tRNA of chironomids provisionally concluded from these data was not biologically tested, however. The 4 S RNA present in hepatoma nucleoli was tested (Nakamura et al., 1968). In this material the different composition and reduced total acceptor activity of nucleolar tRNA compared with cytoplasmic tRNA suggest, as above, that only a few tRNA species are present; the absence of modified nucleotides (other than methylated) in the nucleolar tRNA further suggests that much of it is nascent. These data in the tumor support a nucleolar tRNA synthesis. In conclusion, although direct demonstration is pending, the possibility exists that the distribution of tRNA cistrons may include a few cistrons in the nucleolus organizer region. If that is the case, their arrangement with respect to the rRNA cistrons in the organizer would be of interest to study. It should be recalled in this regard that a large amount of nonribosomal DNA intercalated in the organizer remains still without assigned function. The functional implications of this possible nucleolar synthesis will be considered elsewhere in the text.

Cistrons in the Prokaryote

Phage DNA genomes carry cistrons for tRNAs with a specificity ranging from a few to most amino acids (Daniel et al., 1970; Subak-Sharpe, 1968; Weiss et al., 1968). Together with the fact that the transcription of these cistrons is an early viral function, other evidence discussed in Chapter 5 indicates that these tRNAs are vitally necessary to the virus. On the other hand, the reason for the tRNA molecules, presumed to be of host origin, which are carried by certain large RNA virions such as AMV (Travníček, 1969) remains obscure.

In *B. subtilis* 0.04-0.07% of the DNA hybridizes with tRNA, representing a total of 40 cistrons (Smith *et al.*, 1968). As shown by chromosome density transfer, the majority of these tRNA cistrons are situated at the chromosome origin interspersed with the cistrons for rRNA, 5 S RNA, and ribosomal protein, but with an average map position closer to 16 S RNA. Most of the remainder of the cistrons are at the chromosome terminus, and a few are in between. In conflict with the previous distribution, Bleyman *et al.* (1969), using transcriptional mapping, suggested that all cistrons of this bacterium may occur together as one multicistron (which, as we shall see, creates some difficulty concerning subregulation of cistrons). Similar numbers of tRNA cistrons have been found in *E. coli* (Giacomoni and Spiegelman, 1962), again with a quasi-bimodal distribution (Cutler and Evans, 1967). This is not inconsistent with the mapping in this organism of suppressor genes representing structural tRNA cistrons (Taylor and Trotter, 1967), which reveals five suppressor genes not disposed as a single cluster (von Ehrenstein, 1970). Assuming for the bacterium a total of 40 tRNA species (Yang and Novelli, 1968a) or somewhat more (Muench and Safille, 1968), there seems to be no room in the previous data for a general cistronic multiplicity of tRNA species, although this does not exclude that some species will have a minor multiplicity. It is revealing that the smaller *Mycoplasma* has this minimum number of 40 tRNA cistrons (Ryan and Morowitz, 1969). All these microorganisms still have more than one isoaccepting tRNA species for many amino acids, however. Indeed, the suppressor tRNAs frequently found in the prokaryote are permissible only if isoaccepting species for the particular amino acid, and in some cases perhaps cistronic multiplicity within the species are available. The cistrons of isoaccepting species appear not to be generally linked (Hill *et al.*, 1970). In conclusion, as in the case of rRNA, the prokaryote has relatively more genome involved in tRNA than the higher eukaryote, but the individual cistron multiplicity is one order of magnitude less.

Pathways of Synthesis

Eukaryotic Pathway

The pathway of tRNA synthesis begins with a larger precursor that is submethylated, does not accept amino acid (Sirlin and Loening, 1968), and is not associated with protein (Burdon and Clason, 1969). In HeLa cells the pathway takes at least 40 minutes (Bernhardt and Darnell, 1969). Possible changes occurring during maturation of the precursor will be discussed below in connection with the regulation of synthesis.

Prokaryotic Pathway

In the prokaryote there is evidence to indicate a tRNA precursor (Pace *et al.*, 1970; Waters, 1969). This precursor contains about 40 nucleotides in excess of the mature molecule and distributed to both ends (Altman, 1971). Mutations altering the pattern of base pairing markedly reduce the amount of tRNA that is produced (Smith *et al.*, 1970). This reduction is seemingly due to an effect of the changed tertiary structure of the precursor on its rate of maturation, with the result that less mutant than wild-type tRNA reaches final maturation. The half-life of maturation in the wild-type is 3 minutes at 25°C.

Regulation of Synthesis

The cell content of tRNA, i.e., the ratio of tRNA to DNA, remains relatively constant within a certain range of growth rate. In contrast with 5 S RNA, this means that tRNA and rRNA syntheses are not always coordinated, although for the most part of growth they are. Exceptionally, as in embryonic cells, the syntheses can even be uncoordinated and each can show a marked specificity with respect to developmental stage. The overall rate of synthesis of individual tRNA species is generally of the same order as the rate of rRNA synthesis. Since the cistronic multiplicity of rRNA is at least one order of magnitude higher, this means that the tRNA cistrons in general are utilized more efficiently. Moreover, it should be appreciated that the regulation of many functionally differentiated species as in tRNA is bound to be more complex than for the multiple but essentially similar rRNA cistrons. For this reason, contrary to rRNA, tRNA may be both quantitatively and qualitatively regulated, that is, all species responding alike or differently, respectively. Of these two RNAs, only in the case of tRNA does the potential need exist for a mainly quantitative regulation of obligatory syntheses during cell growth and a mainly qualitative regulation of facultative syntheses during cell differentiation and function. The qualitative regulation may possibly be more significant in the eukaryote than the prokaryote. The regulatory mechanism of tRNA synthesis is less known than that of rRNA in both the prokaryote and eukaryote, hence this brief introduction encompassing the two organisms.

Eukaryotic Regulation

The regulation of eukaryotic tRNA is not understood. Commenting on the possible nature of this regulation, one has to consider first a separate regulation of the individual species of tRNA which may not all be simultaneously in demand, or be all in demand but not to the same extent. In addition, it is not improbable that the individual species have dissimilar cistronic multiplicities. Both of these aspects could contribute to the different copy frequency of the individual species which within limits is what is observed in the cell population of tRNA. It follows from these aspects that it is impossible to distinguish a priori to what extent a genetic dissimilarity is maintained, or, conversely, a regulatory dissimilarity is added onto it. Since at least some degree of regulation must be operative, a word on its possible mechanism is in order, even though evidence on this question is completely lacking. It was suggested in the case of eukaryotic rRNA that the regulation could depend on a translational control resulting in the appearance of a specific repressor activity. In the case of tRNA it is not known whether the cistrons have a specific repressor acting on an operator-like gene, or whether this repressor would be one or many. On the other hand, again by analogy with rRNA, a distinct possibility is that these cistrons as a class have a specific RNA polymerase that recognizes its own promoter. A model of transcription has been proposed by Perry and Kelley (1970) similar to the one considered previously for other stable RNAs. It is based on the partial clustering of tDNA and its high actinomycin sensitivity (equal to the sensitivity of rDNA on a per cistron basis). According to this model, the polymerase would bind to a promoter site (one site per cistron cluster) and then travel unproductively along the template until it encounters an operator site (one site per individual cistron) which, if not repressed, allows initiation of transcription to occur. As will be seen, such a model could help explain the subregulation of individual species which was suggested above.

Given the functional differentiation of tRNA, it is not unreasonable to consider the additional possibility that the cistrons may be integrated into regulatory genomic circuits of the kind previously discussed in detail in connection with eukaryotic mRNA. Indeed, pursuing the analogy, the occurrence of selective activation of tRNA species in response to hormone (see below) lends some credence to this view on genomic circuitry. One has then to consider that together with the need in certain tissues to activate perhaps a few tRNAs for specialized protein synthesis, there is also the need in most tissues to activate at once many or all tRNAs for general protein synthesis. That is, there is the dual need for a qualitative and quantitative regulation. The discussion that follows explores several possibilities of regulation, but the mere duality of regulation just noted serves to remind us that it cannot be unequivocally approached without more basic knowledge than we have at present.

Hormonal stimulation of RNA synthesis in a target organ includes tRNA and rRNA. It was suggested as a possibility in connection with rRNA regulation that this hormonal stimulation could act indirectly via the cytoplasm, although steroid hormone in particular could act nearer to the level of the genome, the precise level remaining undefined. These considerations apply as well to tRNA. As indicated above, under general conditions of growth the need may be for a general tRNA activation, but, owing to the functional differentiation of tRNA, under certain other conditions also for activation of individual species. The latter will be illustrated with several examples from steroid hormone stimulation in Chapter 5. Therefore, the possibility of a more direct stimulation of tRNA, perhaps at or close to the genome level and mostly by means of steroid hormone, deserves consideration. Concurrently with a general stimulation of tRNA, several hormones stimulate the transferase-mediated addition of the CCA terminus (Herrington and Hawtrey, 1971) and the activities of the aminoacyl-tRNA synthetase (McCorquodale and Mueller, 1958) and tRNA methylase (Turkington, 1969). During estradiol stimulation of serine-rich yolk protein synthesis in the avian liver, it was shown that the amount of both serine tRNA, in particular, and methylase increase simultaneously (Beck et al., 1970). Conversely, prior to sex hormone stimulation a tissue can be relatively poor in tRNA, as it is in the case of the hen oviduct (Dingman et al., 1969) and mouse mammary gland (Turkington, 1969). These examples of coordinate net stimulation of tRNA and its enzymes reinforce at first view the case for a rather direct regulation of tRNA by hormone and perhaps at the cistron level. This is because, contrary to rRNA where the hormonal target could be the polymerase, the diversification of tRNA demands a correspondingly diversified target which possibly could be the cistrons themselves. However, this remains in need of proof.

Despite the previous indications for coordination with rRNA synthesis, a certain degree of independence of tRNA synthesis also has to be considered. Independence from rRNA synthesis is shown by the actinomycin-treated L cell fibroblast (Perry, 1962), the mitotic HeLa cell (Pederson and Robbins, 1970b), and certain embryonic systems (Chapter 6), but, in particular, the enucleolate *Xenopus* embryo, all of which synthesize tRNA but not rRNA. (The possibility remains however that part of this tRNA synthesis is mitochondrial.) Synthesis of the two RNAs shows also different sensitivities to many inhibitors, which is readily explainable if the polymerase mechanisms are different. Sometimes, the rate of degradation of the RNAs is also different; tRNA is degraded more slowly than rRNA as the reticulocyte matures (Burka, 1968). The independent mechanism of regulation revealed by these observations should not obscure the fact already mentioned that, in most cells, the two RNA syntheses are considerably balanced with respect to each other over a wide range of cell growth. For instance, there is a common reduction of syntheses upon withdrawal of amino acid (Smulson, 1970). In the case of tRNA, the real significance of this phenomenon is not known, but it might reflect a type of translational control coarser than the ones considered above, which are not principally at the translational level. In any event, the analogy with the RC control of bacterial tRNA is evident.

There is no information available on dosage regulation of tRNA as there is for rRNA, but the moderate cistronic multiplicity engenders the suspicion that some will occur. For example, all known *Drosophila* mutants with clear-cut phenotypic defects ascribable to loss of tRNA, i.e., the "minute" series, appear to involve a deletion of all or most of the cistrons for one tRNA species. This lack of defective point mutants in multiple cistrons suggests a dosage regulation. If that is the case, because of the functional differentiation of tRNA, the dosage regulation might differ for each tRNA species according to its multiplicity, and be such as to contribute to the copy frequency required of each species at any given physiological status of the cell. Such dosage regulation may depend on hormone or on other transcriptional controls discussed below. A different argument in relation to cistron multiplicity is that suppressor tRNAs derived from point mutations are less likely to be effective in the eukaryote than the prokaryote with little or no multiplicity in most tRNA species. The suppressor mutants found in *Drosophila* show this predicted low level of suppression.

Similar to the case of rRNA, the implication of multiplicity in man is likely to be that only massive deletion of cistrons of any tRNA species can result in significant idiopathy. Compared with rRNA, however, the possible occurrence of smaller cistron clusters favors a higher incidence of significant deletions.

To summarize the regulatory possibilities for tRNA, these fall into quantitative and qualitative categories which are not completely separable, and which together represent both an obligatory and facultative genome. Not sufficiently emphasized heretofore for lack of evidence, an aspect to keep in mind is the possible regulatory significance of selective clustering of some cistrons but not others. Earlier, it was proposed that selective activation of individual species could best be explained by integrated genomic circuits of the type envisaged for mRNA. These informational circuits respond to hormone which is also known to activate individual tRNAs and their enzymes, and these are good reasons even to think that certain circuits may include both mRNA and tRNA. The models for mRNA specification illustrated in Figs. 13 and 14 can easily accommodate this possibility. It should be noted that genomic circuitry involves qualitative selection, thus, in principle, representing a facultative genome for tRNA. Hypothetically, there are possible mechanisms of quantitative regulation as well: (a) a promoter with a different efficiency for the polymerase according to the particular cistron cluster; (b) an operator with a different efficiency for a repressor according to the individual cistron, which may or may not be in a cluster. This disposition of the promoter and operator genes comes from the model of Perry and Kelley (1970). Lastly, there could be in conjunction with (b), (c) a translational control operating via a specific repressor. These mechanisms are not mutually exclusive and some of them would concern an obligatory genome for tRNA, while others could be shared with the facultative genome. Of these mechanisms, only the influence of amino acid on transcription has been observed and may represent a translational control as in (c) (which in the case of eukaryotic rRNA appears to be fairly stringent). In addition, the cistronic multiplicity of tRNA offers the possibility of dosage regulation, perhaps more so for those cistrons within clusters, and this type of regulation is strongly suspected even if not proved for lack of sufficient genetic analysis. Such a dosage regulation is however unlikely to be as dramatic as in the case of rRNA, for, in general, the tRNA cistrons are utilized more efficiently. The interesting point is that, since dissimilar multiplicities of individual species also probably occur, their dosage regulation (especially in conjunction with a variably efficient promoter or operator) would contribute to efface any clear distinction between a quantitative and qualitative type of regulation. The transcriptional regulation that was just discussed constitutes the primary regulation of the tRNA pathway. Further opportunities for regulation exist at the posttranscriptional and even functional level, as described below, which rank, respectively, as secondary and tertiary regulations of the pathway. In conclusion, the basic fact to explain concerning the primary regulation is that the copy frequency in the cell population of tRNAs is both different and adjustable. So far as is known, this is achieved by the function of two functionally overlapping genomes, the obligatory and the facultative, in fact these genomes overlapping more than in the case of mRNA. The possibility that there are some exclusively facultative tRNA species is however a definite one. The task now is to find out the extent of the similarity between the mechanistic aspects of the cistronic regulation in these two tRNA genomes.

The rate of tRNA transcription in the HeLa cell, which contains 10^8 tRNA molecules and 2600 cistrons per diploid genome, is such that over one cell generation (1 day) each cistron is, on the average, transcribed 25 times per minute, assuming the molecule is stable. For comparison, this is about one-third of the bacterial rate during rapid growth. In a long-lived cell, such as the mature liver cell, the rate of transcription is much lower, the cistrons being utilized at a small fraction of their total capacity (Quincey and Wilson, 1969). In this cell most tRNA species have a half-life of five days (equal to the half-life of rRNA), which means that most species are transcribed and removed at the same rate (Hanoune and Agarwal, 1970). This quite uniform turnover need not apply to actively differentiating cells.

Synthesis of tRNA precursor depends only partly upon methylation (Bernhardt and Darnell, 1969), suggesting a relaxed control of synthesis as in the case of rRNA. Subregulation of the pathway via modification of the molecule is therefore more likely to occur somewhere at the posttranscriptional level. However, structural maturation of the precursor, i.e., conversion to a molecule with electrophoretic mobility characteristic of the mature molecule, also depends only to some extent upon methylation and pseudouridylation. It follows that subsequent functional maturation of the molecule (as with rRNA) is likely to depend mostly upon posttranscriptional modification; as previously mentioned, the precursor itself is not functional. It is somewhat ironical that the functional importance of specific modification which is so costly energetically, as can be seen here, remains still essentially a postulate. Returning to structural maturation, two possible routes are that (a) the precursor is longer than the mature molecule and a nuclease scission might be cued, for example, by specific site methylation; (b) changes in conformation are involved and these are also probably related to modification. Using the promising analysis of *in vitro* maturation of the molecule, it appears that, as under possibility (a), the precursor is indeed about 30 nucleotides longer (Smillie and Burdon, 1970). Although the molecule could therefore change conformation directly as the result of scission as indicated, this does not eliminate possibility (b), particularly in view of the ample scope for modification in the molecule. In this respect, the evidence in the bacterium is that tertiary structure influences the rate of precursor maturation. In summary, despite the reigning incertitude, it seems probable that (a) conformation of the precursor is cospecified by modification, and that (b) this conformation in turn is essential for a correct cleavage, upon which (c) a last and perhaps most important modification, leading to the functional mature molecule, depends.

Concerning the cell site for posttranscriptional modification, tRNA methylation is found to take place at both presumptive sites of synthesis in the chironomid salivary gland, that is, the nucleolus (Sirlin and Loening, 1968) and chromosomes (Eghyazi *et al.,* 1969), without excluding some additional methylation in the cytoplasm. In contrast, liver methylation was found to occur preferentially in the cytoplasm (Muramatsu and Fujisawa, 1968) but, in view of possible fractionation artifacts, this finding requires validation. In general, however, postsynthetic modifications may be considered unlikely to occur all in the nucleus, if nothing else, because they can serve as a source of functional regulation in the cytoplasm. As will be discussed in Chapter 5, these modifications do have a clear potential for functional tertiary regulation of the molecule superimposed on both the primary transcriptional regulation and secondary posttranscriptional regulation. Evidently, one modification is obligatory for tRNA function, which is the addition of the CCA terminus. In addition to the functional possibilities for tRNA modification in the cytoplasm, it is conceivable that entry of the RNA into the cytoplasm depends on prior modification in the nucleus. So far this point has received little attention in the case of tRNA, but is currently receiving considerable attention in the case of mRNA.

Looking now at the amphibian oocyte as representing a germinal cell, in contraposition to rRNA, the tRNA cistrons are not amplified (Brown and Weber, 1968b). The outcome of this lack of amplification is a low relative tRNA content of the oocyte equal to 1% of total RNA. This means that, relative to the vast amount of rRNA present, little protein synthesis is possible. However, although part of the protein is of maternal origin, in absolute terms the amount of protein synthesis in the oocyte (e.g., ribosomal protein) is large and requires a correspondingly large absolute amount of tRNA. In the oocyte the weight ratio of tRNA to nuclear DNA is equal to 13,000 and extremely high for a single cell. This ratio is due to the rates of synthesis and accumulation of the stable tRNA molecule which are still quite pronounced, even if less than for rRNA, during oocyte growth. The ratio decreases steadily during early development of the embryo until the late blastula stage when it becomes a normal 0.5, at which point tRNA synthesis is resumed. The initial high ratio thus maintains a partial balance of tRNA to rRNA, in order to ensure that protein synthesis remains possible in the oocyte and embryo, in the latter before its own tRNA synthesis begins. The resumption of embryonic synthesis when the ratio is normalized suggested to Brown and Littna (1966b) that the ratio itself is of regulatory importance. This aspect will be further discussed in Chapter 6.

Prokaryotic Regulation

Although generally invariant, the level of tRNA increases when the bacterium is growing slowly (Maaløe and Kjeldgaard, 1966). This increase, apparently unrelated to the needs for protein synthesis (because the ratio of tRNA to ribosome becomes too high) possibly reflects some unknown property of the tRNA regulation. However, as will be seen below, under normal conditions of growth a relationship with protein synthesis—hence with mRNA, rRNA, and ribosomal protein syntheses—is not absent. According to calculations by Manor *et al.* (1969) at full growth the tRNA cistrons are fully loaded with polymerase and are presumably being used at a maximum capacity. Yet these data cannot reveal whether every individual tRNA species behaves similarly in this regard, which, as will become clear, is an important question in regulation that has to be answered.

As discussed in connection with rRNA, stable RNA synthesis is under control of the RC locus whose effects are manifest in the wild-type during deprivation of amino acid. At present, the basic mechanism of control by the locus is assumed to be similar for both stable RNA species. However, the mechanism is complex and not completely understood and particularly so for tRNA; for this species the mechanism remains therefore hypothetical. It appears that stable RNA synthesis is able to proceed in the (stringent) wild-type only for as long as sufficient amino acid is present, which results also, of course, in the production of protein. In relaxed mutants stable RNA synthesis continues during amino acid deprivation, when, of course, no protein is made. The current explanation is that, the more protein synthesis there is in the wild-type, the less GT_4X is formed and this allows the putative ψ initiation factor to bind the core polymerase at the promoter, hence allows transcription to occur. The RC locus is believed to function in the stable RNA regulation during the normal cell cycle of growth, yet which is the particular metabolic function resulting in GT_4X that this locus controls is not known. At any rate, the locus functions in the translational control of transcription. In the case of tRNA it is possible that, since there is only one vegetative polymerase and the ψ factor might be shared with rRNA, the tRNA promoter might be responsible for responding to this control. However, it is not clear how coordinate the RC control is between the two stable RNAs. Since this control depends partly on acylation of tRNA, this may add a measure of self-regulation to the tRNA synthesis. Nevertheless, the general implication is that, because of the functional differentiation of tRNA species, the RC function represents only a coarse mechanism of control. That is, the RC function would effect an indiscriminate reduction or shut-off of all tRNA synthesis as protein synthesis slows down or stops. This is all that is needed in the case of the two monotonous rRNA species, but hardly adequate for the differential regulation of the many discrete tRNA species that must now be considered.

Isoaccepting tRNA species are differentially rather than uniformly transcribed, as revealed by their disproportionate copy frequencies in the cell. For example, three tyrosine tRNAs in *E. coli* (including one suppressor) are present in the approximate ratio of 60:25:15 (Goodman *et al.*, 1968); suppressor tRNAs occur either as major or minor species (Garen, 1968). Many additional examples are given in Chapter 5. Since the cistronic multiplicity is generally unity or close to unity, differential transcription means that a control operates at this level as the result of which synonym cistrons are used at varying capacities, rather than at an even capacity, during the cell cycle. This control amounts therefore to a true fine subregulation of cistrons which is expected to serve to optimize the heterogeneous tRNA population according to the momentary translational needs of the cell, either during the normal cycle or particular metabolic conditions. The mechanism whereby the bacterium achieves this subregulation is not known, yet it must be aimed at the individual cistrons, or, preferably, be part of them. To speculate on the nature of this mechanism, the obvious choice is a differential efficiency of promoters for initiating transcription of their respective cistrons. It can now be seen that, in conjunction with the coarse RC control, this mechanism is capable of achieving simultaneously a quantitative and qualitative regulation of the kind required for tRNA and still depend on translation. A different indication of potential regulatory value comes from the study of certain mutants. They show that tertiary structure may influence the rate of precursor maturation (Smith *et al.*, 1970). However, unless one is willing to accept that the tertiary structure of every tRNA has evolved for the express purpose of subregulation, any subregulation by this means would have to prevail mostly between isoaccepting species, and, even then, its efficiency would be questionable. By comparison, a promoter-mediated subregulation has several advantages in principle. It has no restrictions of the previous kind and therefore is equally probable for any species; it is certainly less onerous to achieve during evolution. A point to note is that any subregulation mediated by promoters would be improbable to apply if, as proposed by some authors, all cistrons were disposed as an inflexible multicistron.

Several miscellaneous observations relate to the regulation of the tRNA molecule in response to metabolic conditions. When one amino acid is moderately limiting for growth, synthesis of the corresponding tRNA is neither repressed nor derepressed (Wong et al., 1969). However, as growth continues to slow down, tRNA in general tends to increase relatively, perhaps as a special property of the RC function. Conversely, as discussed in the following section, for some amino acids the amount of acylated tRNA does influence their synthesis. During repression-derepression of the arginine pathway the affined tRNAs are not altered in their RPC profile (Leisinger and Vogel, 1969), indicating that the species remain largely unchanged. Less than 2% of new total tRNA is formed upon induction of the lactose operon (Smith, 1968), which, of course, involves a considerable formation of specific mRNA. From these examples, the regulation of tRNA is essentially invariant vis-a-vis metabolic conditions, unless they happen to markedly affect the rate of cell growth, that is, the rate of total translation. Also of interest here is the acylation behavior of the molecule during different conditions of growth. Several examples are known of individual tRNA species being affected either quantitatively or qualitatively by growth (Doi et al., 1968; Kwan et al., 1968; Wettstein, 1966). Of three isoleucine tRNAs in *E. coli*, two are completely acylated during exponential growth. A third is not and instead is acylated during leucine starvation, although it remains functionally inactive in protein synthesis (Yegian and Stent, 1969b). In stringent (wild-type RC) strains deprived of leucine, some of the cognate tRNA carries a protecting substance, which is not a normal amino acid, bound to the 2′ or 3′ hydroxyl group. This substance attaches to only one of several isoaccepting leucine species; it is not present in relaxed (mutant RC) strains or exponentially growing stringent strains (Yegain and Stent, 1969a). Some of these observations must somehow relate to general cellular regulatory phenomena involving tRNA as an intermediate, but so far their explanation is not obvious. In addition, a growth-dependent binding of polyamines has been observed (Cohen et al., 1969), which is thought to stabilize the tRNA molecule in its biologically active form (J. R. Fresco, personal communication). Again, whether this has a regulatory or simply structural significance is not clear.

Bacterial Amino Acid Metabolism and Synthetase Function. It has been known for some time that for certain amino acids, i.e., histidine, isoleucine, and valine, the corresponding aminoacyl-tRNA behaves as the corepressor for the synthesis of the biosynthetic enzymes (review by Martin, 1969). The amino acid by itself does not inhibit the synthesis. In the case of histidine it is the degree of acylation of tRNA, not the amount of tRNA, that must increase in order to result in corepression (Stulberg et al., 1969). In the case of a complex biosynthetic pathway, i.e., isoleucine-valine, all the corresponding acylated tRNAs must be present in order to achieve what is called a multivalent repression. In general, it is not excluded that some, rather than all, of the synonym species of the tRNAs implicated in these pathways specifically intervene in the repression.

The initial evidence was therefore indicative of acylated tRNA being the corepressor, but, until its affinity to repressor was tested directly, it remained a possibility that a related synthetase-dependent product was instead the corepressor. Later, it became evident that all the operons for which this type of regulation seemed to apply are, in fact, repressible operons; for these, an aporepressor has yet to be found. Recent work on the *Salmonella* isoleucine-valine operon, which shows multivalent repression, has led to the proposal by Hatfield and Burns (1970) that the cognate acylated tRNAs interact with the first enzyme product of the operon (L-threonine deaminase) and together become the actual aporepressor, indeed a holovalent multirepressor. The basic experimental finding is that the immature deaminase binds specifically leucyl-tRNA, possibly valyl-tRNA, and, by implication, also isoleucyl-tRNA. The postulated repressor activity of this enzyme-tRNA complex on the operon transcription was not demonstrated, however. The complex could therefore act at other levels than the transcriptional one, yet the provocative proposal of Hatfield and Burns (1970) may well have a great heuristic value. Nevertheless, at present, the question is not settled. Work on the *Salmonella* histidine operon has led to an independent proposal essentially similar to the previous one (Kovach *et al.*, 1970), whereas work on the *E. coli* arginine operon has led to the different proposal that the repressive complex acts somewhere at the level of translation (McLellan and Vogel, 1970). In summary, the regulation of end-product amino acid of these repressible pathways is considered to be mediated by an enzyme-aminoacyl-tRNA complex either as a (a) primary transcriptional repression or (b) secondary translational repression. In addition, but perhaps related, there is the well known end-product feedback inhibition of enzymic activity at a (c) tertiary posttranslational level. Turning now to the repressible common aromatic amino acid pathway, a guarded view, at present, is that this pathway is regulated differently from those just described, that is, not directly by acylated tRNA. Even so, reminiscent of an end-product feedback inhibition, phenylalanine-tRNA, eigher charged or uncharged, binds the first allosteric enzyme of the anabolic amino acid pathway (Duda *et al.*, 1968). The significance of this behavior remains to be elucidated.

Aminoacyl-tRNA synthetases are coded by structural genes that are not clustered or related positionally to the structural tRNA genes or amino acid biosynthetic genes. Regulation of all synthetases is to about a constant amount in the cell irrespective of the rate of growth, largely similar to tRNA. Clearly, the common regulation of synthetase and tRNA does not depend on gene proximity. Synthetase regulation is independent of the regulation of the related amino acid, that is, aporepressors other than those mentioned above for regulating amino acid anabolic enzymes are involved. However, the synthesis of some synthetases is not constitutive, but is at least in part transcriptively repressed in the presence of the cognate amino acid. Indeed, this may be a more general occurrence than it appeared at first, because of the fact that some synthetases are irreversibly inactivated during amino acid restriction probably as a nonphysiological effect, hence this effect hinders the observation of derepression of the enzyme (Williams and Neidhardt, 1969). A reciprocal relationship seems therefore to be established whereby the synthetase intervenes (by way of tRNA) to repress the synthesis of certain amino acids as pointed out earlier, and in turn certain amino acids help repress the synthesis of synthetase. In short, the function of synthetase is most directly involved in cell intermediary metabolism.

Mitochondrial and Chloroplastal Transfer RNA and Synthetase

A unique mitochondrial localization of certain tRNA species and their cognate synthetases, distinct from their cytoplasmic counterparts, is found in *Neurospora* (Barnett *et al.*, 1967) and liver (Lietman, 1968; Nass and Buck, 1969). In the case of *Neurospora* mitochondria, tRNA species for at least 15 amino acids have now been shown to differ from the cytoplasmic species using CCD and RPC chromatographic procedures (Epler, 1969). Mitochondrial leucine-tRNA in this organism responds to different codons than the cytoplasmic synonym tRNA (Epler and Barnett, 1967). As shown directly by competitive hybridization of preacylated tRNA, liver mitochondrial leucine-tRNA is homologous with mitochondrial DNA but not nuclear DNA, with which instead cytoplasmic leucine-tRNA is homologous (Nass and Buck, 1969). In *Xenopus* and HeLa cells there is a total of about 12 tRNA cistrons per mitochondrial gemone (Aloni and Attardi, 1971). In accordance with these indications for two genetic systems, one in the nucleus and the other in organelles, the differential inhibition characteristics of mitochondrial and cytoplasmic tRNAs indicate two separate synthetic mechanisms (Vesco and Penman, 1969). Judging by the simplicity of the mitochondrial polymerase described earlier, the synthetic mechanisms should be different indeed. It can be calculated that the genetic information required for 20 precursor tRNAs (3.3×10^4 daltons each) represents 1.3×10^6 daltons of helical DNA; for 20 synthetases (10^5 daltons each) the figure is 3×10^7 daltons. Confronting these amounts of DNA with the total mitochondrial genome of the lower and higher eukaryote, 5×10^7 and 10^7 daltons, respectively, and considering that these genomes not only code for these macromolecules, the conclusion is that at least some mitochondrial synthetases and probably some mitochondrial tRNAs of the higher eukaryote must be coded in the nucleus. With respect to tRNA, this seems to be confirmed by the deficient cistronic complement of the mitochondrion cited above. There is evidence to suggest the restriction applies to the synthetase of the lower eukaryote as well (Gross *et al.*, 1968).

Both mitochondrial tRNAs and synthetases of *Neurospora* react better with the bacterial than with their cytoplasmic counterparts. Moreover, the cytoplasmic synthetase charges bacterial tRNA with a mistaken amino acid. Further in keeping with the predicted prokaryote-like ancestry of the organelle, mitochondria but not cytoplasm of yeast, liver (Smith and Marcker, 1968), and HeLa cells (Galper and Darnell, 1969) possess the initiator formylmethionyl-tRNA characteristic of the prokaryote. Similarly, chloroplasts but not plant cytoplasm contain this initiator tRNA (Smith and Marcker, 1968), together with other tRNA species (Barnett *et al.*, 1969) of their own. In conjunction with the bacterial-like nature of organelle ribosomes, these properties make almost certain that particular mRNAs, irrespective of their nuclear or organelle origin, can only be read inside the organelles.

In conclusion, it is clear that at least some species of the two macromolecules under discussion function preferentially, if not exclusively, within the organelles. Conversely, the data imply that certain species are preferential or exclusive to the cytoplasm, and still others perhaps may be shared. To some extent this compartmentation must reflect the coexistence in the cell of two independent translational apparatuses in relation to two distinct genetic systems in the nucleus and organelles. Other general aspects of organelle autonomy were discussed earlier in this chapter in connection with organelle DNA and rRNA.

CHAPTER 5

Translation

Translation is the final meeting ground for RNA function even if not all species act directly at this level (Fig. 1, p. 10). The polysome, the ultimate translational structure, can be regarded as being economical both in terms of the amount of each RNA species utilized in the production of protein and the concerted regulation of this production. Since informational cistrons are not (or mostly not) present in multiple numbers, this polysomal arrangement still achieves an amplification of translation comparable to that which the basic multiplicity of cistrons, or the further amplification of this multiplicity, alone can achieve for transcription of stable RNA. The fundamental aspects of the polysome, basically similar in the prokaryote and eukaryote, are the theme of this chapter. By comparison with the transcriptional apparatus, except for the ribosome itself, the translational apparatus of the eukaryote is remarkably uncomplicated in relation to the bacterial apparatus—and why this greater constraint resulting in a premature terminal evolution should apply to the translational apparatus is not easy to explain. For these reasons, although not ideal, the presentation of the mechanism of translation that follows combines information from both the prokaryote and eukaryote, but inclines toward the prokaryote about which more is known. The basic diagram of translation is shown in Fig. 19. A few special topics in translation are then discussed, such as some of the structural relationships involved, and, by way of antithesis, certain systems

244 · 5. Translation

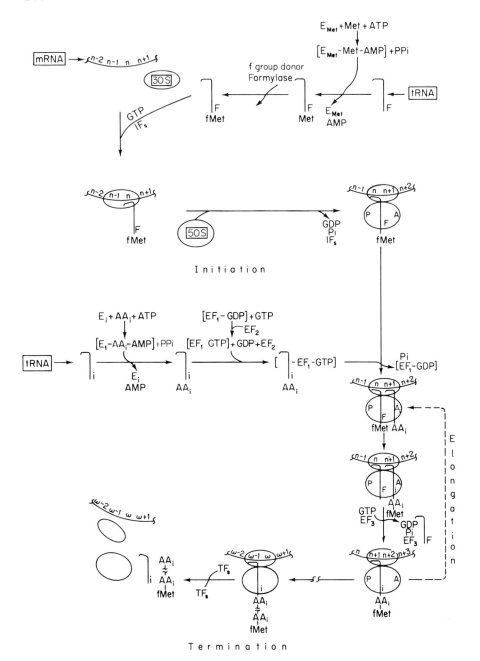

for nontranslational synthesis. Lastly, our attention turns to the regulation of translation which constitutes the main concern of this chapter, and is discussed separately for the two types of organisms. The most salient characteristics of each are considered, with the much greater complexity of the eukaryote becoming obvious. As must be already clear, the present account intends mainly to explore the relationship between RNA and translation and of necessity is only schematic in respect to translation as a field of study in its own right. For further study, articles by Lipmann (1969), Lengyel and Söll (1969), Lucas-Lenard and Lipmann (1971), and several others in *Cold Spring Harbor Symp. Quant. Biol.* 34 (1969) should be consulted. This volume includes a valuable historical review of the subject by Zamecnik (1969).

BASIC MECHANISM OF TRANSLATION

Prepolysomal Events

Transfer RNAs translate the messenger polycodon into a polypeptide. The first step in the process is the binding and activation of amino acid by the amino acid-specific synthetase (Reaction 1, Chart 3). The amino acid is linked by its carboxyl group to the 5'-phosphate of AMP as a mixed anhydride that remains bound to the enzyme. The adenylate-enzyme complex involved is, however, not unquestionably accepted (Loftfield and Eigner, 1969). The previous reaction, somewhat thermodynamically unfavorable by itself, is driven forward by the pyrophosphate split. The carboxyl group retains the high energy status of the terminal ATP phosphate and preserves it through peptidization, although perhaps at a slightly lower energy status after the ensuing esterification described below.

In vitro the activation reaction can be driven backward by excess PP_i with regeneration of ATP (pyrophosphorolysis). In addition, in certain cases, excess tri-

---

Fig. 19. Schematic outline of events during initiation, elongation and termination of translation in *Bacillus stearothermophilus* (adapted from Lengyel and Söll, 1969). The events are described in the text. Symbols and abbreviations for initiation are: $n-2, n-1, n, n+1, n+2$, a series of adjacent codons in mRNA; n, initiation codon; "bottomless bracket with F," initiator tRNA; F met, methionyl-tRNA synthetase; IFs, initiation factors. The oval shapes represent ribosomal subunits. Binding sites (A and P) for tRNA depicted in the large ribosomal subunit are probably shared with the small subunit. The formylation reaction for initiator tRNA is simplified in the outline. Symbols and abbreviations for elongation and termination are: EFs, elongation factors; "bottomless bracket with i," tRNA accepting amino acid AAi; $w-2, w-1, w, w+1$, a series of adjacent codons in mRNA; w, termination codon; TFs, termination factors. Other symbols are analogous to those for initiation. Interruption of the sequence at the bottom of the figure allows for growth of the previous dipeptidyl residue attached to tRNA into a polypeptidyl chain. The current nomenclature of translation factors is arbitrary and is not used in the figure.

phosphate nucleoside (ATP) brings about a nucleophilic attack by it on the AMP phosphoryl group with formation and release of AT_4A, a tetraphosphate dinucleoside. Any triphosphate nucleoside other than ATP gives rise to a hybrid tetraphosphate dinucleoside (AT_4X) (Zamecnik and Stephenson, 1968, 1970). Since this type of molecule was looked for and found in the mammalian cell, the reaction might be of some physiological value. As mentioned in Chapter 4, there are strong indications in bacteria that compounds of this type mediate the regulation of stable RNA.

(Activation) 1. $aa_z + ATP + synthetase \xrightleftharpoons{Mg^{2+}} (aa_z - AMP + synthetase) + PP_i$

(Acylation) 2. $(aa_z - AMP + synthetase) + tRNA_z \xrightleftharpoons{Mg^{2+}} aa_z tRNA_z + synthetase + AMP$

Reactions 1 + 2. $aa_z + ATP + synthetase + tRNA_z \xrightleftharpoons{Mg^{2+}} aa_z tRNA_z + synthetase + AMP + PP_i$

(Peptidization) 3. $aa_z tRNA_z + (polysome - aa_n tRNA_y)$

$$\xrightarrow[K(NH_4)]{\substack{Mg^{2+} \\ GTP}} \left| \begin{array}{l} \text{peptidyl} \\ \text{transferase} \\ \text{factors} \\ R-SH \end{array} \right.$$

$(polysome - aa_{n+z} tRNA_z) + tRNA_y$

Chart 3

The second step mediated by the synthetase is the acylation of tRNA specific to the amino acid (Reaction 2). This is achieved by a nucleophilic attack by tRNA on the amino acid carboxyl group, and results in the esterification of the amino acid to the $2'$- or $3'$-OH of the tRNA terminal AMP residue; the equilibrium favors $3'$-OH (Feldmann and Zachau, 1964). The combined reaction (Reactions 1 + 2) is visualized as a four-stage catalysis cycle in which the release of aminoacyl-tRNA limits the overall rate of acylation (Yarus and Berg, 1969). However, the presence of amino acid, even before activation, affects the combined reaction. The thermodynamic aspects of the reaction were considered by Watson (1970). In this reaction the synthetase undergoes conformational changes which can be recognized physicochemically. This allosteric behavior indicates a functional substructure of the enzyme in the form of subunits separately carrying the recognition sites for amino acid, ATP, and tRNA. In cases where the synthetase consists only of a single chain (Arndt and Berg, 1970), this means a virtual rather than actual substructure. The aminoacyl-tRNA ester resulting from the reaction is stable at neutral or acid pH, and the peptidyl-tRNA that will form as the polypeptide grows is even more stable. The normal rate of acylation is 30% of the maximum possible rate, but can vary enormously under different physiological conditions (Williams and Neidhardt, 1969).

In vitro the combined reaction is highly sensitive to ionic strength (optimum I, below 0.2), Mg/ATP ratio, pH (Shearn and Horowitz, 1969), and free base available (Loftfield and Eigner, 1969; D.W.E. Smith, 1969), in which capacity one of the terminal ribose OH groups may function. The optimal conditions of reaction vary however, sometimes quite markedly, with each aminoacyl-tRNA. How much of this variation applies within the cell is difficult to tell. Also, the general stimulatory effect of polyamines on the reaction is becoming increasingly evident (Takeda and Igarashi, 1970). Once again the physiological implications are not clear. An example of the power of biological adaptation is shown in halophilic bacterial systems which require an ionic strength equivalent to 3.8 M KCl for optimal activity (Griffiths and Bayley, 1969).

Stereospecifically, the combined activation-acylation step is the essence of translation since subsequently the tRNA will transfer the amino acid in response to a codon that recognizes only tRNA. Of these two reactions, acylation is the more discriminatory, and the binding of amino acid to synthetase that precedes activation is less discriminatory (Owens and Bell, 1968). Activation error involving related amino acids such as valine and isoleucine is not infrequent, but the wrong amino acid is usually released from the enzyme complex at the acylation step (Baldwin and Berg, 1966), which prevents fixation of the error. The overall mechanism of acylation is therefore self-correcting (the magnitude of the error was discussed in Chapter 1). Moreover, in the presence of cognate tRNA the synthetase makes less mistakes to begin with (Loftfield and Eigner, 1969); with certain synthetases the activation is possible only in the presence of the cognate tRNA. In other cases, the enzyme is stabilized by both the amino acid and tRNA, yet this synergism does not depend on their covalent bonding as aminoacyl-tRNA (Mitra *et al.,* 1970).

Heterologous reactions between synthetase and tRNA from different organisms can reveal certain aspects of the mutual recognition as yet not well explored. These heterologous reactions are generally less specific than the homologous ones, the lesser the specificity the greater the phylogenetic distance (review by Jacobson, 1971a). In exceptional cases, however, the specificity is retained over wide phylogenetic distances (Anderson, 1969a). Often a heterologous synthetase is more discriminatory against a subnormal, e.g., submethylated, tRNA than a homologous synthetase is (Peterkofsky, 1964), possibly due to the correspondingly subnormal tertiary structure of tRNA. Often a heterologous synthetase is also more discriminatory between normal synonym tRNAs. It is likely that the nonuniversality of heterologous reactions revealed by these observations responds partly to trivial reasons such as different optimal requirements. However, there are indications that these reactions fail more often when the codons of the synonym tRNAs being compared are different (von Ehrenstein, 1970). This might suggest a connection between acylation and codon, but much further work is needed to establish what the connection may be.

Synthetases

The molar ratio of synthetase to cognate tRNA in the bacterial cell is generally between 0.5 and 1 (Novelli, 1967), but can be less than 0.2. Perhaps this enormously high ratio functions, in part, to maintain a critical concentration of charged tRNA in the vicinity of the polysome, below which the concentration of tRNA would become limiting. Attesting to their functional importance, the synthetases amount to 10% of all enzymes present in the bacterial cell.

The same bacterial synthetase acylates more than one synonym tRNA species, e.g., serine tRNAs with entirely unrelated anticodons (Sundharadas et al., 1968). This means that the enzyme is degenerate. However, there may be differences in the rate of acylation of synonyms. Eukaryotic synonym tRNA species, on the other hand, tend to be acylated by more than one synthetase in the cytoplasm (Kull and Jacobson, 1969). In addition, there are different synthetases to acylate synonym species which are exclusive to the mitochondria and chloroplasts (Chapter 4).

Because of the advantages of tRNA for physicochemical study, it might be correctly assumed that much correlative work of this nature is proceeding on the synthetase. It is not attempted to do justice to this work in this brief summary. The synthetases have molecular weights of approximately 10^5 daltons, but some of them are larger; many of them consist of two or more subunits. On a weight basis the enzyme is therefore four times larger than tRNA. Related to its allosteric properties, the enzyme can reversibly adopt an inactive configuration (Lee and Muench, 1969). Both active and inactive configurations are stabilized by the presence of the cognate tRNA. A yeast synthetase has now been crystallized (Rymo and Lagerkvist, 1970), which opens the general possibility of a much needed high-resolution physical analysis of this important enzyme.

Synthetases can reflect heterogeneity in their structural genes as the result of mutation. In general, depending on the particular synthetase and mutation, replacement of one or a few amino acids renders the enzyme thermosensitive or alters one or more of its binding constants (Printz and Gross, 1967), whereas multiple replacements totally inactivate the enzyme. Some-

Basic Mechanism of Translation 249

times, conditional temperature lethals form a structurally different synthetase at growth-restrictive temperatures, suggesting enhanced mistranslation of the enzyme under such conditions. At other times, the temperature-sensitive enzyme is physiologically more prone to dissociating into subunits.

Implicit in this discussion was the idea that, because the interaction between synthetase and tRNA is almost an enzyme-to-enzyme one, it could prove to be very special. However, it is not clear at present whether factors additional to those mentioned earlier are involved in the interaction. A recent account (Mehler, 1970) conveys the special status of this interaction very clearly, while at the same time stressing the different behavior shown by different synthetase-tRNA pairs. The K_m values of synthetase for tRNA are generally in the range of 10^{-4} to 10^{-6} M. The absolute order of the catalytic steps mediated by one synthetase species was described by Allende et al. (1970), an order which should be viewed in connection with the catalysis cycle proposed earlier (Yarus and Berg, 1969). However, from the differences just noted, it is likely that the steps will vary in detail with the particular synthetase-tRNA pair. General aspects of synthetase function were reviewed by Neidhardt (1966), Novelli (1967), and Peterson (1967). Recognition between synthetase and tRNA, in particular, was discussed by Yarus (1969).

The Ribosome

The ribosome, a name coined by Roberts (1958), is discussed here as representing four classes. The ribosome is (a) 70 S (about 2.5×10^6 daltons) in bacterial cells and (b) twice that size, 80 S (about 4.5×10^6 daltons), in the cytoplasm of plant and animal cells (Spirin and Gavrilova, 1969). Two additional classes of ribosomes were described in Chapter 2: (c) the mitochondrial 60 S ribosome in the higher eukaryote which is the smallest so far known and (d) the mitochondrial 70-80 S ribosome in the lower eukaryote. These four classes of ribosomes have been reviewed comparatively by Wittman (1970). Returning to classes (a) and (b), which are the ones considered in this chapter, the two subunits which compose the bacterial 70 S ribosome are 30 and 50 S, and the subunits of the eukaryotic cytoplasmic 80 S ribosome are 40 and 60 S. The rRNA species contained in these subunits are 16 and 23 S, and 18 and 28-30 S, respectively; in addition, the large subunits contain 5 S RNA. (It should be apparent from the variable size of rRNA that the S value of the 60 S subunit in particular is nominal. Moreover, the larger the RNA, possibly the larger the number of proteins.) The subunits of the 70 S ribosome comprise about two-thirds of rRNA and one-third of protein; the subunits of the 80 S ribosome comprise about one-half each of rRNA and protein. Compared with the eukaryotic cytoplasmic ribosome, the bacterial ribosome is more dissociable at low

Mg concentration (1m*M*), possibly in relation to its lesser protein content; it is sensitive to chloramphenicol while the eukaryotic ribosome is sensitive to cycloheximide (except perhaps the lower eukaryote ribosome); and it may have different requirements for initiation, as discussed below (review by Schlessinger and Apirion, 1969). In addition, the eukaryotic ribosome is less prone to mistranslation. A different specificity of the bacterial and eukaryotic ribosome is further suggested by the fact that their subunits are not interchangeable (see later), and, similarly, translation factors are not completely interchangeable (Ciferri and Parisi, 1970). Within the eukaryotic cell, noninterchangeability may also be true of the mitochondrial and cytoplasmic ribosome.

Ribosomal proteins include acidic and basic proteins—the latter distinct from histones—which tend to be small proteins (average 20,000 daltons) and differ between the prokaryote and eukaryote. The bacterial 30 S subunit contains about 20 proteins; the 50 S subunit approximately 40 proteins, most of them present only in one copy (Moore *et al.*, 1968). The number of copies of each protein is therefore invariant in regard to the functional state. However, the absolute number of proteins in the 30 S subunit varies, possibly representing different compositional states of the subunit in relation to the various steps in translation (review by Kurland, 1970). The proteins which are always present are termed unit proteins and those which are not always present are fractional proteins. Of these, the unit proteins are located centrally in the subunit and correspond with the core proteins described below, whereas many of the fractional proteins occur superficially. Although a similar analysis of the proteins in the bacterial 50 S subunit is not yet available, it is probable that the outcome will be similar to that for the 30 S subunit. Thus, in general, the heterogeneous population of ribosomal subunits may reflect the transient presence in the subunit of factors for initiation, elongation, release, etc., during the corresponding step of translation. In turn, each compositional state may correspond, according to some estimates, to one of a dozen possible subtly different conformational states of the subunit. In the eukaryotic ribosome, the exchange of ribosomal protein mentioned in Chapter 4 may be indicative of similar phenomena. The eukaryotic ribosome is envisaged to contain about three times as many protein species as the bacterial ribosome, and their characterization is also under way (Welfle *et al.*, 1969).

Clearly, the ribosome is an extremely complex organelle, representing as it does a noncovalent assembly of a few RNAs and many proteins. Important advances are being made concerning the unravelling of this complexity (reviews by Kushner, 1969; Schlessinger, 1969). It has been possible to achieve *in vitro* a dismantling and amazingly precise spontaneous reconstitution of the constituents of the 30 S *E. coli* subunit. A temporospatial mapping of the assembly, thus far the most complex of such aggregations achieved *in vitro*, has been published (Mizushima and Nomura, 1970). In brief, the reconstitution leading to a functional subunit is sequential and cooperative. It is sequential in the sense that internal or core proteins are assembled prior to the proteins occupying the surface, and it is cooperative in the sense that the number of sites in the 16 S rRNA molecule which can bind the proteins independently of one another is very small compared with the number of proteins. The reconstitution is first order in respect to the functional subunit: a rate limiting, probably unimolecular, reaction represents the structural rearrangement of an obligatory intermediate in the reconstitution. A simplified description of the reconstitution, neglecting secondary associations, is that the rRNA binds mostly the internal proteins and these in turn mostly bind the superficial proteins. In their pleiotropic effects, defective ribosomal mutants show the same interdependence of proteins. It is revealing that the intermediate found during reconstitution is seen to accumulate in certain mutants. All the above indications are that each protein has a unique topographical relationship to the other proteins and to rRNA, which, in turn, is dynamically interactive with the proteins. It was further shown that the functional reconstitution *in vitro* requires mature 16 S rRNA but fails if this is replaced by its precursor. This finding, however, cannot exclude that the *in vivo* assembly requires the precursor. The general implications are that the functional subunit requires both mature rRNA and the complete protein ensemble.

The plausible rationale behind the reconstitution work is, of course, that the orderly assembly *in vitro* should bear a considerable parallel to the ordered assembly of the nascent native subunit (Chapter 4). In this respect, despite the assembly beginning during rRNA transcription, data presented in Chapter 2 indicate that the proteins do not markedly affect the higher order structure of rRNA; this structure remains grossly unchanged after purification from protein. A possibility exists that the clustering of ribosomal protein cistrons common to *E. coli* (Schlessinger and Apirion, 1969) and *B. subtilis* (Smith *et al.,* 1968) serves to stipulate, perhaps by gene dosage, that the individual proteins are made close in time to one another yet in the order permitting their fastest association. An additional fact behind this possibility is that, because of the coupling with transcription and the location of their cistrons, these proteins are made right next to the rRNAs. Recently, also the 50 S subunit of *B. stearothermophilus* was reconstituted *in vitro* (Nomura and Erdmann, 1970). Whereas the main conclusions given above on the reconstitution of the 30 S subunit may apply to this other subunit, a new phenomenon was revealed. It appears from these experiments (and from work on *E. coli in vivo*) that the assembly of the 50 S subunit somehow depends on the simultaneous assembly of the 30 S subunit, yet the reverse is not true. Since no ribosomal protein is shared between these subunits, the element which causes this dependence is not known. From these spectacular advances it would not come as a surprise if the complete ribosome can be reconstituted in the near future.

In addition, the function of hybrid ribosomes is being investigated. Hybrid ribosomes, consisting of a small protozoan subunit and a large rat subunit have the same potential for translation in response to poly U as the indigenous protozoan ribosome (Martin and Wool, 1969); this does not hold if the origin of the subunits is the reverse. Except for the large protozoan subunit, all tested combinations between both subunits of different eukaryotes are active. By contrast, whatever the source of prokaryotic and eukaryotic subunits, these do not reassociate with one another.

Apart from peptidyl transferase in the large ribosomal subunit and possibly translation factors in both subunits, no function is known for individual ribosomal proteins. Yet, that individual mutant ribosomal proteins can affect translation shows their direct involvement in the process (see Effects of Antibiotics, p. 263). It might well be that most ribosomal proteins function cooperatively, and the sophisticated structure of the ribosome is optimized for the total rather than individual function of these proteins. Also, the motion of all proteins relative to one another and the pliable RNA backbone must be taken into consideration for the conformational transitions of the ribosome. While all this makes the functional analysis an arduous one, it is also difficult to see how it can eventually succeed without a real understanding of intrinsic RNA function in the ribosome. The structural importance of the major rRNA species for the physical integrity of the subunits is obvious enough, but subtler functions in translation have been suggested, in particular for the minor rRNA species, which remain unsolved. No doubt a badly needed understanding of structure-function relationships to account for the great number of reactions known to occur in the ribosome is forthcoming from research in this area, one of the most active in molecular biology. Henceforth, the recently discovered crystallization of the ribosome *in situ* under conditions of hypothermia (Barbieri et al., 1970) paves the way to high-resolution physical analysis.

As will be clear in what follows, the ribosome is unquestionably to be regarded as the functional, modular unit of translation. Its structure is, however, asymmetrical with respect to function, as revealed by (a) the single copy representation of most ribosomal proteins and (b) the region where translation occurs appearing to be localized to the intersubunit crevice. In this regard, mRNA is currently envisaged to run in the interior of the ribosome along the crevice, past the binding sites for tRNA located at the inner subunit surfaces of the crevice. It bears stressing, however, that the inner structure of the ribosome remains uncharted. Why the size of the eukaryotic ribosome should be twice as large, apart from indicating a greater complexity, also remains unexplained.

Polysomal Events

Translational events at the polysomal level are known in general outline and differ only in detail for the prokaryote and eukaryote. These events in

Basic Mechanism of Translation

the prokaryote were reviewed recently by Lengyel and Söll (1969) and in the eukaryote by Munro (1970). All translational elements considered, the polysome is structurally complex enough to be classed as a quinternary structure. Two points worth mentioning in this regard are that (a) no polysomal event occurs by itself but, instead, is rigidly controlled by an enzyme or factor and (b), more particularly, these enzymes and factors recognize higher order structure more than discrete chemical groups in the participant macromolecules. Concerning tRNA, for example, its correct tertiary structure is every bit as important for interaction with these enzymes and factors as it was for the previous interaction with synthetase. However, using the same example, the structure of the tRNA also exposes certain chemical groups for interaction: one such group is the common GTψCG sequence in arm IV which is necessary for ribosome binding.

Messenger RNA is read from the 5' to 3' terminus as the protein chain grows in the direction from the N to C terminus. Since mRNA is also made 5' to 3', this codirectionality of transcription and translation is explained by the ancestral coupling between these processes, still used by the bacterium. As described in Chapter 1, first the bacterial initiator F-Met-tRNA binds to appropriate factors and GTP and forms an obligatory initiation complex which then reacts with the small ribosomal subunit in the presence of 5' terminal (or near terminal) initiation codons in the mRNA. The initiation factors possibly recognize both the formyl group and acylation-dependent tertiary structure of initiator tRNA, thus discriminating against nonformylatable Met-tRNA (Rudland *et al.*, 1969). This specificity of function denotes special affinities of factors and ribosomal subunit toward initiator tRNA, and also permits free exchange of subunits.

Initiator tRNA enters the 30 S subunit at the A binding site, and, when the 50 S subunit joins the complex, it translocates to the P site in the 50 S subunit, during which GTP is split. However, there are two views in conflict with this scheme: one is that initiator tRNA enters directly at the P site, because the initiation factor has a unique specificity to do this; the other is that there is a third site, perhaps a subsite of A, for entry of tRNA (Culp *et al.*, 1969). In either event, the templated ribosome has assumed a more compact conformation (Schreier and Noll, 1970) and is now ready to form the first peptide bond.

The elongation reactions that follow are best exemplified by the liver system (Moldave *et al.*, 1968). There is a different nomenclature for enzymes and factors in other systems, but their functions are similar to those in this system. Contrary to the initiation reactions, the elongation reactions are obviously not specific to any aminoacyl-tRNA. Multiple forms of a transferase I interact with the second incoming aminoacyl-tRNA (by which they are stabilized) specified by the codon operative at the time, but probably also screened by the ribosome; the transferase promotes binding of this aminoacyl-tRNA to the A site. As observed in *B. stearothermophilus,* the elongation factors (collectively termed transferase I in the liver system) first bind aminoacyl-tRNA and GTP to form a free ternary intermediate which then releases aminoacyl-tRNA to the A site. Contrary to the initiation factors, these factors specifically exclude initiator tRNA, which helps prevent misreading of the internal GUG codon as Met instead of Val (Ono et al., 1969). Elongation factors also exclude nonacylated or denatured tRNA and nonformylated initiator tRNA. Each of the initiation and elongation reactions have therefore their own fail-safe devices.

Returning to the liver system, next a peptidyl transferase, forming an integral part of the 50 S ribosomal subunit, catalyzes a nucleophilic attack by the free α-amino group of the second incoming aminoacyl-tRNA on the esterified carboxyl group of F-Met-tRNA at the P site. A cyclic diester involving the two cis-OH groups in the terminal ribose of tRNA may act as an intermediate in this peptidization reaction (Rich, 1968); however, other functions have been proposed for the 2′-OH group in this reaction (Zamecnik, 1962). Peptidization results in the formation of dipeptidyl-tRNA at site A and displacement of free initiator tRNA from site P (Reaction 3 in Chart 3). As expected, the peptidyl transferase functions solely in the complete templated ribosome. The function of the enzyme requires the CCA terminus of tRNA at site P but apparently only the CA portion at site A (Monro et al., 1968). Thermodynamically, the high energy in the ester bond ($\Delta F = -7$ kcal/mole) suffices to drive peptidization (-3 kcal/mole) irreversibly.

A transferase II (translocase) then moves dipeptidyl-tRNA to site P, thus freeing site A for the third incoming aminoacyl-tRNA. The ribosome simultaneously moves one triplet relative to the mRNA, except when the ribosome is immobilized at the endoplasmic reticulum membrane, in which case the actual movement is supposedly by mRNA. In either case, stepwise translocation ultimately is a measure of codon-anticodon binding, i.e., tRNA "telling" the ribosome when to move one triplet. After the codon occupied by this third aminoacyl-tRNA, no new enzymes or factors are required (Erbe et al., 1969). The entire cycle is repeated from then on, except for peptidyl- instead of aminoacyl-tRNA at site P, until a termination codon appears in the mRNA.

Attesting to its functional prominence, the translocase involves 2% of the total E. coli soluble protein. One molecule of translocase seems to be active per ribosome (Leder et al., 1969). Two steps specifically requiring GTP, not a nonhydrolyzable analog, are the terminal stage of binding of tRNA before peptidization and its translocation, from which it is assumed that at least two GTP molecules are consumed per unit elongation of the polypeptide. Since the initial stage of binding of tRNA and the peptidization reaction can use a GTP analog, the question arises as to whether GTP as such is required at these stages for an allosteric purpose only. However, since there is enough energy for peptidization in the peptidyl donor, the more general question is: what is the extra energy liberated from GTP required for? It could be for driving the ribosome along the mRNA, either by means of a high energy intermediate or again allosterically by causing a conformational change in the ribosomal subunits. A different element controlling translocation at least in bacteria is cAMP: its level may also determine whether the translocation-dependent RNase V (Chapter 2) becomes active in degrading mRNA (Kuwano and Schlessinger, 1970).

To exemplify polysomal events with the *E. coli* tryptophan operon discussed in Chapter 3, a cluster of 20 ribosomes travels behind the transcribing polymerase. Hence, in this operon there may be a definite stoichiometry in translation with respect to informational transcription. Each ribosome occupies about 200 Å (a subunit diameter) of extended mRNA. Since the extended internucleotide distance is about 7 Å, this means that each ribosome encompasses a length of mRNA of about 30 nucleotides; still, other data indicate that the center-to-center distance between translating ribosomes may be two to three times this number of nucleotides. The ribosome tandem allows multiple translations to proceed simultaneously, each ribosome translating essentially independently of the rest. Yet, the ensemble as a whole is coordinated as discussed below. Because in this bacterial system transcription and translation are coupled, they have the same absolute duration. Because of the coding ratio, however, the maximum rate of translation can only be one-third that of transcription, or about 60 msec per amino acid at $37°C$. Since this overall rate is actually observed (Manor *et al.*, 1969), the indication is that translation is not physiologically limiting.

Specific factors involved in chain termination and release were discussed in Chapter 1. It was mentioned that peptidyl transferase appears to play a part in these processes. Termination requires the transfer by esterification of the completed chain from the last translating tRNA to a water molecule. Since peptidyl transferase seems capable of catalyzing this esterification as an alternative to the normal amide formation during peptidization (Fahnestock *et al.*, 1970), it might be that a termination factor-determined allosteric transition in the enzyme could influence termination, thus making termination a partly ribosome-catalyzed reaction. The present view is that dissociation of the ribosomal subunit concurrent with release from the polysome is not spontaneous but is again mediated by a specific factor. This would mean that dissociation is regulated as opposed to its being incidental to the intrinsic stability of the ribosome. Moreover, since a factor acts stoichiometrically, it could regulate dissociation very rigidly. Related, if not the same, factor(s) have been proposed to control translational events at both ends of the polysome, initiation and release, thus bringing forth a true regulatory cycle. However, the fate of the ribosome between falling off the 3' terminus of mRNA and reentering at the 5' terminus to reinitiate another round of translation is not clear. About 10-20% of the bacterial ribosomes appear to occur as free subunits in between rounds of translation and to exchange freely. Whether there are, in addition, free ribosomes in the bacterial cell is debated. In contrast, both free subunits and ribosomes are likely to occur in the eukaryotic cell. This question is discussed in more detail later.

Some Characteristics of Eukaryotic Translation

Eukaryotic translation conforms with the previous general scheme but fewer details are known than for bacterial translation. The present purpose is to mention briefly some translational similarities as well as differences between the two types of organisms. A primary difference with profound consequences is, of course, that eukaryotic translation is not coupled to transcription, except presumably in cytoplasmic organelles.

Concerning initiation of translation, one of the most studied proteins is globin. In this case, it was not clear whether initiation involved N-acylated aminoacyl-tRNA as in bacteria (Laycock and Hunt, 1969), acylated tRNA (Arnstein and Rahamimoff, 1968), deacylated tRNA (Mosteller *et al.*, 1968), or no tRNA at all, but a factor. However, as in many bacterial proteins, it seemed fairly certain that in globin the N-terminal amino acid (methionine) was excised to expose another amino acid (valine) at the terminus. According to recent work on different proteins, the eukaryotic initiator may be nonformylated Met-tRNA responding to AUG and possibly GUG as in bacteria (Smith and Marcker, 1970), or *N*-acetyl aminoacyl-tRNA (Liew *et al.*, 1970; Polz and Kreil, 1970). Thus, pending further work, it appears that the initial amino acid, if at all modified, may not be formylated as it is in bacteria. Although not yet isolated, all components of the bacterial initiation complex including factors are present in the eukaryote (Heywood, 1970). Much less can be said regarding any mechanistic differences of chain propagation and termination in the eukaryote. Indeed, chain propagation in liver was used as one of the general model systems in the preceding presentation of translation. Likewise, chain termination in the eukaryote resembles bacterial termination, including the pattern of codon recognition and the presence of termination and release factors. However, the number of these factors may or may not be the same as in bacteria (Chapter 1). Furthermore, a point of general importance concerning all eukaryotic translation factors (not just the last mentioned) should be noted here: it is not known whether the factors as a whole always behave stoichiometrically with respect to the translational apparatus as they appear to do in bacteria. If they do not behave stoichiometrically, their regulatory capacity may be qualitatively more significant than in bacteria.

The eukaryotic ribosome is characterized by a greater size of rRNAs and also a larger number of ribosomal proteins. The complication in respect to the latter is therefore in the number of different proteins rather than in the number of copies of each. Although this structural complexity suggests a greater functional (rather than mechanistic) complexity compared with the prokaryotic ribosome, little is definitely known about this. However, almost certainly this complexity results in a greater capacity for interaction with cytoplasmic elements in general, and, typical of the eukaryote, endoplasmic reticulum membranes and hormones, in particular. It is perhaps also significant here that the eukaryotic ribosome mistranslates *in vitro* less than the bacterial ribosome (Weinstein et al., 1966). As in bacteria, on the other hand, a differential protein content (Henshaw and Loewenstein, 1970) and phosphorylation (Kabat, 1970) of the ribosomal subunits may indicate conformational changes in the translating ribosome to correspond with the various steps of translation.

Polysome size is a reflection of peptide chain length, which indicates a uniform spacing of ribosomes on mRNA. For example, globin (17,000 daltons) is made on a polysome with 5 ribosomes, tropomyosin (35,000) on a polysome with 5-9 ribosomes, actin (70,000) on a polysome with 15-25 ribosomes, and myosin (200,000) on a polysome with 50-60 ribosomes (Heywood and Rich, 1968). This constant numerical relationship further indicates that monocistronic mRNAs are an obligatory characteristic of the eukaryote (Kuff and Roberts, 1967). As shown *in vitro*, as the proteins just mentioned are made they assume their native quaternary configuration, involving association of subunits.

The rate of mammalian translation *in vitro* at 37°C was found to be less than 1.5 minutes for globin and 6 minutes for the ferritin subunit (25,000 daltons). This rate is of the order of 0.5 to 1 second per amino acid, or 15-30 times slower than the full bacterial rate. Germane to what follows, the Q_{10} for translation is 2.5. Certainly a major difference with the bacterium is the wide range of organismal size and life-span of the higher eukaryote. In this regard, a larger size in mammals is accompanied by a lower rate of protein synthesis (as indicated by the renewal of albumin), and this is apparently achieved by a lesser number of ribosomes active per cell rather than by changing the actual rate of translation (Munro, 1970). Interestingly, therefore, fish liver at 22°C shows the same rate of translation as the mammal, but on acclimation to a lower temperature the rate increases by increasing the rate of elongation; the compensation is partly due to increased transferase I activity (Haschmeyer, 1969). By contrast to the homeotherm discussed later, changes in the rate of translation might then be a more common phenomenon of poikilotherm regulation.

Further aspects of translation typical of the eukaryote are discussed in the next section, and again later in connection with the regulation of translation.

SPECIAL TOPICS IN TRANSLATION

Models of Structure-Function Relationships

In first thinking about polysomal function, the analogy was made between the GTP-driven translation of ribosomes along mRNA and the ATP-driven displacement of myosin on actin during muscle contraction. According to a model of the ribosome proposed by Spirin (1968), a GTPase-dependent intrinsic pulsation of the subunits would regulate the translocation of all translational components at or near the ribosome. This model has gained acceptance in that it is now believed that the ribosome undergoes a cycle of contraction and expansion with each amino acid added. Moreover, it has now become clear that the translational functions presented in the previous section occur mostly inside the ribosome. Thus, according to Spirin's model, the various translational components would interact with one another inside the ribosome and between the subunits, the polarity of translocation being ruled by the intrinsic asymmetry of the ribosomal subunits. A second model of translocation proposed by Lengyel and Söll (1969) envisages that the ribosomal subunits continuously dealign and realign with respect to each other. A different proposal is that the ribosome directs allosterically the pairing of each tRNA with the appropriate codon (Woese, 1970a), and, in assisting in this function, the elongation factors promote translocation. These models are probably oversimplified but all concentrate on explaining the still unknown force that drives the ribosome along mRNA, or vice versa. It is clear from them that the mechanistic connotation in the name "translational apparatus" is an apt one. As pointed out, whatever the model, there is no question that the ribosome must be regarded as the functional unit of translation. Nevertheless, a working understanding of ribosomal function depends on elucidation of the overall quinternary structure of the polysome.

Hence the function of the ribosome must comply with the structural requirements of translation, chief among which are those pertaining to polysomal structure. As one example of these structural requirements, one may ask which is the minimum functional length of mRNA? Granting that one ribosome spans about ten amino acids' worth of mRNA and that two to three times this length lapses between consecutive ribosomes, a minimum polysome with a stable number of two functional ribosomes would require an informational length of mRNA equivalent to about 50 amino acids. This is the range of the largest protamines. However, since several protamines and biological peptides have a shorter chain length than this, the question arises as to how these shorter molecules are manufactured. They may be made by (a) an informationally shorter mRNA capable of carrying only a single ribosome (the actual length of mRNA may be greater depending on the additional length of noncoding sequence), but, apparently because of structural reasons, this arrangement is exceptional (Kuff and Roberts, 1967). Alternately, they may be made by (b) an informationally longer mRNA in a true polysome first as a larger precursor protein which is then cleaved into smaller fragments, a mechanism which is generally known to occur. Lastly, they may be made by (c) a mRNA-less oligopeptide synthesis as discussed later; for this type of synthesis the maximum length of product is not known, but is believed to be 15-20 amino acids (Lipmann, 1971).

Another example of structural requirements of the ribosome concerns the disposition of tRNA within the ribosome. In regard to this aspect the following points are worth noting. First, given that the binding to mRNA is at the center of the tRNA molecule where the anticodon is located, a standard axial length of the adaptor (80-90 Å) becomes mechanically essential in order for peptidization to occur at the 3' terminus. Second, the peptidyl transferase being an integral part of the large ribosomal subunit, the 3' terminus of tRNA must come close to the subunit in order for the enzyme to reach it (Monro et al., 1968). Finally, and as speculation, since the transfer of amino acid seems to depend on prior binding of codon and anticodon, it could be that this binding alone elicits a conformational fluctuation in the tRNA or ribosome which makes the transfer sterically possible. However, the specific recognition between ribosome and tRNA is probably also sterically important. According to the model of codon-anticodon interaction proposed by Fuller and Hodgson (1967), mRNA can recognize two adjacent anticodons by forming a kink while the axes of the two tRNA helices remain 22 Å apart. In this model, the two anticodon arms can form H bonds with each other using a ribose OH group and a phosphate group oxygen, respectively. The final structure of tRNA and ribosome upon which the mutual steric relationship between adjacent tRNAs depends is not known.

In the eukaryotic ribosome the large subunit sits against the endoplasmic reticulum membrane (Sabatini and Blobel, 1970). A central channel is believed to traverse the large subunit through which the nascent chain forming inside the ribosome moves the length of some 30 amino acids either on its way into the cisternal space, in the case of the attached polysome, or into the cytosol, in the case of the free polysome. The newest part of the chain is thus protected by the ribosome. The question remains as to what drives the chain out of the ribosome: a possible candidate is the pulsating ribosome itself. [It is not clear whether or not the large bacterial subunit has a central channel (Lubin, 1968).] In the case of the attached ribosome, it is unlikely, however, that the chain itself holds the subunit to the endoplasmic reticulum membrane, for which specific surface sites on the ribosome and membrane are, instead, more probable. It was proposed (Sabatini and Blobel, 1970) that a hydrophobic N terminus of the growing chain, presumably encoded in mRNA, could initially seek out these membrane sites and promote their attachment to the corresponding ribosome sites. This view assumes that the complete, translating polysome joins the membrane. Contrary to this view, it is also proposed that the large subunit first joins the membrane and then the initiation complex (small subunit, mRNA, tRNA, factors) seeks out that bound subunit (Baglioni et al., 1971). In this case, polysome formation and translation depend on prior membrane binding.

The endoplasmic reticulum-bound polysome is currently viewed as an export protein manufacturer and the free polysome as supplying the cell's own proteins (Ganoza and Williams, 1969). The first class of polysome predominates in organs such as liver and pancreas, and the second class in actively multiplying cells. This distinction in terms of product between free and bound polysome may perhaps not be an absolute one, however. Nevertheless, as it stands, the distinction says that, according to the two views given above, either mRNA or the initiation complex carries the specificity for the polysome to become free or bound, on which will depend the destination of its product. In different ways both views stress essentially a binding specificity of mRNA, which seems a logical extension of its informational specificity. A corollary of this binding specificity of mRNA is that the matching sites in the ribosome and membrane, or any additional factor as may intervene, are necessary but not sufficient conditions in order for the polysome to bind. Hence, in reality, the crucial mechanism by which a cell directs its product is not quite decided. This conclusion is valid even if other mechanisms to facilitate this direction are still possible. To cite an example, if the ribosomes in the free and bound polysomes were not interchangeable with one another (i.e., belonged in different pools), they might help each class of polysome find its destination in the cell. In this respect, there are indications that some ribosomal proteins may differ between the two classes of polysomes (Fridlender and Wettstein, 1970), yet unfortunately, this could either be the cause or because of the polysomal classification. The fact is that, apart from the membrane association, no significant differences have been found between the two polysomal classes.

Somewhat more speculative views bear on aspects of the bound polysome. As considered in Chapter 3, a unit polysome-endoplasmic reticulum package was proposed to be produced in response to hormone (Tata, 1968), which implies a segregation of the reticulum. In addition, based on the instability of structural mRNAs for enzymes and the altered appearance of bound polysomes in hepatoma, Pitot (1969) has argued that the typical rosette disposition of the polysome on the endoplasmic reticulum membrane reflects a specific relationship of the polysome to a mosaic of matching attachment sites in the membrane. As seen above, attachment to the membrane is by the ribosome or the nascent chain, not by mRNA. However, it is fair to mention that the rosette is adopted as the polysome forms on newly made mRNA at the outer nuclear envelope, so that a purely mechanical explanation is not discounted. Be that as it may, Pitot (1969) goes on to argue that, by interacting with the internal cell environment, the membrane would control the stability of mRNA. The final implication is that the anomalous changes that seem to occur in the endoplasmic reticulum of hepatoma are formally analogous to those occuring during normal liver differentiation. In conclusion, Pitot considers that a "membron" unit in the rough endoplasmic reticulum would function somewhat as the equivalent of a cytoplasmic operon, a genetic (autonomous?) endowment of endoplasmic reticulum providing for regulation. [Of interest, Pitot also reviews the evidence for a primitive rough "endoplasmic reticulum" in bacteria,

Special Topics in Translation   261

since bacterial ribosomes may be weakly bound to membranes (Hendler, 1965). It is also clear that the bacterial cell membrane itself has important functions in relation to macromolecular syntheses.] Despite its prokaryote oriented overtones, this proposal focuses on the structural relationships in the polysome and membrane, which is quite reasonable in attempting to understand translation insofar as the bound polysome is concerned. Yet, from all that was said regarding this polysome, it is evident that increased knowledge of the mechanics of polysome assembly vis-a-vis membrane assembly is urgently needed. Perhaps some of this knowledge may come from the study of hormonal action on the membranes. For example, this time with social overtones, it is beginning to look as if female sex hormones stimulate the *in vitro* binding of polysomes in the male while male sex hormones stimulate binding in the female (Sunshine *et al.*, 1971). The precise functional implication *in vivo* of these findings is unclear, for, as pointed out before, any hormonal action ought to be secondary to the mRNA specification of the binding. The basic question which, surprisingly, it still apparently unsolved is whether membrane-bound polysomes can occur in cells cultured in the absence of hormone.

Implicit in these considerations, there is hardly any doubt that the total architecture of the cell is not only functionally but also regulatorily significant at all stages of translation. This must in general be so since translation relies on the cytoplasmic environment as much as transcription does on the nuclear environment. However, what is in mind here is not the organization of membranes discussed above, but, in particular, the organization of the cytoplasmic matrix, which is in fact the sole one interacting with the free polysome. Viewed in this way, one important aspect to cytoarchitecture is compartmentation of function in terms of active cell sites (including membranes) as well as the metabolic support for these sites. This compartmentation need not represent a static condition, but, quite the contrary, a dynamic condition maintained by an equally dynamic architecture representing very high molecular specificities. As will become clear, without this organization of cytoplasm the regulation of eukaryotic translation would be difficult to imagine. Yet little is known of this organization because it cannot be preserved when the cell is disrupted, almost a sine qua non of biochemical analysis. It is a well known fact that all cell-free protein synthesis works at a marginal fraction of the *in vivo* capacity. Some of these architectural aspects will be discussed further in connection with the regulation of eukaryotic translation.

Quantitative Aspects of Transfer RNA during Translation

Regulation of synthesis of tRNAs and ribosome was discussed in Chapter 4. Here we shall focus on their final quantitative relationship, which is one important parameter of translation and definable enough to merit separate treatment. As expected, within the physiological range, the concentration of tRNA is rate limiting for translation (Anderson, 1969b). Misreading *in vitro* is greater at a low rather than a physiological tRNA concentration. In log phase bacteria the ratio of tRNA to translating polysome-bound ribosomes (the majority of bacterial ribosomes) is about 15 and more than 100 in slow growing bacteria (Maaløe and Kjeldgaard, 1966): clearly, at the highest ratio few tRNA molecules can be translationally active. In the eukaryote the ratio of tRNA to total ribosomes (not all translating) is 12 to 20 as shown by the hatching *Xenopus* embryo (Brown and Littna, 1966b), adult rat liver cell (Quincey and Wilson, 1969), and HeLa cell (Table 4). Whether higher ratios than these occur in the eukaryote is not known. However, in the opposite direction, possibly lower ratios are found in immature eukaryotic organs (Chapter 4).

At the log phase ratio of tRNA to ribosome about one-tenth of bacterial tRNA is ribosome bound. The ratio of a single species of Phe-tRNA to ribosome was estimated to be 0.5 in log phase *E. coli* (Anderson, 1969b). Probably similar values obtain for certain tRNA species in the eukaryote (Hoagland, 1969). These low specific ratios clearly show that a very efficient mobilization of aminoacyl-tRNA is mandatory to meet the normal rate of protein chain growth, otherwise any tRNA becomes rate-limiting as indicated above. This is also probably one reason for the high concentration of synthetases and the great affinity of translation factors for the ribosome. In addition, an optimal concentration gradient of aminoacyl-tRNA may be maintained by the cytoplasmic matrix (or membranes) in the proximity of the polysome. Pointing in the same direction, the degree of tRNA acylation is high in the functional cell. Both tRNA and aminoacyl-tRNA are capable of ribosome binding, so that their ratio competitively affects the rate of translation (Levin and Nirenberg, 1968). Thus about 70% of the bacterial tRNA is acylated irrespective of the rate of growth (Yegian *et al.*, 1966). Even starvation for any one of several required amino acids does not result in complete deacylation of its cognate tRNA. In yeast, however, the acylation may decrease during exponential growth (Giege and Ebel, 1968), apparently due to a reduction of intact 3' acceptor termini. In rat liver, again almost all tRNA is fully charged even after a 24-hour fast (Allen *et al.*, 1969). Moreover, as expected for an effective regulation, both synthesis and acylation of eukaryotic tRNA (Chapter 4) are under a joint hormonal control.

Evidently, the absence of aminoacyl-tRNA would cancel synthesis of a protein requiring the amino acid if no synonym tRNA capable of reading the pertinent codon were available. This situation is found in a lethal amber-suppressor strain discovered by Söll and Berg (1969). Because of a mutation in position I of the anticodon, a major glutamine tRNA species (CAG responding) can read amber (UAG) as glutamine, and, as a consequence, not enough of a minor synonym species is available to read the original CAG codon. More particularly, a single tRNA species within a multiple synonym set could serve to read a hypothetical regulatory codon, as shown for arginine

Special Topics in Translation 263

in an *in vitro* model system (Anderson, 1969b). What is implied in this paragraph is that the distribution in mRNA of synonym codons may be, apart from structural considerations, of regulatory value. From what is known, this is more likely to occur in the eukaryote.

The tRNA complement of an organism can be adjusted according to functional needs. In the eukaryote certain tRNA activities predominate in an organ according to the composition of its typical protein product, again presumably because without such increase those tRNAs would be strongly limiting. One example is the silk gland (Garel *et al.*, 1970). This situation in general is likely to reflect not only quantitative but also qualitative differences in the transcriptional regulation of synonym tRNA species; the genetic and hormonal mechanisms of this differential or selective regulation were discussed in Chapter 4. Also in the bacterial cell the rate of transcription varies greatly for each synonym species, as shown by their disparate relative abundance in the tRNA population (Goodman *et al.*, 1968). We shall return later to the functional implication of this variation.

Between organisms, in addition to the regulatory differences just mentioned, examples are known of numerical cistronic variation with respect to individual tRNA species. Different organisms differ in the number (and, as mentioned, relative abundance) of tRNA species as well as in the codon-anticodon combinations more frequently used for a given amino acid. As hypothesized above, some organisms may contain no tRNA at all to read certain codons (Caskey *et al.*, 1968), which would then not be represented in their genomes. For example, six amino acids in each *E. coli* and mammalian liver command a total of 16 tRNA species, of which seven codons are common to both systems, seven codons unique to liver, and five codons unique to the bacterium (two species in the bacterium and four species in liver are redundant). The general implication of this variation is that the representation of tRNAs varies widely between genomes. Likewise, the possible fact that certain bacteriophages as well as lower eukaryote mitochondria carry the information for a complete set of tRNAs indicates coding specificities different from the bacterial and eukaryotic host cell, respectively. The aspects referred to in this section show that pronounced variations occur between organisms in the tRNA cistrons and their multiplicity, and between cells in the frequency of tRNA species within the population. Some of these aspects, especially the latter, are basic to the consideration of tRNA as a regulator of translation discussed at the end of this chapter.

Effects of Antibiotics on Translation

Antibiotics are used as a tool to probe the mechanism of translation. As an illustration, the *in vitro* effects on translation of one of these, streptomycin, are briefly described here. These subtle effects are probably separate from the drastic cessation of translation that streptomycin causes *in vivo* and which is believed to be by interference with the translation initiation complex. For a general coverage of the subject, including many other antibiotics, the reader may consult the reviews by Weisblum and Davies (1968) and Petska (1971).

Streptomycin and related aminoglycoside antibiotics cause an ambiguity of translation formally analogous to informational suppression, that is, a defectively coded protein is read correctly; conversely, a correctly coded protein is read defectively. Different mutant ribosomes misread differently in the presence of streptomycin. Neomycin even allows denatured or single-stranded DNA to function as mRNA (Bretscher, 1968b). By using circular DNA in these experiments, incidentally, it was shown that free ends in this artificial template are not requisite for translation.

Streptomycin binds to one ribosome site, but other antibiotics bind to many (Davies and Davis, 1968). It is now known that streptomycin resistance involves mutation of a core protein in the 30 S ribosomal subunit, and that removal of split (fractional) ribosomal proteins does not increase the ambiguity of translation. The mutant core protein therefore contains the binding site for streptomycin. The same protein has been somehow implicated in the binding of initiator tRNA (Ozaki *et al.*, 1969).

The mechanism of antibiotic action of interest here is the resulting distortion of normal ribosome structure that alters the codon-anticodon specificity to produce the ambiguity described above. Also of interest, the streptomycin-dependent ribosome, the product of mutations at the streptomycin locus, is also distorted and consequently translationally inactive, yet can be "cured" by the further distorting effect of the antibiotic. On the other hand, the mutant streptomycin-resistant ribosome is one-tenth as ambiguous *in vitro* in the absence of antibiotic as the sensitive ribosome, indicating that ribosomal structure is directly involved in the fidelity of translation. Other examples with a similar indication are a 30 S subunit mutation that suppresses *in vivo* all three nonsense codons (Rosset and Gorini, 1968) and still other mutations that, in fact, mimic the effect of streptomycin. This implies that the ribosome discriminates between incoming tRNAs, either normal or suppressor. Curiously enough, several antibiotics that prevent chain elongation, hence protein synthesis, do not prevent the movement of ribosomes along the mRNA (Curgo *et al.*, 1969); this unproductive translocation requires the usual translational components such as tRNA and GTP but presumably not all translation factors. As described in Chapter 4, this phenomenon has been instrumental in the analysis of the mechanism of bacterial stable RNA regulation.

Nontranslational Oligopeptide Synthesis

This section deals briefly with certain systems with unique characteristics which may well represent successful living fossils, as it were, of primitive protein synthesis (Chapter 1), that is, before the need evolved for the translation of larger products from templates. Thus short biological peptides are synthesized by means of an aminoacyl-tRNA synthetase-like enzyme without intervention of ribosome, tRNA, or mRNA. The synthesis is therefore not RNA templated. This is the case for glutathione (Mooz and Meister, 1967) and small peptide antibiotics such as gramicidin (Gevers et al., 1968). The specificity of the enzyme, which is related to but different from the synthetase, alone determines the sequence in the oligopeptide. In this system, ATP-activated amino acids form an intermediate enzyme-bound phosphorylated complex. Hence the system resembles a translational one, except (a) that the amino acid is bound to an enzyme (not tRNA), and (b) the order of several enzymes (a polyenzyme system but not mRNA) determines the order of amino acids in the oligopeptide; like a polypeptide, this grows in the N to C direction. Not surprisingly, therefore, the system is effective only for short products. Nonprotein amino acids such as D-amino acids are found exclusively in these oligopeptides, which, of course, is essentially because the synthesis does not have to comply with the genetic code.

Akin to the previous system, complex multidimensional polymers of the bacterial cell wall contain polyamino acid links, the amino acids being transferred by tRNA but still without ribosome or template taking part (Lipmann, 1969). There is no explanation of why tRNA is used in this but not the previous system, except that the reason must somehow lie in the history of the system; not infrequently, tRNA is found at the root of atavistic synthetic systems, as already indicated in Chapter 2. In any event, certain tRNA species are exclusively devoted to such nontranslational role. For example, one species serves to insert glycine in the peptidoglycan of *S. epidermidis* whose total glycine is as abundant as in all cell protein combined. The Gly-tRNA which functions exclusively in the synthesis of this polymer appears to possess thiouridine as its only modification (Stewart et al., 1971). This may suggest that the system continues to use ancestral-like (unmodified) species of tRNA, as opposed to most other species which have evolved (been modified) jointly with the translational apparatus that they serve. Most revealing, in addition, the exclusive Gly-tRNA of *S. epidermidis* responds to none of the four codons for the amino acid. In all, therefore, these nontranslational systems bring into prominence the "code within the code" which operates at the level of tRNA and synthetase (or synthetase-like enzyme) as was explained in Chapter 1.

While these systems suggest the possibility that oligopeptides exclusive to the eukaryote may be similarly made, it is probable that some of them are not. For example, oxytocin and vasopressin, two hormone-like nonapeptides consisting of L-amino acids, appear to derive from a large protein precursor (Sachs, 1969) whose synthesis is, of course, translational. An explanation for this ostensibly wasteful synthesis is believed to be that the remainder of the precursor serves as the carrier protein, neurophysin, for the oligopeptides.

A role for tRNA not requiring template, which is different but perhaps related to the previous roles, has come to light in bacterial and mammalian extracts. This is the direct transfer of amino acid to the N-terminal position of acceptor proteins such as ribosomal protein (Leibowitz and Soffer, 1970). A soluble enzyme, tRNA-protein transferase, achieves a correct peptide linkage again without benefit of ribosome, Mg, or GTP. Presumably the transpeptidation is used to modify protein to a functional purpose.

REGULATION OF TRANSLATION

Regulation in the Prokaryote

Because translation is coupled to informational transcription, many aspects of bacterial translation have already been discussed in Chapter 3. Due to this physical coupling, much of the regulation of translation takes place, in effect, at the level of transcription, ensuring the rapid translational response typical of the bacterium. Moreover, the overall kinetics of this response is very nearly symmetrical as a consequence of the short life of mRNA, that is, both the onset of translation during induction and the discontinuation during repression are rapid. Given that bacterial translation is thus not independent of transcription, the emphasis in this section is therefore on the more intrinsic regulatory aspects of translation. However, since these aspects require further substantial investigation, their present discussion can only be brief and serve for no more than orientation. The principal interest is in bacterial translation, but extensive reference to viral translation is required for comparative purposes—hence the inclusive title of this section.

A first consideration is that, because of the coupling to transcription (and the almost constant rate of chain propagation), the overall rate of translation is a direct function of the rate of formation of mRNA. It also turns out that the functional duration of mRNA is partly dependent on (the duration of) transcription. The overall rate of translation is equally a function of the number of polysomes at work: as the rate of translation increases so is the number of newly made ribosomes that enter the polysome fraction. These rules need not, however, be taken to apply literally in every instance, and indeed what is explored in the following discussion are some of the possible points of departure from these rules. A second consideration is that most constituents of the translational system are themselves subject first to some sort of transcriptional control and then to translational control. As a result of this comprehensive control, the function of the translational system becomes dependent on the metabolic status of the cell and the rate of translation is adjusted to the rate of cell growth. Collectively, the translational aspects represent a second level of regu-

Regulation of Translation 267

lation additional to the primary regulation at the level of transcription. Yet these two levels of regulation may be and most probably are integrated in a formidably complicated network. We shall now look into the regulation of the steps of translation in the order of chain initiation, elongation, and termination.

By analogy with transcription, initiation of translation is probably more closely regulated than elongation or termination. By comparison, but for exceptional circumstances, both elongation and termination proceed at a relative constant rate. It is not clear how the rate of initiation is controlled, however. As far as is known, even though translation is as a rule initiated soon after transcription, initiation is not controlled by transcription except indirectly for the provision of initiation codons in mRNA. Nevertheless, from now on, in practice it is only the coupled transcription-translation system that we have to consider. In this system, the intrinsic elements that govern initiation of translation are (a) the irregular periodicity of initiation inherent in the transcription promoter gene (discussed in Chapter 3), and (b) the availability of initiation factors for both transcription and translation. Extrinsic elements of regulation are the inducer or corepressor required for interaction with repressor, another intrinsic regulatory element. Mechanistically, it follows, of course, that all participants in transcription and translation must be operative in order for translation to commence, namely, promoter genes, polymerase, cAMP, ribosomal subunits, initiator tRNA, initiation factors, GTP, and substrates. The translation initiation complex in particular, which provides the complete specification for the process and therefore has an obvious potential for regulation, was discussed in detail in Chapter 1. (Polarity mutants offer a clear but rather special example of control of initiation, namely, internal initiation in polycistronic mRNAs.) However, at some stage in translation, but probably beginning with initiation, positive and negative feedbacks become important, namely, (a) catabolite repression at the transcriptional and translational levels; (b) end-product inhibition; and (c) effects and availability of substrates at the translational level. Admittedly vague, this is all the information we have about the important control of initiation of translation.

To digress for a moment from bacterial translation, translation initiation factors are introduced in the bacterial cell by infecting DNA phage (Hsu and Weiss, 1969), evidently to replace the bacterial initiation factor. As a result of this replacement, T4 phage infection alters the translational ability of the bacterial ribosome toward host and unrelated viral templates, while enforcing the ability toward its own templates. These phage-specified factors should therefore be considered part of the general subversive strategy against bacterial host function. This strategy was discussed in Chapter 3 in connection with the informational transcription of phage.

Certain recent findings in RNA phages may also be relevant to mention at this point. From this work, which is considered in some detail later, it appears that there may be specific ribosome binding factors distinct from the factors intervening in the translation initiation complex. Preliminary views are that these factors would positively recognize certain sequences in the viral RNA preceding any of several initiation codons, on which they form a binding site for the ribosome. Thus several different ribosome binding sites have been demonstrated in viral RNA which might mean as many binding factors. The interaction of these binding factors with the translation initiation factors, if any, is still unknown. Superficially at least, the analogy of binding factors with transcriptional σ factors in both bacterium and bacteriophage is clear. However, in bacteria the status of binding factors is more speculative than in the case of viruses. The reason for this is because, contrary to the partly replicative purpose of viral RNA, the discreteness of bacterial translation and its coupling to transcription may obviate the need for the specification of ribosome binding. Nevertheless, with or without the need for such specification, it should not be forgotten that intercistronic sequences in polycistronic bacterial mRNAs function rather similarly to the binding sites in viral mRNAs in that they are capable of regulating internal reinitiation.

Returning to the bacterium, the highly ordered structure and function of the polysome permits one to visualize possibilities for further regulation at many translational steps after initiation. There is as yet insufficient evidence from molecular genetics to provide specific examples of this area of regulation, but some circumstantial evidence in its favor is available. First, the factors involved in elongation lend themselves for regulation. These factors are present in a stoichiometric ratio to the ribosome at all stages of cell growth (Gordon, 1970), from which it is easy to imagine that under certain conditions they may become limiting for growth. The enzyme-mediated translocation is another possible step for regulation and, as mentioned, is susceptible to the level of cAMP. Codon composition demonstrably affects the *in vitro* rate of chain elongation (F. R. Kramer, quoted by Davidson, 1968). A preferential use of synonym codons, partly exploiting the wobble recognition, could thus influence the rate of elongation *in vivo* which at any time represents the grand average of the rates of translation of each individual codon. An alternate possibility is that the rate of elongation may be adjustable directly by the ribosomal subunits, since the subunits appear to undergo phase-specific conformational transitions and these could come under the influence of physiological conditions in the cytoplasm. There is also the possibility, to be appraised later in the light of information on viral translation, that the attainment of tertiary or even quaternary structure by the proteins being synthesized likewise affects the rate of elongation. Moreover, it is conceivable that this rate is somewhat dependent on the higher order structure of mRNA, and this could vary regionally according to the sequences in each mRNA. There seem therefore to be several opportunities for regulating elongation. Chain release depends for its rate on the availability of termination factors and this may well be an area of terminal control, although not the most attractive one to consider. Dissociation factors control the recycling of ribosomal subunits through the polysome.

In reviewing all these possibilities it should be borne in mind that the rates of elongation and termination do not vary nearly as much as the rate of initiation, or as much as the number of ribosomes at any time committed to translation. For example, as mentioned already, translation proceeds during normal growth at the maximum rate permitted by the rate of transcription, which means that the rate of elongation, in particular, is not limiting. (At slow growth the rates of transcription and elongation decrease jointly, so that under this condition elongation again need not be limiting.) Hence, postinitiation steps are, in general, likely to be a minor source of adjustment of translation. However, the ensemble of steps from initiation to release may still, as a whole, create some sort of homeostatic equilibrium of translation at any given rate of cell growth, although evidence to support this statement is outstanding.

After release the chain may, in cases, be subject to postpolysomal regulation, similar in principle (if not in scope) to that discussed later at some length for the eukaryote. Although again no specific evidence on this point is available, several possibilities can be listed. First, the acquisition of quaternary structure by enzymes may be instrumental in timing their function. Second, the topography of the polysome in the cell could influence the outcome of translation. Suggestive of this influence, for example, sometimes translation is delayed with respect to informational transcription, as in the case of penicillinase mRNA. It is not known how the exceptional stabilization of this mRNA is effected, but it might be by its association with membranes. Similarly, a correlation of the locale of translation in the cell with that of the substrates for the enzymes that are made was considered in Chapter 3 to be another possibility. (Conversely, it is interesting that, as in the eukaryotic cell, the ribosome, hence translation, is excluded from the bacterial "nucleus.") Third, metabolic conditions in the cytoplasm might be important in regulation. For example, the protracted expression of early sporulation mRNAs depends on the status of cytoplasm. The second and third of these possibilities bear a considerable formal analogy and this could mean that they are coextensive, if not actually identical, with each other.

In the growing bacterium the ribosomal subunits dissociate after release from the polysome. The subunits then recycle directly into the polysome without reforming free ribosomes (Phillips *et al.,* 1969). Subunit dissociation is under control of a stoichiometric factor, which perhaps offers another opportunity for regulation. Under normal conditions of growth the balance is therefore between ribosomal subunits and polysome, and, as far as is known, the reassortment of subunits is random. A fraction of the ribosomal subunits is not active in protein synthesis, however. This can be seen in shift-up experiments where within 5 minutes there is an increase of as much as 50% of polysomes (Harvey, 1970), which cannot all derive from newly made subunits. On the other hand, under conditions of slow growth, an excess of ribosomal units accumulates. Contrary to normal growth, in this case it appears that the subunits do reassociate to form ribosomes which virtually represent a reserve or inactive pool of subunits. That is, on renewed growth, these ribosomes dissociate and the subunits thus made available reenter the functional polysomal cycle. The most likely conclusion from these data is that the ribosomal subunits are not causally involved in translational regulation. Perhaps whether they remain active or inactive, or enter a reserve pool, is determined by a balance between dissociation and initiation factors. Thus it is significant that, as mentioned in Chapter 1, these factors are probably one and the same.

The remainder of this section is concerned with aspects of higher order structure of mRNA that may be of consequence to translation. (According to the definition in the footnote to p. 14, the structure to be considered is a low order tertiary structure. It is described as a secondary structure by most workers in the field, and this designation is retained here to avoid confusing the reader.) So far, work in this area has been entirely in the small RNA phages which contain three cistrons: in the 5' to 3' direction, for maturation protein, coat protein, and replicase. An example of intrinsic secondary structure comes from the partial sequence analysis of the R17 phage coat protein cistron, which folds on itself with a high degree of base pairing, hence in a helicoidal fashion (Adams *et al.,* 1969). The phasing of codons in this cistron is such that the third (wobble) bases are not opposite one another but displaced by one base position, thus permitting two-thirds of the base pairs to change the secondary structure of the RNA molecule by mutation without altering the amino acid sequence of the resulting protein. (This freedom of secondary structure with respect to informational content could then be one reason for the code degeneracy. Historically, however, it was first thought that, to the contrary, the role of degeneracy in regard to the mRNA molecule was to prevent autocomplementarity; Rich, 1968.) Other evidence in R17 phage (Steitz, 1969) indicates the presence of relatively long nontranslatable sequences preceding and in part different for each cistron, which include the ribosome binding sites. The inference is that these sequences are recognized by as many binding factors, and, consequently, that not all viral translation must begin 5' terminally. A different possibility disclosed by this work is that, to become effective, the initiation and termination codons have to be exposed whereas the ineffective codons remain buried in the secondary structure. The implication now is that these codons may be suitably exposed or buried by a changing secondary structure of the RNA, which in any case must somehow be transiently altered during translation. For example, *in vitro* translation of f2 phage replicase depends on partial translation of the preceding coat protein cistron. This can be explained if the helicoidal structure of the coat protein cistron changes during translation and in the process helps expose binding sites and initiation codons of the replicase cistron (Lodish, 1968a). Also in R17 phage it appears that the RNA structure controls the binding of ribosomes to the replicase cistron (Jeppesen et al., 1970). All this suggests that, apart from ribosome binding proper, the intercistronic sequences may also be involved in the structural changes aimed at translation (and later perhaps at encapsulation). In summary, it is beginning to look as if the secondary structure of the viral RNA molecule has been decidedly exploited to a translational purpose during evolution.

Whereas the previous examples provide some idea of how secondary structure can influence the expression of viral mRNA, extrapolation as to how much the structure can affect the expression of bacterial mRNA remains hypothetical. A reason is that, because of the tight coupling of translation to transcription, which is first secured by the translation initiation complex, bacterial mRNA might need to be less internally regulated for translation than viral mRNA. Moreover, contrary to RNA phage mRNA, the true polycistronic function of bacterial mRNA does impart coordination to translation as required. If there were to be any secondary structure to bacterial mRNA after all, a first possibility would be that synonym codons are distributed to achieve this structure much as they are in phage. A second possibility would be that operator sequences are transcribed covalently with mRNA and form some sort of rate limiting device for ribosome entry, a device which, according to this view, would depend on secondary structure. Such device could be particularly useful for independent translational control of constitutive functions. A third possibility would be that intercistronic sequences are present, although much shorter than in phage and at most a few nucleotides. As before, the coupling of translation to transcription militates against these sequences serving as specific ribosome binding sites as in phage. Moreover, if specific, any number of sites commensurate with the number of bacterial informational cistrons would be impractically high. These intercistronic sequences would then only provide some sort of spacer or higher order structual region, although perhaps a very important one; for example, they appear to regulate internal reinitiation. To be sure, except perhaps for the last, thus far there is no evidence regarding these possibilities in the bacterium.

Of interest to mention are other effects of protein on the expression of mRNA, again only known in small RNA phages. As implied above, these effects provide for the self-regulation of viral translation. In f2 phage, translation of late mRNA (the nascent plus RNA strand) is eventually blocked by one of its own products, coat protein, starting from the coat protein cistron onward (Eggen and Nathans, 1969). This blockage reduces ongoing translation of the distal replicase cistron, of which by then no more product is required. This effect on the replicase cistron is the opposite of that which earlier in the cycle made its very translation possible (see above). On the other hand, translation of the maturation protein cistron is independent of the coat protein cistron: instead the limited production of maturation protein may be due to a lower affinity of the ribosome for the binding site of this cistron (Lodish, 1968b). Finally, as the few finished plus RNA strands become free, their subsequent translation is prevented by encapsulation within viral protein to form the progeny virions, a process in which RNA folding is necessary. To recapitulate, the three viral proteins are translated separately and their very different amounts controlled in reality by a variety of mechanisms; contrast this economy with the obligatory total translation of certain mammalian RNA viruses (Chapter 3). These general characteristics of viral RNA regulation oppose the view of an operonlike linear translation of their small genomes. Even though the replication of the viral genomes is linear and complete as in a bacterial operon, this is only incidental to the replication subserving the reproduction of the virus. Whether some product effects of the viral type occur restrictedly during translation of bacterial polycistronic operons remains possible (perhaps at the intercistronic spaces), but has not yet been documented.

In summary for the bacterium, as a consequence of coupling, translation normally follows immediately upon transcription. Rather against the usual emphasis on transcriptional control prompted by this coupling, it is now becoming apparent that there may be a measure of independent translational control (Bock, 1969). In that case, the two levels of control are necessarily coextensive with each other, and, in addition, both systems are attuned to cell metabolism and nearly equally exposed to its products. Unfortunately, a complete description of the functional integration of transcription and translation with cell metabolism—one of the challenges for the future—requires a greater understanding of bacterial cell organization than we have at present.

Regulation in the Eukaryote

Together with a much more complex genomic regulation than the prokaryote, the eukaryote relies much more on translational control of cell activities and this control is also at its most complex in the eukaryote. As a result, most of the day-to-day regulation of eukaryotic translation is at its own level. This is not to deny for a moment that during cell differentiation or cell stimulation new informational and stable RNAs must be made available or that, even in quiescent cells, many of these RNAs must be continuously replenished, thus making transcriptional control of cell activities an equal requirement. With respect to translation, this primary transcriptional control therefore stands as a qualitative control in the sense that it alone can make new genetic information available. Despite being a primary control, informational transcription must also be subject to some sort of internal feedback from translation; although its nature is obscure, this feedback means that the regulation of transcription cannot be totally independent of translation. On the other hand, the typically eukaryotic regulation of translation indicated above as being largely independent of transcription can be either semiqualitative or quantitative. It can be semiqualitative by virtue of withholding informational transcripts from reaching translation, and it can be quantitative by controlling the actual rate and extent of release of these transcripts for translation. The regulation of translation comprises several sublevels which are the subject of this section, but which, for reasons of scope, can be no more than summarily evaluated. For a more elaborate discussion and survey of the subject, the reviews by Munro (1970) and Lucas-Lenard and Lipmann (1971) are recommended.

Disregarding cytoplasmic organelles, most eukaryotic translation occurs in the cytoplasm and is, of course, physically separated from transcription which occurs in the nucleus. A realistic biological appreciation is however that a separate cytoplasm represents an innovation concomitant with rather than contingent to the appearance of a nucleus, with the implication that the establishment of each of these two compartments is of equal importance to the cell. The fact remains that in the eukaryote a physical coupling of transcription and translation is impossible (except in cytoplasmic organelles), which by itself dictates the need for an independent translational regulation. A different consequence of the segregation of translation is that it is perhaps more directly exposed to cell metabolism, suggesting a closer interrelationship with metabolism, of which a few examples will be given later.

The point was made earlier that eukaryotic translation is in general mechanistically similar to bacterial translation but its regulation is enormously more complex. To a certain extent the complexity of regulation derives from the different functional organization of the eukaryotic cell, yet it also suggests that corresponding subtle mechanistic differences remain to be found. One such mechanistic difference could apply to translation factors. In contrast to what is believed for the bacterium, in the eukaryote these factors might vary their concentration, hence their overall stoichiometry, with respect to the translational apparatus. This qualification affects all translational steps where factors intervene, namely, from chain initiation to release, and has clear regulatory implications. A second mechanistic difference, also with regulatory implications, might be reflected in the accumulation of ribosomes. Whereas few ribosomes accumulate in the bacterial cell, many ribosomes accumulate in certain eukaryotic cells to the point where distinct ribosome pools are formed.

Eukaryotic informational transcription leading eventually to protein synthesis was considered in Chapter 3. From that discussion, there are grounds to believe that extensive prior regulatory transcription is obligatory for informational transcription, which means that there are two discrete levels to the transcriptional process. Contrary to the prokaryote, moreover, informational transcription is, as a rule, not immediately followed by translation, although one probable exception are the histones. In the prokaryote we saw that translation follows automatically as mRNA is made, in fact while it is being made; the twofold implication is that transcription of mRNA limits translation and the main regulation is that of transcription. In the eukaryote the delay of translation ranges from perhaps a minimum of one-half hour to a maximum of days, but it can be months in the case of eggs or even millennia in the case of plant seeds. On the whole, therefore, translation does not depend momentarily on transcription and the supply of mRNA need be less instantaneously limiting, both aspects creating an opportunity for a more independent regulation of translation. It was shown in yeast, and presumably applies to the eukaryotic cell in general, that the relative delay between DNA replication and the functional replication of enzyme potential achieved some time after translation is much greater than in the bacterial cell (Mitchison and Creanor, 1969). Thus, in effect, the biosynthesis of eukaryotic protein is regulated significantly at the translational level, a point which was made in the introduction to this section. A corollary of this point is that many participants in this biosynthesis are themselves likely to be directly or indirectly under extensive translational control, thus enforcing the truly pervasive nature of the regulation. As presently discussed, this regulation involves several sublevels. Evidence is also available for an intermediate (posttranscriptional) level of regulation of translation located between the two transcriptional levels and the translational level, namely, during the processing of mRNA which was considered at some length in Chapter 3. In contrast again with the automatic use of bacterial mRNA, this processing appears to include an important phase of selection of the eukaryotic molecule which contributes to part of the observed translational delay.

The onset of the transcriptional response of the eukaryote is almost as rapid as in bacteria. For example, DNA-like RNA is synthesized within a matter of minutes in response to hormone (Kidson and Kirby, 1965; Tata, 1968). Most of this RNA is labile RNA whose uncertain regulatory versus informational status was dealt with in Chapter 3. To the extent that the RNA has a regulatory function its transcription would precede that of mRNA, resulting in the discrete pretranscriptional level of regulation of protein synthesis referred to earlier. A rapid transcriptional response therefore need not mean, and usually does not mean, a rapid informational response. A consequence of this regulatory transcription is that it somewhat reduces the kinetic asymmetry of transcription-translation resulting from the long life of eukaryotic mRNA (Fig. 20). To expand on this question, the asymmetry of transcription-translation depends on such factors as: (a) the relative anticipation of regulatory transcription in regard to informational transcription; the extent of both (b) the posttranscriptional

Regulation of Translation

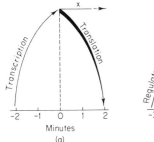
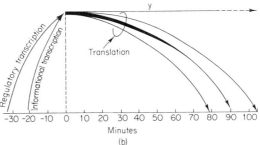

Fig. 20. Representation of symmetry relationships in transcription and translation. The ordinate (not drawn) represents a single transcriptional-translational event from beginning to end in the direction of the arrow and with the duration of time shown in the abscissa. Bacterial transcription and translation (a) have about the same duration without a substantial delay in between, so that either is representative of the half-life of mRNA; an average duration of 2 minutes was chosen. The overall process of bacterial transcription-translation is represented as being very nearly symmetrical, except for cases (x) where translation is delayed (e.g., penicillinase and spore mRNAs). In the higher eukaryote (b) the overall process is represented as being asymmetrical to varying degrees because of the presumed relatively rapid transcription but variably delayed and lasting translation; the asymmetry is reduced to the extent that regulatory transcription precedes informational transcription. In cases (y) the asymmetry is increased because of longer delay in translation (e.g., maternal embryonic mRNAs). The duration of the transcriptional and translational processes in the eukaryote given in the figure is a bona fide approximation, but note the change of scale relative to the bacterial processes.

regulation and (c) the pretranslational regulation discussed below; and (d) the active half-life of mRNA under translation. The reader will find a diagram of this sequence of processes in Fig. 21. Although it is difficult to assess the inclusive duration of any of these processes in the eukaryote, it is appropriate to say that, by and large, the kinetics of transcription-translation will differ substantially from the bacterial cell. This is the central fact which Fig. 20 intends to convey.

The comprehensive regulation of eukaryotic translation involves all six levels from the pretranscriptional to the posttranscriptional illustrated in Fig. 21, thus considerably more than translation proper. As shown in the figure, even between transcription and translation many more things happen to mRNA than in bacteria. All these levels are complex—being difficult to tell which is the most—and the theories that eventually ought to account for their dynamics and organization will be as complex as any biological theory that we have at present. These levels will now be considered in order, starting with the pretranslational, and certain representative aspects of each level will be discussed. Regulation involves both

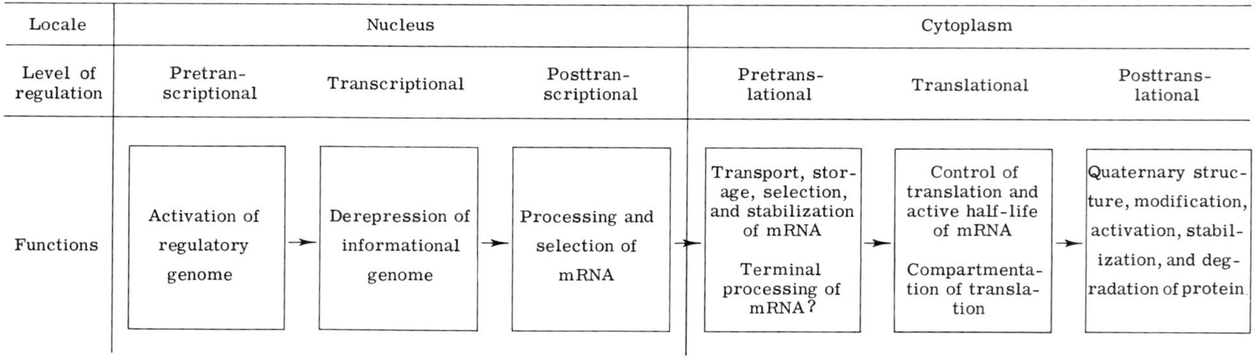

Locale	Nucleus			Cytoplasm		
Level of regulation	Pretranscriptional	Transcriptional	Posttranscriptional	Pretranslational	Translational	Posttranslational
Functions	Activation of regulatory genome	Derepression of informational genome	Processing and selection of mRNA	Transport, storage, selection, and stabilization of mRNA Terminal processing of mRNA?	Control of translation and active half-life of mRNA Compartmentation of translation	Quaternary structure, modification, activation, stabilization, and degradation of protein

Fig. 21. The integral regulation of higher eukaryote translation. The speculative diagram is schematic and only for the purpose of orientation. Early functions up to and including some of the functions under the pretranslational level of regulation were discussed in Chapter 3; the remainder of the functions in the present chapter. The first level of regulation involves transcription but is considered pretranscriptional with reference to the informational transcription beginning at the subsequent level. Certain functions are illustrated at the level at which they are believed to occur most typically, but may extend into the preceding or following level.

the manufacture and degradation of protein, but mostly the first process will be discussed throughout this section and only a word will be said at the end about the second.

Beginning with the prepolysomal (or pretranslational) level of regulation, the first question to consider is the stabilization of mRNA—a process which comprises several aspects. In contrast with the prokaryote, stabilization of mRNA is very common in the eukaryote and exaggerates even further the effects of uncoupling between synthesis and translation of mRNA, resulting in the asymmetry depicted in Fig. 20. This stabilization was already considered in Chapter 3 in relation to the posttranscriptional specification of mRNA. It apparently involves the association of mRNA with approximately equivalent amounts of nonribosomal (informosomal) protein which accompanies mRNA into, not just to, the polysome. Within the general context of selection of mRNA, this protein may regulate the functional quantity of mRNA by controlling both the stability and availability. For example, translation is less delayed in regenerating than normal liver (Bucher, 1963), suggesting such a regulation. Concerning two other aspects of stabilization, informosomal protein could, first, contribute to specifying whether a mRNA will be translated as a free or attached polysome, an important discrimination whose mechanism remains unsolved. Second, informosomal protein might affect the secondary structure of mRNA in the polysome, if such order of structure exists in the eukaryote. (The functional structure of eukaryotic mRNA remains an aspect potentially important for regulation, but is not pursued here for lack of basic evidence; consider, however, the spiral or rosette shape of the free polysomes in the cell, which suggests some kind of higher order structure.) Despite all that is said in this paragraph, the general function of the informosome is not as well defined as the importance which one assumes for it would demand. Informosomal function remains therefore in need of much investigation. Two further forms of mRNA stabilization are discussed below, namely, by retention in inactive polysomes and by association with endoplasmic reticulum membranes. It is difficult to distinguish functionally these two forms of stabilization from the previous one by informosomal protein if this protein is supposed to remain associated with mRNA in the polysome. Yet all three forms of stabilization may occur either separately or combined. Pointing in this general direction, disaggregation of polysomes in dividing HeLa cells does not result in a complete loss of cytoplasmic mRNA. Inasmuch as some mRNA stabilization depends upon membrane association, the process would also come under the indirect control of hormone. Very likely, however, the stability of the free polysome falls in part under this control.

Other possible aspects of prepolysomal regulation are as follows. The posttranscriptional processing of mRNA which began in the nucleus may not be completed until the molecule arrives in the cytoplasm. This terminal maturation of the molecule may again be in relation to informosomal protein and perhaps correspond with the stabilization discussed in the preceding paragraph. Inasmuch as this maturation would be necessary in order to obtain a functionally mature mRNA, it would have an indirect regulatory character. However, a true pretranslational regulation operates at the level of tRNA. This aspect will be reviewed as a special topic in the following section.

Recent findings indicate that eukaryotic mRNAs contain repetitive poly A tracts and also more typical redundant sequences. Some of the original evidence for the homopolymer was presented in Chapter 2. A-rich sequences of up to 50-70 nucleotides (or possibly longer at times) are associated with reticulocyte mRNA and the size of these sequences correlates with that of the polysome (Lim and Canellakis, 1970). This globin mRNA should be approximately 440 nucleotides long but instead behaves electrophoretically as if it were 650 nucleotides (Gaskill and Kabat, 1971). The poly A appears to be at the 3' terminus of mRNA (Burr and Lingrel, 1971) and possibly as the result of a posttranscriptional covalent addition in the nucleus (Darnell et al., 1971). Because the poly A in nuclear RNA is recovered in functional cytoplasmic mRNA perhaps quantitatively, its presence in mRNA may be obligatory, although not the same length in all mRNAs. Poly A seems to occur at or near redundant protions of mRNA which, like the homopolymer, are presumed to be noncoding. Whether these redundant sequences are transcribed with the mRNA, as it could be expected, is not known. Questions that arise concerning these noncoding sequences are therefore whether they are found in all mRNAs and, in the case of the redundant sequences, whether they are specific to particular mRNAs or shared between many mRNAs. According to Darnell (1971), poly A is absent from histone mRNAs; according to Kedes and Birnstiel (1971), if noncoding sequences are present in these mRNAs, they would appear not to be molecularly redundant or shared with other mRNAs. Thus at least for poly A it is clear that it may be obligatory to some but not all mRNAs. What is the function of poly A in these mRNAs? Although several functions are possible (and will be considered later), one of them is that the homopolymer functions as a handle or cue in order for mRNA to be saved and transported to the cytoplasm; convenient for this purpose, poly A is very insensitive to nucleases. Based on this view, homopolymer addition would be the outcome of the complex (and unknown) specification in the nucleus that decides which mRNAs are to be translated. This role of poly A could be once again in relation to the informosome—hence the reason for mentioning this topic here.

Before discussing the polysomal level of regulation, some attention should be given to the functional distribution of ribosomes in the cell. This is best seen in certain cells, part of whose ribosomes are, as it were, poised for translation rather than actually performing translation. However, it must be insisted from the beginning that, even if this condition is probably general to eukaryotic cells, its extent may vary greatly between different cells and some cells may not show it at all. Thus, in regenerating liver the 30% of newly made ribosomes do not seem sufficient to account for the enhanced translation, the bulk of which seems to be carried out by 30-60% activated polysomes that are normally inactive (Hoagland, 1969). A possible implication is that there is an inactive pool of polysomes that is preserved for homeostatic reasons. Clearly, though, in moving from a stationary into a growing phase the cell also requires new ribosomes. Such a pool would have to be actively preserved, because, otherwise, nonfunctional polysomes such as in liver under certain dietary conditions (mentioned later) are degraded. These inactive polysomes behave therefore like informosomes even though they are not. To account for his observations in liver, Hoagland proposes that a "translation package" which consists of polysomes, endoplasmic reticulum membranes with loosely bound tRNA (about 50% of total tRNA; Hoagland *et al.*, 1968), translation factors, metabolic adjuvants, etc., responds unitarily to the stimulus for regeneration. This certainly is a view concerned with the broader functional organization of cytoplasm. According to Kabat and Rich (1969), in embryonic muscle there is a constant pool comprising as much as 90% of free ribosomes (about 20% of total ribosomes) that do not participate in protein synthesis, or at least not as actively as the ribosomal subunits recycling through the polysome. This ribosome pool is typical of early embryonic cells (Chapter 6), which, in addition, contain inactive polysomes. Similarly, during slime mold differentiation there is a preferential removal of ribosomes from the so-called monosome pool and a preferential renewal into the polysome pool (Cocucci and Sussman, 1970). Although no firm conclusions are possible at this time, the collective implication of these data is fairly obvious, namely, that translationally active and recycling ribosomal subunits bypass inactive ribosomes or polysomes, which, however, can be called upon on demand. This high incidence of inactive ribosomes may be correlated to the general difficulty of dissociating the eukaryotic ribosome *in vitro*. These active and inactive ribosomes may differ in their proteins but not in their rRNAs. They may differ in the number of their ribosomal proteins, in specific transpeptidation of amino acid to these proteins (see Nontranslational Synthesis), or in site-specific phosphorylations of these proteins (Kabat, 1970). Because phosphorylation of ribosomal proteins is influenced by cAMP, it, in particular, has an obvious potential discriminatory value.

The question is whether an inactive ribosome is really a nontemplated ribosome. Present views tend to favor that, unless subunit dissociation is prevented during release from mRNA, subunit reassociation into a ribosome depends upon reentering a complex with mRNA. However, it was seen that "polysomes" with one ribosome cannot occur permanently. The simple answer may therefore be that a nontemplated ribosome is the result of an interplay of factors in response to metabolic and hormonal conditions in the cell, and these factors also determine when the ribosome will reenter the polysomal cycle. By any mechansim, a reserve pool of ribosomes or polysomes contrasts with the only exceptional accumulation of ribosomes in the bacterium; this organism generally meets the demand for more protein synthesis with the immediate synthesis of more ribosomes, which thereafter will recycle as subunits.

These data might also suggest the existence of a topographical segregation of ribosomes within the cell space, in principle, irrespective of whether they are free or attached. This segregation might be not only with respect to inactive-active ribosomes as just discussed, but also with respect to ribosomes differing in their translational product. This latter possibility has been raised ever since the functional dichotomy of free and bound polysomes became apparent. The answer to this question of a topographical segregation, which could be of great physiological importance, still remains elusive. Pertinent to this general subject, recent evidence indicates that the cell membrane is a site for specialized ribosome function. In the reticulocyte (without a recognizable endoplasmic reticulum) 18% of the ribosomes are in polysomes bound to the cell membrane where they engage in nonglobin protein synthesis (Bulova and Burka, 1970), in contrast with the remainder of (free) polysomes engaged in globin synthesis. Cell membrane bound fibroblast polysomes likewise appear to differ in their translational function (Glick and Warren, 1969). The related question of whether ribosomes devoted to a particular translation may represent qualitatively different classes of ribosomes will be considered again later.

A considerable degree of regulation is manifest at the next translational level, the polysome. It is obvious that as mRNA becomes available, any increase in protein synthesis must involve several general parameters such as the number of ribosomes, their proportion in the polysome, and the activity of the polysome. In addition, hormones control the quantitative expansion of the translational apparatus as well as its attachment to endoplasmic reticulum membranes. By more specific mechanisms hormones should also be able to control translation qualitatively. Having set these general conditions, we shall now separately consider the polysomal stages of initiation, propagation and termination. (The reader may consult the previous section on the prokaryote for general discussion of several possibilities for regulation at these three stages). Initiation is probably the most important point of translational regulation, either qualitative or quantitative. Initiation is likely to operate by an active mechanism possibly involving the informosome and certainly the translation initiation complex. Of these, the informosome was already discussed but the way in which it may interact with the initiation complex is totally obscure. Also discussed were all elements similar to those of the bacterial initiation complex, including factors, present in the eukaryote. Although it is clear that the complex suffices to explain the initiation of translation—as it does in bacteria—our understanding of the function, let alone regulation, is, however, rudimentary by comparison. We must content ourselves with speculating along certain plausible models. Distinct initiation factors or special ribosomal proteins could ideally recognize a family of messengers and such a protein-specific control would be admirably suited to the needs of tissue differentiation. The distinctive initiation sequences in mRNA that this model implies were first hypothesized by Rich (1968). Later, a run of detachable 5' terminal mRNA sequences or "tickets," each worth one ribosome, was proposed by Sussman (1970) to control the extent of ribosome loading. The implication of this proposal would be that there are noncoding 5' terminal sequences in mRNA (hence other than initiation codons) whose function is within the general area of initiation, but these sequences remain to be shown. What they bring to mind is something like the 3' terminal sequences previously mentioned which clearly could function in termination. In conclusion, even if we may be justified in thinking that the initiation of translation is more complicated than in bacteria, at least as far as regulation goes, no special structure in eukaryotic mRNA has yet been found in support.

Experimental observations possibly related to the regulation of initiation are (a) an interferon-dependent modification of ribosomal subunits that allows translation of cellular but not viral mRNA (Levy and Carter, 1968), in fact the reciprocal of the virus-specific translation initiation factors mentioned earlier; and (b) the inhibited ribosome of the mitotic cell discussed in Chapter 6. Another possibility of regulation at initiation depends on whether (c) the posttranscriptional inactivator which intervenes in adaptive modulation (Chapter 3) somewhere between the pre- and posttranslational levels, acts at the 5' end of mRNA. An insulin-dependent stimulant of protein synthesis which in muscle acts on the large ribosomal subunit (Wool *et al.,* 1968) may involve either neutralization of this posttranscriptional inactivator, or, alternatively, specification of a positive translation initiation factor. (d) A prosthetic group, such as heme, seems to enhance initiation of globin translation (Marks *et al.,* 1968), but whether it does so directly is not known. According to one model of rough endoplasmic reticulum formation presented earlier (Baglioni *et al.,* 1971), (e) the occurrence of reticulum is a prerequisite for initiating the synthesis of export proteins.

Concerning the process of chain propagation, the few determinations available in reticulocytes do not seem to establish whether the rate remains constant in general or varies to suit the physiological status of the cell. By comparison, in bacteria the rate of propagation is relatively invariant except at slow growth, which means, of course, that this rate is less determining of the amount of protein to be made than, in the first place, the rate of initiation or the total number of initiations (dependent on the number of polysomes at work). However, several potential instances of regulation of the rate of chain propagation can be cited: (1) the hydroxylation of proline and lysine in collagen which is concurrent with translation and necessary for its completion (Schubert, 1969); (2) the glucosamination of immunoglobulins during translation (Sherr and Uhr, 1969) which could be similarly necessary; (3) a heat-labile microsomal membrane factor which in liver inhibits transferases and is antagonized by GTP (Scornik *et al.*, 1967); (4) the stimulation of fish liver transferase I activity during cold acclimation (Haschmeyer, 1969), indeed one of the few instances where an increased rate of elongation has been shown. It will be noticed that (1) and (2) suggest the possibility that the incipient secondary or tertiary structure of the protein may influence translation. As already mentioned, however, propagation (and termination to be considered next) is as a rule, at least for the homeotherm, probably less significant for regulation than initiation.

Concerning chain termination, lastly, in liver polysomes this is decreased by catabolite (glucose) repression, whereas cAMP has the opposite effect (Khairallah and Pitot, 1967). This reciprocal condition is also found in the bacterial promoter, where a normal level of cAMP mediates an unspecific stimulation of transcription and appears to do so by activating a binding protein directly responsible for the stimulation (Chapter 3), whereas both the cAMP level and the stimulation are lowered by catabolite. In the case of liver it is not known whether cAMP increases chain termination directly or via a protein, or whether it acts on termination codons or factors. Most likely, this metabolite has in addition many other translational effects as a conveyor of signal from hormone (Kenney, 1970). The possibility also exists that the $3'$ terminal poly A sequences mentioned earlier play a role in termination. It is necessary to remember that this is a truly metabolic process (Chapter 1) and its regulation could very well be more elaborate in the eukaryote.

Several examples of regulation have not yet been allocated to any of the above phases of translation. These are: (a) a growth hormone-dependent stimulant of liver translation acting on the small ribosomal subunit (Barden and Korner, 1969), perhaps akin to the insulin-dependent stimulant listed under regulation of initiation; (b) a serum factor that contraarrests an intracellular inactivator of translation (another possible posttranscriptional inactivator) which, interestingly, is absent in tumors (Amos, 1967); and (c) an unstable RNA that in HeLa cells promotes polysome aggregation in compensation for the opposite effect of supranormal temperature (McCormick and Penman, 1969). These and the previous examples show that a variety of substances interact with the translational apparatus, a point whose implications in terms of cell organization will come to the fore later. A question of interest is whether any nonpolysomal RNA (excluding tRNA) intervenes indirectly during translation. The cytoplasmic presence of nonpolysomal polydisperse RNA of unknown function (Chapter 2) makes this a possibility to keep in mind.

Chemically and cytologically, the next or postpolysomal (posttranslational) level of regulation is a protean one on account of the variety of reactions and cell locales that are involved. Most H bonds in protein are formed during assembly of the chain. After assembly, amino acid residues can be covalently substituted with groups which affect the tertiary structure and function of protein, although some substitutions have already begun during assembly. Indeed, many proteins are strictly dependent on this kind of substitution for their function. Examples of protein substitutions are: sulfation which occurs mainly at the Golgi apparatus; glucosylation which is very ubiquitous in the cell; phosphorylation; acetylation; and addition of lipid. A modification, rather than substitution, with strong stabilizing effects on structure is disulfide bridge formation. Also at this level of regulation some proteins undergo physiological cleavage, sometimes extracellularly, not infrequently the cleaved portion being as large as the functional one. In the case of hormones and enzymes, this cleavage conveniently serves to convert an inactive into an active form and sometimes to provide it simultaneously with a carrier protein or moiety, respectively, derived from the cleaved portion. In addition, fragments of one polypeptide chain might conceivably be transpeptidated to another, but so far this phenomenon is documented only for single tRNA-bound amino acids (see Nontranslational Synthesis, p. 264). The quaternary structure of a protein is based on association of subunits by H and covalent bonds, and the association can occur during or after assembly. However, although it remains a possibility, there is no good evidence for a coordinated translation of the subunits that are to join, e.g., α and β chains of hemoglobin. What is positively known is that failure of subunit association, or lack of availability of a prosthetic group in other cases, can lead to a feedback inhibition on translation. These feedback effects are then another important aspect of post-translational regulation. For example, synthesis and/or release of protein hormones is inhibited by the protein product of the target cell. Possibly also the specific relationship of differently programmed polysomes to cytoarchitecture, and certainly the flow, storage, and concentration of protein (including complexing) through the cytoarchitecture and in particular the vacuome have feedback effects on translation. Clearly, at this postpolysomal level the regulation of translation involves the whole economy of the cell and, frequently, the organism.

The pervasive nature of eukaryotic translational regulation is perhaps nowhere more clearly shown than in the fluctuation of polysome aggregation. This aggregation is an index of the intensity of protein synthesis and depends on the supply of amino acid. The diurnal cycle of polysome aggregation in rat liver is determined by the cyclic ingestion of protein and shows a maximum aggregation at dark (Fishman *et al.*, 1969). *In vitro* any amino acid deficiency leads to polysome breakdown, but *in vivo* only tryptophan does. The reason is because tryptophan is the least abundant amino acid and cannot be sufficiently rapidly replaced in the cell pool, whereby its cognate tRNA becomes limiting

(Allen *et al.*, 1969). In a sense, this makes Trp-tRNA responsible for integrity of the polysome, but any physiological condition other than diet leading to exhaustion of any amino acid pool could create a similar situation. Irrespective of the amino acid, why its deprivation determines disaggregation of the polysome, instead of merely interrupting chain elongation, is not clear but certainly attests to the dynamic nature of the polysome; perhaps the cell destroys what it does not need. Concerning the energetic aspects of polysome stability, glucose, but not an isocaloric intake of fat, restores to normal the starved polysome profile *in vivo* (Whitman *et al.*, 1969). Hence, although energy is mandatory, not simply energy is involved in polysome stability. In addition, the residual polysomes in starved animals incorporate *in vitro* less amino acid per unit weight of RNA (Sox and Hoagland, 1966), possibly in relation to the presence of inhibitory translation factors (von der Decke, 1969). Presumably analogous to these observations in the mammal is the one in plants where polysomes are rapidly built up in response to light (Clark *et al.*, 1964). Concerning the level of acylation of tRNA with respect to diet, in rat liver this remains almost complete even after a 24-hour fast (Allen *et al.*, 1969). It was therefore surprising to find that single amino acid deficiencies in the diet reduce the charge of the corresponding tRNA, indicating a remarkable homeostatic balance. As expected, tryptophan deficiency showed the greatest effect.

To sum up so far the variety of effects on translation—and the previous listing was by no means exhaustive particularly in respect to hormones—it is evident that a systematized picture of the regulation of translation has not begun to emerge. Perhaps some day these effects may fit into some sort of "fields of control" (an elusive yet justified biological concept), organized in a hierarchical order and simultaneously affecting various areas of translation. To put this notion graphically using Fig. 21, the linear axis connecting regulatory blocks in the figure will in that case bend back on the page so that nonlinear fields of control overlap two or more of the blocks; for example, one field of control will cover translational and posttranscriptional regulation and another field will control translational and pretranslational regulation. Some of these fields then fold back from cytoplasmic translation onto the nucleus, a point which was already evident in Fig. 1 (Chapter 1). What these fields of control may mean in chemical terms is not obvious, or, more appropriately, not as obvious as the implication that the organization of the cell is directly involved. Given the complexity of eukaryotic cytoplasm, only this organization can supply each translational molecule with its proper chemical environment at the right place.

Thus it is difficult to imagine that the architecture of the eukaryotic cell, representing as it does a true internal differentiation of the cell, will not have an essential role in regulation. To repeat the underlying concept, this role is because of the cytoarchitecture providing to the molecular processes attendant upon protein synthesis the same infrastructure and compartmentation, metabolic interaction, and metabolic support that, say, membranes provide to enzymic pathways inside cytoplasmic organelles. For instance, the highly discretionary stimulatory effects on translation of many protein hormones reaching their target via unspecific cAMP (Kenney, 1970) suggest that the balance of regulatory specificity must come from somewhere in the immediate environment of the polysome. It is particularly clear in the case of membranes that they may provide the polysome with such a special microenvironment [enriched among other things for tRNA and synthetase (Bandyopadhyay and Deutscher, 1971); see also the proposal by Hoagland (1969) above], and they may also help segregate different functional classes of bound polysomes [see the proposal by Pitot (1969) below]. It is equally clear that some of these properties need not be exclusive to membranes but may also apply to the highly organized cytoplasmic matrix which is the sole one accessible to the free polysome. Despite all this, however, not much is positively known about the role of cytoarchitecture in the regulation of translation compared, for example, with what is known of its role in secretion. For a fact, much of the above mentioned analytical difficulties regarding translational regulation are often the result of the polysome being tested under *in vitro* conditions where all architecture has been lost.

Proposals by Tata (1968) and Pitot (1969) were mentioned earlier which are related to this cytological aspect of translational regulation. To recapitulate the first proposal, Tata envisages that a unitary complex of mRNA, rRNA, and endoplasmic reticulum is coordinately synthesized in response to growth hormone. It makes perfect sense, of course, to build up the whole translational apparatus in response to a strong stimulus for protein synthesis (the bacterium does it). If Tata's views are for a continuous functional complex in the cytoplasm, they implicitly demand either nonexchangeability of ribosomal subunits (which he considers not to be differentiated), or, alternatively, their full loading of mRNA while still in the nucleus. So far these views lack a clear-cut experimental basis. Speaking against the first implication, although not necessarily condemning it, exchangeability of ribosomal subunits has been shown to occur in yeast (Kaempfer, 1969) and muscle (Kabat and Rich, 1969) and, therefore, is probably a general occurrence. A remaining possibility is however that Tata's complex may function as such only once in the cytoplasm (see below). In the second proposal, Pitot argues that the endoplasmic reticulum is a mosaic in terms of specific attachment sites for polysomes and by which, in turn, polysomes are stabilized. This view implies a functional segregation of polysomes. The important point is that this and the previous proposals by Hoagland (1969) and Tata (1968), in common, recognize some form of endoplasmic reticulum differentiation. Whatever their merit—and they are only speculative—these proposals are the first to take into account cytoarchitecture as a possible means of coordinating translation, for which reason alone they deserve attention. This coordination, moreover, can now be recognized as being intimately related to the membrane-dependent physiological aspects of translational regulation referred to above. Endoplasmic reticulum was singled out for comment here, but it should be stressed that cytoarchitecture means considerably more than just this membrane system. A complex infrastructural organization must be general to all cytoplasm, yet this membrane system is not.

Since the discovery in 1962 that the nucleolus is the source of ribosomal subunits, this has been considered to be its major contribution to the process of translation. In the light of recent work this view on the nucleolus may have to be expanded to include a more active participation in translation. Sidebottom and Harris (1969) found in heterokaryons that genetic information is expressed only after nucleolar rRNA synthesis is patent, and also that, contrary to previous observations in normal cells, the appearance of almost all new RNA in the cytoplasm is dependent on this nucleolar synthesis. Related to these observations, Ringborg *et al.* (1970) reported in insects that the terminal processing of 18 and 28 S rRNAs takes place in the chromosome. According to this report, which awaits confirmation, the chromosome would then be an intermediate station between the source of rRNA and its final cytoplasmic location. At face value, taken together these data might suggest that the ribosome promotes rather than merely assists in the cytoplasm the expression of mRNA, and it does so perhaps by facilitating the release from the chromosome and/or the subsequent transit of mRNA. For background, prior to this suggestion the general model had become (after some considerable initial dispute) that the ribosomal subunits and mRNA interacted with one another not before reaching the cytoplasm. As of the present, the mechanism of this putative ribosomal promotion of mRNA would remain to be clarified, but the formal similarity with Tata's proposal in the previous paragraph should not pass unnoticed. Relevant to this context, Tata (1968) concluded that when cells embark on a new differentiative career (in his system, metamorphosis), the old rRNA is replaced by new, which is rapidly distributed to endoplasmic reticulum membranes. This conclusion suggests that a new class of ribosomes may be required for the new cell career as much as new mRNAs obviously are. Similar conclusions were arrived at by Cocucci and Sussman (1970) in their study of slime mold fructification. These conclusions have a bearing on the question of a possible segregation of ribosomes previously considered on several occasions in the text.

The important difference introduced by a ribosomal promotion of mRNA is that the segregation may involve different classes of ribosomes in reference to their templates. In the simplest case imaginable, one class of ribosomes would read a family of functionally related mRNAs. However, it could still be that the ribosomes are identical and only temporarily classified as they promote or convey a particular mRNA, to regain their identical status as they recycle through the polysome. (This last possibility would mean that Tata's complex mentioned earlier functions as such only once in the cytoplasm.) In either case, were the occurrence of different classes of ribosomes to be substantiated, it would represent a typical eukaryotic phenomenon. There are facts against this idea that mRNA is conveyed by the ribosome, however: (a) in early amphibian and echinoderm embryogenesis new mRNAs appear which are translated in the absence of new ribosomes; (b) in mammalian cells DNA-like RNA is transported to cytoplasm in the absence of concurrent rRNA synthesis (Perry and Kelley, 1970). Although it is possible to dismiss arbitrarily these facts as exceptional or to argue liberally (as will be done in Chapter 6) as to how they might fit the previous idea, neither dialectic course can prove or disprove it. Critical experimentation on the points touched in this and the previous paragraph is therefore yet to come. Assuming for the time being that ribosomes do indeed convey mRNA, which questions does this raise? First, the question whether the proteinated rRNA-mRNA complex which is necessary to postulate is found in the nucleus; there have been claims in this direction but the evidence is, to say the least, confusing. Second, there is the question of the relationship of this complex to the informosome. Third, evidence presented in Chapter 4 suggests the presence of a few tRNA cistrons in the nucleolus. This remains to be established, but, as pure speculation, it might turn out that to provide tRNAs of specific function (initiation?) in relation to a ribosomal conveyance of mRNA is another aspect of the nucleolar participation in translation.

An aspect of posttranslational regulation that has purposely been left for consideration until last is the degradation of protein. Degradation of protein is as carefully regulated in the eukaryote as is synthesis. Eukaryotic proteins, particularly enzymes, are intrinsically labile as, it turns out, an inbuilt clearing-off mechanism which is also fully exploited regulatorily. Generally, enzymes need to be stabilized by their substrate, cofactor, etc. (review by Schimke, 1970), otherwise they break down. The same is true of certain proteins stabilized by specific ligands, e.g., apoferritin is stabilized by iron. This phenomenon mimics end-product activation and a consequence is that increased enzymic activity does not always directly depend on translational activity. By contrast, prokaryotic enzymes are usually stable and when unwanted are inactivated or more often simply outgrown by cell proliferation. (It seems that in the particular case of defective prokaryotic enzymes, e.g., nonsense terminated, these are preferentially degraded.) Pursuing the comparison, the typical mature eukaryotic cell relies more on stabilizing its protein metabolically or controlling it by the modulatory type of regulation discussed in Chapter 3. In this regulation, the inactivating genes postulated to control either translation or stabilization of enzyme would represent another eukaryotic innovation.

If anything, this account of the regulation of eukaryotic translation--at which level all macromolecular syntheses are finally integrated—should serve to emphasize our precarious understanding of the regulation. A personal impression is that whenever all the relevant pieces of evidence are fitted together into a working pattern, the regulation of translation will turn out to be both complex and deeply integrated with cell metabolism. In sharp contrast with the mechanistic similarity to prokaryotic translation, in any cell of the higher eukaryote translation is, in reality, an organismal business: energy, substrates, signals, and feedbacks all come from the organism. (This is true of course of eukaryotic replication and transcription as well, even though, for different conceptual reasons, in the case of transcription some emphasis was put on the intrinsic capability of the cell to direct it.) The pressing question to explain is therefore not so much the details by which eukaryotic translation differs from prokaryotic translation but the manner in which it is regulated in the cell within the context of the organism. Probably the best known example of the regulatory complexity of translation is found in the family of hemoglobins because of the genetic, structural, and metabolic evidence which is available. Whereas little of this evidence is as yet clearly interpretable in terms of regulation, however, the monograph on this family by Weatherall (1965) does give the reader a feeling for the fundamentals of the problem.

TRANSFER RNA AS A REGULATOR OF TRANSLATION:
A REVIEW

The Hypothesis: Modulation of Translation

The involvement of tRNA in cell regulation as is known for the prokaryote was considered in Chapter 4. This included transcriptional control of stable RNAs and amino acid anabolic enzymes. Irrespective of organism, the mass effects of tRNA on translation, as befits one of the principal reactants of the system, were outlined earlier in this chapter. The question arises as to whether the functional properties of the tRNA molecule warrant a consideration of its participation in the regulation of translation with emphasis more on these properties than on the mechanistic relationship of the molecule to the translational system, although evidently the two aspects must be intimately connected. This review of regulatory tRNA function is organized in subsections according to various types of organisms, neoplastic cells, and tRNA-related enzymes. Two recent compilations of data may be useful to the reader seeking further information on the subject (Sueoka and Kano-Sueoka, 1970; Borek, 1971).

By way of providing a conceptual framework to the problem, the views by Ames and Hartman (1963) and Kano-Sueoka and Sueoka (1966) will be singled out for comment here. As the central point, both these views stress how a few limiting or modifiable tRNA species may modulate translation. Of the two views, the Sueokas' view is most particularly relevant to cytodifferentiation as it does not depend on genomic changes. In short, the modulation hypothesis

(Kano-Sueoka and Sueoka, 1966) contends that a single tRNA species with a unique pattern of codon recognition will slow down or stop translation as its concentration decreases, and a species modified either in the anticodon or synthetase or ribosome recognition sites will translate mRNA in a qualitatively or quantitatively different manner. Thus a true regulation is in mind, rather than an adaptation of the tRNA population to the composition of mRNA (as, for example, in the silk gland mentioned earlier). Basically, the hypothesis requires changes in individual tRNA species in order to modulate translation. Model experiments by Anderson and Gilbert (1969) supporting the hypothesis were previously described. In and of itself, moreover, the occurrence of redundant tRNA species (reading the same codon) argues, in principle, for a regulatory potential. Later to be considered as an extension of the original hypothesis, changes in related enzymes rather than tRNA itself can bring about a formally similar regulation.

Other basic facts underlying a possible regulatory function of tRNA are the widely different concentration of individual species in the cell, the broad range of base modification in the molecule, and its exquisitely sensitive tertiary structure. Still other properties favorable to a regulatory function were noted in the discussion of the molecule (Chapter 2) and translation. In particular, the ensemble of modificatory enzymes, by profoundly altering the tertiary structure of tRNA, undoubtedly has a potential for regulation. The same is true of subtle structural rearrangements in tRNA because of primary structural change or higher order structural interaction with the intracellular environment, e.g., local pH or other molecules in this environment. In this last regard, certain molecules (e.g., hormones) may activate, regulate, or inhibit tRNAs specifically, but, for lack of basic evidence, this remains a passing comment.

Two levels of regulation principally invoked in making tRNA capable of a regulatory role are: (a) the transcriptional level for the synthesis of new modulatory species, and (b) the posttranscriptional level for modification and/or structural rearrangement of extant species that then become modulatory. Either level of regulation will introduce the qualitative changes in the tRNA population which are necessary in order to regulate translation according to the modulation hypothesis. Of these two levels, transcriptional regulation (a) is not at the level of tRNA but at the level of the genome (Chapter 4). Posttranscriptional regulation (b) is at the level of tRNA but is equivalent to a nongenetic alteration of the genome; in fact, it represents an extension of the facultative genome (with the exception of the CCA terminal addition which virtually represents an extension of the obligatory genome). Moreover, posttranscriptional regulation involves modificatory enzymes functionally related to intermediary metabolism. Hence, according to these levels, translational regulation by means of tRNA can be either a genetic or metabolic regulation whose venue encroaches upon and supplements the prepolysomal regulation discussed in preceding sections. In summary, from concepts developed in Chapters 2 and 7, the reasons that make tRNA a good candidate for regulation go back to its ancestral functions reflected today in a central function in translation, which is affected by a variety of enzymic reactions. Although there are therefore some solid grounds for a regulatory tRNA function, the task is now to ascertain to what extent this is borne out by the evidence. Being that tRNA is an obligatory auxiliary in translation, this regulatory function can however only be expected to apply to a portion of the tRNA population: there is a clear parallel here to the suppressor function of tRNA.

In the type of work to be discussed the individual species of tRNA are usually recognized within the mixed population extracted from the cell by their amino acid radioactivity in a chromatographic elution profile. Out of the few radioactive peaks that are commonly observed for any amino acid, each peak may in principle be ascribed to one tRNA species, although in actuality, as noted below, this is certainly not always the case. At any rate, the interpretation critically depends upon the power of resolution of the techniques employed. Frequently, but not always, a new tRNA species or a modified species will show radioactivity peaks in the elution profile different from that of preexistent synonym species. To determine whether the change in activity profile corresponds to a new or modified species, additional independent criteria are mandatory. Without them the only information is that the tRNA population is altered. As will be seen, however, with few exceptions these criteria are not generally provided, and seldom have the individual species been isolated on which to prove differences in structure or, more important, function. The overall evidence remains therefore of a preliminary and decidedly phenomenological character. Another reservation concerning the interpretation of the data is that the elution profile most used, i.e., after *in vitro* acylation, strongly depends upon the extent of synthetase purification (Wettstein, 1966; Yegian and Stent, 1969a) and therefore requires confirmation *in vivo*, also not generally provided. Further uncertainties are introduced by the possible aggregation and denaturation of tRNA, pyrophosphorylase effects, etc. The artifacts derived from these sources are evaluated by Novelli (1967). Ultimately, the acid test of the rationale behind a regulatory function of tRNA in translation rests with demonstrable differences under cell-free conditions; this has not yet been achieved. Only this demonstration can prove that any variation observed in tRNAs is for regulating translation and not something else, or perhaps, just trivial.

Bacteria

To begin with bacterial tRNA, changes in activity are documented during growth. We shall first examine *B. subtilis*. During growth in poor medium, of several species tested only one of three Ser-tRNA peaks disappears. Whether the disappearance is for lack of acylation or synthesis is not decided (Goehler *et al.,* 1966). Tyr-tRNA gives different elution activity profiles during the exponential and stationary phase of the culture. No biological explanation has been found for this difference (Arceneaux and Sueoka, 1969), but the possibility exists that two synonym species are differentially transcribed in relation to the cell cycle of growth. Val-tRNA shows two activity peaks whose ratio changes early during sporulation and step-up or -down conditions, because of increased amino acid acceptance in one peak (Doi et al., 1968). The proportion of peak I increases during exponential growth relative to stationary growth, while that of peak II remains constant (Heyman et al., 1967). Late sporulation also leads to accumulation of one of two Lys-tRNA peaks depending on the culture medium (Lazzarini and Santangelo, 1967).

Turning to *E. coli, in vivo* acylation of Phe-tRNA differs between growth in minimal and enriched medium, other tRNAs showing no change. *In vitro* acylation of three Leu-tRNA peaks differs markedly depending on whether the RNA is recovered from the cell supernate or ribosome. Judging from its rare occurrence in the ribosome, the major Leu-tRNA species is less in use by the cell (Wettstein, 1966). Minor changes in Ile-tRNA activity profile are observed between anaerobic and aerobic cultures (Kwan et al., 1968). Of the three Ile peaks, peaks I and II are completely acylated *in vivo* during exponential growth as well as during Leu starvation. Peak III is not acylated during exponential growth, but it is acylated during starvation although, in addition, it is in an inactive, probably (thiol) reduced state (Yegian and Stent, 1969b). Several instances are therefore known of tRNA activity dependent on the oxidative status of the cell.

A protector substance is carried by *E. coli* Leu tRNA I solely in Leu-starved (repressed) stringent strains, as discussed in Chapter 4 in connection with stable RNA regulation. The substance is absent during exponential growth (Yegian and Stent, 1969a), when, as described above, the species seldom comes in contact with the ribosome. Since the same species is also broken down during phage infection (see later), this species might be generally involved in regulation. To compound with this involvement of Leu-tRNA, starvation of relaxed strains for Leu results in the appearance of new Leu-, His-, and Arg-tRNAs (which do not appear during His or Arg deprivation). It is argued that the amino acid itself may have specific metabolic effects on tRNAs whose codons begin with C (Fournier et al., 1970).

Additionally, tRNAs are partly charged with as yet unidentified amino acid derivatives (Yegian et al., 1966). Although the metabolic meaning of this remains a mystery, it could be in relation to sundry functions of tRNA in a number of intermediary pathways (Chapter 2). By contrast, plain scavenging of these derivatives seems an unlikely function for tRNA. Lastly, among other factors the tRNA molecule may be implicated in senescence, e.g., in aged cultures tRNA is not as fully modified as it is in vigorous cultures (Wettstein and Stent, 1968), but whether or not the implication is a causal one is not clear. In conclusion, the copious but fragmentary evidence in bacteria cultured under various metabolic conditions does not say much about the participation of tRNA in regulation under these conditions, beyond suggesting that it may possibly occur.

Viruses

Viral infection introduces foreign translational requirements in the cell thus providing a system, historically the first, in which to search for new tRNA activity. These translational requirements involve either obligatory or facilitatory changes of the host cell machinery of synthesis and changes that accompany the transition from early to late viral protein synthesis. As will be seen, the virus-dependent changes of the cell machinery are widespread and affect tRNA, modificatory enzymes, and synthetases. However, a necessary distinction often lacking in this system is between virus-specified function and virus-modified host function, the latter presumably mediated by virus-specified enzyme. Acceptable evidence for virus-specified tRNA is as follows: T4 phage infection drastically alters the ^{35}S MAK elution profile of host tRNA, this alteration depending on protein synthesis. The presence of T4-specific Leu- and Pro-tRNA in a third ^{35}S elution region was shown by direct annealing of acylated tRNA to phage DNA (Weiss *et al.*, 1968). Leu-tRNA is actually found in the hybrid product (Daniel *et al.*, 1970). Recent evidence shows that certain T-series phages can code for their own complete set of tRNAs. This suggests that the phage provides for its use within the cell those synonym species which the cell lacks. Yet, as Daniel *et al.* (1970) point out, the function of these tRNAs may never be solved unless it can be first shown that *E. coli* is the natural host of the phage in question, for, if not, such function may not be demonstrable.

Less rigorous evidence for herpes virus-specified Arg- (and Ser-) tRNAs in mammalian cells was presented by Subak-Sharpe (1968). From analysis of the frequency of GC doublets in the viral and mammalian DNAs, the virus would seem a priori to require a different set of Arg codons. By this analysis, a general difference in codon requirements (not only Arg) in respect to the mammal was also found in other viruses whose genome is larger than 23×10^6 daltons, but, as expected from the lower informational content, not in viruses whose genome is less than 5×10^6 daltons (Subak-Sharpe, 1969). It might therefore be after all that the herpes virus-specified tRNAs, as perhaps the phage-specified tRNAs above, are obligatory for viral translation because of their absence in the host, in which case their function is strictly not regulatory. However, as a reminder of possible technical difficulties, the findings in herpes virus could not be confirmed (Morris *et al.*, 1970). In addition, adeno-2 virus, which requires exogenous Arg for growth in mammalian cells as does the herpes virus, produces no detectable virus-specific Arg-tRNA (Răska *et al.*, 1970). Clearly, Subak-Sharpe's proposal remains to be settled. In the meantime, is there any indication for a regulatory function of viral tRNA?

In answer to that question, the nonsense (*ochre*) suppressing activity in *E. coli* Su-4+ strains disappears after T4 phage infection (Brenner et al., 1966) and this was interpreted to mean a phage-induced modification of host tRNA. However, whether this modification is aimed at canceling host translation, in this restricted sense being regulatory, remains to be shown. By far the most thorough analysis of this question is that by Kano-Sueoka et al. (1968). After T2 phage infection changes are found only for Leu-tRNA which are T-even specific and depend on synthesis of phage DNA and protein. Briefly, *E. coli* Leu-tRNA shows two peaks of activity in MAK chromatography, peak I being the predominant one. One minute after infection a new and faster eluting Leu-tRNA (peak F) appears and then increases by 2 minutes, during which time peak I decreases and peak II increases. By 8 minutes peak F has disappeared. In conformity with these changes, using heterologous (yeast) synthetase a new peak is observed by 2 minutes in front of peak II; since yeast synthetase discriminates only peak I before and after infection, the new peak appears to be related to peak I. Five Leu-tRNA species were then characterized in the normal cell by further subdividing the MAK profile according to codon response, which in increasing order of elution are: Leu-tRNA$_1$ (CUG), Leu-tRNA$_2$ (poly UC), Leu-tRNA$_3$ (CUU, CUC), Leu-tRNA$_4$ (poly U), and Leu-tRNA$_5$ (UUG). A correlation by change in codon response and activity profile was established for the infected cell: by 1 minute after infection the newly appearing Leu-tRNA$_F$ responds to poly UG, while simultaneously Leu-tRNA$_1$ decreases markedly; by 8 minutes, when Leu-tRNA$_F$ is no longer observed, significant increases occur in Leu-tRNA$_3$ through Leu-tRNA$_5$. Reverse phase chromatography (RPC) also resolves five Leu-tRNAs in the normal cell and these remain partly resolved after MAK rechromatography (Kan et al., 1968). In accordance with the MAK data, at 8 minutes after infection the RPC Leu-tRNA$_1$ (CUG) decreases, while Leu-tRNA$_5$ (UUG) increases. It is believed that RPC Leu-tRNA$_1$ through Leu-tRNA$_3$ are the same species in both normal and infected cells, but RPC Leu-tRNA$_4$ and Leu-tRNA$_5$ are different (corresponding approximately with the slowest moving MAK fractions). Concerning the tRNA of immediate interest, tRNA$_F$, this is considered to derive from the normal Leu-tRNA$_1$ after being split in half by the action of a phage-induced nuclease (Kano-Sueoka and Sueoka, 1968) with a consequent change in codon recognition. The splitting takes place during acylation, or without acylation if the tRNA is H bond disrupted. As expected under these circumstances, tRNA$_F$ is incapable of any response to enzyme. An interpretation of these data is now possible from the fact that the biologically active Leu-tRNA$_1$ is found in association with the polysome and therefore is presumably used by the cell, but disappears from the polysome as it breaks down to inactive tRNA$_F$. In view of the additional fact that the serviceable Leu codon (CUG) is not used by the phage, the conclusion by Kano-Sueoka and Sueoka (1969) is that the phage destroys this Leu-tRNA to disable cell translation at a prepolysomal level. They do not, however, imply that the complex shut-off of host function is entirely accomplished by this means. Yet, as pointed out before, Leu-tRNA$_1$ might be rather generally involved in regulation and might thus be critical to host function. Leu-tRNA$_1$ has now been sequenced (Dube *et al.*, 1970) making possible further analysis of this interesting system. Hence, the system represents phage modification, or better, destruction of cell tRNA for a purpose which, in a certain sense, may be considered to be regulatory.

The molecular nature and biological reason of the other extensive early changes described above which Sueoka and his collaborators reported in T2 phage (Kan *et al.*, 1968) is not known. For example, the reciprocal quantitative variation of RPC Leu-tRNA$_1$ and Leu-tRNA$_5$ could indicate either conversion of tRNA$_1$ to tRNA$_5$ or *de novo* synthesis of the latter. According to Waters and Novelli (1968) using a different RPC system, however, Leu-tRNA$_1$ is only partially inactivated rather than having disappeared. Recently, Kan *et al.* (1970) extended their observations at 8 minutes after infection: the normal codon response of RPC Leu-tRNA$_1$ and tRNA$_5$ remains unchanged but the latter's response changes quantitatively. Unfortunately, no complete codon assignment of tRNA$_5$ was workable and the question is still open. At any rate, the resulting fivefold relative increase in UUG recognition is bound to have quantitative effects on translation. Later in the infective cycle, at 20 minutes, there is evidence for the onset of T4 phage-specified tRNA synthesis corresponding approximately with the MAK peak II region of the elution profile (Weiss *et al.*, 1968). This tRNA synthesis is probably related to the transition from early to late phage protein synthesis. Further changes possibly related to late phage protein synthesis are suggested by two qualitatively new Leu-tRNA peaks appearing at 120 minutes after T2 phage infection (Waters and Novelli, 1967). These late tRNAs are phage-induced since they have no counterpart in the normal cell, but there is no evidence as to whether they are phage-modified cell tRNAs or new phage-specified tRNAs. In conclusion, it is clear that considerable production and/or modification of tRNA accompanies the infective cycle, but, except for the early Leu-tRNA$_F$, the function remains undefined.

A different type of change of biological activity was found in Pro-tRNA after Qβ phage infection of *E. coli* (Hung and Overby, 1968). Although the MAK elution profile is no different, the tRNA from infected cells binds two to three times less to the ribosome in the presence of poly C (a reminder that changes may go unnoticed by chromatography). Significantly, in the few surviving cells the Pro-tRNA is normal. This change is not observed after infection with the related MS2 phage, nor does it occur in Phe-tRNA with Qβ. Since phage specification is all but excluded, in this case, from the small size of the viral genome, the reason behind the change remains unclear. All the work on phage so far discussed was done using synthetase from normal cells, thus excluding interference from a possible concomitant variation in enzyme during the infective cycle.

Plants and Animals

What could well prove to be a demonstrable modification of tRNA along the lines of the modulation hypothesis has become known in plants. Cytokinin activity, as the result of postsynthetic modification of tRNA bases, is present in a small but significant proportion of several tRNA species (but not rRNA) from many sources including plants and animals (Armstrong et al., 1969). In plants cytokinins act as growth factors to promote cell division, the N-6 substitution of the adenine residue being essential for this action. In animals the function is still elusive. Two examples of cytokinin substitution are: N^6-isopentenyladenosine and 6-(γ,γ-dimethylallylamino) purine, both derived from mevalonic acid. The modification occurs immediately next and to the 3' side of the anticodon of the tRNA species. The apparent rule for the presence of the modification is that U must be the first letter in the codon recognized by the tRNA (Peterkofsky and Jensensky, 1969) and this involves many, if not all, amino acids with U_I codons. The relationship is indirectly confirmed by the observation that many tRNAs with U_I are nonpolar (Wehrli and Staehelin, 1971), which is compatible with their possessing a lipophilic modification such as cytokinin. Because of its large size the substitution may influence the AU base pair formed with the codon by the other (5') end of the anticodon. Possibly by this means the substitution reinforces the binding of codon-anticodon (Gefter and Russell, 1969; review by Lengyel and Söll, 1969), ultimately influencing the effectiveness of translation. It also appears that enzymes can introduce the cytokinin residue in toto from free cytokinin-PP_i into certain A residues in selected tRNA species (Kline et al., 1969), which might portray the mode of origin of the substitution. Thus a regulation of translation based upon posttranscriptional modification of tRNA can be envisioned in this system. The reader is directed to a comprehensive review of the subject by Hall (1970b).

Observations dealing with changes in the tRNA population have been made in animals. In a broad survey of organs within and between mammalian species, only Tyr-tRNA revealed a characteristic major difference in profile between fibroblasts and epithelia or normal organs (Taylor et al., 1968). Minor differences were also observed for Gly- and Ser-tRNA between mammalian organs. These differences observed *in vitro* were also found *in vivo*. Since synthetase variation was excluded in the survey, this clearly was one of the first examples of cell lineage differentiation at the tRNA level. It suggests a differential structural gene activation rather than a secondary posttranscriptional modification of tRNA, presumably in correspondence with new mRNAs appearing during lineage differentiation. On the other hand, between birds and mammals the fibroblast-type Tyr-tRNA remains unchanged but noticeable differences in profile were found in Leu- and Tyr-tRNAs from the same organs. Recently, however, with the higher resolution of RPC, different tRNA species for each of several amino acids were found between mammalian organs (Hatfield and Caicuts, 1969).

Preliminary evidence on the microheterogeneity of the α-hemoglobin chain has been interpreted as indicating the presence of infrequent codons ambiguously served by minor tRNAs that occasionally insert an unusual amino acid (von Ehrenstein, 1966), in what virtually amounts to a "programmed" ambiguity within a cistron. This possibility was first suggested by Itano (1964) and further elaborated upon by Ingram (1964). More convincing, if not probative, evidence is now available. The α chain of rabbit hemoglobin contains several Leu residues. Using a cell-free reticulocyte system and tRNA fractionated by CCD, Galizzi (1969) found that the Leu codon at residue 48 is read only by a minor tRNA species, while the majority of the Leu codons can be read by two major species and some codons by only one species. Different mouse strains vary in the amino acid sequence of the α chain. In this case, it was shown in a cell-free system that a Val codon exclusive to the C_3H strain is read by a tRNA that is absent in the $C_{57}BL$ strain (Yang et al., 1969b). Similarly, certain synonym tRNAs are required for the α but not the β chain of rabbit hemoglobin (Anderson and Gilbert, 1969). It is easy to see how these minor tRNAs, particularly in the example with rabbit Leu-tRNA, could function regulatorily. In general, this type of explanation may apply to much of the known genetic heterogeneity in other proteins. It has been proposed, for example, for normal immunoglobulins (R. J. Campbell, 1967; Mach et al., 1967), without, however, gaining a wide acceptance (discussed by Jerne, 1967).

Concerning the effects of hormonal stimulation of tRNA, in rat liver cortisone increases the acylation of all tRNA species without changes in the elution profile for at least several that were tested (Agarwal et al., 1969). However, rat liver MAK profiles of Lys- and Phe-tRNAs show changes after thyroidectomy which are followed by partial restoration to normal soon after administration of thyroxine (Yang and Sanadi, 1969). In rooster liver, estrogen causes the production of Ser-rich phosvitin. Correspondingly, of four Ser-tRNA peaks on BD-cellulose chromatography the major peak is relatively increased (Maënpää and Bernfield, 1969). Because the effect of estrogen is only on Ser-tRNA, this is probably an example of selective stimulation of synthesis. Other examples of hormonal control of tRNA, including posttranscriptional modification, were given in Chapter 4.

Although this review is not exhaustive, but selective, the evidence presented for variation in tRNA activity in the higher eukaryotes so far examined is not greater than in bacteria. This might seem surprising at first glance, but it must be recalled that most of the evidence in eukaryotes deals with mature cells and not with cells exposed to drastic changes in growth as is the case of bacteria. In fact, in neoplastic cells discussed below the evidence is probably more plentiful. Considering the variety of differentiative pathways that healthy eukaryotic cells undergo, the conservative picture that emerges is therefore probably significant and not a sampling artifact. This indication for the involvement of a minor fraction of the tRNA population in regulation stems directly from the general high cistronic multiplicity of tRNA in the higher eukaryote. Hence only minor tRNA species, presumably with a lesser multiplicity, are likely to be involved in regulation. This is also the argument for explaining why high levels of informational suppression are rare in the higher eukaryote. Nevertheless, a final evaluation of the modulation hypothesis as concerns the higher eukaryote should be deferred until more high resolution studies on different systems are available. A definite possibility, as yet not firmly on record, is that of specific loss of tRNAs (and synthetases) with progressive cell maturation, primarily as the result of transcriptional repression.

Neoplastic Cells

Apart from the special interest in neoplastic cells, it was hoped that their variety of growth patterns would allow a crucial test of regulatory tRNA function. The evidence is as follows. Ehrlich mouse ascites cells show main shifts of activity peaks on MAK for Phe- and Ser-tRNAs and extra peaks for Gly- and Tyr-tRNAs compared with normal mouse organs (Taylor *et al.,* 1968). The difference in Phe-tRNA does not involve modification of the anticodon (Taylor, 1969). Certain tumor and adenovirus transformed cells alone contain both Tyr-tRNAs found in normal cells, one characteristic of fibroblasts and the other of epithelia. The elution profiles remain abnormal whether using normal or neoplastic enzyme, again implicating tRNA itself. The question remains unanswered as to whether the changes represent a primary event in carcinogenesis or sequelae attendant upon the long-term passage of the neoplastic cells in culture (HeLa and transformed cells) and the peritoneal cavity (ascites cells). A more definite answer to this question comes from the ethionine-induced hepatoma referred to below. Baliga *et al.* (1969) described new slow moving activity peaks on MAK for Asn-, His-, and Try-tRNAs of Novikoff hepatoma compared with rat liver, as well as differences in the elution profile for seven other amino acids. Mouse leukemia Asn-tRNA possesses one additional activity peak on RPC which, it is suggested by Gallo *et al.* (1970), may be related to the sensitivity to L-asparaginase as an antileukemic drug. Between serum-free grown L-M cells and tumors induced after they are injected into mice, mainly four tRNA profiles show change out of many tested on RPC (Yang *et al.,* 1969a). When the tumor cells are grown serum-free, they revert to the original profile but tend to relapse upon addition of serum. Either nutritional factors or differentiation of the L-M cells in the host are proposed as the explanation for this interesting phenomenon.

Plasma-cell derived tumors have attracted experimental attention because of the ease of identifying their protein products. A plasmocytoma lacks one activity peak each of Leu-tRNA (total three peaks) and Thr-tRNA (two peaks) on MAK compared with liver or other plasmocytomata reacted with liver enzyme

(Mach *et al.*, 1967). With the higher resolution of RPC the differences between various plasmocytomata, producing either myeloma immunoglobulin A or G, were circumscribed to Ser-tRNA which shows alterations in activity in three out of four peaks observed (Yang and Novelli, 1968a). It is interesting in this regard that Ser is one of the few variable amino acids in the active fragment of immunoglobulins. These changes are due to tRNA but as usual it was impossible with the data on hand to decide whether they represent synthesis or modification of the molecule. A survey of Leu-tRNA was conducted on RPC in various plasmocytomata all secreting the same light immunoglobulin chain, but with single amino acid differences (Mushinski and Potter, 1969). Of five Leu-tRNA peaks in mouse liver, some plasmocytomata were very deficient in peak 3 and/or 5 and others had more of peaks 4 and 5; synthetase variation was reasonably well excluded in this work. Depending on the particular plasmocytoma, an independent transcriptional regulation of the cistrons coding for each putative Leu-tRNA species is suggested.

Clearly, there is no simplification of the neoplastic cell in terms of tRNA, ruling out dedifferentiation of the neoplastic cell at this molecular level. Rather, although neoplastic cells may retain undifferentiated traits as in the case of Tyr-tRNA, they often proceed to a diversification of tRNA for which there is no normal equivalent. Judging from the information so far obtained, the study of tRNA modulation in neoplastic cells has just about lived up to its original promise: the challenge is now to relate the modulatory behavior of the molecule to the etiology of the neoplastic process. An interesting possibility is that carcinogens may first act by chemically modifying a few tRNA species in the population (Agarwal and Weinstein, 1970). For example, most of the alkylation introduced by these agents is now known to be in tRNA rather than in DNA as was originally thought. Together with the additional possibility that all spontaneous carcinogenesis could be due to alterations in minor tRNA bases (review by Zamecnik, 1971), this area of biomedical research might become one of the most important as far as tRNA is concerned.

Enzymes Related to Transfer RNA

Enzyme variation is less well documented than that of tRNA, and, in those cases where it is, the changed site or mode of interaction with tRNA remains largely undefined. However, for the present question of regulatory modulation of translation, alteration of enzyme function is just as important as that of tRNA. This says that the effects of enzyme and tRNA with respect to regulation are formally interchangeable, thus placing enzyme effects as an extension of the original modulation hypothesis. Indeed, part of the variation discussed for tRNA in the preceding sections may be the result of overt or covert enzyme variation. Enzyme variation can take two forms. First, there is authentic enzyme alteration reflecting a different primary and/or quaternary structure, ultimately traceable to newly activated structural cistrons. Of some importance, because of the probable uniqueness of these cistrons for enzymes (or their subunits) in contrast to the tRNA cistrons, a functional restriction dependent on cistronic multiplicity as discussed for tRNA may not apply. This difference would tend to make regulation by enzyme more stringent than by tRNA. However, this argument may not be true for eukaryotic synthetases whose cistronic multiplicity cannot be excluded as probably it can in prokaryotic synthetases. Second, there is modification of extant enzyme involving physiological posttranslational change and in this case any new cistron involved is not the enzyme structural cistron. By comparison, fluctuation in enzyme levels would be of relatively less consequence than the two previous forms of variation. Last, but not least, the real possibility that covariation in enzyme and tRNA is determined by separate mechanisms makes the translational regulation all the more complex and intriguing.

Synthetases. Because of the special relationship with tRNA, variation of synthetase in particular will result in quantitative or qualitative changes in translation which are very similar to those introduced by tRNA, and are, therefore, of equal regulatory value. Minor activity changes in synthetase were indeed pointed out in several of the previous examples for tRNA. That this key enzyme is prone to have regulatory functions annexed to its primary translational function was clear from the discussion of bacterial amino acid metabolism in Chapter 3; there is no reason to suppose that these regulatory functions (even if poorly known) are absent in the euakryote. The relative importance of synthetase variation in the prokaryote and eukaryote might be different, however, by virtue of the apparent fact that a synonym set of tRNAs may be served by one enzyme in the prokaryote and by several synonym enzymes in the eukaryote. Moreover, in the eukaryote certain synonym synthetases (together with the corresponding tRNAs) function exclusively in either the cytoplasm or organelles. Eukaryotic synthetases thus tend to behave like isozymes.

A substantial modification is induced exclusively in *E. coli* Val synthetase during early T-even phage infection. As shown by density label, the host synthetase is totally converted to a new more stable form, possibly dimeric, whose appearance depends on protein synthesis (Neidhardt *et al.*, 1969). The working hypothesis calls for an addition of a phage-specified protein to the host enzyme. The mystery surrounding this conversion is that under laboratory conditions it may be gratuitous to the phage, that is, the conversion does not demonstrably affect phage growth. One is again reminded of the possibility mentioned earlier that *E. coli* may not be the natural host.

In the rabbit, Strehler *et al.* (1967) found that heart muscle possesses three Leu synthetases separable by DEAE-cellulose, two of which differentially attach Leu to their respective tRNAs. In addition, reticulocyte and liver synthetases differ quantitatively in the pattern of acylation. Of two Ala-tRNAs separated on RPC, one is acylated by kidney, reticulocyte, and liver synthetase, whereas the other is acylated by reticulocyte and liver, but not kidney, synthetase. If representative of the *in vivo* condition, this evidence then suggests that synonym synthetases are distributed to different tissues in a nonexclusive, partly overlapping pattern. Goldman *et al.* (1969) compared Novikoff ascites tumor with rat liver. On RPC, ascites Phe-tRNA lacked the first of three activity peaks when acylated wtih ascites synthetase, but showed it with the liver synthetase. However, the ascites synthetase partially recognizes liver peak I. Thus, both ascites Phe-tRNA and synthetase appear to be modified. This situation may be common to many systems and shows the difficulty to ascribe a functional change to one or the other molecule.

Methylases. It has been argued that many of the changes in tRNA activity listed above reflect the activity of modificatory enzymes influencing translation through their widespread effects on the higher order structure, hence function, of tRNA. A good example of modification was the plant cytokinin substitution affecting structural aspects of codon-anticodon recognition. This concept of structural modification—whose significance extends to tRNA function in general and not just the modulation of translation—stems from the pioneer work of E. Borek on methylases. The great importance of these enzymes is that, by recognizing different regions of the tRNA structure which they modify, they bring the whole structure into play for regulation, in addition to the anticodon whose properties in turn they influence. Since all substitutions introduced by the enzymes are potentially regulatory and there may be hundreds of enzymes involved, the scope for regulation at this level is, indeed, enormous. Moreover, by virtue of the nature of these substitutions, a close gearing of their formative pathways to intermediary metabolic pathways is established. In particular, this gearing to metabolism makes tRNA modification very sensitive to internal conditions in the cell, however, also very costly in terms of energy expenditure. The two families of enzymes to be discussed, methylases and ethylases, may act on tRNA at any time starting with its synthesis.

Transfer RNA methylases consist of several distinct species that, for example, in the eukaryote are base, organ, and species specific. In the bacterium, within 10 minutes of T2 phage infection, there is a dramatic increment of methylase activity that is not uniform for all the base-specific enzymes (Wainfan *et al.,* 1965). The substrate may be a host or phage tRNA whose modification is presumably required for phage function. A requirement of this nature is apparently not universal since no activity increment is found during T1 phage infection. To the contrary, after λ phage induction there is a marked decrease of methylase activity, again with differences in base-specific enzymes, followed later by a resumption to normal levels. A dialyzable methylase inhibitor is detectable in freshly induced cells (Wainfan *et al.,* 1966). These data suggested to Wainfan (1968) that methylases participate in the regulation of the λ phage repressor (discussed in Chapter 3), but this remains to be proved. Both in the case of T3 phage infection and of cultured mammalian cells infected with foot-and-mouth disease virus, there is another methylase inhibitor whose function probably is to destroy the immediate methyl group donor. The large magnitude of the inhibition suggests a restraining influence by phage on the bacterial translational machinery. This may be true of the mammalian virus also since it inhibits host translation (Ascione and Woude, 1969). On the other hand, in the case of virus-transformed mammalian cells an increase of methylase activity was reported with SV-40 virus, but could not be found with polyoma virus (Breier and Holley, 1970). A word of caution by Kaye and Leboy (1968) concerns the adequate control of the ionic environment in these *in vitro* enzymic assays, which otherwise may lead to spurious interpretation. A second word of caution is that the viral effects on the enzyme or the methyl group donor could well be the unavoidable result of some loss of cell function rather than directly caused by the virus. Nevertheless, taking the data above at face value, the conclusion is that methylase activity can change in any direction depending on the system.

Increase and base-specific shift of methylation are extensively documented during mammalian carcinogenesis. The differences can sometimes be almost qualitative, as, for example, the appearance of U methylase activity which is minimal in healthy mammalian cells (although common in bacterial cells). These changes are partly due to the removal of methylase-inhibiting proteins but partly also to variation in the methylases themselves, including their altered specificity toward substrates. Although the causal relationship of this methylation pattern to carcinogenesis remains to be proved (review by Borek, 1969; Craddock, 1970), it is highly suggestive that several carcinogens all increase the methylation of precancerous tissue. According to Riddick and Gallo (1970), an increased activity and different activity pattern of methylases is characteristic of neoplastic cells as well as immature normal cells, including fetal cells. Once the activity reaches its maximum in these cells it does not continue to increase, irrespective of any further increase in the cell growth rate. These findings have led to the interesting notion that the neoplastic process is equivalent to the unmasking of embryonic genetic information, at least as far as tRNA modificatory pathways are concerned. The great prospective importance of this area of research was pointed out earlier in connection with tRNA.

Moderate differences of methylase activity between several normal mammalian organs are reviewed by Borek (1969). However, whereas these differences certainly evince tissue differentiation at the level of methylase, their regulatory significance remains to be seen. As an example, according to Tidwell (1970), the temporal pattern of methylation during liver regeneration suggests that translational regulation depends on this pattern. In this respect the effect of estradiol on the pig uterus is noteworthy; ovariectomy reduces the methylase activity and alters the MAK elution profile of Ser tRNA, both of which are restored to normal by administration of the hormone (Sharma and Borek, 1970). Of course, uterine stimulation involves considerable histodifferentiation and may be properly regarded as a model of differentiation. As will be seen, in the traits just mentioned, this model resembles the developmental systems discussed in Chapter 6.

Ethylases. Ethylases appear to be implicated in the initiation of the ethionine-induced hepatoma. Ethylation is from a cell donor other than S-adenosylethionine (Ortwerth and Novelli, 1969). It involves all or most tRNA species (Axel *et al.*, 1967), presumably by competing for physiological methylation sites in the molecules. Liver Leu-tRNA shows three activity peaks on MAK responding to poly UC and poly UG, but major differences are found in hepatoma. Concerning ethylation itself, before the hepatoma appears, ethylated liver tRNA shows only one Leu activity peak responding to poly UC; this indicates a deficiency for the UUG codon. By the time the hepatoma is established after months, the Leu-tRNA has returned to normal (as there is no longer dietary ethionine available for ethylation). Thus, the neoplastic condition persists and indeed develops after the tRNA is restored to normal, in agreement with the notion that promotion and maintenance of tumors are by mechanisms separate from those for initiation. However, the manner in which ethylated tRNA is related to tumor initiation, and whether or not it is a causal relationship, is not known. In this respect, although ethylation spares nucleic acids other than tRNA, it does, however, affect proteins (Craddock, 1970), whose involvement in this type of tumorogenesis is equally unknown. Nevertheless, the increased methylation of precancerous tissue caused by carcinogens (see above) adds to the suggestion that ethylation has a role in early tumorogenesis.

Evaluation

In conclusion, an extensive descriptive evidence suggests a diversified role for tRNA in the modulation of translation. This role is probably more important for cell differentiation and cell function than for basic cell growth, but in addition, as an aspect of cell differentiation, it may be important for neoplastic growth. As described in the introduction, this modulatory role depends on selective synthesis or modification of individual tRNA species, either resulting in a qualitative variation of the RNA population. Of these possibilities, synthesis in particular is worth discussing with regard to its implications. In differentiating cells, modulation by selective tRNA synthesis would imply the function of a facultative genome whereby individual species are regulated independently of other species, including their own isoaccepting species. Presumably, this facultative genome would have been initially under control of a developmental transcriptional program until it eventually came also under control of hormone. In neoplastic cells, modulation by tRNA synthesis presumably means a departure from the original program in a normal cell.

Thus far, the suggestion is that a translational regulation by tRNA may be biologically widespread, but the greater implications in the eukaryote should not be overlooked. Not unexpectedly, from the number of modified tRNA activities observed in any system, the impression is that regulatory tRNAs are minor components of the population. That is, within certain synonym tRNA sets there may be major species required for general translation and minor species facultatively used for modulation.

In the ultimate analysis, however, causal proof and definition of the mechanism of translational regulation by tRNA are still lacking. Although commonly associated with translation, it needs recalling that any variation in tRNAs could also be connected with regulation of many other metabolic pathways. At present, the most accomplished analyses are those in bacteriophage infection and plant cytokinins, and, from the standpoint of human biology, the most promising ones are those in mammalian hemoglobins and, particularly, neoplastic cells. It could well be that with further work these systems completely fulfill the conceptual promise of the original modulation hypothesis. Meantime, until a regulatory explanation is found for them, the qualitatively different distribution of tRNA species which is found in cell organelles, phage genomes, organs and tumors remain interesting examples of molecular differentiation of the translational apparatus. In the event it were substantiated, a modulatory function for tRNA would mean that Nature has put to use the multiplicity and degeneracy of the molecule once more to a functional (sometimes dysfunctional) purpose.

PART III
Biological and Evolutionary Aspects of RNA

Chapter 6 surveys the function of eukaryotic RNA as seen at the level of a complex organelle, the nucleolus. Also considered under this function are two inclusive aspects of cell life, namely, the cell cycle and embryonic development. Chapter 7 recapitulates from an evolutionary standpoint the most important traits of the RNA species as they pertain to their role in genetic function. Contrasted, these two chapters, one exemplifying the diversity of RNA function that can be appraised at different biological levels and the other searching for the evolutionary norms that are discernible behind this function, aim at the broader picture of the involvement of RNA in matters of cell biology.

CHAPTER 6

Topics of Interest to Cell and Developmental Biology of the Eukaryote

The topics covered in this chapter concerned with the eukaryote—the nucleolus, the cell cycle, and embryonic development—attempt to provide a better biological persepctive of the RNA molecules than was possible in preceding chapters dealing with the more molecular aspects. By confronting living rather than biochemical situations, these topics will also inevitably show how much there is yet to be learned. Unfortunately, however, because of the unavoidable omission of other important biological parameters, the approach to each topic cannot be as balanced as one would have desired.

Two other very active areas of molecular biology which do not yet warrant inclusion in a general monograph of the present type, but doubtless soon may, are the formation of antibody (review by Edelman, 1970) and nerve cell function (reviews by Glassman, 1969; Agranoff, 1971). Significantly perhaps, although at different orders of complexity, both these areas deal with memory-like cellular phenomena. In the case of antibody-producing cells, what is lacking is the mechanism, if that is what it is, whereby a competent cell manages to match a particular antigen with a suitable anticopy, the antibody. Recent reports have suggested the specific involvement of noninformational RNA in the handling of antigen, and it might well be that a major conceptual surprise lies in waiting in this problem. The reader, therefore, is best referred to an evaluation by Edelman (1970) of the genetic mechanisms currently envisioned to be behind antibody formation. In the case of nerve cells, it is not possible at present to distinguish the RNA metabolism connected with specific function from the intimately related metabolism connected with basic maintenance. Evidence for specific neuronal proteins has nevertheless begun to appear (review by Shooter and Einstein, 1971). The relationship of these proteins with neuronal electrophysiological activity is not at all clear, nor is it too clear how we shall attempt to relate them. Adding to this gap, a revealing finding is that much more of the mammalian unique genome, representing mostly the informational genome, is transcribed in the brain than viscera (Brown and Church, 1971). The complexity of the brain transcript is therefore equivalent to a very large number of proteins, i.e., 2.5×10^5 different proteins with an average length of 330 amino acids. If it really occurs, such a complexity of product would reflect the complexity of brain function, but it must not be forgotten that this function is based on a heterogeneity of cell types each of which is functionally subdifferentiated.

THE NUCLEOLUS

It took 180 years to establish the central function of this typically eukaryotic organelle discovered around 1780 by Fontana. On present knowledge, this function is the assembly of ribosomal subunits (Perry, 1962) and involves several separate but interrelated aspects: (a) making and processing a common RNA precursor, and (b) packing it with protein derived from a different cell compartment. Of these aspects, the common precursor may or may not be exclusive to the eukaryote, whereas processing of a precursor is not exclusive but only more highly developed in the eukaryote. As a consequence, the eukaryotic ribosomal pathway can be separately regulated, first, at the level of synthesis and, second, at the level of processing and proteination of the transcript, and the two regulatory levels show a greater degree of autonomy than in the prokaryote, e.g., hormone will affect one and/or the other level. The regulation comprises several subregulatory stages discussed in Chapter 4 which are probably contingent to the compartmentation of the pathway inside the nucleolar organelle. Only the eukaryotic pathway has these distinct subregulatory stages as well as a pool of large ribosomal subunits. Related to its more complicated manufacture, the eukaryotic ribosome is larger than the prokaryotic ribosome mainly because of the greater number of ribosomal proteins. In all probability, however, the higher order structure of the ribosome is much more elaborate than what the greater number of proteins might suggest. Therefore, the eukaryotic ribosome is not only larger but probably considerably more complex. This structural complexity suggests a greater functional complexity:

the domain of function which is involved is not known, but it probably has to do with the capacity for general interaction or, in particular, with cytoplasmic membranes and hormones. Returning to the nucleolus, it will be seen that its molecular complexity is also very great and perhaps only second in the cell to the chromosome. This complexity may be necessary for other nucleolar functions, in addition to the central ribosomal function, which are as yet unknown. This complexity is reflected in the fact that the nucleolus is the only part of the eukaryotic chromosome which is commonly considered to be a distinct synthetic organelle. Hence it may take more than the usual reductionist attitude of the experimenter to define the nucleolus functionally, toward which goal the purpose of this section is directed.

Synthesis of eukaryotic rRNA is, with very few exceptions, manifest in the presence of a nucleolus. The converse need not be true since at times a nucleolus may be inactive. In general, however, both nucleolar size and activity correlate well with the cell metabolic status (review by Sirlin, 1962). In the growing insect spermatocyte, for instance, the amount of nucleolar RNA (mainly rRNA) is directly proportional to the nuclear volume (Schrader and Leuchtenberger, 1950), itself an index of cell function. At the same time, if there is no fusion of nucleoli, for an equal metabolic status nucleolar size is related to the number of rRNA cistrons which occur within each nucleolus organizer. (Fused nucleoli contain more than one nucleolus organizer and this simple relationship can no longer apply.) This number of cistrons per organizer may vary because of chromosomal mutations, with the result that the total number of cistrons per cell cannot be deduced from the number of organizers which are present. On the other hand, in nondividing, nongrowing cells, such as liver and nerve cells, nucleolar activity reflects rRNA replacement rather than net synthesis. A question which is still pending is whether all the rRNA made in these nongrowing cells reaches maturity in the cytoplasm; some evidence indicates that it does not (Attardi *et al.*, 1966).

The Ribosomal RNA Cistrons: Natural History

As a base line for the present discussion we may recall that in the prokaryotic chromosome the major rRNA cistrons occur as clusters that are disposed in two regions where the 5 S RNA cistrons, the majority of tRNA cistrons, and many ribosomal protein cistrons are also located. This physical linkage of the translational apparatus as a whole was largely lost during evolution toward the eukaryote. As far as we know, except for the rRNA cistrons themselves, most of these other cistrons have separated from one another in the eukaryote. This separation suggests that a more dynamic regulation of the translational apparatus has evolved in place of the original regulation requiring a physical proximity of cistrons. Thus to date the only known nucleolar templates at the DNA level are rDNA and perhaps some tDNA. Of importance, however, stating that these are the only nucleolar templates is based solely on the lack of evidence to the contrary.

Almost certainly the rRNA cistrons will become the first in the eukaryote whose regulation is understood. The general validity of the molecular theory then to come will depend on the extent to which it explains not one ribosomal system, but all of them, that is, the natural history of the cistrons. A good start in this direction has been the isolation of a putative repressor for rRNA synthesis which was discussed in full in Chapter 4 in connection with regulation of this RNA species. The point was made that such a regulation based on a common repressor for all the repetitious rRNA cistrons is the simplest one possible at the chromosomal level to permit an immediate and cooperative response of these cistrons. How the repressor itself is ultimately controlled is, however, not known.

The best characterized cistrons in *Drosophila* and *Xenopus* are clustered as a single block at the nucleolus organizer locus of each haploid complement and essentially only one rDNA strand of the locus is transcribed. In these animals there is a maximum of one nucleolus per complement or two nucleoli per cell, although frequently the number is reduced to one by nucleolar fusion. In certain insects and plants there is more than one cistron block or organizer locus per complement, and many more are found in mammalian cells. Correspondingly, in these cells there is a plurality of nucleoli, the exact number again depending on the extent of fusion. These low and discrete nucleolar numbers are representative of most cells, but certain cells, mostly in the lower eukaryote, have more complex or even variable nucleolar patterns, of which two examples follow. Ovarian nurse cells of *Drosophila* have a single discernible nucleolus at an early stage, which, because of nucleolar fusion, probably indicates one nucleolus organizer per complement. At later stages, however, large masses of nucleolar material become abundant. A rather similar condition prevails in the larval salivary gland of sciarid insects. In these chromosomes, an organizer produces a nucleolus during early stages which later tends to disperse; at this stage, numerous "micronucleoli" appear and often are in what seems to be an intimate association with chromosomes (Sirlin and Schor, 1962a,b). That these micronucleoli are involved in rRNA synthesis was demonstrated by cytological hybridization (Pardue *et al.*, 1970), but had been previously anticipated from their ultrastructure (Jacob and Sirlin, 1963). These two conditions in the insect seem to predominate in cells such as these nurse and salivary cells with hypertrophied function and correspondingly hypertrophied chromosomes, i.e., endopolyploid and polytene, respectively. As can be seen clearly in *Drosophila*, and possibly also the sciarid, nonhypertrophied cells in other tissues of the animal have only one or two nucleoli.

What concerns us here is the origin of the nucleolar masses and micronucleoli present in the hypertrophied cells. Basically, the question is whether these nucleolar materials are the product of the organizer or the unmasking of "hidden" cistrons located elsewhere than in the organizer. In the case of nurse cells, there is a strong possibility that the nucleolar masses originate from organizer rDNA that has undergone amplification. This exceptional somatic counterpart of germinal amplification would be due to the fact that the nurse cell supplies rRNA to the oocyte, and, accordingly, *Drosophila* may lack germinal amplification. The origin of micronucleoli in the sciarid salivary cell is less clear. The Gall group (Pardue *et al.*, 1970; Gerbi, 1971) favor the interpretation that the micronucleoli represent ramifications of organizer rDNA, on the basis that (a) such ramifications are observed and (b) nonorganizer chromosomal regions with associated micronucleoli contain no detectable rDNA. The evidence under (b) is, however, not compelling because, for example, in *Drosophila* salivary glands what is initially the locus of the chromosomal organizer for some reason later shows no detectable rDNA (see below). It remains possible, therefore, that some micronucleoli originate from scattered rDNA, as originally interpreted by Jacob and Sirlin (1963) on the basis of ultrastructural evidence. As will be discussed, this "hidden" rDNA is not uncommon. In addition, Gerbi (1971) found no rDNA amplification in the sciarid salivary cell. Of interest, she goes on to suggest that the heterochromatic organizer may remain underreplicated during growth and the production of micronucleoli might be to maintain the diploid amount of rDNA. This suggestion is equally valid in the event that micronucleoli originate from nonorganizer rDNA.

The preceding ribosomal systems indicate that the pattern of clustering of cistrons in different organisms is variable, ranging from one to several major clusters or organizers, and in certain organisms some cistrons may be scattered in smaller clusters. The latter could be the case, for example, in yeasts (Schweizer *et al.*, 1969) and perhaps also in crabs (Skinner, 1969). In addition, it will be seen that some organizers are not always active, and, from the previous evidence, the small clusters may be active only in certain tissues. In moving to the higher eukaryote, e.g., mammals, it would seem that the reason for the increased number of major nucleolus organizers cannot simply be for the more dispersed cistrons to be able to escape basic regulation by a common free repressor. The increased numbers of organizers could instead be, for example, to permit a modulated or more plastic response of the separate cistron clusters to the repressor under different cell circumstances. Perhaps then this separate arrangement has a greater adaptive value for the most advanced eukaryote. However, in order to arrive at any generalization, other ribosomal systems have to be considered.

A most remarkable functional adaptation to the increased demand for rRNA is the amplification of rDNA typical of certain oocytes. According to Brown and Dawid (1968), this germinal amplification is manifest whether there are many free nucleoli, as in the amphibian oocyte, or only one nucleolus that remains attached to the chromosome, as in the clam oocyte. According to Vincent *et al.* (1969), however, the rDNA is not amplified in the single nucleolus of the starfish oocyte. Pending confirmation, this last observation may serve as a reminder that there is no real reason why amplification should be obligatory in all oocytes rather than just in those which require a particularly large amount of rRNA. In oocytes of certain dipterans the amplified rDNA occurs within an enormous heterochromatic chunk which sits on the nucleolus (Lima-de-Faria *et al.*, 1969) and sometimes shows evidence of breaking up. Hence, broadly speaking, there is a range of distribution of rRNA cistrons in different cell types with, at one end, the typical paucinucleolate somatic cell, and, at the other end, the multinucleolate oocyte. Within this range, increasingly numerous cistrons tend more to detach from the main organizer, in effect becoming discrete organelles at the oocyte end. However, such a detachment of cistrons once again does not guarantee evasion from a common basic regulation by repressor, and in fact the repressor activity is nonexistent at the time of maximum rRNA synthesis in the oocyte. Rather, the detachment may be mechanically necessary to permit the amplification finally resulting in a myriad oocyte nucleoli; or it may be related to the unique regulatory properties of the cistrons which permit the necessary incoordination of rRNA and protein synthesis in the oocyte to occur. As mentioned above, this phenomenon of amplification, usually regarded as germinal, may apply as well to certain somatic cells such as the insect ovarian nurse cells which contain an excess of nucleolar material; it is possibly in these cells, not the oocyte receiving rRNA from them, that the rDNA is amplified.

Some of the examples in cells with polytene chromosomes suggested that there may be "hidden" clusters of rRNA cistrons. By extension, this suggests that there may be major nucleolus organizers also hidden. This is, in fact, known for several types of cells with typical (nonpolytene) chromosomes where the number of such hidden organizers is of the same order as of organizers which are normally expressed. For example, not all organizers in the plant or cultured mammalian cell (Phillips and Phillips, 1969) normally result in a definitive nucleolus. It is well authenticated by breeding experiments in plants that an organizer can be present but not expressed (Keep, 1962). Human-mouse hybrid cells exclusively produce murine 28 S RNA, and possibly exclusively murine 18 S RNA, suggesting that the homologous human cistrons, although present, are repressed in this particular combination where, therefore, the murine organizer prevails (Eliceiri and Green, 1969). There is no such prevalence in the case of the mouse-hamster combination, yet it is clear that the man-mouse combination mimics the physiological situation in other cells. From data of this kind, it has been considered long ago that there is a competition for expression favoring dominant or "strong" organizers (e.g., Longwell and Svihla, 1960). A mechanism to explain this dominance is unknown; it could reside in the condition of DNP in the "weak" organizers or in genetic elements outside these organizers, but having effects on them. Of these two possibilities the latter places the number of active organizers (when this is above the diploid two) under genetic control, an area of research which lies entirely in the future (see Flaherty et al., 1972). At any rate, the phenomenon of nucleolar dominance is contingent upon an excessive number of cistrons being present over the minimum required for viable function.

Toward the higher eukaryote there has been a remarkable conservation, in fact a consolidation in respect to the prokaryotic condition, of one or a few highly multiple cistron blocks. This arrangement can have a regulatory reason which, as explained, could be that the presence of multiple cistrons in tight blocks facilitates their joint regulation. In contrast, to make the argument clear, were protection of the rDNA complement from chromosomal accidents such as deletions to have been a prime evolutionary consideration, this would have favored a universal dispersal of at least half the number of cistrons, which is the minimum number dictated by the dosage rule for normal function (see below); the single major organizer in the insect and amphibian shows this not be the case. For correcting deletions there is now instead (as established in the insect) an active mechanism in the form of a transient compensatory replication, namely, the magnification discussed in Chapter 4; this mechanism can act in the opposite direction to correct chromosomal duplications as well. As pointed out, moreover, a moderate numerical increase of organizers could be of some adaptive value in the most advanced eukaryote. In part this could be a protective value, since, for example, a subviable individual (with less than half the number of cistrons) is less likely to occur in man with five organizers than in other animals with only one. Hence, in general, a typical nucleolus in the higher eukaryote is the outcome of the massing of closely related cistrons onto which other aspects of ribosomal function (and perhaps nonribosomal functions) have been built.

We may conclude that there has been a great plasticity in the evolution of the pattern of distribution as well as utilization of the rRNA cistrons inherited from the prokaryotic ancestor, where presumably they were disposed in one or a few blocks of moderate cistronic multiplicity much as they are in the present prokaryote. In its range the evolutionary pattern of somatic cistrons is no wider than that which was noted between different cell types, irrespective of evolutionary scale, when both somatic and germinal cells were taken into consideration. Perhaps only animals that possess polytene chromosomes show the entire range of possible cistronic variation. We have not asked which mechanism could be responsible for the evolving pattern, but, perhaps, the constitutive heterochromatin that is usually located in the vicinity of the nucleolus organizers has played a physical part.

In any balanced genome the nucleolar numbers (before fusion) are in proportion to the number of intact organizers present, irrespective of ploidy. The question arises concerning the nucleolar numbers in unbalanced genomes, particularly diploid genomes, although not many have been examined in this regard. In the simplest case of genomes with an organizer deficit such as those in mutants of *Drosophila, Xenopus, Chironomus,* and maize, one organizer out of two suffices to maintain the basic rRNA synthesis. This is the dosage compensation rule of regulation which states that, given the correct arrangement, one-half of the diploid number of cistrons suffices for normal synthesis. This rule achieves a functional homeostasis whereas magnification (as is known in the insect) adjusts the physical number of cistrons, so that, in a sense, these two transient equilibrating tendencies counterbalance each other. According to the dosage rule, therefore, organisms with either one-half of a diploid organizer or a complete haploid organizer are as viable as those with a diploid organizer. A double-sized nucleolus is formed by the organizer in these organisms, suggesting a functional compensation of synthesis and/or a certain critical nucleolar mass. In a comparable situation, the oocyte of the *Xenopus* mutant with only one organizer amplifies its rDNA normally and thus forms a normal nucleolar complement. In the next case to consider, that of genomes with an organizer surplus, the response seems to differ between plants (e.g., maize; Lin, 1955) and animals (e.g., *Drosophila*; Schultz and Travaglini, 1965). The plant responds to the organizer surplus with the formation of excess nucleoli and excess rRNA synthesis whereas the animal, adhering to the dosage rule, does not respond. The reason for this difference is not known. Taken together the data suggest a functionally redundant rDNA, as is now confirmed by all recent molecular evidence. Unhappily, these data cannot clarify whether transcription in the normal animal is limited to a finite number of cistrons or blocks of cistrons, or occurs in all cistrons available at a variably compensated rate; this ambiguity is not resolved by the molecular evidence. The reader is referred to a full discussion of these questions in Chapter 4 from the standpoint of the regulation of rRNA synthesis.

The occurrence of a polycistronic rRNA precursor in the eukaryote would run counter to a general evolutionary trend only in the event that mRNA has lost the polycistronic precursor that it frequently had in the prokaryote. This question was discussed in Chapter 3 and is not solved. As they mature, both eukaryotic rRNA and mRNA function, of course, monocistronically. If the prokaryote lacked a polycistronic rRNA precursor, which is also not solved, then together with the consolidation of multiple cistron blocks, this polycistronic precursor would be another way by which the eukaryote has coped with an increasing complexity of ribosomal function.

Nonribosomal RNA Species in the Nucleolus

A biochemist's view looking at the total nucleolar RNA is that rRNA represents the only high molecular weight RNA pathway in the organelle (Penman,

1966) and of which nothing in the end is retained by the organelle. However, because of the design of experiments, lack of clear-cut evidence for other pathways of high molecular weight RNA should not be taken as binding. For example, recently a minor nonribosomal 45 S RNA rich in AU was reported in the nucleolus during actinomycin inhibition (Choi and Busch, 1969). Messengerlike RNA has also been reported in normal nucleoli (Brentani and Brentani, 1969). This claim, as well as the preceding one, require confirmation. Yet, because of the diversity of nucleolar funciton, the presence of some mRNA, in particular, whether or not coded there, is not an impossibility. In addition, small polydisperse, but not large polydisperse, RNA is present. The Mn-activated polymerase responsible for the synthesis of these AU type dRNAs has been found in the nucleolus (Jacob et al., 1968). Autoradiographical studies in various types of cells have consistently shown a very stable nucleolar RNA fraction (e.g., Amano et al., 1965) whose identity is not clear. As considered later, an equal or greater possibility is that some of these RNAs represent chromosomal contributions, as is the case for 5 S RNA which will now be discussed.

As opposed to high molecular weight rRNA, 5 S RNA is an extranucleolar contribution to the nucleolus that offers a good example of a genetically independent but coordinate regulation with rRNA. (Although this RNA is included in this section, it is also considered ribosomal.) As discussed in Chapter 4, the coordination takes the form of a dependence of 5 S RNA synthesis on the presence of rDNA rather than a coupling to rRNA synthesis. The result is that 5 S RNA can be synthesized provided rDNA is present but irrespective of whether or not it is functional. However, despite the lack of coupling, the two RNA syntheses are delicately balanced against each other in the cell. The mechanism responsible for such balance remains to be explained. Some of the genes for 5 S RNA were known to be clustered and others to be dispersed in the chromosome. Elucidation of their topography in the genome in relation to rDNA was therefore awaited with interest. According to cytological hybridizations by Wimber and Steffensen (1970), in dipteran polytene chromosomes there is no noticeable accumulation of 5 S RNA genes in the proximity of the nucleolus organizer. In contrast, the genes in the prokaryote are close to the 23 S RNA genes in particular, and, according to some views discussed in Chapter 4, may even share common polycistrons. This would mean that there has been a break in the ancestral proximity or linkage on the part of the eukaryote, and, in fact, the reverse of the consolidation undergone by rDNA. Interestingly, therefore, is that in the typical eukaryotic interphase nucleus the functional topography of 5 S DNA may still be in relation to the nucleolus (to which we shall return in connection with the nucleolus-associated chromatin). Although evidence is lacking, this coming together of 5 S DNA and rDNA in the nuclear space is not excluded in the specialized dipteran salivary nucleus.

Concerning nonribosomal low molecular weight RNAs, most tRNA cistrons are not nucleolar. Evidence in chironomid salivary chromosomes presented in

Chapter 4 does nevertheless indicate that a few cistrons may be nucleolar. As a further example, certain insect oocytes possess a secondary nucleolus that becomes labeled only in soluble RNA (Halkka and Halkka, 1968). Nucleolar tDNA remains, however, to be established and generalized. In any case, the majority of tDNA has separated from rDNA compared with their close proximity in the prokaryote (although also in the prokaryote a little tDNA is separate from rDNA.) In the eukaryote this general drift away from rDNA raises the question that any tRNA cistrons that may have remained in close association to rDNA, as mentioned above, might serve some special function in relation to the ribosomal pathway, as already considered in Chapter 5. Indeed, this possibility adds to the interest of any native nucleolar tRNA.

Several low molecular weight RNA species of unknown function, but not ribosomal, occur in the nucleolus. Apparently some of them are exclusively present in the nucleolus (Weinberg and Penman, 1969). Whether this exclusive occurrence and other evidence discussed in Chapter 2 indicate a nucleolar origin of these species is not clear. These exclusive species are completely stable, and since the nucleolus disaggregates during mitosis, this may mean that the species reenter the nucleolus during successive cell cycles. Moreover, of these species, the 8 S and U3 RNAs are transiently H bonded to 28 S RNA only while this is moving through the nucleolar compartment (Prestayko et al., 1970), the meaning of which with respect to the function of 28 S RNA is not known. Clearly, the immediate interest is to find whether these small species have any role in the regulation of the ribosomal pathway. However, as in the case of mRNA or tRNA, there is no shortage of nucleolar functions in which they could possibly intervene. It hardly needs adding that the number or topography of the cistrons for all these possible nucleolar nonribosomal species have not been worked out.

Nucleolar Protein

Protein can make up to 90% of the nucleolar dry weight, or the equivalent of up to 25% of the wet weight. This concentrated mass of protein is plainly responsible for the bulk of the organelle—and immediately suggests that some protein is specific to the organelle. Isolated liver nucleoli contain 30% soluble protein, 30% histone, and 40% residual acidic protein. The acidic proteins are the most active metabolically. For example, in sarcoma cells they increase to 60% of total nucleolar protein (Muramatsu and Busch, 1967). Part of the soluble and acidic proteins are enzymes. The enzymes involved in the rRNA pathway alone include polymerase, several base-specific methylases and pseudo-uridylases, and probably more than one each endonuclease and exonuclease. Other enzymes involved in nucleolar metabolism are reviewed by Sirlin (1962). Another part of the soluble and acidic proteins are ribosomal proteins destined to combine with the rRNA. However, comparison of nucleolar and ribosomal proteins of the starfish oocyte by electrophoresis revealed no striking similarities, and furthermore, that only part of the proteins are shared in common by the nucleolus and ribosome (Vincent et al., 1966). Two implications of these findings are, first, that most of the acidic protein is indeed not ribosomal in nature and, second, that the nucleolus does not supply all the ribosomal protein to the cell either directly or indirectly (see below). However, in these studies the basic ribosomal proteins have not been well characterized. The nucleolus may also contain migrating cytonucleoproteins (Kroeger et al., 1963) whose assignment to any of the previous classes of protein is at present undecided.

Depending on cell type, nucleoli show variable amounts of protein synthesis as judged by autoradiography. The "specific activity" is, however, generally less and the total activity much less than in cytoplasm (review by Hay, 1968). Correspondingly, it was shown biochemically that much of the nucleolar acidic and basic protein is transferred from the cytoplasm by a process dependent upon RNA and ATP production (Kawashima et al., 1971). Without these observations signifying that there is no nucleolar protein synthesis at all, they agree with the presence of few, if any, recognizably mature nucleolar ribosomes (Monneron and Bernhard, 1969).

The nucleolus of the HeLa cell contains a balanced pool of ribosomal proteins representing 5-10% of the total ribosomal protein in the cell (Warner, 1966). Biochemical data in this cell indicate that 60% of this protein is added to the ribosomal subparticle as this forms in the nucleolus. It is not unlikely that the remainder of the protein is subsequently added to the particle in the cytoplasm where it exchanges with other (soluble) ribosomal protein. It was mentioned in Chapter 5 that this exchange may, in part, reflect conformational transitions of the translating ribosome. The exchange further suggests, as the other data above, that much of the ribosomal protein is made elsewhere than in the nucleolus.

Definitive evidence concerning the cytoplasmic origin of ribosomal protein is that in liver the protein is made *in vitro* by cytoplasmic polysomes (Terao et al., 1968). However, cleaving eggs and the enucleolate mutant of *Xenopus* which make little or no rRNA, produce very little ribosomal protein (Hallberg and Brown, 1969). Assuming that most ribosomal protein is coded outside the nucleolus organizer—which is a plausible view but still needs to be proved—this finding indicates a nongenetically linked coregulation of the syntheses, much as in the previous case of 5 S RNA. Again as in this case, L cell ribosomal protein continues to be made during actinomycin inhibition of rRNA synthesis (Craig, 1970). By analogy with 5 S RNA, the data indicate that ribosomal protein synthesis is coordinated with but not coupled to rRNA synthesis and also depends on the presence of rDNA (Chapter 4). By which mechanism this coordination of protein and RNA syntheses is achieved is even difficult to guess, although a possibility is that this mechanism operates somewhere at the level of translation. It is clear that such overall coordination of syntheses is necessary in order to secure the stoichiometry between rRNA and ribosomal proteins. When it is considered that in the higher eukaryote these proteins alone represent a vast and near equimolar ensemble of about 100 molecular species, their synthesis obviously requires a tight balance with that of rRNA. It should be noted at this point that ribosomal proteins and 5 S RNA alike build up a pool in the cell, most of which is in the nucleolus.

If it is accepted that most ribosomal proteins are produced in the cytoplasm, as seems likely, then this serves to illustrate the principle of differential amplification of protein synthesis and stable RNA synthesis mentioned in Chapter 5. For example, in the amphibian oocyte with cistrons expressly overamplified to produce a large amount of rRNA, there is no indication in favor of the ribosomal protein cistrons being similarly amplified to match the rRNA with a correspondingly large amount of these proteins. Instead, as in somatic cells, the amplification of ribosomal protein production in the oocyte is at the level of transcription-translation. So far the present argument applies to histone, whose production also requires amplification to supply the numerous nucleoli in this oocyte. However, in addition to the argument, it remains possible that the ribosomal protein cistrons have a basic somatic multiplicity carried into the oocyte which is similar to that of the histone cistrons now to be discussed.

Evidence for the cytoplasmic site of histone synthesis is given later in connection with the cell cycle. Previously, the possibility was considered that the histone cistrons were located near the rDNA. However, a circumstantial argument against this particular possibility being applicable to all histone cistrons was that 5 S RNA and possibly most ribosomal proteins are coded outside the nucleolus organizer. It has now been found by hybridization of echinoderm mRNA representing three histone fractions that the cistrons are massed together and not in proximity to rDNA (Kedes and Birnstiel, 1971). These cistrons are in multiple numbers, approximately 400, and possibly all cistronic sequences are nearly identical, yet this does not exclude that each histone species is less cistron multiple. In their high multiplicity, togetherness, and possibly near identical repetitiousness, these informational cistrons are therefore remindful of rDNA itself. Moreover, presumably because of their high GC content and close proximity, these cistrons band as a satellite like rDNA. A related question of present interest, of which nothing is known, is how much the cistronic regulation of histone resembles that of rDNA. These similarities with rDNA, which functions very much like it does in the prokaryote (Chapter 4), cannot but also remind us of the view that histones are primitive proteins. Consonant with this view are their simplicity, low antigenicity, near universality, that much of their translation is coupled to replication and shows no delay (therefore, is short lived), and, perhaps related to the last point, that their mRNAs lack both the homopolymer and redundant sequences that are found in other eukaryotic mRNAs. A guess is that if ever an exceptionally simple regulation, processing, or transport is disclosed for eukaryotic mRNAs, these will include the histone mRNAs.

Concerning the metabolic regulation of histone, if the separateness of histone and rRNA cistrons in the echinoderm can be generalized, the lack of two lysine-rich histones by the enucleolate *Xenopus* mutant (Berlowitz and Birnstiel, 1967) might suggest another nongenetic coordinate regulation with rRNA synthesis. Such a coordination would probably occur during the S period of the cell cycle since histone synthesis is, in general, strongly coupled with replication. During this period, rRNA synthesis also increases but in a less clearcut pattern. This dissimilarity of pattern, together with the fact that histones do not show a stoichiometrical relation with respect to rRNA, implies that their coordination with rRNA is less strict than in the case of ribosomal proteins or 5 S RNA. Indirectly in line with this, there is no large pool of histones in the cell as there is for these other molecules. Moreover, as will become apparent when discussing the cell cycle, any coordination need not necessarily involve all histones. Contrasted with the similarities to rDNA in the previous paragraph, the present difference perhaps is basically because histone transcription is not dose compensated during the cell cycle, whereas rRNA appears to be. At the metabolic level, however, the difference does reflect an uncertain relationship between histone and the ribosomal pathway.

Certain polyamine-producing enzymes are also missing in the *Xenopus* mutant (Russell, 1971). This suggests even another coordinate regulation with rRNA synthesis, perhaps for those amines that become attached to rRNA.

Curiously enough in view of the preceding relationships, rRNA polymerase is present in normal amounts in the *Xenopus* mutant (Roeder *et al.,* 1970). It is, therefore, not coregulated with rRNA.

Ultrastructure and Organization of the Nucleolus

This overview of the nucleolus has dealt mainly with its ribosomal function, which, to remind the reader, disregards possible nonribosomal functions of which we may not as yet be aware. It is indeed a moot question whether all that has been described can be uniquely ascribed to the ribosomal function. Until such time when the interplay between the several possible nucleolar functions is understood, and the nucleolar ultrastructure to be discussed becomes interpretable molecularly in unequivocal terms, any attempt at correlating them, as the one that follows, must inevitably remain less than fully meaningful.

The nucleolar ultrastructure consists of matrix, fibers, and particles (review by Hay, 1968). The matrix is protein in nature and its function is unknown. The fibers and particles are RNPs and, in part, represent the various maturation steps of ribosomal subparticles. That the RNP in fibers and particles is in part ribosomal agrees with the observation that there are no free ribosomal subparticles in the nucleolus, but are all interconnected by fibers (Miller and Beatty, 1969b). The fibers and particles may also in part represent elements of the nucleolar architecture proper, that is, nonribosomal elements. The manner in which histones and acidic proteins are represented in these nonribosomal elements is, however, not clarified. First, concerning histone, it is not decided whether all histone is associated to rDNA or to any other DNA included in the nucleolus for that matter, or whether some histone participates independently of DNA in the nucleolar architecture. In this respect, in any large nucleolus the small mass of DNA per se is bound to have insignificant structural effects on the architecture, but, of course, as a binder of protein the effect could be enormously greater. Second, concerning acidic proteins, these have enzymic functions as described above and possibly also have other functions (loosely termed structural) in relation to the architecture. Again, however, the disposition with respect to rDNA is not known. This general incertitude with regard to rDNA proteination is not alleviated by the ultrastructural observations suggesting that this DNA is the most denuded of the mitotic chromosome (Hsu et al., 1967), although it can be argued that the nucleolar structure has disappeared by this stage. Part of the acidic proteins do, however, represent the nucleolar pool of ribosomal proteins. It is again not excluded that these proteins as a class contribute to the architecture, particularly in view of their considerable mass and aggregative properties. How the various nonribosomal RNAs are represented in the ultrastructure is speculative. According to Narayan et al. (1966), some of these RNAs are present in the fibers. In summary, it is not feasible at present to distinguish ultrastructurally between the ribosomal pathway and the intrinsic morphology of the organelle, about which in effect we are practically entirely ignorant. All we know is that this massive organelle (about 1 μ^3) is held together by the interaction of the three principal structural elements just discussed, and, it will become clear, this interaction is somehow conditional on the functional pathway.

Despite the current poor interpretation of structure, all the indications above (and more to come) are that the nucleolus is a subtly organized organelle. More than any other chromosomal derivative, the nucleolus develops a considerable epigenetic organization both at the structural and metabolic levels (Sirlin, 1962, 1963). The persistent ability of the organelle to fuse with like is further indicative of self-organization. As already stated, these are the reasons why of all chromosomal derivatives we think only of the nucleolus as an organelle. The organization depends as much on products of the nucleolus organizer as on products of cytoplasm, hence indirectly on other chromosomal loci as well; as we shall see, this is an important point when it comes to defining how the organelle is controlled. An instance of organization at the metabolic level is given by the nucleolar orthophosphate pool. In maize the amount of nucleolar P_i is 30-50% of the total cell P_i and its concentration is three to five times that of macromolecular organic P mostly in RNA (Libanati and Tandler, 1969). It is possible that the resulting exceptionally high P_i concentration, up to 0.7 M, helps maintain the solubility of the concentrated nucleolar protein. In addition, this P_i behaves as if enclosed in a relatively closed pool for it shows a unique delayed kinetics of entry into nucleolar RNA (Tandler and Sirlin, 1964). Other examples of nucleolar organization are the high concentration of cations (Tand-

The Nucleolus

ler *et al.*, 1970) and of Fe linked to polysaccharide (Robbins and Pederson, 1970); in particular, this nucleolar Fe may move into the condensing chromosome preparing for mitosis. Still another, if less impressive, example of organization is the presence of nucleolar vacuoles which are purported to be related to RNA synthesis (Johnson, 1969), but just as likely are related to some other function.

These examples clearly indicated that the nucleolus represents an intranuclear compartment with respect to the distribution of certain minerals. A compartmented nucleolar pool of triphosphate nucleoside precursors for RNA synthesis was similarly proposed on the basis of the differential sensitivity of the nucleolus to compounds affecting that synthesis (Sirlin and Loening, 1968) and the nucleolar structure (Monneron *et al.*, 1970). This proposal had a precedent in the abovementioned unique kinetic behavior of nucleolar P_i destined for RNA synthesis. If correct, the proposal would afford another instance of metabolic nucleolar organization. It remains to be shown, however, that the pharmacological effects of these compounds are indeed on pools and not having to do with a special sensitivity of the nucleolar rRNA polymerase.

During nucleologenesis the organelle is assembled in relation to the onset of ribosomal function, yet the intriguing manner in which this functional coassembly occurs cannot even be guessed at. It follows that the two aspects previously discussed, ribosomal pathway and nucleolar structure, are intimately interdependent functionally. As we shall soon see, maintenance of the nucleolar organelle thereafter continues to depend on sustained ribosomal function. All this means that nucleolar form and function are in fact codependent. Hence the functional nucleolus is at all times a dynamic structure maintained by the flux of product and possibly also many intrinsic components, few of which may really be static. All four nucleolar pools so far mentioned, namely, large ribosomal subparticle, ribosomal protein, 5 S RNA, and minerals, are in a state of flux. In the present sense a discrete nucleolus is therefore a "statistical" structure, much as the chromosome itself is. Not unexpectedly, the organelle possesses no physical limiting membrane.

The particular events leading to nucleologenesis are as inadequately understood as the nucleolar structure. As shown using actinomycin, these formative events are absolutely dependent on the organizer function, i.e., rRNA synthesis. They may also, however, depend on contributions (or interactions) from the rest of the chromosomes which, for all we know, could be equally important (Tandler, 1966). The hypothetical contributions to the nucleolus to be considered include protein and, because of its immediacy, especially RNA. Any contribution of this sort from the chromosomes would be significant in that it could place the organizer function in the context of a direct support from the rest of the genome, in other words, under genetic control. In order to explore this our main interest is in direct chromosomal contributions and, in the case of RNA, in nonribosomal RNA. To proceed in the direction from the general to the particular, we shall consider first what evidence there is for this genetic control of formative events and later examine what is known of these events themselves. An obvious point is that the control need not be confined to these events which are of present interest to us, but may also extend into the functional nucleolar phase. A second point is that the more general this genetic control is, the more the nucleolus qualifies as a true organelle as opposed to a local chromosomal annex.

To begin with the formative nucleolar phase in regard to genetic control, chromosomes in organizer-less nuclei form a number of RNP bodies that have been interpreted, although not undisputedly, as an abortive contribution from the chromosomes to the nonexistent nucleolus (Swift and Stevens, 1966). Normally, according to this interpretation, the bodies would become confluent in the nucleolus forming anew at late telophase. Some evidence suggests that the process depends on an RNA synthesis whose nature remains unprecise (Stockert et al., 1970), but which evidently could correspond with the RNA present in the bodies. A protein, perhaps a nucleolar matrical protein, was tentatively proposed to organize these extranucleolar contributions at the nucleolar site (Sirlin, 1963). This proposal of an organizing protein has the virtue of being experimentally testable and there may even be some evidence in favor of it (Stevens and Prescott, 1969). As we turn to the functional nucleolus, the evidence for chromosomal contributions (other than 5 S RNA) becomes even more meager. The sole evidence for these contributions is in fact that, after different regions of chromatin are spot-irradiated with ultraviolet, the effects seem to reach rRNA synthesis (Perry et al., 1961). Since nonspecific effects of irradiation are almost impossible to eliminate in experiments of this kind, moreover, the evidence is no more than suggestive. In sum, both the forming and functional nucleolar phases remain precariously examined with regard to any direct macromolecular contribution from the chromosomes that might possibly be related to a genetic control. The unfortunate conclusion at present is therefore that the status of this control is not known.

Trivial to mention, on the other hand, indirect chromosomal contributions—via the cytoplasm—to the nucleolus either forming or functional must be legion. Because the controlling influence of these contributions would tend to be correspondingly indirect and merge imperceptibly with the purely structural requirements of the nucleolus, these contributions are not at the moment a favorable ground for exploring a genetic control. It is worth noting, however, that in the case of the forming nucleolus these contributions provide an alternate explanation of the RNP bodies, although this interpretation is not necessarily mutually exclusive with the previous one—that the bodies represent prenucleolar components derived from the cytoplasm. For many years the indications have been that RNA and protein constituents of the nucleolus in the previous cell cycle appear to enter the newly forming nucleolus during telophase by way of the cytoplasm (review by Sirlin, 1963). In light of the recent evidence the amount of this preformed RNA may be too much to represent the scanty stable low molecular weight RNAs or other nonribosomal RNAs mentioned earlier, but could correspond to 5 S RNA or to the stable nucleolar RNA fraction of nature unknown which is detected by autoradiography. The RNA could also include 45 and 32 rRNAs left over from the previous nucleolus, this time inside fibrils attached to the metaphase chromosomes (Fan and Penman, 1971), which subsequently appear to reenter the nucleolus at telophase and much as described for the bodies above. Irrespective of the interpretation, the large amount of protein contributed by these bodies is clearly important for the reformation of the bulky organelle.

Looking now at the formative events in the nucleolus itself, it first forms as a pars fibrosa. This pars consists of RNP, as described above, and may be looked upon as a nucleation centre for the nucleolus. This pars also appears in the above mentioned abortive bodies, if, for lack of an organizer, a definitive nucleolus does not form. A pars granulosa then generally appears peripherally to the pars fibrosa of the nucleolus as rRNA synthesis gets first under way (review by Bernhard and Granboulan, 1968). The pars granulosa represents mainly maturing ribosomal subparticles. Even the organizer-less *Xenopus* mutant, with only a trace of nondefinitive nucleolus (hence termed a pseudonucleolus) capable of all but a vestigial rRNA synthesis (Brown and Gurdon, 1964), initially goes as far as forming particles that disappear later (Hay, 1968); however, the ribosomal nature of these particles is not established. At some stage the protein matrix (pars amorpha) is added onto the two pars to complete the nucleolar structure. The present description notwithstanding, it was explained that what controls or guides these events into forming a nucleolus—as opposed to a passive dispersion of the ribosomal product—is not elucidated. Not only formation but also structural integrity of the nucleolus is function dependent since, as mentioned, the organelle is maintained by the flow-through of substance. Interference by antibiotic or antimetabolite will result first in segregation of the basic nucleolar elements (matrix, fiber, particles) into separate zones, then in disappearance of particles, and eventually in total dissolution of structure. Because of this functional dependence, the ultimate pharmacological effects on nucleolar ultrastructure are bound to be the same whether the action is at the template, transcriptional, or metabolic levels. Deleterious agents for the nucleolus have therefore more than RNA synthesis as their target. A segregation of nucleolar components similar to the ones obtained experimentally can be observed in the early amphibian embryo and in the residual nucleolus of the *Xenopus* mutant, two physiological instances of subnormal nucleolar function.

The organization of nucleolar DNA is currently assumed to resemble that of a puff, that is, complete DNA loops emerging from the chromosomal axis (Rodman, 1969). In polytene chromosomes, for example, the rDNA can be shown by cytological hybrid to lie outside the chromosome, yet remain connected to bands in the chromosome (Pardue *et al.*, 1970). The same holds true of the chironomid nucleolus (Jacob and Sirlin, 1964) described in detail below, where the rDNA corresponds to only part of the chromosomal organizer and most rDNA may be outside of it. It is therefore likely that chromosomal proteins are more tenuously associated with organizer DNA than with the rest of chromosomal DNA, even though these two DNAs are physically continuous with each other. Thus in metaphase mammmalian chromosomes the nucleolus organizer consists of DNP filaments which are 50 to 70 Å thick as against 150 to 200 Å in the remainder of the chromosome (Hsu *et al.*, 1967). In addition, the rRNA polymerase requires less ionic strength for activation *in vitro* than the nucleoplasmic polymerase, as if less inhibition by protein had to be overcome. In summary, these observations indicate that the chromosomal organizer locus is relatively denuded of protein or is less supercoiled. Reduced supercoiling is in turn suggested in amphibians (Miller and Brown, 1969) and dipterans (Ritossa *et al.*, 1966a) from the fact that the organizer locus is disproportionately long relative to the amount of rDNA. Reduced supercoiling is however not incompatible with denudation, and may even depend on it.

As usually observed with autoradiography, the progression of nucleolar RNA label from the pars fibrosa (containing the rDNA) to the pars granulosa parallels the conversion from precursor rRNA particle to derivative ribosomal subparticles in transit to the cytoplasm. The progression of nucleolar protein label does not clearly follow the same physical direction, reflecting the various levels at which protein enters and leaves the pathway. In most interphase animal cells the pars fibrosa, where the first RNA label is observed, is central, irregular, and contains interdigitations of chromosomal DNA entering from the nucleolus-associated chromatin (see below). The DNA entering the nucleolus includes but may not exclusively consist of organizer rDNA. The nucleolus of the polytene salivary chromosome of insects is almost diagrammatic of the condition just described (Fig. 22); this nucleolus is disposed radially around the organizer locus in the chromosome and has a central pars fibrosa and a peripheral pars granulosa. In this nucleolus, as rRNA matures, the RNA label flows centrifugally from the inner pars fibrosa into the outer pars granulosa (Jacob and Sirlin, 1964). On the other hand, free nucleoli in the amphibian oocyte have a circular piece of rDNA. At one stage of growth these free nucleoli themselves are annular and label in RNA uniformly along their length, because the pars fibrosa is likewise continuous and circular (Lane, 1967).

In conclusion, it is not soon that a theory of the nucleolar organelle—not just the ribosomal pathway—will emerge, but its basic elements are discernible.

Nucleolus-Associated Chromatin

Interphase nucleoli usually possess, depending on cell type and status, a variable but sometimes quite large amount of peripherally associated chromatin which

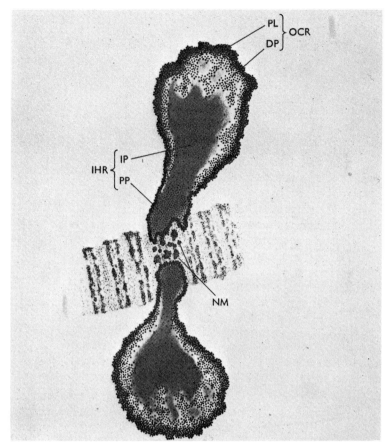

Fig. 22. Structural organization of the nucleolus. The diagram summarizes the pattern of organization of the nucleolus in the larval salivary gland of the chironomid. The parts shown are: NM, nucleolar material within the organizer locus in the chromosome; PP, proximal part of the nucleolus; IP, intermediate part; DP, distal part; PL, peripheral layer. PP and IP correspond with the pars fibrosa described in the text, DP and PL with the pars granulosa. According to recent information, the rDNA is contained in the pars fibrosa (including NM) rather than throughout the organizer locus in the chromosome (from Jacob and Sirlin, 1964).

in all well-analyzed cases is continuous with the intranucleolar chromatin. This nucleolus-associated chromatin should therefore not be equated with organizer rDNA, although in certain cells part or even all the rDNA may be contained therein. The associated chromatin does include constitutive heterochromatin from several adjacent chromosomes, which helps explain its frequent condensation. For example, in the murine interphase nucleus the associated chromatin includes the centromeric DNA which is part of constitutive heterochromatin.

However, despite being condensed, the associated chromatin can be quite active in RNA synthesis, and sometimes as in embryonic cells (Karasaki, 1968) be even more active than the nucleolus itself. The paradox of a condensed but transcriptively active chromatin was tentatively explained in Chapter 3 in terms of microstructural heterogeneity; by definition, the paradox means that euchromatin is present in the associated chromatin. That this chromatin is heterogeneous is further suggested by its content in early replicating DNA (Comings and Kakefuda, 1968) which is another typical feature of euchromatin. Other indications for heterogeneity are given below.

The question arises as to what is the special function, if any, of this associated chromatin. If certain euchromatic regions of the chromosome are supposed to interact with or donate molecules to the nucleolus, would they be preferentially located in this chromatin so intimately in contact with the organelle? The question cannot be properly answered. In the narrow terms of general topography, however, the answer might be yes. Each chromosome appears to occupy a fixed relative position in the interphase nucleus (Harris, 1959), a view which receives indirect support from the finding that each chromosome may establish a number of specific contacts with the inner nuclear envelope (Comings, 1968). A rationale would therefore be that certain chromosomes, in fact a substantial part of all chromosomes, are preferentially associated with the nucleolus, but the question remains to what extent the association is a truly functional rather than a fortuitous one. During amphibian embryogenesis, the nucleolus is initially compact and becomes looser as it grows, while concurrently the associated chromatin becomes less conspicuous (Jacob, 1969), at which point nucleolar RNA synthesis increases markedly. Again, the dilemma is whether the observed decompaction of associated chromatin, presumably indicative of derepression, is or is not in any causal way connected with the incipient nucleolar synthetic activity. Perhaps if, as suggested by Amaldi and Buongiorno-Nardelli (1971) from their cytological hybridizations, 5 S DNA is located in the associated chromatin, a model for the specific effects of this chromatin is forthcoming.

Recent work has refocused attention on the relationship between the nucleolus and the synaptonemal complex. Initially, Moses (1968) had speculated subjectively that the typical synaptonemal complex might provide a function for molecular DNA redundancy. An atypical complex present in the cricket spermatocyte is associated with the nucleolus (Sotelo and Wettstein, 1966). In the cricket oocyte, containing an enormous body of DNA which functions as an expanded nucleolus, a similarly atypical and multiply concentric synaptonemal complex is found attached to the body (Jaworska and Lima-de-Faria, 1969). The number of individual concentric elements in the complex agrees with the low degree of gametic rDNA amplification in this animal, from which a connection between the complex and rDNA amplification was considered possible. The situation is obscure, however. A nuclear DNA satellite present in the oocyte contains only 1% of DNA that behaves as rDNA by hybridization. A possibility was therefore that the synaptonemal complex represented the association of nucleolar rDNA with an extraneous DNA of unknown function. According to Gall (1970), the extraneous DNA is highly redundant, and, in agreement with the previous data, not rDNA. It will now be interesting to find out if these two redundant DNAs are in the same or different chromosomes. In any event, the status of the nucleolar body—generally regarded as a classical example of heterochromatin—is not a qualitatively simple one. In short, this body is, on the one hand, annexed to the nucleolus and contains rDNA (presumably an amplification of organizer rDNA) and, on the other hand, is associated with a synaptonemal-like complex and contains a second type of redundant DNA. This is indeed a good example of structural and molecular heterogeneity combined, which brings to mind the many possible roles of constitutive heterochromatin (discussed in Chapter 3). To a variable degree, this combination will probably be found to be common to many somatic cells when they are examined in detail.

THE EUKARYOTIC CELL CYCLE

Introduction

The bacterial cell cycle has been considered separately in Chapter 3. In review, the discussion emphasized the timing and relative coupling of replication (including chromosome segregation), transcription, and translation during the bacterial cycle. In the bacterium these processes take place more or less continuously, not in a discernible multiphase cycle as in the eukaryote. The amount of DNA per bacterial cell varies above the haploid value within a range dependent on the rate of growth. In addition, depending partly on this rate, cytokinesis can be out of phase with DNA replication.

The value of understanding the eukaryotic cell cycle becomes clear when it is considered that most macromolecular aspects of the cell are eventually referrable to discrete permissive events occurring at some point in the cycle. As a result, the production of many key macromolecules is rather accurately timed during the cycle. Bearing in mind the macromolecular complexity of the cell, it seems likely that a considerable expenditure of genomic potential is used in the form of regulation just to secure this orderly timing. However, despite what this might suggest, it needs stressing that the cell cycle is much more than a temporal bracket of cell activities. It represents, first, the inclusive program for obligatory transcriptional regulation of the cell discussed in Chapter 3, and, second, any independent translational program as may be in operation, as discussed in Chapter 5. The transcriptional program is, of course, the primary program, but, in finally specifying the cycle of cell activities, both programs rank as of equal importance. In essence, and a point which cannot be emphasized enough, this means that the cell cycle cannot be better understood than the basic obligatory regulation of transcription and translation. Given the considerable incertitude surrounding these two fundamental programs, since many of the integrating elements are still missing, it can hardly surprise anyone that the molecular biology of the cell cycle can at present be only superficially described. The presentation to follow deals only with macromolecules and, of necessity, in a general way. However important for the progression of the cell cycle, nonmacromolecular aspects such as precursor pools and energy reservoirs are beyond our present scope.

The interphase period of the cell cycle is divided into a replicative phase S which is preceded and followed by a gap, G_1 and G_2, respectively. Interphase comprises most of the cell cycle with the relatively short period of mitosis, from prophase to telophase, overlapping the end of one cycle and the beginning of another. Mitosis is soon followed by cytokinesis. The relative and absolute duration of each phase varies enormously with cell type, however, except for mitosis and cytokinesis which vary the least. The balance between phases is shifted according to the physiological or developmental condition of each cell.

The rate of cell division is therefore widely variable depending on cell and tissue type and their physiological condition. Cytodifferentiation proceeds *pari passu* with the establishment of an apparent rate of cell division and eventually with the rate of cell growth in the final resting G_1 period. By these means, cell reproduction is virtually restricted to comply with fixed tissue or organ limits, this being an important aspect of organismal homeostasis, which still remains a challenge to understand. To cite an example, within epithelia it is considered that positional effects in relation to the epithelium as a whole govern the cycle of each cell, and this is facilitated by each cell being ionically coupled to its immediate neighbors; the nature of these positional effects is not known but it is suspected that at least in part they involve cell membrane phenomena. Division means the doubling of the synthetic potential of the cell, for eventually the cell mass is doubled, even if not simultaneously, across all its components. Yet, after the actual doubling of synthetic potential which occurs sometime between interphase (probably the G_2 period) and cytokinesis, the concerted function of the increased cell mass is maintained. The path then ahead of any cell is either to further proliferate or to age, sometimes while still undergoing differentiation, and finally to die (Bullough, 1968). The point here is that cytodifferentiation is no alternative to aging.

Considering the entire cell cycle, it is possible to say that the overall rate of transcription tends to increase uniformly, although less so than the rate of translation. It is possible, however, that with greater analytical resolution transcription may prove to follow a biphasic course with minima at the times of replication and mitosis. This may be illustrated with examples taken from different cell systems. Transcription occurs everywhere in the macronucleus of ciliates (the metabolic nucleus in these organisms) except within a discrete band undergoing DNA replication (Prescott and Kimball, 1961), in agreement with a physical limitation of template to one of these functions at a time. Similarly, in embryonic life the overall rate of transcription reaches a minimum in the most rapidly cleaving cells which replicate their DNA the fastest (Gurdon, 1968). In naturally synchronized cell populations, as in slime molds (Mittermeyer *et al.*, 1964), two peaks of transcription are discernible; the first occurs early in the S period (there is no G_1 period in these organisms) and the second in G_2. In artificially synchronized cell populations, as in cultured hamster (Crippa, 1966) and HeLa cells (Scharff and Robbins, 1965), the rate of transcription increases monotonously throughout the cycle and falls at the end of the G_2 period. The ostensible contrast between these cell populations, one naturally and the other artificially synchronized, is probably due to the different degree of synchrony and/or individual cycle of replicons in each population. In the slime mold the superimposition of the first transcription and replication maxima at the S period is therefore possibly not indicative of a coupling between these processes. Hence, although the multireplicon nature of the chromosomes makes the evidence difficult to obtain, there is no strong indication generally in favor of any coupling of replication and transcription as there partially is in the case of bacteria. This difference could very well have to do with the more complex organization of the eukaryotic chromosome.

The major RNA species are simultaneously and continuously transcribed during all of interphase (Bello, 1968; Pfeiffer, 1969). According to some observations, their rate of synthesis has doubled after replication, suggesting perhaps a gene dosage effect at the root of the doubling of the cell synthetic potential mentioned earlier. Other observations are contradictory, however, and the preceding view cannot be unquestionably upheld (Enger and Tobey, 1969). Thus, for tRNA the rate of synthesis is constant throughout interphase (Tidwell and Stubblefield, 1971). Also uniform is the rate of phosphorylation of acidic proteins which appears to correlate with the release of informational transcription (Chapter 3); it should be recalled that these proteins phosphorylate very specifically during selective stimulative conditions. Under these circumstances, it is perhaps safer to say that there is no clear-cut and sustained increase of transcription of major RNA species associated with replication which would indicate that DNA is limiting. For example, many of the sequences of polydisperse RNA made throughout interphase seem to be informationally equivalent (Pagoulatos and Darnell, 1970). In contrast, a detailed study by Clason and Burdon (1969) of the minor stable nuclear RNA species in the 4 to 7 S range, a total of approximately 25 species, has revealed that those at the low side of the range are transcribed continuously during interphase and those at the high side are transcribed beginning with the replication period (this is not a dosage effect). This duplicity of pattern could bear on the specific function of these nuclear species which is still undecided, although the apparent relationship between molecular weight and time of transcription, in particular, is probably a trivial one. Of greater importance, the pattern within this group of small RNAs is a reminder of possible differences in pattern within other RNA species, in this regard none too well studied.

All cells in an organism will carry an obligatory transcriptional program responsible for the basic regulation of stable and informational RNAs necessary for their own survival and reproduction. This obligatory program is, therefore, in control of the basic cell cycle and represents a series of quantitative and qualitative regulations depending on RNA species, as seen in the patterns just described. (RNA polymerases are probably not limiting, but cannot be excluded from contributing qualitatively to these regulations.) In addition, upon entering differentiation, the cells will develop a facultative program characteristic of each cell type. This program will then be responsible for the essentially qualitative subregulation of individual species, in the first place mRNA and tRNA. Many mRNAs, the "luxury" mRNAs, may be regulated exclusively by this program. Yet, as seen in Chapters 3 and 4, there needs to be a certain overlap rather than a total exclusion between the obligatory and facultative programs involving mRNA and tRNA. Thus cytodifferentiation is primarily reflected in the molecular population of these two functionally differentiated species, but mainly mRNA. Provided the cell continues to divide, each cell cycle will now obey both the obligatory and facultative programs. That is, as a dividing cell it will obey the first program, and, as a differentiated cell, the second. It is possible, but not certain, that differentiation also entails changes in the obligatory program for stable RNAs, since different cell types show different patterns of stable RNA syntheses, and in certain terminal cell types they are eventually abolished. However, it is more probable that these changes in obligatory program during differentiation are also the result, for example, of hormonal action (which, of course, is a most important controller of the facultative program). An interesting but unsolved question which touches on the reversibility of the differentiated state, and which received some attention in Chapter 3, is thus how much of the exclusively facultative genome remains functional in a cell line maintained artificially without hormone. Finally, the outcome of the obligatory and facultative genomes is reflected at the level of translation which will now be discussed.

Looking at translation, the overall rate increases throughout the cycle and then decreases toward the end. It is not meaningful, however, to compare these kinetics with those of the transcriptions previously described. The relevant comparison would be with the kinetics of informational transcription which unfortunately is not known, since only data for total DNA-like RNA are available. Even if a direct comparison is not possible, it is still possible to relate the behavior of translation to informational transcription by considering two additional aspects which are not altogether dissociable: one aspect is that translation represents an enormous variety of proteins which appear at different times after their mRNAs; the other is that mRNAs may appear at most times during the cell cycle. This description predicts the observed cumulative increase in the rate of translation, for this rate now becomes a compounded rate relative to that of transcription. As expected, this increased translational activity is paralleled by the increase of stable RNAs and ribosomes toward the latter part of the cycle. Interestingly, on the other hand, the increased translational activity is heralded by the highest methylation of tRNA at the midcycle (Tidwell and Stubblefield, 1971). Now to focus attention on individual proteins, certain "housekeeping" proteins and enzymes are produced more or less continuously and others at rather discrete periods in the cell cycle; histones, considered to be in this second category, are discussed later. "Luxury" proteins also show a variable timing: collagen is made by fibroblasts throughout the cycle (Davies et al., 1968), but γ-globulin is made by myeloma cells mostly during the early S period (Byars and Kidson, 1970). However, because of asymmetry considerations depicted in Fig. 20, it is not possible from this alone to deduce when the informational transcription for globulin is occurring.

Biological and biochemical aspects of the cell cycle were reviewed by Becker (1969), Monesi (1969), Mueller (1969), and Prescott (1969). Perhaps a pertinent comment here is that this number of reviews for one year is proportional to the interest in the subject, but hardly a measure of the progress made in understanding it.

The G_1 Period

The length of the cell cycle is mostly regulated at the G_1 period. In relative duration this period is the most variable, being shorter the faster the rate of cell division. It is totally expendable in the lower eukaryote. At the other extreme, cells that become divisionally arrested either by contact inhibition or to begin their differentiation, do so at the G_1 period. Most of the typical differentiated function of a cell occurs therefore at this period. Differentiation does not however necessarily exclude further cell division, and, in fact, early stages of differentiation require it. Most cell fusion also occurs at this period.

During the G_1 period a cell commits itself or not to replicate its DNA, of which, in practice, the first alternative leads directly to further division. (Evidence that the commitment to divide is taken during the G_2 period for a small proportion of cells will be presented later.) This point of decision is thus responsible for the extent of tissue growth. An example is given by certain tumor cell lines which grow faster, not because of a shorter cycle (it is in fact longer), but because more of the cells divide, that is, the trigger to replication is more active. On the other hand, when a cell withdraws from the proliferative pool, it enters what is conventionally termed a G_0 period. This can be a period for cells either to complete differentiation or to remain until they reenter the cycle as they are again called upon to divide. Many quiescent stem cells are thus poised at G_0.

The nucleolus reforms at the early G_1 period during postmitotic reconstruction of the nucleus. The organelle manifests the initial burst of rRNA synthesis during the cell cycle. A nucleolus is not present when this synthesis is not scheduled in the cycle, as during the early fast cycles in the amphibian embryo, where there is practically no G_1 period. Later in development of this embryo, when rRNA synthesis is scheduled, there is still a question as to its actual initial rate (which in the case of the echinoderm embryo appears to be maximal). In any event, the cumulative rate of rRNA synthesis increases markedly only as a lengthening G_1 (and G_2) period allows the nucleolus to attain a certain functional size.

From a biochemical point of view, the complexity of syntheses in preparation for a new cell division or for the alternative of differentiation rank about equal at the G_1 period. However, during most of the G_1 period no specific translational requirements are apparent. Toward the end of the period it is believed, nevertheless, that there is a control point for the cell cycle, as was described above. The nature of this control is not clear but it could be at the translational level, for example, by the production of a replication initiation factor (Prescott, 1969). In cells with a typical G_1 period this synthesis of a protein factor would depend on transcription (of a mRNA?) earlier during the period. In cells with a short or with no G_1 period, two alternatives are that production of the factor is entirely controlled at the translational level, i.e., is not dependent on transcription during the same cell cycle, or that the factor is directly inherited from the preceding cell cycle. In general, therefore, the need for transcription during a typical G_1 period is partly to build up or refurbish the cell machinery for macromolecular syntheses and partly to compensate for macromolecular losses during the cycle. It should be recalled that a resting cell in the G_0 period undergoes the greatest turnover of RNA and protein.

Positive cytoplasmic factors whose nature or target are obscure intervene during these preparations for syntheses. These factors are recognizable during microorganismal parabiosis, plasmodial fusion, etc. (review by de Terra, 1969). Transcription factors are recognized by embryonic and adult amphibian nuclei after being transplanted into the oocyte, indicating that similar factors operate in the donor cells with a typical G_1 period. However, the factors in the unfertilized amphibian egg are of a negative character, that is, they suppress transcription in the transplanted nucleus; one such factor could be the putative rRNA repressor. As the recipient embryo reaches the stage of cleavage, positive factors (initiation?) begin to appear during the now expanded G_1 period, and, contrasted with the factors in the egg, these promote transcription in the transplanted nucleus. Functionally, therefore, because of these factors, the transplanted nucleus has assumed the characteristics demanded of a nucleus in the new cell environment. Thus each RNA species begins transcription at the correct developmental stage as the result of the intervention of factors. However, no matter how important the function of these factors obviously is, it has to be viewed within the total context of selective programming of the genome as was discussed in Chapter 3.

The S Period

The S period is the period of DNA replication. The putative replication factors involved appear during the late G_1 period, or presumably during the late G_2 period when the former is absent. The presence of these factors was first suggested by the work on transplanted amphibian nuclei (Gurdon, 1968). Several obligatory replication factors have now been indicated in HeLa cells (Mueller, 1969) and they appear in the cytoplasmic supernate only during the S period. The fusion of even several cells which are not in this period to any cell which is, results in their joint replication. These factors are believed to control initiation of replication by interacting with both DNA polymerase and initiation sequences in the template. However, these views are inferred secondhand from inferences in the bacterium and therefore require substantiation; indeed, the same is true of the initiation factor itself which needs to be demonstrated by isolation and direct testing. Nevertheless, it is believed that while the presence of polymerase is obviously obligatory for replication and perhaps involved in its regulation, it is unlikely to act as a master controlling element. Evidently, these putative initiation factors, together with the elements pertaining to their regulation, could play an important part in securing the homeostasis of tissues. This is, in general, because, although it is a necessary but not sufficient condition, replication in fact almost always leads to cell division.

It is currently assumed that replication of a DNA segment precludes its simultaneous transcription. However, because of the asynchrony of replicons in any nucleus, transcription still continues during the S period. Moreover, transcription occurring during the early part of this period is required for further replication. Hormones acting on the chromosome appear to affect the cell mostly during this period, which may be taken to signify that a partial denaturation of template, or, more probably, new formation of the chromosome is required for changing the regulatory program of transcriptions. To explain the latter point, hormone may more readily influence the pattern of proteins involved in the regulation of chromosomal function when these proteins reassemble in the newly forming chromosome (perhaps also suggesting a direct hormonal effect on translation). This important prerequisite for chromosomal rather than just DNA replication may be true, and for similar reasons, in all cells beginning to differentiate irrespective of hormone. It is clearly not true of differentiated, nondividing cells whose transcription is selectively but restrictedly activated whether by hormone or other means, that is, without entering a new differentiative program.

The relative duration of the S period varies according to the total duration of the cell cycle. Exceptionally, as in rapidly dividing embryonic cells, the period occupies practically the whole cycle (Gurdon, 1968). In this situation the cells are committed to little else but producing molecules for basic metabolism and division. According to Comings and Mattoccia (1969), all categories of rodent DNA sequences, redundant and unique, have replicated equally at all times during the S period. This question remains to be settled for the most repetitive (satellite) sequences whose replication may be restricted to certain parts of the S period, particularly the latest. The same uniform replication is reported for rDNA in mouse cells where this DNA is divided between several nucleolus organizers each present in a different chromosome (Watson, 1969). At the peak of the S period almost all of the oxidative energy of the cell goes into macromolecular syntheses (Robbins and Morrill, 1969), important among which is, of course, DNA synthesis.

Replication of DNA in general depends strictly on translation (as is also true in the bacterium). One reason for this must be that not just DNA but the entire chromosomal apparatus has to be replicated and the complex array of chromosomal proteins requires a minimum of coordinate replication. A second reason may be the need for proteins or enzymes required in connection with DNA replication itself. Hence with regard to all these proteins, chromosomal and enzymes, at least their translation must be coordinated with DNA replication. Precisely this is found for total nuclear protein in the mouse fibroblast (Auer et al., 1970). As expected, therefore, synthesis of histones occurs preferentially during the S period. The evidence in this respect for acidic chromosomal proteins is less clear-cut. These protein syntheses will now be discussed.

Histone synthesis and accumulation in chromatin is coupled in time with DNA replication and depends upon it. Some histone can, however, accumulate even when replication is inhibited (Gurley and Hardin, 1968). A histone mRNA (8 S) appears in HeLa cells in small cytoplasmic polysomes and is translated during the S period (Gallwitz and Mueller, 1969b). This mRNA is definitely not translated in the nucleus, which constitutes evidence for the cytoplasmic origin of histones. The coupling of histone synthesis with DNA replication is at the level of transcription, for the appearance of histone mRNA in the polysome depends on both replication and transcription. Yet it is not known when actual synthesis of this mRNA occurs, and thus the coupling could instead be by a short-lived factor, although such a possibility is disclaimed by Pederson and Robbins (1970a). It should not be construed from this description that the coupling of histone synthesis and DNA replication is absolute. In effect, different histone fractions in the HeLa cell differ in their degree of dependence on replication (Sadgopal and Bonner, 1969), and it was suggested that the degree of dependence could vary with the cell status. Some histone synthesis takes place even during the G_1 period, and much must occur in cells like neurons during the G_0 period. Posttranslational modification of histone, on the other hand, occurs at almost all periods of the cell cycle. Nevertheless, if one assumes for a moment that (a) the normal complement of histones is not vastly in excess of what is functionally required and (b) that histone sites in the chromatin are not interchangeable, then it follows that a prerequisite to prevent residual transcription during replication would seem to be: by the time a segment of DNA has replicated, all histones associated with that segment must also have replicated. This argument applies equally to any other protein whose main function, as in most histones, is to repress transcription, unless, of course, there is some alternative mechanism that obviates such need for repression by proteins particularly during replication. In support of the argument, to the writer's knowledge, no synthesis of histone during the G_2 period has been reported, that is, the full histone complement is achieved during the S period. In any event, the temporal coupling of histone synthesis with DNA replication helps to explain the near stoichiometry between these two types of molecules.

Acidic chromosomal proteins are not synthesized in marked coordination with DNA in HeLa cells (Mueller, 1969). Like most other nuclear proteins, the bulk of these proteins is made in the cytoplasm (Stein and Baserga, 1971). Their synthesis, which begins in the G_1 period, increases quantitatively during the S period (Stein *et al.*, 1970). Perhaps the lack of a temporal coupling of these acidic proteins with DNA replication in contrast to most histones is not very surprising if one considers the functional diversity of acidic proteins, their flux in and out of the chromosome, and especially their unknown form of association with the DNA. At this point, it bears recalling that the multimolecular organization of the chromosome as a whole is as yet not understood. It follows that chromosomal replication in terms of any particular protein cannot be understood in detail.

The specificity of cell DNA replication during the S period is made clear by the finding that viral DNA replication can occur in the nucleus even during the G_1 period. This also indicates that the onset of the S period is not determined by the availability of DNA polymerase. The multireplicon pattern of the individual eukaryotic chromosome which becomes apparent during this period is based on DNA molecules whose continuity over the entire chromosome is presumed but not established. This pattern is cell type specific and each replicon is replicated once per cell cycle. The overall regulation of replication is at the chromosome (multireplicon) level but the replicons are somewhat separately subregulated, as concluded from the study of autosome translocations (Stubblefield, 1966). By analogy with the bacterial chromosome (Chapter 1), it is possible that certain factors, which are perhaps integrated with replication complexes and act on initiation sites in each replicon, are in immediate control of the replication. Yet what controls the multireplicon pattern to which these factors would be respondent is not known. In this regard, some evidence has led to the suggestion that the late replication of heterochromatin is due to the fact that its replicons replicate more in unison (and later) than those of euchromatin (Comings, 1970). At present, attention is being focused on the question of the possible attachment of replicons to the inner nuclear envelope. Presumably, the attachment would be for a structural and/or regulatory purpose in connection both with somatic chromosomal replication and meiotic chromosomal pairing. The basic findings in this respect are that (a) the first replicated DNA is immediately adjacent to the envelope (Comings and Kakefuda, 1968) and (b) the synaptonemal complex involved in the early phase of meiotic pairing is first formed at the envelope. Pursuing the bacterial analogy, such an association of replicons with the envelope would indeed suggest a replication complex.

A small fraction (3%) of HeLa cell DNA with a distinct isopycnic banding pattern, i.e., discrete composition, is replicated at the end of the S period and is subsequently required for the cell to enter mitosis (Mueller, 1969). Autoradiography shows this DNA fraction to be distributed to all or most chromosomes. Replication of this DNA depends on both transcription and translation, again perhaps suggesting the need for a specific initiation factor. This late replicating DNA might be centromeric in function. Similarly, a DNA fraction (0.3%) with distinct composition replicates belatedly during the zygotene stage of plants and is required for meiotic chromosomal pairing (Stern and Hotta, 1969). Replication of this DNA, probably also occurring in all chromosomes, is coupled to translation and the resulting protein binds firmly and specifically to this DNA. The question that remains is what is the relationship of this DNA synthesis to the formation of the synaptonemal complex. In addition, a small DNA fraction of average composition is replicated during the pachytene stage of meiosis. This DNA is recombinatory in nature and represents strand repair.

In a speculative vein it has been argued that in differentiated cells there could be a cell-specific early order of replication of the genes concerned with cell differentiation. According to this argument, these genes could then profit from an early dosage effect at the level of transcription, supposing that this were indeed an advantage in view of the possible additional control during translation. At any rate, the factor determining transcription would be the cell-specific pattern of replication mentioned earlier, which is a novel concept. An allied concept was proposed more mindful of the eukaryote. This is the "linear reading" of genes whereby transcription would occur only during a fixed period of the cell cycle and follow the gene order in the chromosome (review by Mitchison, 1969). This concept implies an adaptive gene allocation characteristic of each cell type, but gene expression now depends on the pattern of transcription to follow the gene order rather than on the pattern of replication. Thus, but for the linearity, with respect to selective sequential transcription this is essentially not a novel concept. In short, these two concepts intend to explain what determines specific protein synthesis at discrete points in the cell cycle, namely, replicative or transcriptive order.

The G_2 Period

In mammals the G_2 period is short and constant. In the lower eukaryote without a G_1 period, it is long and variable. The protein spectrum at this period is different from any other period (Kolodny and Gross, 1969). This new protein synthesis is presumably directed to mitosis, such as for chromosome condensation and formation of the mitotic apparatus, and it is required almost to the end of the period. The mitotic apparatus, in particular, must engage a considerable fraction of the translation potential of the cell. Yet there is no firm indication that the translation for the apparatus started in the G_1 period, as might have been expected from the knowledge that the microtubular components of the apparatus preexist in a nonstructured form (Mazia, 1966). This then suggests that these components are made in the G_2 period and conserved by the succeeding daughter cells. On the other hand, in ciliates a labile protein which is required for division accumulates during the first two-thirds of the cell cycle, but is no longer labile or required during the last third (Zeuthen, 1964). The equivalence of this protein to spindle protein has not, however, been established.

The mitotic spindle contains about 5% of RNA, whose nature and function is not known (DuPraw, 1968). A possibility remains that this RNA may be a product of centriolar DNA, whose transcriptional ability—assuming it has one—remains uncharted. Evidently, mention of this possibility suggests means by which the centriole goes about its spindle organizing function. At any rate, even though the spindle proteins may be conserved, new spindle RNA may be made at each G_2 period.

There is evidence for transcription during the G_2 period that may relate to the translations described above. Particularly clear in this respect are the studies with actinomycin in slime molds (Mittermayer *et al.*, 1964). They reveal that inhibition of early G_2 transcription delays the oncoming mitosis and cancels the second following mitosis, whereas inhibition during late G_2 only delays the second mitosis. Hence information necessary for two consecutive mitoses is released during this period, which agrees with the conservation of microtubular components suggested above. Similarly, it is known that in cells lacking a G_1 period an anticipatory preparation for mitosis takes place during the G_2 period of the preceding cycle. The full sequence of events during the G_2 period is clearly illustrated in mammalian cells *in vitro* (Prescott, 1969). In these cells, it appears that early translation is needed for subsequent transcription, which in turn is needed for late translation, which again is needed for mitosis. It is not taxing the imagination to rationalize this sequence in terms of limiting initiation factors, etc., but the precise significance is not known.

Part of the replication of the chromosomal apparatus may continue after the S period, and from arguments given before this is more likely for nonhistone than for histone proteins. Further data on this point are presented in the next section. Perhaps related to this residual chromosomal replication is the observation that certain blocks of heterochromatin look cytologically different at the G_2 period from what they did before the S period (Nagl, 1970). If that is the case, the link between DNA replication during the S period and the remainder of chromosomal replication during the G_2 period remains undefined.

Evidence has been obtained for a small proportion of cells in animal and plant tissues that are indefinitely arrested at the G_2 period instead of the usual G_1 (Pederson and Gelfant, 1970). These cells respond immediately with mitosis to any emergency such as compensatory growth. The population of these cells consists of subpopulations responding differently with mitosis to different physiological stimuli, and no explanation is known for the heterogeneity. Interestingly, these cells can be "superinduced" by actinomycin to enter mitosis (Candelas and Gelfant, 1970), suggesting the function of a translational inactivator of the type proposed for modulatory regulation in Chapter 3. The general implication would be that a certain "mitosis mRNA" is normally held in check by the inactivator function and actinomycin suppresses this RNA-dependent function, thus permitting translation of the mRNA and mitosis to occur. Transcription of this putative mRNA could well be what was prevented in the actinomycin experiments in slime molds described above.

Mitosis or M Period

Most macromolecular chromosomal syntheses required for the chromosome to enter mitosis will have taken place during the preceding periods of the cell cycle. That is, with respect to chromosomal syntheses the M period is a quiescent one. Concerning the molecular basis of the ensuing chromosome maneuvers during this period our ignorance is almost complete (DuPraw, 1968). Taking into account all the equipment needed for these maneuvers, they must use a goodly amount of genetic information, although, as mentioned, part of this might be independent information stored in centriolar DNA. As discussed in Chapter 3, the histone to DNA ratio remains largely unchanged from the S through the M periods while the acidic chromosomal proteins increase relatively by a factor of about 2.5. By itself, this suggests that some acidic proteins have a general structural role concerned with chromosome condensation. Clearly, any such role could be critical to the behavior of the mitotic chromosome. In addition, one particular histone containing SH groups during interphase has these groups converted to -SS- bridges during metaphase (Bonner et al., 1969), definitely pointing to a structural role for this histone in chromosome coiling (yet not exclusively by means of these bridges). It has been known for some 40 years now that thiol groups in nuclear nonchromosomal proteins behave in an opposite way in terms of time to those in chromosomal proteins such as this histone, but the functional meaning of this reciprocal relationship remains as unclear as ever.

Chromosome condensation is mediated by certain factors as indicated by experiments involving the fusion of nonmitotic and mitotic HeLa cells (Rao and Johnson, 1970). If the fused nonmitotic cell is in the G_2 period, its chromosomes condense immediately in the normal way; if the cell is in the S period, "pulverized" or precociously condensed chromosomes appear instead; if it is in the G_1 period, its chromatids condense. It was mentioned that Fe stored in the nucleolus may be transferred to the condensing chromosome. Cell tonicity also appears to be implicated in the cellular events occurring during mitosis. Thus hypertonicity alone (1.5 to 2.8 times the normal) brings about chromosome condensation, nuclear envelope dispersion, polysome disaggregation, and inhibition of transcription, which are all typical of the M period (Pederson and Robbins, 1970b). Lastly, cell division or cytokinesis takes place toward the end of this period and is attendant to mitosis, sometimes both processes occurring almost simultaneously. However, cytokinesis is dissociable from mitosis under exceptional circumstances such as ploidization, formation of syncitia, and artificially induced cleavage of enucleated eggs. Cytokinesis depends on cell surface activity and perhaps on the dynamic infraskeleton of the cell, but it follows from the previous remarks that it does not depend on a normally functioning spindle apparatus, on which instead chromosomal maneuvers depend (at least in higher eukaryotic cells). As in the case of these maneuvers, the molecular basis of these other phenomena of the mitotic cell is not understood.

During the M period macromolecular syntheses in the cell are at their lowest ebb. It was mentioned before that all macromolecules in the mitotic chromosome have been made prior to M. Transcription reaches a minimum during this period and may be practically nil, clearly because of the extreme condensation of the chromosome. One of the first syntheses to decline appears to be that of labile RNA. However, tRNA may be exceptional in this respect in that it continues to be made by HeLa cells synchronized in the M period, at about 10% of the interphase rate (Pederson and Robbins, 1970b). This conclusion depends strictly on the degree of synchronization achieved between cells as well as on the exclusion of mitochondrial contamination. Also 5 S RNA would be exceptional for, according to Penman *et al.* (1970), it is synthesized during mitosis at 50% of the interphase rate. At these rates, the two mitotic syntheses combined amount to only 4% of the average rate of total RNA synthesis during interphase, so that the total mitotic synthesis is still very small. On the other hand, the nucleolus has disappeared during late prophase, and in those cells where it persists (as in certain plants) it is in any case no longer functional in making rRNA. Some of the 45 and 32 S RNAs made prior to the nucleolus disappearing remain within nucleolar remnants attached to the mitotic chromosomes, and then resume maturation as these remnants reenter the newly forming nucleolus at telophase (Fan and Penman, 1971). As described in Chapter 4, this preservation is not found for 5 S RNA after actinomycin treatment, presumably because in this case the nucleolus has been totally destroyed. Nevertheless, from the data above, it appears likely that some 5 S RNA is preserved in the normal mitotic cell.

Relative to interphase, translation is reduced by 60-70% during the M period, and, correspondingly, cytoplasmic polysomes are largely degraded. The cell amino acid pool is, however, not radically affected. The ribosome becomes inhibited during this period by a trypsin-sensitive material which appears to act by blocking initiation of translation (Salb and Marcus, 1965). This proteinaceous material is removed from the ribosome by certain viruses, such as poliovirus, thus restoring ribosomal function. Obviously, this inhibitory material must be released physiologically at the end of telophase, but what determines the release is not known.

Despite the extensive polysomal degradation, about half of the preformed functional mRNA in HeLa cells persists through the M period (Hodge *et al.*, 1969; Steward *et al.*, 1968). This is compatible with the average half-life of mRNA in this cell (Chapter 2) being longer than the 30 to 45 minutes that it takes to complete mitosis. Hence the cell's inability to translate mRNA is primarily due to the above mentioned ribosomal blockage. As the new cell cycle begins, polysome reconstitution takes place on preexisting mRNA and not necessarily in physical relation to the reformed nuclear envelope. The cell does not therefore avail itself of the opportunity offered by mitosis to wipe out old mRNA, presumably because this mRNA is still functional. Nevertheless, polysomes also reform on fresh mRNA which is about half of the postmitotic mRNA in HeLa cells. Polysome formation, in this case, is in relation to the pores of the nuclear envelope, as beautifully illustrated in plant microspores by Mepham and Lane (1969).

Some General Points Concerning the Cell Cycle

In summary for the previous sections, the cell cycle consists of three metabolic periods during interphase from G_1 through G_2, and one period mainly for mechanical tasks of the chromosome during M. The G_1 period is in preparation for DNA replication, or, alternatively, for functional differentiation in this period. Replication almost inexorably leads to cell division. Differentiation may be preliminary and lead to further differentiation in succeeding cell cycles, but eventually the cell enters terminal differentiation at some G_1 period. However, at this terminal stage, most mature differentiated cells are still capable of modulating their functional response at both the transcriptional and translational levels. If the cell will no longer divide it enters a G_0 period, where it rests or continues to function, as the case may be. The S period is concerned with replication of the whole chromosomal apparatus even though the didactic emphasis is routinely on DNA replication alone. By itself, DNA replication is limiting with respect to somatic (diploid) cell division, but essentially not limiting in a regulatory sense with respect to macromolecular cell syntheses beyond the S period. Moreover, during DNA replication, there is no firm evidence for dosage effects involving either noninformational RNA or protein synthesis. The G_2 period is one of preparation for mitosis in the current and probably also the following cell cycle. Yet this does not necessarily imply interrupting any cell function (other than for replication) that may persist from the G_1 period. If a G_1 period is absent, its functions are taken over by G_2, which then becomes the combined preparatory period for replication and mitosis as well as, in reality, the period for most cell macromolecular activities. However, unless the G_1 period is characteristically absent, most cells differentiate during G_1 (or G_0). The M period involves the partition of the presently functionally arrested chromosomes into two daughter nuclei. Finally, cytokinesis apportions the doubled cell mass to two daughter cells.

As forewarned in the introduction, many details are still needed to round up the picture of regulation across the cell cycle in macromolecular terms. This is hardly surprising given the comprehensiveness of the subject and the lack of effective genetic analysis. Consequently, for the most part, the picture presented here consisted of generalizations on what goes on at the level of macromolecules during the cycle, although the effort was made to relate these aspects to regulatory principles derived in preceding chapters. On the more positive side, the idea is taking root that the progression of the cell cycle depends on an orderly sequence of transcription-translation events, each round signaling the onset of the next (Sachsenmaier, 1966). Of all periods, this sequence is most clear in G_2. It reveals a certain immediacy in the control of the transcriptional-translational events which are involved. The relatedness to the notion of sequential gene activation, first derived from chromosomal puffing and later incorporated into current models of gene regulation described in Chapter 3, is too obvious to need elaboration. Two further general points are as follows. First, the cell's program for transcriptions during the cell cycle must be open to external influence, indeed essential in a multicellular organism, of which some examples from the effects of hormones and factors are listed below. Second, much more clear than in the prokaryote except perhaps for replication, in the eukaryote there is a marked dependence of cell activities during the cycle on translational control.

By now virtually every macromolecule and enzyme associated with their structural transcription have been shown to be stimulated (or in cases inhibited) by hormone, indicating that, in principle, hormones have total control over the cell cycle. In contrast to this, it is not known to what real extent the transcriptional regulation of the basic cell cycle (as opposed to facultative cell differentiation) actually depends on hormone. Nevertheless, it is probably true to say that under many circumstances local factors in tissue or systemic factors in the organism are just as important in regulation, and probably more so, than hormones. It is a fact that cells in culture can multiply in the absence of hormone (Price et al., 1967). The endogenous or obligatory transcriptional regulation of the cell cycle vis-a-vis external influences being referred to here was discussed in Chapter 3, which the reader is advised to consult. An additional level of hormonal control over macromolecules during the cell cycle is, of course, at translation. The previous considerations on hormonal dependence at the transcriptional level apply to this level as well. A basic regulatory program of transcription and probably a similar program of translation thus appear to be the most important intrinsic determinants of the cell cycle.

A few external factors acting on cells have been preliminarily recognized, although none has been positively identified so far. (Intracellular factors, particularly for initiation of replication and transcription, were described in previous sections.) Serum contains circulating factors that stimulate cell division as well as transcription of all RNA species (Todaro et al., 1967). This stimulation has been confirmed in several recent reports. Conversely, contact inhibition, which involves membrane phenomena (thus presumably not factors), reduces all RNA species proportionately. Recent views are for a triggering balance between retarder and promoter growth substances involved primarily in the net flux of electric charge in the cell (Szent-Györgi et al., 1967), obviously important for any membrane phenomena. A serum factor controlling translation was mentioned in Chapter 5. Mitotic activity in liver is affected by circulating inhibitory factors (review by Swan, 1958). In this regard, a promising line of work is that on certain tissue-specific proteins, termed chalones, which may act as endogenous inhibitors of mitosis (Bullough, 1968), although, as will be seen, there is also an external element involved in their action. The concept is that, by circulating freely in the body, chalones would be able to control organ homeostasis. An obvious role for them would seem to be inhibition of replication initiation factors. The theoretical framework behind the concept is that (a) chalone consists of a general inhibitor moiety and a moiety responsible for tissue specificity; and (b) chalone is acted upon by hormones, being either stabilized or destroyed according to the particular hormone. Some chalones have been partly purified. The potential functional importance of proteins of this kind for normal and aberrant cell growth needs no emphasis. What is necessary is now to prove that their function accords with the advanced concept.

RNA may be a peripheral component of many types of cells (Weiss, 1969). This was suggested by observations not depending on cell fractionation, thus independent at least of fractionation artifacts. Such RNA could account for 10-20% of the electronegative charge density in the cell surface. The higher values correspond to cells with a faster rate of growth, which by itself suggests a certain lability of the superficial RNA. (Another major electronegative cell surface component is sialic acid.) Any rôle of RNA in the many membrane-dependent functions in the cell is, however, not known, unless it would be entirely concerned with charge density, which is unlikely. Apparently it is also not involved in the flux of ions. As proposed by Smythies (1970), RNA forms part of the receptor site for transmitters in the neuronal synapse. The molecular nature of these putative RNAs in the cell surface is not known.

It is a simple expectation that the synthetic potential of an interphase cell will be somehow correlated with the size of this cell. The quantitative correlations that emerged from an autoradiographical survey of several rodent organs (Schultze and Maurer, 1967) are remarkable in the degree to which they bear out this expectation. These correlations are as follows: (1) amino acid incorporation into nuclear protein is proportional to nuclear volume, or in first approximation to nuclear dry mass. As predicted from this, accumulation of nuclear protein is directly correlated to the fraction of DNA which has completed replication (Auer et al., 1970). (2) Incorporation into cell protein is linearly proportional to nuclear volume taken as the percentage of cell volume, and the same holds in terms of dry mass. (3) As measured per gram of tissue per unit time, the turnover of whole cell protein is 23-fold higher than that of RNA, but only twofold higher in the nucleus. The turnover of nuclear protein is sixfold higher than that of chromatin and nuclear sap RNA, but only 1/30th as high in the nucleolus. This last figure reflects the exceptionally high rRNA turnover of the organelle.

Finally, cell aging is the inevitable fate of an otherwise healthy cell unless it rejuvenates itself in preparation for reproduction. Differentiated function, in particular, is no alternative to aging. Some basic data on the nucleic acids of aging cells are as follows: (a) their DNA is more difficult to extract, which suggests a greater cross-linking of chromatin; (b) they contain relatively more heterochromatin and, correspondingly, less DNA-like RNA including possibly mRNA, in line with a generally restricted genomic activity (Harbers, 1968); and (c) their stable RNA is partly degraded, in accordance with a reduced translational capacity. Although the functional implications of these changes are evident, a broader view on cell aging is that it is also the result of unchecked metabolic and molecular deterioration, that is to say, entropy. More precisely, Orgel (1963) suggested that aging is accelerated by eventual mistranslation of enzymes involved in macromolecular syntheses such as polymerases, synthetases, and methylases, which, because of the nature of their function, would exponentially enforce further mistranslation. The notion behind this proposal is that general mistranslation eventually spells malfunction. Some experimental evidence for this notion was adduced both in fungi and *Drosophila* (Holliday, 1969). On the other hand, mistranslation and somatic mutation could both lead to autoimmune conflict. These stochastic views on cell aging resulting in cell death for certain do not exclude a nonspecific genetic component like in most other biological systems. Against these stochastic views, however, is another calling for a genetically directed or even programmed cell death to maintain, for example, the equilibrium of tissue (review by Bullough, 1968) and the course of morphogenesis (review by Glücksmann, 1951). Yet, when all is said, for most eukaryotic cells what kills them is not their own wearing out—whatever its cause—but the death of the organism.

EMBRYONIC DEVELOPMENT: A REVIEW

Because development encompasses such diverse cellular aspects as multiplication, determination, differentiation, growth, and function, it could serve—were it sufficiently understood—to recapitulate all that was said up to this point concerning the eukaryote. In actuality, however, as previously noted in the case of the cell cycle, the very vastness of the subject results in many of these aspects not having as yet been approached in molecular terms. For these reasons, and as far as the data permit, development will be reviewed with the emphasis on the integrative and regulatory aspects of eukaryotic macromolecular syntheses considered in preceding chapters. This approach also provides a frame of reference by which to interpret the developmental data. By way of introduction, we begin by discussing, in general, a few concepts and attributes of embryonic systems.

General Aspects of Development

Coordination of the various processes leading to a mature organism is the very essence of development. Unfortunately, it is not clear whether a unitary

account of these developmental processes, which relies on the basic principles for macromolecular syntheses presented in previous chapters, is conceptually possible. There are three categories of developmental processes to consider here, which in their order of compliance with these macromolecular principles are: (a) cellular, (b) supramolecular, and (c) supracellular. The implication with respect to genetic information is that categories (b) and (c) are more dependent on remote or second-order products. [The order of biological complexity of these processes changes, respectively, to: (b), (a), and (c).]
(a) Cytodifferentiation adheres to these principles and is primarily a cellular process (even if triggered from outside the cell) involving the flow of information from DNA to protein, hence typical templates for macromolecular syntheses. All known cases of cytodifferentiation include an actinomycin-sensitive step, and, therefore, transcription at some time or another. Cytodifferentiation encompasses the steps responsible for the final characteristics of cells including control of their size. The outstanding question remains, however, whether cytodifferentiation is triggered from outside the cell by signals involving only molecules or also the supracellular stimuli which are considered below; this question will be discussed later in a section on induction. (b) The character of developmental processes changes once we start considering patterning of macromolecules outside the cell and with effects on themselves and other cells, such as the deposition of layers or aggregates of mucoprotein (a nonfibrillar protein) and collagen (a fibrillar protein). One argument is that properties resident in these substances themselves are responsible for the pattern. This second category of processes would then depend on proteins or other substances acting, in general, as matrices (not enzymes), from which it follows that these processes have acquired a supramolecular character.

In the third category (c), exemplified by processes such as morphogenesis or the vectorial emergence of form, we simply do not know which are the operating principles, this being as true of a tissue as of an organ or organism; these processes are supracellular. The character of this category seems sufficiently distinct from (b), but it is important to appreciate that they may be functionally related; for example, a supramolecular territory (basal lamina or juxtacellular matrix) contiguous to a cell may be essential in order for supracellular effects from that cell on other cells to take place. It is also plain to see that these supracellular processes must depend and be congruent with those in the previous categories, but this does not tell us how their molecular nature is related. More simply, the supracellular processes must depend upon the informational flow from DNA to protein, but are they implemented by using this flow or its terminal products? The previous comments imply that the question is not answered. Concerning the far simpler question of signals to release these processes, Gustafson and Toneby (1971) have suggested that the appearance of embryonic collagen could lead to morphogenesis—another example of the interrelationship between categories—yet there is no reason why small molecular or macromolecular signals in the cellular category should not equally occur. At present, therefore, because of the fundamental incertitude in this category, a theory of positional information is emerging in search of basic supracellular principles (Goodwin and Cohen, 1969; Wolpert, 1969). One aspect is whether or not these principles include nongenetic information of the type described in Chapter 1. All that is clear is that some sort of cell communication via junctional complexes or other properties of membranes must be available. Hence the interest in this area, perhaps one of the most intriguing and far-reaching in cell biology, lies, on the one hand, in the nature of the linkage between macromolecular assembly and membrane phenomena, and, on the other hand, in the possible self-replicating nature of the membrane phenomena with a view to intercommunication between cells. Without wishing to point to anything more than a formal resemblance, the first of these two aspects is also the present interest in neuronal behavior.

This review deals with development up to early embryonic differentiation. Following the pattern of previous chapters, the review covers the topics of replication, transcription, and translation, in general, and then centers on induction in a later section. These topics are relevant to the (a) templated cellular activities and, derivatively, to the (b) patterned supramolecular activities considered in the preceding paragraph. It must be realized that the immediate way in which these topics relate to the (c) supracellular activities is not obvious.

Embryogenesis sets out to achieve a dynamic spatial segregation of cell types in the embryo through selective cell division and migration, for the classical description of which the reader is referred to the textbooks by Waddington (1956) and Balinsky (1965). A frequently overlooked aspect attendant upon this segregation is, however, the equally important selective elimination by cell death (Glücksmann, 1951). Three main phases of embryogenesis are: (a) a preparative phase from oogenesis through cleavage stages, (b) a succeeding morphogenetic phase starting with gastrulation, and (c) an organogenetic phase that starts as the previous phase nears completion and ends gradually in early adulthood. Of these phases, the morphogenetic one, in particular, remains an enigma to developmentalists, quite simply because of the supracellular phenomena which are involved.

The future three-dimensional order of the embryo is already latent in the one-celled egg by dint of ooplasmic localizations with more or less defined potentialities to direct subsequent differentiation of the different parts of the embryo into one or another direction, in part meaning a control of this process by the differentiative status of the cytoplasm. The essential point is that the various ooplasms are discretely localized in the egg and each has a specific developmental significance, from which it follows that this cytodifferentiation of the egg is very meaningful in terms of development. These ooplasms are imprinted by the maternal organism, in particular the ovarian follicle cells, during oogenesis.

Hence oogenesis becomes the beginning of development as well as the biological bridge between generations. As will be discussed, this maternal imprinting is best viewed as the creation of a developmental program in the egg cytoplasm, reminding one conceptually of the genomic program written in a different molecular language, because the present one may partly be a supramolecular language. Beginning with fertilization, extraovarian life is the cumulative realization of these ooplasmic potentialities as opposed to a departure from any preexisting pattern. As already mentioned, what we have is a segregation of ooplasms into distinct embryonic regions which are becoming more and more committed to their future differentiative pathways. This realization is, however, not a simple process but a multireciprocal process involving a myriad of interactions with the embryonic genome. During enactment of the ooplasmic potentialities a genomic program, presumably also going back to oogenesis, is being deployed and further evolved through continuous interaction with the ooplasmic program. Unfortunately, beyond their necessary occurrence, little is known of these programs. The molecular nature of their interaction, resulting in the creation of developmentally significant territories, remains one of the weakest links in the understanding of development. Perhaps what is included are informational molecules released by the genomic program and then selectively distributed to these territories (or actually "selected" by them), these molecules in turn being reciprocated by factors (or other influences) released by the cytoplasmic program. Eventually, the combined nucleocytoplasmic program dictates the orderly appearance of RNA species in the embryo, which in turn controls the morphogenetic and organogenetic phases of development. Cell-to-cell membrane interactions during these phases probably constitute an important feedback on this program.

A basic concept in the preceding paragraph is the remarkable directionality of the developmental program shown in the timely provision for relevant molecular and cellular mechanisms. Because this concept is important, we may recapitulate the nature of this program. Nucleocytoplasmic interactions create a common developmental program starting with two separate programs. One of these programs is genomic and is passed down the germinal lineage. Another program is ooplasmic and equally continuous from generation to generation. Inasmuch as it is probably moulded by both the oocyte and maternal genomes, this program in a sense alternates between somatic and germinal lineages. Both programs depend on genetic information and together in turn generate cellular phenomena, derivatively supramolecular phenomena, and eventually supracellular phenomena. In addition, it appears that supracellular information of a nongenetic nature and unclear specification may be handed down the female germinal line to subsequent generations. This latter aspect of development could be as profoundly significant as it is at present poorly understood.

Eggs have been traditionally regarded as belonging in a mosaic or determinate type (mollusc, ascidian) or a regulative type (echinoderm, amphibian) according to whether the basic embryonic pattern is fixed firmly or loosely during the first cleavage divisions. After it is interfered with, the pattern can be reorganized or regenerated in the regulative egg, but less or not at all in the mosaic egg. There are no absolute types, however, for certain regions in an embryo may tend more to one type, whereas other regions may tend more toward the other. In addition, the regulative egg becomes fixed in pattern as it develops. The distinction should therefore best be considered a temporal, not a fundamental one, and in fact has no phylogenetic basis. What is clear is that whatever molecular mechanism is determining the difference, it runs in advance in the mosaic egg; hence in this type of egg chemical differentiation, or its implementation, precedes morphological differentiation the most. As already mentioned, ooplasmic territories with defined potentials are established first in the egg and become manifest later in the multicelled embryo; as a rule it is these territories that are more inflexible stage by stage in the mosaic than in the regulative embryo. One striking example of these territories is the germ plasm governing the chromatin elimination described in Chapter 3 (see DuPraw, 1968, for discussion). More generally, as originally proposed by A. Weismann in his classic "germ plasm theory," this sex plasm bestows the ability to direct gonadal differentiation upon an indifferent nuclear population as this is parceled off into cells containing the plasm. Yet later, as development proceeds, it can be seen that ooplasmic differentiation becomes clearly a matter of two-way nucleocytoplasmic interactions. Historically, this complication of development brought about the introduction of the allied but more sophisticated concept of "morphogenetic fields." These formally resemble physical fields (e.g., gravitational) but with the added dimension of time; for their discussion the reader is referred to Waddington (1966).

A real difference between embryos is, of course, their natural environment. At one extreme, the amphibian embryo thrives in nutritionally indifferent surroundings (except for oxygen, etc.) and entirely depends for survival on its yolky reserves. Being a closed system, this embryo can at best maintain a constant mass of RNA and protein; its reserves are converted, so to speak, into informationally rich molecules. (Some of these reserves are, however, protein.) At the other extreme, the mammalian embryo prospers in a nutritive environment which allows it to grow and gain mass; this embryo represents an open system. With little reserves of its own, this embryo soon diverts much of its structure to drawing foodstuffs from the placental circulation. Formation of this considerable amount of structure also means that macromolecular syntheses must become totally engaged very early in the embryo. The nonplacental embryo becomes an open system as it begins feeding.

Whatever their postovulatory career, during ovarian life all eggs represent open systems deriving not only nutrients but also macromolecules from the broader maternal environment. Most commonly, as in the amphibian and mammal, the egg itself is a powerful producer of macromolecules. These eggs have a functional genome, sometimes with lampbrush chromosomes, and active nucleoli during oogenesis. In addition, in these eggs the ovarian follicle cells directly contribute substances of their own making as well as passively contributing other substances arriving in the bloodstream from maternal organs, such as yolk proteins made in the liver. Some of these substances may also be informational macromolecules destined for the localized ooplasms mentioned earlier, but their nature or form of transfer to the ooplasms remains hypothetical. (The follicle cells also form the membranaceous chorion of the egg. This acellular layer is traversed by mutually contacting cytoplasmic processes derived from both the follicle cells and the egg, obviously representing a direct means for exchange of substances.) The opposite situation is exemplified by the meroistic insect egg in *Drosophila*. This egg is synthetically dormant during oogenesis and presents a nonfunctional genome lacking lampbrush chromosomes or nucleoli. In this egg most of the necessary macromolecules are synthesized in the ovarian nurse cells from where they pass into the egg, although the ovarian follicle cells continue to contribute as in the amphibian. Parenthetically, lest it may seem that the meroistic egg is a retrograde system, the insects possessing it are among the most advanced evolutionarily; the advantages of this system will be explained below. Certain rodent eggs, such as the mouse, appear to lack lampbrush chromosomes, and it is considered that their genome is less active than in other rodent eggs which have this type of chromosome. It has been proposed that mouse eggs are united in an ovarian syncytium whereby surviving eggs are fed macromolecules by eggs that become atretic (Davidson, 1968), in which case the latter eggs would be functioning as nurse cells. The modalities of ovarian growth described in this paragraph are not clearly correlated with either the relative amount of yolk in the egg or the mosaic versus regulative type of egg. These modalities do however impose a different rate of egg growth: the dormant eggs nourished by nurse cells grow, in fact, much faster than the active eggs left to fend for themselves. This must be a reflection of the higher gene dosage in the combined nurse cells (4000-8000C) compared with the 4C genome in the active eggs (Jacob and Sirlin, 1959). Indeed, the nurse cell functions as an amplifier of macromolecular syntheses—this is the advantage alluded to earlier.

The participation of regulatory factors is very apparent during development. Cytoplasmic factors for replication and transcription of the intracellular type presented in connection with the cell cycle, and which are recognizable since early embryonic stages, are usually included in this topic. In the axolotl, a protein factor, the product of a now identified gene, is first apparent in the germinal vesicle and later is obligatory for gastrulation (Briggs and Justus, 1968), but is not detectable beyond the gastrula stage. A similar factor was described in *Rana* (Smith and Ecker, 1969). In the sea urchin, the addition of vegetal micromeres restores to normal both the tendency toward animalization and excessive transcriptional activity characteristic of separated animal cells (Markman, 1961). In this system, the micromere factor(s) responsible is insensitive to actinomycin but sensitive to lithium, suggesting that the factor is transmitted via intercellular junctional complexes which are known to be sensitive to this cation. Factors operating during early embryonic induction will be considered separately in a later section. Reminiscent of the modulatory type of regulation discussed in Chapter 3, certain metabolites have propitiatory effects on differentiation, e.g., phenylalanine acting on neuroectodermal pigment cells (Wilde, 1961) and vitamin A acting on epithelia (Fell and Mellanby, 1953); however, the truth of the analogy remains to be seen. Nerve growth factors (Levi-Montalcini et al., 1972) and nerve trophic factors (Dresden, 1969) are also well known, the first being a nonspecies-specific diffusible protein. A low molecular weight fraction allows the multiplication and differentiation of chondrocytes and melanocytes (Coon and Cahn, 1966). An epidermal growth factor is a small basic protein (Taylor et al., 1970). The mode and place of action of this variety of factors, some intracellular and others extracellular, is yet to be converted to molecular terms. The point to remember is that they may, and in cases certainly do, represent the programmed function of regulatory genes during development.

Many of the previous factors are extracellular and their transmission does not require cell contact. However, all tissues of the early embryo are interconnected by junctional complexes which in the later embryo are retained only between similar tissues and transiently shared between dissimilar ones (Hay, 1968b). These complexes permit cell-to-cell communication via electrical coupling and exchange of small molecules (Lowenstein, 1968). These intercellular potentials and molecules may represent a flow of signals not only for metabolic homeostasis or concerted cell movement but also for coordination of macromolecular syntheses, which is a most essential aspect of development. In addition, macromolecular syntheses at the beginning of development, i.e., egg maturation and ovulation, are under extrinsic control by maternal hormone. Thereafter, extraovarian animal development up to organogenesis does not rely upon the action of typical hormones. The accent here must be on "typical" because some embryonic inducers of great importance in development may have to be regarded as atypical hormones. In contrast, plant embryogenesis depends on hormones, though here again many of them act more like growth and cell division factors.

Cell membranes, or more correctly, cell surfaces, have important roles in various communal cell activities manifest during development. Examples are contact inhibition, cell migration and aggregation, sorting out of cell types, and morphogenetic movements. These surface activities involve recognition and motion, and, except perhaps contact inhibition, include the supracellular component considered earlier. What kind of molecular processes and again whether or not nongenetic or second-order genetic information is involved in these activities is not clear. Because much information may depend on or be transmitted by these activities, they represent another of the major missing links for understanding development. Membrane activity must somehow be communicated to the cell interior, from which the concept of a transducer function of membranes originated (Wolpert and Gingell, 1969). In addition, this communication affects the internal motile apparatus of the cell represented in the microtubular system (Granholm and Baker, 1970). Probably also, the communication eventually reaches the genome and affects its regulatory program with a view to coordinating differentiation. As a model only, one can think here of something like the internalization of a protein hormone signal via cAMP dependent pathways. In reality, many aspects could be involved in the internalization we are considering. This remains therefore a "black box" of membrane biology, and yet another developmental area whose understanding is left for the future. Naturally, membrane activity must ultimately be communicated to other cells, in which aspect glycocalices, extracellular matrices, and again junctional complexes (see above) are believed to intervene.

The cell cortex of ciliate protozoa, that is, the cell membrane together with a subjacent layer of cytoplasm, has a distinct developmental significance. This says that the inheritance of cortical characteristics is associated with the cortex itself and/or its kineties, including part of the microtubulature, rather than with the deeper lying plasmatic areas. This idea of a cortical inheritance arose from experiments on *Paramecium* (Sonneborn, 1967) and *Stentor* (Tartar, 1961). They showed that, for example, the pattern of kineties is self-replicating. According to Sonneborn, however, although these kineties contain DNA, this does not seem to be directly implicated in the inheritance. An explanation by Sonneborn is that whereas proteins are encoded in DNA their arrangement is not, but instead could be "encoded" or primed by a preexisting arrangement of proteins. Such a system was referred to as a nongenetic informational system in Chapter 1. Here an analogy with something subtly in between the supramolecular and supracellular phenomena is apparent and a certain relationship to membrane phenomena in general is evident. Of interest, according to Curtis (1965) working on amphibian eggs, the notion of cortical inheritance may have a wider significance for eukaryotic morphogenesis in general. However, which substances are responsible, he does not divulge.

The following review of development is selective rather than comprehensive and is based primarily on the amphibian and echinoderm embryos which are the best studied biochemically, but when practicable or informative makes reference to the mammalian and other embryos. The intention will be to indicate, superimposed on the pattern of DNA replication, first, the appearance in the embryo of auxiliary RNAs, then, the presumed regulatory and informational RNAs in relation to the previous RNAs, and lastly, the course of translation vis-à-vis the total RNA machinery. Reviews are available on the ultrastructural facets of development (oogenesis, Nørrevang, 1968; through cytodifferentiation, Sherbet and Lakshmi, 1967), although the correlation between most ultrastructural and biochemical aspects of development has yet to be worked out. In the oocyte, in particular, it would seem that the ultrastructural complexity that is peculiar to this cell is compatible with the great amount of information stored. Suffice it to add that thereafter the complexity of more typical cytoarchitecture increases in the embryo roughly in parallel with the major phases of development. This might not seem particularly illuminating, but it recalls some of the points made in Chapter 5 regarding the increase in the organization of translation in parallel to the degree of cytodifferentiation. Plant material is reviewed by Galston and Davies (1969) and Loening (1968b); it is dealt with here only in passing.

Despite the vastness of the subject of development, the progress made during the last decade in the description of its molecular aspects has been very real. One result of this general progress has been that the basic rules for interpreting developmental data are beginning to stem from developmental systems rather than from bacterial systems. A second and inevitable result has been that the ignorance concerning the conceptually more important aspects of regulation and general coordination is now becoming only too clear. So far, in this respect, developmental systems have not provided more causal information than perhaps other established cell systems. It is these developmental regulatory mechanisms that should, therefore, receive the most attention in the coming years. Meantime, following the present thinking of many molecular biologists, a basic regulatory concept stressed here is that of developmental programming. A developmental program means the existence of regulatory circuits responsible for the orderly activation of gene ensembles resulting in new patterns of development. The basic concept is coextensive with that formulated for functional rather than developmental changes of the eukaryotic transcriptional apparatus in Chapter 3. From the viewpoint of biological organization, to secure an harmoniously developing embryo is as complex as to maintain an harmoniously functioning organism, and probably more so. In fact, the first notions of program were put forward in the 1950's to explain development in genetic terms, and these notions stand conceptually as forerunners of the more molecularly oriented models of program described in Chapter 3. During development such program is caught in the dramatic and long-drawn phase of transcriptions leading progressively from a totipotent cell to stable and rather irreversible cell types. Yet, relevant to the continuity of the system, when these cells mature functionally they continue to undergo similar, if less dramatic, changes in transcription. However, before proceeding, it should be admitted that this regulatory concept of development requires a good deal of work to give it substance. Whatever the work, as witnessed by the concept of "oncogenesis as a blocked ontogenesis" put forward by Potter (1969), the subject is of more than academic interest.

Replication during Development

The idea entertained in the past was that development could be explained by directional changes in DNA, or gene conversion, although positive evidence in favor of these changes was never provided. A recent version of this idea is that the changes could consist of a scrambling of genes resulting in qualitatively new DNA sequences (Stent, 1969), by a process similar to what is believed might happen during antibody formation according to the clone selection theory. The problem again is that to explain development such scrambling of genes must be highly directional. These possibilities seem unlikely when it is considered, first, that no directional mechanism for changing DNA is at hand, and, second, that development is based on progressive determination of cell types significantly more than on selection of cell types. Hence the current view is that DNA sequences do not change greatly (if at all) during development, that is to say, genome differentiation is not at the root of cytodifferentiation. By extension, barring somatic mutation, the DNA in all cells of an organism is informationally identical. Stated in this form, this is now recognizable as one of the basic principles of modern cell biology. By implication, it is the expression of genetic information which varies from cell type to cell type during development. Although no technique is available to put this view of a qualitative DNA constancy to an absolute test, much indirect evidence and certain biological experiments in particular are in support of it. For example, a small number of late embryonic or early adult nuclei are capable of taking an enucleated egg through normal development (Gurdon and Woodland, 1968); a whole carrot plant can be reconstructed starting from a single phloem cell (Steward et al., 1958) (this being also an example of cytoplasmic totipotency). Above all, the view of genomic invariance has conceptual economy in its favor, for otherwise one highly accurate mechanism would be required to alter the DNA and a second equally accurate mechanism to specify which part of it is expressed.

However, there are departures from the rule of genomic invariance. One example is the diploid diminution of insect chromosomes described in Chapter 3 by which specific loss of chromosomes creates a differential gene content. A second but less dramatic example is the widespread restriction of certain genes to the Y chromosome. Not by chance, these two qualitative genomic variants are connected with sex determination. Nevertheless, accepting the informational constancy of the genome as a first approximation (it can be no more in the absence of proof), one has only to look at the changes occurring in heterochromatin to realize that the chromosome rather than the DNA does indeed differentiate during development. Biochemical evidence for these changes in the chromosome will be given later. They will possibly reflect a changing genetic program resulting in differential gene expression. As will become clear, this amounts to saying that not all DNA is transcribed at any one time, and, it follows, there must be a mechanism to decide what is to be transcribed. What should be noticed is that this differential gene expression is almost the operational equivalent of the previously refuted qualitative changes in the genome.

DNA is methylated at the postreplicative level and the activity of the base-specific methylases involved in this modification varies during development. However, contrary to some understanding in the bacterium, the function of eukaryotic DNA methylation is entirely unknown and it would be unjustified to assume that it has a causal role in differentiation. It could well represent a phenomenon, so far unexplored, attendant to selective transcription or some other aspect of DNA as a template. Scarano and his associates (1967) presented evidence that they nevertheless interpreted as possibly indicating a permanent directional change in the DNA sequence during early echinoderm development; methylation of cytosine followed by deamination to thymine would result in an altered TG base pair. Further work is required to validate this interpretation but, briefly, once again the nuclear transplant evidence cited earlier is against it. A more conservative proposal is therefore that the conversion to thymine, which was confirmed in hepatoma cells, acts to promote replication or transcription by causing template destabilization (Sneider and Potter, 1969). Presumably, such effect would occur at the initiation sequences for these processes in the template and in keeping with the views that were mentioned.

To begin with the description of replication during oogenesis, this takes place during the preceding oogonial interphase at which time the 4C DNA value is attained. An exception is rDNA that is subsequently amplified during meiotic prophase of the oocyte. Oocyte maturation includes two reductional divisions that bring the DNA value down to the 1C gametic level. In the amphibian, entry into this stage follows rupture of the germinal vesicle, a process activated by maternal gonadotropic hormone. As shown with actinomycin and puromycin, maturation depends upon both transcription and translation (Gurdon, 1968). Although the actinomycin-sensitive transcription could reside in the ovarian follicle cells, this inclusive dependence reveals an immediate overall control of maturation. It was also shown that the mitotic and cleavage arrest characteristic of the mature oocyte is caused by a cytostatic factor present in the endoplasm (Masui and Markert, 1971). This factor will be neutralized at the time of fertilization by something in cortical cytoplasm, which then becomes very active developmentally. As the germinal vesicle of the oocyte ruptures, a second regulatory factor appears in the cytoplasm which will later promote replication through early cleavage stages (Smith and Ecker, 1969); its appearance at this time depends upon translation only. Because of this factor, nonreplicating amphibian nuclei will replicate their DNA when transplanted into cytoplasm of oocyte stages after vesicle rupture but not earlier. By this test the factor is therefore not active within the unruptured vesicle. In any case, it does not become active until after fertilization, that is, when the first blocking factor has been neutralized.

During cleavage, cell division soon attains its peak rate in the life cycle and nearly as high as the bacterial rate. The word "cleavage" derives from the lack of cell growth between cell divisions, that is, the fertilized egg is progressively parceled down to smaller cells. In the case of the amphibian, for example, some 12-15 rounds of division during cleavage of the egg result in some 20,000 embryonic cells. Since the egg had increased by more than a 100,000-fold in volume during oogenesis, these embryonic cells are still considerably larger than adult cells. In keeping with their frequent division, early cleavage cells have no G_1 and little G_2 period and most of the interphase is spent in the S period. Accordingly, these cells make few macromolecules other than those required for division. As development proceeds and cell division slows down, the M period increases in duration very slowly, G_1 and S increase more or less together, and G_2 increases most rapidly (Gross, 1968). Replication continues at a high rate throughout embryonic life, but toward the midpoint the cell cycle assumes more of the characteristics in adult dividing cells: it becomes regulated at the G_1 period. Development is also unique in that the ratio of DNA to dry cell weight is continuously increasing. Clearly, in nonmammalian embryos the cell cycle must be regulated endogenously, although this is not by internal precursor pools or in indirect response to the environment, except for temperature.

Replication during cleavage depends on preparation during the G_1 period, or, alternatively, during the G_2 period when G_1 is absent. This is shown by transplantation experiments. Amphibian nuclei transplanted into cleavage cells which are in the S period must be in the G_1 period to be able to replicate their DNA. If the nuclei are transplanted when they are in the G_2 period, mitosis and reentry into the G_1 period are required before replication is again inducible. By contrast, in transplants between amoebas, most transplanted G_2 nuclei will replicate their DNA immediately, that is because this animal has a very short G_1 period and much of the preparation for replication is completed early in the G_2 period. Probably the same result would have been obtained if amphibian nuclei from early embryonic stages (with no G_1 period) had been used in the previous experiments. Often the chromosomes in transplanted nuclei derived from late embryonic stages and induced to replicate in this fashion show abnormal numbers, presumably because of an original asynchrony of replication in the individual chromosomes.

To recapitulate, the maximum rate of DNA replication is attained during embryonic cleavage. In the amphibian the total DNA per embryo, when plotted logarithmically, increases abruptly until midgastrulation and from then on increases more slowly. This means that, starting from gastrulation, DNA accumulates at a decreasing rate per total DNA content. In the endoderm the rate of replication is lower than in the remainder of the embryo, the difference becoming greater in the gastrula and neurula stages (Woodland and Gurdon, 1968). In contrast to the amphibian, the mammalian embryo, active in syntheses since very early during cleavage, develops slowly and with a typical cell cycle (including a G_1 period) from the beginning; it therefore forms nucleoli. The mammalian condition is, however, not exclusive to the homeostatic uterine environment for it also occurs in certain free-living lower animals.

Early cleavage cells contain considerable amounts of cytoplasmic DNA, often surpassing the amount of nuclear DNA, of which most is found in mitochondria and possibly a little in yolk. The relative excess of cytoplasmic DNA disappears after cleavage because of the great increase in nuclear DNA. Although its transcription is enhanced in the sea urchin from the blastula to the gastrula stage (Hartmann and Comb, 1969), almost nothing is known of the early replication of this extranuclear DNA (Gross, 1968). According to data by Pikó (1970), it is possibly not replicated in the cleaving mouse embryo. At present, there is no indication that this DNA participates in directing the course of embryonic differentiation, even though it is essential for supporting mitochondrial function. However, in no embryo whatsoever has this participation in differentiation been excluded, nor can it be excluded on general grounds, even if, admittedly, because of the low informational content of mitochondrial DNA this participation seems very unlikely. Cytoplasmic DNA not contained in organelles such as the species specific poly dAT in insect eggs (Travaglini *et al.*, 1968) has no known function. (This cytoplasmic dAT should not be mistaken for nuclear dAT, which is part of the highly redundant chromosomal DNA in the insect.) The ratio of the cytoplasmic copolymer to nuclear DNA remains constant during early development, which perhaps suggests a storage function. Likewise, the function of the DNA that from cytochemical observation moves to the cytoplasm in certain eggs is not known.

Concerning the mechanism of cell replication during cleavage, this does not immediately depend upon transcription, as shown by experiments using actinomycin which permits survival up to the blastula stage. In contrast, puromycin and cycloheximide arrest cleavage (Davidson, 1968). At least part of the effect of these drugs is on the making of spindle proteins rather than directly on DNA replication. In effect, DNA replication can be induced in amphibian nuclei transplanted into early embryonic cytoplasm whose protein synthesis has been practically suppressed (Gurdon, 1968). Cytoplasmic protein which enters the nucleus just prior to the S period during cleavage might represent a factor or an enzyme involved in inducing or assisting DNA replication in some manner. From the previous observation, this protein is likely to last in the cell for some time after it is made. Whether it corresponds with the amphibian factor described earlier which appeared at the time of germinal vesicle rupture is not clear. By contrast, DNA polymerase is known to be present in the cytoplasm of cleaving cells, but it is unlikely to serve directly in the control of replication. The above enzymes or factors concerned with early DNA replication probably are maternally supplied but very soon are made anew in the embryo. On the other hand, the maternal pool of precursors for DNA (as well as RNA) persists at about constant levels in the embryo, therefore, it is neither limiting nor controlling replication. For embryos with an extended cleavage stage, such as the amphibian, the regulation of replication is the single most important regulation during early development. It is clear, however, that the fine mechanism of this regulation is no better understood than in the adult.

Cell differentiation, in general, requires a suitable rate of cell division. In many differentiating cell types this means a slow down and frequently an eventual cessation of division. In a basic sense, however, cell differentiation (including induction and regeneration) depends on cell division and more particularly on replication of the chromosomal apparatus. This is equally true of hormonal and carcinogenic action and of the antibody response. Presumably, the dependence of differentiation on chromosomal replication relates to the acquisition of new patterns of transcription being made possible during chromosomal replication, which is the explanation given before for hormonal action. Further, during the ensuing mitosis, regulatory proteins may gain preferential access to the chromosome and participate thereafter in the reprogramming of transcription (Gurdon and Woodland, 1968).

In conclusion, during early embryogenesis the genome multiplies rapidly, and, on present evidence, is distributed to prospective cell progenies without any specialization of DNA. In the amphibian embryo, with a predominant cellularization but little genomic activity, these cell progenies differentiate according to the cytoplasm they receive and which their genomes then proceed to modify. This cytoplasm was organized for differentiation under the direction of the nuclear genome during oogenesis, probably not only the oocyte but also the maternal nuclear genome. In the mammalian embryo, which grows rapidly and has an active genome from the beginning, the cytoplasm is more of an original creation of the cell progenies. To a different degree in the two embryos, a feedback between genome and cytoplasm is progressively established. The rate of replication decreases during gastrulation as the morphogenetic pattern begins to emerge, so that cellularizaiton preceded and created a basis for overt differentiation.

Transcription during Development

Beyond question, because of the large number of underlying processes and the need for their close integration, development is the paradigm of the complexity of eukaryotic regulation, and this is perhaps nowhere more manifest than in the regulation of developmental transcription. It can therefore hardly come as a disappointment that this account does not set out to solve this regulatory complexity. Yet, given some of the limitations pointed out in the introduction, the need is plain for a basic principle to guide discussion of the tendentious ensemble of processes which, after all, is development. The choice of principle soon narrows to one already introduced in Chapter 3, namely, "differential gene activation," for which further documentation, though not proof, is provided in this section. This differential activation of genes depends on the mechanism of selective derepression of the genome, which thus becomes the master switch for the release of genetic information at the root of the control of development. As we shall contend, the specification of this release of information reveals a dynamic developmental program inscribed in the chromosome.

Differential gene activation is possibly reflected in the changes which occur in the embryonic chromatin. Changes are observed in the acid lability of amphibian DNA from gastrulation onward (Brachet et al., 1968) and may indicate a changing type of association of DNA with chromosomal protein. Evidence was presented to suggest that, in the presence of exogenous RNA polymerase, more of frog chromatin becomes progressively available for transcription during development. This increase in template availability seems to correspond with a decreasing amount of protein in direct contact with the DNA (Kohl et al., 1969), despite the almost certain probability that the total chromosomal protein is increasing. Similarly, compared with the blastula stage, sea urchin chromatin at the pluteus (late gastrula) stage contains more template activity, chromosomal (derepressor) RNA, and acidic protein, but less histone (Marushige and Ozaki, 1967). Some histones show quantitative and qualitative variation during early development. All these changes indicate a gross differentiation of the chromatin which must somehow correspond with the fine changes occurring in embryonic transcriptive activity.

It should be clear by now that differential gene activation is not the same as "differential gene composition," this representing a genomic difference negated by the concept of genomic invariance put forward earlier. On the other hand, without compromising the principle of differential gene activation, "differential gene content" is known to occur in two cases of amplification, namely, rDNA during oogenesis and nonribosomal DNA in somatic cells of certain insect larvae. These two cases of DNA amplification may be possible only because of the specialized chromosomes in these cells, namely, lampbrush and polytene, respectively (Chapter 3). A developmentally programmed nonribosomal gene amplification in the more general case of embryonic cells with ordinary chromosomes cannot be judged on present evidence, but it remains a possibility (for which, again, a mechanism mediated by reverse transcriptase is entertained by some authors). It does not occur, however, in the histone cistrons during sea urchin development

(Kedes and Birnstiel, 1971). Lastly, somatic cistronic multiplicity as in the case of rDNA and 5 S DNA does not result in differential gene content for this multiplicity remains stable during development. By contrast, the previous germinal amplification of rDNA creates a transient differential gene content. As will be explained later, the somatic multiplicity does allow a "differential utilization of genes" or gene dosage regulation over that possible for nonmultiple cistrons.

Let us now attempt to define what cytodifferentiation, as an aspect of gene activation, is chemically all about at the final level. All we can do concerning this definition is to restate on faith that the molecular basis of differentiation is represented by the protein endowment of each cell, and leave it at that. As Jacob and Monod (1963) put it, "two cells are differentiated with respect to one another if, while they harbor the same genome, the pattern of proteins which they synthesize is different." Hence differentiation means those events conducive to the acquisition of the final cell protein endowment. In defense of this doctrine the reader should consult the writings by Davidson (1968) and Gross (1968). Reservations that were already made in the introduction, as to whether this doctrine based on macromolecular principles suffices to explain morphogenesis, need not be belabored here.

In previous discussion of derepression of the genome in Chapter 3, the closing remarks concerned self-programming of the genome. The salient feature of this programming is a progressive activation of genetic circuits in such a way as to ensure the timely release of information appropriate to the ongoing trend of function or differentiation. In the case of the embryo, such progressive release of information will allow it to develop initially with a minimum of realized information. Gene activation is preordained genetically within the chromosome but depends continuously on events outside the chromosome. This is clear as early as the genetic activation of the male pronucleus, which is totally dependent on the egg cytoplasm. Later, these events are well illustrated in early development where the progressive nucleation of local cytoplasmic territories results in reciprocal chromosome-cytoplasm interactions that feed back on the chromosome until its expression is canalized into explicit differentiative pathways. A second illustration of the action of cytoplasm comes from the frequent need to retransplant several times into enucleated eggs older nuclei first transplanted into similar eggs, before these nuclei can sustain long-term development of the embryo. All this can be interpreted to mean that an accruing differentiative status of cytoplasm is continuing to remould the initial transcriptional program traceable back to oogenesis, and this in turn is affecting the status of cytoplasm. (Notice for reference later that an early functional preeminence of the maternal genome is implied.) The important point now is that development depends on a continuous transcriptive response to these cytoplasmic events, otherwise it soon aborts.

The period during which these nucleocytoplasmic interactions occur covers roughly from cleavage to the beginning of any particular organogenetic induction. At the molecular level the interaction results in the striking stage and regional specificity of early transcription and the concurrent adjustment of embryonic translation. Starting with gastrulation, as the organ primordia begin to be defined, each developmental system will increasingly resemble quantitatively and qualitatively the adult system. However, the phrase "ontogeny recapitulates phylogeny" reminds us that the basic programs tend to be preserved among developmental systems.

The long-term developmental progression just described is marked by the activation of the auxiliary genome for stable RNAs prior to gastrulation, and, beginning with gastrulation, the insidious establishment of the facultative genome both regulatory and informational. In particular, the stage specificity of transcription brings out the distinct regulation of the individual RNA species now to be discussed, beginning with the stable species. More than a significant departure from somatic cells, this distinctiveness of regulation is because of the disparate need for augmenting the different types of RNA at different developmental stages, as a consequence of the inbalanced prenatal endowment in respect to each type. The stage specificity of individual species excludes the possibility that the regulation is determined by precursor pools; as measured directly, these pools do not limit transcription. The same is true of RNA polymerases (Roeder *et al.*, 1970), although their complex structure, distribution in the nucleus, and certainly the presence or absence of transcriptional factors, can influence regulation. The possible nature of the general regulation affecting the various RNA species has been considered by Gurdon and Woodland (1968). We discussed the regulation of each RNA species in specific terms in Chapters 3 and 4, and this will not be recapitulated here unless necessary. The following discussion of transcription leans heavily on the best studied embryonic material in this regard, the amphibian.

Ribosomal RNA

The mature amphibian oocyte is an enormous cell (0.5 mm^3 in *Xenopus*) which, on a weight basis, contains an amount of total RNA within normal somatic cell limits. Almost all of this RNA is, however, rRNA. This preembryonic buildup of rRNA during oogenesis was discussed in Chapter 4. The sources of this rRNA in the oocyte are the originally multiple chromosomal rDNA, and, quantitatively the most important, the additionally amplified rDNA occurring in the form of multiple free nucleoli. This amplified rDNA is believed to derive from the chromosomal rDNA, but Wallace *et al.* (1971) have put forward the alternative suggestion that it could derive from "episomal rDNA" which is perpetuated nonchromosomally along the germ plasm. Another quite different view (Crippa and Tocchini-Valentini, 1971) is that the rDNA is amplified from chromosomally produced rRNA by means of a reverse transcriptase, that is, an RNA-primed DNA polymerase like the one recently discovered in oncogenic RNA viruses.

The amplified rDNA of the oocyte is the best known example of differential gene content. Because of it, the buildup of rRNA proceeds at a rate several times higher than in somatic cells, which results in the large oocyte being practically filled with this RNA. Of greatest interest is the peculiar regulation that permits this extraordinary accumulation. Whereas a mature cell produces ribosomes in relation to the rate of protein synthesis and free ribosomal subunits never accumulate beyond about 10% of ribosomes, the oocyte stores ribosomes of which relatively few can be engaged in protein synthesis and up to 60% of which occur as free subunits. (It should not be construed from this that the total amount of protein synthesis in the oocyte is negligible, for, quite the contrary, rRNA alone has to be matched with an equimolar quantity of ribosomal proteins.) One early clue for the "anomalous" regulation in the oocyte seemed to be that, since most rDNA occurs as detached nucleolar "episomes," these could lack a regulatory gene and thus escape the normal regulation. The possibility of control by ribosomal protein itself was unlikely since this protein is regulated jointly with rRNA. Recently, however, the finding that a putative repressor activity is absent at the time of maximum synthesis in the oocyte suggests that the detached rDNA has, after all, no basic regulation from which to escape. Instead the disposition of rDNA inside multiple nucleoli may be for other purposes of regulation, chief among which are the high rate of rRNA synthesis and the consequent partial uncoupling of this from protein synthesis. Perhaps an additional reason is that these multiple nucleoli are more efficient than a few oversize nucleoli with an equivalent total mass.

Using the amphibian as the norm, the embryonic synthesis of rRNA does not become apparent until gastrulation (review by Brown, 1967). The rDNA amplified during oogenesis remains present in the embryonic cytoplasm but is not further replicated or transcribed, either because of its location or its content of methylcytosine (Chapter 4), or both. Until gastrulation, however, cleavage of the oocyte with a very high relative amount of rRNA for a single cell has resulted in many embryonic cells also with a large amount of rRNA. It follows that the intense synthesis of protein during the stages immediately following gastrulation, which is when embryonic rRNA and ribosomes begin to be made, depends largely on maternal ribosomes. The enucleolate mutant embryo subsists on these maternal ribosomes until the time of its death at the free swimming stage. Hence maternal ribosomes last until basic morphogenesis is more or less completed. Concurrently, rRNA synthesis becomes increasingly active from the time when the embryo begins to grow after gastrulation. The new synthesis does not, however, contribute appreciably to the total rRNA content of the embryo before the swimming stage but, as the result of this contribution, by feeding stages the rRNA content has doubled. By this time the embryo has developed an additional requirement for exogenous Mg in its food which is necessary for the integrity of the newly formed ribosomes.

Of all RNA species, only the synthesis of rRNA is not initially related in amounts to the total DNA amount in the embryo. Instead, rRNA continues to accumulate rather uniformly relative to DNA on a per cell basis. The rate of rRNA synthesis, at first slow during gastrulation and very rapid during neurulation, becomes nearly constant relative to DNA by the hatching stage (Gurdon, 1967). By then, even though the rate of synthesis is faster, the ratio of accumulation of rRNA relative to other RNA species becomes also constant, as demanded by their final proportion in the cell. Because the multiplicity of the somatic rRNA cistrons remains constant for life, changes of multiplicity do not determine the great increase of rRNA during this period. However, since all cistrons are not utilized, it is not known whether the increase depends on a larger number of cistrons being transcribed, or, alternatively, on a greater overall rate of transcription of the same number of cistrons; this aspect was discussed for the adult animal in Chapter 4 but there are reasons to suppose that the same considerations apply to the embryo. The increase of rRNA synthesis corresponds with the lengthening of the G_1 and G_2 periods of the embryonic cell cycle, as is true for most other RNA species during development. At the cytological level the first appearance of rRNA synthesis corresponds, of course, with the appearance of nucleoli. The first nucleoli to appear were small because of their limited growth during the short cell cycles, and their rRNA production was consequently small. Later, with the lengthening of the cell cycle, the nucleoli have time to grow larger and produce more rRNA. At this point, it is important to mention that in the echinoderm the rate (not amount) of rRNA synthesis per nucleus measurable during early cleavage is already as high as that found later during the gastrula stage. Since, until recently, in the echinoderm the rRNA synthesis during cleavage had been obscured by the great accumulation of DNA-like RNA, this should serve as a caution that the amphibian pattern of rRNA synthesis described in this paragraph might be due for some revision (even though the interference by DNA-like RNA may be less). As we shall see, if it can be generalized that DNA-like RNA is masking early rRNA synthesis, then this is the converse of what happens in the oocyte.

As in adult cells, in the embryo both 18 and 28 S RNAs are transcribed stoichiometrically in a common 45 S RNA precursor. This precursor originates from the somatic (chromosomal) rDNA in the embryo, not from the amplified (free) rDNA remaining from oogenesis that, as mentioned, is now not functional. Since it appears that this latter rDNA has become nonfunctional partly because of its low methylcytidine content compared with somatic rDNA, this suggests an as yet undefined specific mechanism in relation to rRNA polymerase capable of discriminating rDNA sequences which are not sufficiently modified. There is also some eivdence to indicate that part of the first made precursor may never reach maturation (Brown, 1967). Another explanation could simply be that maturation is delayed. In any case, in the just activated embryonic nucleolus synthesis appears to predominate over processing of rRNA. Maturation of rRNA will become normal as development proceeds. Either of the previous possibilities concerning the earliest rRNA synthesis in turn perhaps suggest that the regulation of different aspects of the pathway varies at different periods of development.

At most stages of development 5 S RNA is transcribed coordinately, hence stoichiometrically, with the major rRNA species. Like these species, 5 S RNA synthesis begins at gastrulation, yet, whether they are coordinated right from the beginning or some time after, is not known. Eventually, the coordination must devolve from a coregulation of these RNA species, since their cistrons are in separate regions of the chromosome. This coordination also occurs in late oogenesis when the 5 S DNA, being somatically highly multiple in the amphibian, is not further amplified germinally as rDNA is (Chapter 4). In contrast, uniquely during early oogenesis, that is, prior to rDNA amplification, the coregulation did not apply, for then synthesis of 5 S RNA (and tRNA) far exceeded that of rRNA (Mairy and Denis, 1971). This imbalance between 5 S and rRNA is redressed later during oogenesis as the syntheses become coordinated. These stoichiometry relationships suggest, in addition, that the reason for the exceedingly high somatic gene dosage for amphibian 5 S RNA is to match the overamplified rRNA synthesis in the germinal cell. This would be because, whereas a permanent somatic dosage of 5 S DNA representing only 0.05% of total DNA is tolerable, a permanent dosage of rDNA representing 200% of DNA would certainly be intolerable. Consequently rDNA is transiently amplified just during oogenesis. Nevertheless, the heavy somatic load of 5 S DNA also means that, except during oogenesis, this DNA is marginally and much less utilized than rDNA during life. In the cytoplasm, 5 S RNA is always associated with ribosomes and thus the amount per ribosome (1.5-2% of total ribosomal RNA) remains constant during life. As mentioned, the situation in early oogenesis may be different.

There are differences between the rRNA transcription in different regions of the amphibian embryo. In the endoderm, in correspondence with the late appearance of nucleoli, transcription begins several hours later than in other regions. After neurulation, however, the rate of transcription per cell or per unit of DNA synthesis has become greater in the endoderm (Woodland and Gurdon, 1968). The greater synthetic rate appears to be related to the lower rate of cell division (Flickinger *et al.*, 1970a,b) more than to any other known property of the endoderm. This enhanced synthesis, surprisingly against the traditional view of an inert endoderm, is borne out by other RNA species to be discussed.

The time at which rRNA transcription starts in other embryos is not necessarily the same as in the amphibian described above. Even in embryos for which there is no biochemical analysis, initiation of this transcription is always recognizable microscopically thanks to the rule of thumb that it coincides with the first formed nucleoli. As measured biochemically in the nematode *Ascaris*, rRNA transcription happens very prematurely before cleavage, and, what is exceptional, only in the male pronucleus; it then ceases promptly to be resumed at the four-cell stage, this time by the combined maternal and paternal rDNA (Kaulenas et al., 1969). In molluscs the nucleoli first appear during cleavage, in ascidians very belatedly at metamorphosis. In echinoderms, nucleoli appear from cleavage to gastrula stages depending on the species. In echinoderms with cleavage nucleoli it was shown that almost from the beginning rRNA synthesis attains the same rate per nucleus as in much later stages (Emerson and Humphreys, 1971). It was pointed out that this difference with the course of synthesis in the amphibian, where the initial rate reported is lower than the subsequent rate, indicates the need for a careful revision of the amphibian data. In fishes, and probably also birds, during blastulation the yolk sac contributes ribosomes to the overlying embryonic disc. In other words, the gap between fertilization and the onset of rRNA transcription can be shorter or longer than in the amphibian embryo. More than anything else, these temporal differences depend on the balance between the amount of ribosomes that the embryo inherits maternally through the oocyte and that which it needs stage by stage. During the phase of cellularization (cleavage) many embryos, including the amphibian, do not need more ribosomes than those supplied maternally. Conversely, other embryos such as the mollusc (and the mammal described below) do not receive a sufficient maternal supply and must add to it from an earlier stage.

The mouse oocyte contains an amount of rRNA equivalent to 100 adult cells. Rather surprisingly, on a cell volume basis this amount is several times higher than in the much larger amphibian oocyte. Each cell in the 20-cell mouse embryo contains the rRNA equivalent of only one adult cell (Ellem and Gwatkin, 1968). The rRNA first appears in its precursor form by the four-cell stage and in its mature form by the eight-cell stage (Woodland and Graham, 1969). The 5 S RNA also appears at the four-cell stage. Taken together the observations suggest that the mammalian embryo starts with a store of maternal ribosomes which rapidly dwindles on a per cell basis as the embryo divides, but soon the embryo checks this decrease and starts providing for its early massive growth with its own production of ribosomes. Although this rRNA synthesis is premature in terms of development compared with the amphibian and echinoderm, it is not so on the basis of time. The four-cell stage takes 2 days in the mouse, or longer than it takes these other animals to reach the late blastula stage when rRNA synthesis starts.

Concerning the presence of a repressor of rRNA synthesis, a low molecular weight inhibitor was first reported in *Xenopus* from studies of the RNA in dissociated cell monolayers and analysis by MAK chromatography (Wada et al., 1968). The inhibitor activity was found to be inversely proportional to the *in vivo* rate of rRNA transcription, that is, maximal at the midblastula and in the vegetal part of the early gastrula stage, from which it was assumed to represent a physiological activity. However, in embryos dissociated within their vitelline membrane and whose RNA was analyzed by gradient sedimentation, these observations were not confirmed (Landesman and Gross, 1968). These authors view the discrepancy between the studies as arising from either the more injurious monolayer technique, in which case the inhibitor would not be physiological, or the analytical procedure (i.e., MAK chromatography does not completely resolve rRNA from polydisperse RNA), in which case the inhibitor might not be for rRNA. The existence of an inhibitor was nevertheless clear from nuclear transplant experiments. Thus, after transplantation into enucleated egg cytoplasm, tadpole endoderm nuclei cease their rRNA synthesis (and their nucleoli regress), to resume synthesis at gastrulation like normal embryonic nuclei (Gurdon and Woodland, 1968). From these observations, the function of the inhibitor may be stage dependent or the inhibitor initially present may be progressively used up. The latter possibility is the least likely in view of the fact that some kind of regulation must persist into adult life. As described in Chapter 4, a putative repressor molecule has finally been identified by Crippa (1970). The molecule was found to be protein rather than low molecular weight as proposed earlier by Wada et al. (1968). Except for this, however, their biological interpretation has proved to be substantially correct.

Autoradiographical evidence in many organisms indicates that the nucleolus-associated chromatin becomes activated just preceding or concomitantly with activation of the nucleolus, e.g., in sea urchin (Karasaki, 1968) and mouse (Mintz, 1964). As discussed in connection with the nucleolus, a causal correlation between the function of this chromatin and nucleolar rRNA synthesis remains possible, but, if it does exist, is far from understood.

In summary, synthesis of rRNA begins in different embryos at a time depending on the duration of the maternal supply of ribosomes and in relation to the overall course of embryonic translation for which they are destined. Because of its highest single proportion in the cell, synthesis of this auxiliary RNA becomes the most important quantitatively since early embryonic life. Yet, as will be seen, it is generally not the earliest RNA synthesis to appear in the embryo.

Transfer RNA

The amphibian oocyte begins its career with a tremendous relative amount (80%) of low molecular weight RNA, of which at least part is tRNA (Thomas, 1970) and part 5 S RNA, as described above. However, during the remainder of oogenesis the rate of tRNA synthesis becomes much lower than that of rRNA, to some extent no doubt because tDNA is not amplified. As a result, the final amount of tRNA in the mature oocyte, although high for a single cell, is low relative to rRNA (Brown and Littna, 1966b). Thus the egg contains less than 1% of total RNA as tRNA, or on average less than one molecule of tRNA per ribosome, not because the molecule is unstable (which it is not in this case) but because much more rRNA is made. About 10% of the tRNA present in the oocyte is associated with the ribosome, or less than 0.1 molecule per ribosome. When compared with the corresponding figures in late embryonic or adult cells, which are over one order of magnitude higher, this signifies a major imbalance in the machinery for protein synthesis such that not all this machinery can be of use to the oocyte itself. As in the case of rRNA, the actual beneficiary of much of the oocyte tRNA is, in fact, the embryo.

It is not known by what mechanism tRNA is regulated during development, or, in particular, what allows the egg to accumulate so much, although the latter has a parallel in rRNA. The ratio of tRNA to nuclear DNA is about 13,000 in the egg and becomes 0.5 later in embryonic development. Synthesis starts again when the ratio is 0.4 to 1 (Brown and Littna, 1966b). The regulatory implication of this critical ratio is not clear, however. From the sudden rise in the rate of tRNA synthesis during the short period of the early gastrula stage when DNA is increasing relatively little, Gurdon (1968) has argued that the ratio per se has no special regulatory significance. Instead, he suggests a causal correlation with the changing status of the embryonic translational apparatus at that period, in turn perhaps related to the synthesis of DNA-like RNA during the preceding late blastula stage. This explanation based on translation comes close to previous considerations on tRNA regulation in Chapter 4. Clearly, no definitive explanation is at hand.

During amphibian development the synthesis of tRNA starts at the late blastula stage. Previous to this and during oogenesis there is an exclusive terminal CCA turnover of the molecule. The precise significance of this turnover is not known, but it also occurs in the echinoderm where the molecule is known to be functional already in the unfertilized egg (Gross, 1968). After cleavage, tRNA accumulates in a constant proportion to DNA, that is, when the whole embryo is looked at, the amount of tRNA doubles with each cell division. However, from data given below, this is not so in all cells. The actual rate of synthesis is initially higher than for DNA until neurulation, declining later to a constant relative rate by the hatching stage (Woodland and Gurdon, 1968). In view of the constant accumulation of tRNA, this probably means that until neurulation the molecule is turning over *in toto* and the rate of synthesis is regulated to maintain the accumulation. What relation this total turnover has with the initial terminal turnover, if any, is not known.

By the swimming stage, tRNA constitutes approximately 15% of total RNA, which is nearly a 30-fold increase relative to the proportion in the egg. By the hatching stage, tRNA attains the normal ratio of 15 molecules per ribosome, of which two molecules are associated with the ribosome. At all embryonic stages, including the mature but not the early oocyte, about 10% of tRNA is associated with the ribosome. Once attained this normal ratio is maintained by the greater accumulation of rRNA than tRNA. In summary, in the amphibian an earlier and faster embryonic synthesis of tRNA redresses the imbalance with respect to rRNA created in the egg, and, thereafter, a normal balance is maintained.

In the endoderm the synthesis of tRNA starts at the same time as in the rest of the amphibian embryo, but proceeds at a rate five- to ten-fold higher per cell or per unit of DNA synthesis (Woodland and Gurdon, 1968).

In dipteran (Endy and Dobrogosz, 1967) and sea urchin embryos (Comb and Silver, 1966) synthesis of tRNA starts at the onset of gastrulation, either earlier or at the same time as rRNA. Quite differently, in the mouse embryo newly made tRNA precursor appears during the four-cell stage and mature tRNA during the following stage (Woodland and Graham, 1969); this was also the case with rRNA.

The general rule with respect to stable RNAs is therefore that, in embryos derived from large eggs full of reserves (including stable RNAs) such as the dipteran, echinoderm, and amphibian, synthesis of stable RNAs is delayed. Depending on the imbalance among these RNAs, their initial rate of accumulation can differ, as appears to be shown by the amphibian, but eventually full coordination is attained. In embryos derived from small eggs, such as the mammalian, stable RNA syntheses are little delayed and are more fully coordinated from the beginning.

Subsequent to stable RNA syntheses attaining full coordination sometime during early development, the regulation of tRNA may eventually become more complex than that of rRNA and 5 S RNA. As discussed in full in Chapter 4, tRNA appears to require a dual quantitative and qualitative regulation. This dual regulation may start during embryonic life sometime after the onset of cytodifferentiation. Before cytodifferentiation the coordinated regulation of tRNA was mainly quantitative, that is, as an obligatory genome in control of all tRNA species and resulting in their dissimilar but basically stable proportions in the cell tRNA population, which is presumably optimized for the translational needs that exist at this stage. This obligatory genome, therefore, represents part of the developmental transcriptional program and regulates tRNA uniformly across the embryo. As a quantitative regulation it probably remains the basic one for the entire life of the organism. However, as discussed for translational modulation later in this chapter, with the onset of cytodifferentiation it is possible that the tRNA population is readjusted qualitatively and perhaps species added or subtracted. The reason for this is, of course, that as embryonic differentiation proceeds there is a corresponding differentiation of translation, and this may be reflected more and more at the level of the tRNA population. Eventually, this population will adjust to the functional demands of each developing organ. The developmental program in operation by this time includes a facultative genome for tRNA in addition to the basic obligatory genome. A facultative genome means therefore one by which individual tRNA species can be subregulated independently of one another, including their own isoaccepting species. Possible regulatory strategies for creating a facultative genome for tRNA were considered in Chapter 4. In summary, an early obligatory program for tRNA regulation is supplemented by a facultative program as embryonic development proceeds past the phase of cytodifferentiation. In principle, it is possible that distinct facultative genomes differentiate in each tissue or organ, each of these genomes standing for a differently balanced tRNA population. As hormones are introduced much later in development, these facultative programs may all be coordinated at the level of the embryo. By this time, indeed, the embryo has begun to resemble the adult organism.

The Status of the Evidence Concerning Nucleolar Synthesis. Some evidence in favor of a nucleolar synthesis of tRNA, additional to the predominant chromosomal synthesis, was presented in Chapter 4. The insensitivity of current molecular hybridization techniques to detect this possible nucleolar synthesis was mentioned. However, a lack of developmental correlation between the appearance of the nucleolus and tRNA synthesis has been posited as evidence, in general, against this nucleolar synthesis. Also, the observed correlation between the appearance of the nucleolus and rRNA synthesis, while strengthening such already well established nucleolar contribution, has further been used as an argument to preempt any other nucleolar contribution and, in particular, tRNA. Close scrutiny reveals that the case against tRNA on the basis of the embryonic data as they stand is unfounded, as can be seen in what follows.

In *Xenopus* rRNA synthesis starts at an apparently low rate, from which it is customary to state that it begins at gastrulation (e.g., Brown and Dawid, 1969) when the rate is already appreciable. Extrapolation of the data to the origin indicates however that rRNA synthesis starts at the late blastula stage (stage 9 in Fig. 5, Brown, 1965; Fig. 12-8, Gurdon, 1967). Recent data by Knowland (1970) are still not sensitive enough to pinpoint accurately the first appearance of rRNA synthesis; as shown by Emerson and Humphreys (1971) in the sea urchin, this appearance may be obscured by the intense DNA-like RNA synthesis taking place at the time. Synthesis of tRNA starts in *Xenopus* also at the late blastula stage but at an apparently much higher rate than rRNA (Brown, 1965; Gurdon, 1967). Contrary to the case of rRNA, this initial stage of tRNA synthesis is stated correctly (e.g., Woodland and Gurdon, 1968). The conclusion in *Xenopus* is therefore that, as far as the data reveal, there is no real differential stage specificity of rRNA and tRNA syntheses.

Hence, in regard to the evidence in the amphibian, the question becomes whether the cytological data on the stage of appearance of the nucleolus are compatible with the biochemical findings above pointing to a joint onset of rRNA and tRNA syntheses at about the late blastula stage. The answer may be yes. In the amphibian this question has to be decided without the help of autoradiography, a particularly useful technique for these studies, since, to the writer's knowledge, the earliest observations using autoradiography are at gastrulation in the newt (Karasaki, 1965). The critical information is then that definitive nucleoli are found with the electron microscope in some *Xenopus* cells as early as the midblastula (stage 8) (Hay and Gurdon, 1967), which is when approximately both stable RNA syntheses start. In other animals the evidence is not as indicative, but the consideration that the nucleolus is present at the start of stable RNA syntheses seems also valid. For example, in one species of sea urchin, tRNA synthesis begins at the onset of gastrulation (Comb and Silver, 1966), and in another species with a similar timing of transcription, active nucleoli are observed with autoradiography at the same stage (Karasaki, 1968). Similarly, in the dipteran both stable RNA syntheses occur during gastrulation (Endy and Dobrogosz, 1967); in mouse both syntheses are detected at the four-cell stage and may begin even within the same cell cycle (Woodland and Graham, 1969). In the dipteran and mouse, definitive nucleoli are present at these stages, respectively. These examples show that the times of appearance of the nucleolus, rRNA, and tRNA are indistinguishable. They should therefore correct the impression that the embryonic data are against a nucleolar contribution of tRNA. In fact, as they stand, these examples could equally well be construed to support this contribution. For reasons that will be clear, this is not the present intention.

The study by Hay and Gurdon (1967) of the enucleolate *Xenopus* mutant, which shows a normal tRNA synthesis but a vestigial rRNA synthesis, should equally correct the notion that the data on this mutant, in particular, are against a nucleolar contribution of part of the tRNA. These authors found, as Jones (1965) did earlier, that the principal nucleolar lesion of the mutant is in the formation of the pars granulosa which is implicated in rRNA, but not in the formation of the pars fibrosa which contains an undefined RNA. Hence these cytological observations, while agreeing with the presence of most tRNA cistrons outside the nucleolus, cannot exclude the presence of a few cistrons in the nucleolus. Quantitatively, this interpretation is consistent with the autoradiographical observations by Wallace (1966) showing that, within experimental error, the total RNA made in the whole enucleolate nucleus (including the nucleolar pars fibrosa) is no different from that in the extranucleolar nucleus of the wild-type. In conclusion, the general occurrence of a nucleolar tRNA synthesis must in the future be judged experimentally rather than, as it has been, on the basis of correlations. More pertinent, without knowing at which embryonic stage this nucleolar synthesis might start, any correlative cytological-biochemical evidence is all but irrelevant to decide whether or not it occurs.

DNA-like RNA

This designation of RNA is adopted here because of its prevalent use by workers in the field. Two RNA species are included under this heading, labile RNA and mRNA. Initially, they will be discussed jointly, avoiding any preconception regarding their possible causal relationship, namely, that part of labile RNA is a precursor to mRNA. This relationship was considered in Chapter 3 and for it, in fact, there is some suggestion in the present material. As the discussion proceeds, the RNA will be treated as labile RNA when it has been characterized by one but preferably more of several criteria: polydisperse sedimentation, composition, turnover rate, and hybridization behavior under standard conditions. Messenger RNA shares some of these characteristics, but it will be referred to as such only when the RNA is found associated with the polysome; when its template activity is demonstrated in a cell-free system (in practice, the activity more than the product is specific to the RNA) or when other biological aspects compellingly suggest this template activity; and when the RNA is homologous to unique DNA. A serious difficulty is that seldom are a sufficient number of these criteria simultaneously provided to allow a reasonable distinction between the RNAs. Nevertheless, by the best possible use of these criteria a developmental correlation between the RNAs will be attempted with their possible functional significance in mind.

A basic point remains, however, that since the functional relationship between these RNAs is not as yet elucidated in any biological material, we cannot expect to see it elucidated in the present one. Whatever this relationship, we will adopt the premise that dRNA includes regulatory as well as informational RNAs. On this basis, in dealing with dRNA we are dealing presumably with (a) the function of the regulatory circuits mentioned in the introduction, representing a basic ingredient in any future comprehensive theory of development; and (b) the release of new genetic information in the form of mRNA dependent on these circuits and resulting in new patterns of development. These aspects being among the most fundamental for the guidance of development, no more need be said about the importance of dRNA.

Prior to discussion, a renewed word of caution concerning the interpretation of the dRNA-DNA hybrids which will be frequently mentioned is in order. First, present standard procedures to allow detection of these hybrids (saturation experiments) involve mostly reiterated rather than unique DNA sequences (Chapter 3). Second, as a consequence of derivative reiteration in the transcript, these hybrids (competition experiments) involve familial or imprecise matching rather than precisely homologous matching; the competition is therefore between rather than within families of reiterated transcripts. dRNA with this hybridization behavior includes or is mostly composed of labile RNA. On the other hand, using stringent procedures, a fraction of dRNA was shown to hybridize with unique DNA

(Davidson and Hough, 1971; Gelderman et al., 1971). This unique RNA may include mRNA and will be considered as such. Hence the evaluation of dRNA at present is far from being quantitatively absolute or even qualitatively oversignificant, but, with these limitations firmly in mind, a profitable comparison of this RNA between embryonic stages is possible.

Amphibians. During the lampbrush chromosome stage of midoogenesis about 2% of the total RNA is dRNA homologous to 3% of DNA, assuming asymmetrical transcription (Davidson et al., 1966). The sedimentation behavior of this oocyte dRNA cannot be studied in the presence of the excess newly made rRNA, but it is believed to be polydisperse. About one-half of the total DNA transcribed in the oocyte is transcribed into highly redundant RNA copies (Davidson, 1968). Since about half of the *Xenopus* genome is highly redundant, this means that dRNA is transcribed from a maximum of 6% of the redundant genome. As seen in Chapter 3, this figure approaches the proportion of active DNA in mature cell types. Qualitatively, however, the relative frequency of different dRNA sequences is very disparate in the oocyte. In this section all the RNA studied by molecular hybridization represents this type of redundant transcript.

At least half of the dRNA made during the lampbrush chromosome stage is still present after several months in the mature oocyte, to about 1% of the total RNA, as shown by hybridization. Presumably this dRNA is protected by protein as, or as soon after, it is transcribed, as suggested by the abundance of newly made RNP on the chromosome loops. Assuming a proportional representation in the dRNA of all redundant DNA components, the calculated complexity of dRNA still present in the mature oocyte is equivalent to 300 cistrons each of 10^3 base pairs (Davidson and Hough, 1971). This moderate informational content must have been repeatedly copied to account for the 1% of cell RNA that dRNA represents, although, as mentioned, the copy frequency of individual sequences varies widely.

After oogenesis three main bursts of dRNA transcription occur at ovulation (as the result of pituitary hormone stimulation), fertilization, and the midblastula stage. At ovulation an amount of dRNA equivalent to 0.02-0.1% of total RNA is made (Brown and Littna, 1966a), or considerably less than was made during the lampbrush chromosome stage. By the early blastula stage the ratio of dRNA to DNA synthesis is at a minimum per cell, which is possibly related to the lack of G_1 and G_2 periods of the cell cycle during this part of development (Gurdon and Woodland, 1969). At the midblastula stage the rate of transcription begins to rise again and climbs sharply during the gastrula stage; the steady-state amount of dRNA doubles together with the total amount of DNA, as in the case of tRNA but faster than rRNA (although the increase of rRNA may be obscured by the more important increase of dRNA). The rate of dRNA accumulation behaves as an inverse function of the number of cell divisions elapsed between developmental stages (Flickinger et al., 1970b). By the swimming stage, when all RNA species approach their ratio in the mature organism, the content of the embryo in dRNA is about 2%. As will be seen, these temporal changes in the amount of dRNA are accompanied by marked changes in the sequences represented.

The dRNA from the lampbrush chromosome stage deposited in the ovulated *Xenopus* egg is conserved until at least the early blastula stage, but one-third is lost by the late blastula (Crippa et al., 1967). In one amphibian species with synchronous oogenesis the dRNA was found to be retained through gastrulation (Davidson and Hough, 1969). Concerning the dRNA made in *Xenopus* at ovulation and just after fertilization, nothing can be said about its duration for it does not appear to have been adequately tested. Moreover, it is not clear whether its synthesis is nuclear or cytoplasmic (mitochondrial?) (Gurdon and Woodland, 1969). Denis (1968) reports that none of the dRNA sequences detectable in the early gastrula stage or later, were present in the preceding stages from unfertilized egg to late blastula stage. However, his hybridization competition assays were carried out under nonsaturating conditions, and, for this reason, he tested for the change in distribution of the most frequently represented sequences. Davidson et al. (1968), on the other hand, find that newly made mid-to-late blastular sequences (at the onset of increased dRNA transcription), although not represented in any previous sequence even as close as the early blastula stage, have a minor but significant representation in the next early gastrula stage. Davidson's competition assays were carried out under saturation conditions and thus measured a wider spectrum of sequences, which may account for the discrepancy with Denis' data. Thereafter, Denis (1968) observes that most dRNA sequences present in the gastrula are also present in later stages, when some sequences may even be more abundant, although apparently a small proportion are no longer present. Tail-bud sequences all persist in the more differentiated tadpoles. Whole adult dRNA competes well against gastrula dRNA, but adult heart or liver dRNA competes considerably less well, indicating a progressive organ specificity of the sequences.

From the previous data it is clear that, beginning with the egg, several distinct waves of dRNA transcription herald the progress of early differentiation, namely, at the lampbrush chromosome stage, ovulation, fertilization, mid-to-late blastula, and gastrula stages. From the gastrula stage onward a more parsimonious release of information takes over from the previous more stage-dependent series of transcriptions and then gradually converts into the adult pattern of transcription. All this progressive release of information is, of course, controlled at the transcriptional level, but it is unfortunate that there are no data on possible translational influences on this transcription. The cumulative DNA coding for dRNA increases correspondingly during this period: 0.3% in the egg, 2% in the gastrula stage, and 8.5% in the tadpole (Denis, 1968), even though these values may be somewhat overestimated because of the hybridization conditions. Assuming single-strand transcription the last value becomes 17%, which is still less than half of the one-half redundant genome scanned in these experiments. Considering the DNA that is transcriptively active at any developmental stage (Denis, 1968), up to 90% of the redundant genome is not being utilized (Davidson, 1968). The largest fraction of DNA eventually used for dRNA synthesis is used already by the gastrula stage, in agreement with the maximum relative output of dRNA at this stage. These observations indicate an extensive molecular differentiation that precedes the overt morphological differentiation to follow later; not until well past the neurula stage will cells become microscopically more interesting.

As mentioned already, the amount of dRNA is inversely related to the duration of the embryonic cell cycles. A predictable finding was therefore that there will be more dRNA when development is caused to slow down by lower temperature (Flickinger et al., 1970b). It is of interest that embryos raised at lower temperature show an earlier determination to differentiate. The question arises whether this developmental prematurity is related to the accrued dRNA, as seems likely. If so, the implication would be that the differentiative program follows an intrinsic rhythm somewhat independent of the cellularization of the embryo.

The dRNA molecules become gradually stabilized as molecular differentiation continues. The half-life of dRNA during the gastrula stage is only a few hours (Denis, 1968). Contrasted, however, after the tail-bud stage both unstable dRNA (as before) and also now stable dRNA are present, the latter having been made by genes which are no longer active. Recalling that the total amount of dRNA increases at an equal rate with DNA starting with the mid-blastula stage, the curve for DNA capable of hybridizing readily with stable dRNA has the same shape as that of total dRNA, but trails behind it. The greatest difference is again at the gastrula stage where, of 40% of pulse-labeled dRNA, only 10% hybridizes readily (Denis, 1968). Brown and Gurdon (1966) examined further these aspects of dRNA in the enucleolate *Xenopus* mutant. Classifying dRNA as heavy (>20 S; >10^6 daltons) and light (10-20 S), the first comprises 75% of dRNA at all developmental stages. However, although mostly heavy dRNA accumulates until the tail-bud stage, thereafter, some of the heavy dRNA can be quantitatively chased into light dRNA, in agreement with the data by Denis (1968) mentioned earlier. From this time onward light dRNA is stable (as stable as tRNA) and accumulates to 2% of total RNA (or one-fourth of tRNA) by the swimming stage, which is the terminal stage for the mutant. The presumed conversion therefore implies partial stabilization of the dRNA molecule. The efficiency of conversion compared with that of rRNA precursor would be only 10% and takes hours rather than minutes, the remainder of dRNA being rapidly degraded mostly in the nucleus. Brown and Gurdon concluded from their data that, if some heavy dRNA is indeed converted into light dRNA, then this convertible heavy type must be made after the neurula stage. Instead, if light dRNA were to be made entirely *de novo*, this could only be much later than in the neurula stage. This potential stabilization of dRNA may have an important functional significance as discussed later.

Regional differences of dRNA synthesis in the amphibian embryo are revealed by autoradiography. The first nuclei in *Xenopus* to commit themselves to RNA synthesis are those in the presumptive mesoderm (particularly the anterior mesoderm) and endoderm, starting abruptly with the midblastula stage (Bachvarova and Davidson, 1966). Insofar as autoradiography reveals total RNA, these observations combined with the preceding biochemical data suggest a highly localized dRNA synthesis in the embryo. This is significant when considering that the anterior mesoderm plays a major organizational role during the gastrular movements which begin at the blastoporal region. It was first observed in the newt that the early blastoporal nuclei are the most committed to RNA synthesis (Sirlin, 1960). Activation of presumptive ectoderm nuclei then follows by the midgastrula stage (Bachvarova and Davidson, 1966). Postgastrular endoderm accumulates dRNA at a faster rate on a per cell basis than the rest of the embryo (Woodland and Gurdon, 1968). Animal and neural halves of the blastula have dissimilar dRNA sequences, the animal half possessing a greater variety. The axis and belly of the tail-bud stage also differ in this respect, the belly being richer in sequences (Flickinger *et al.*, 1966), which agrees with the faster accumulation of dRNA in the endoderm. These data indicate a marked regional differentiation of the embryo in terms of dRNA, supposedly representing regulatory and/or informational sequences.

Other Animals. The pattern of dRNA synthesis in the sea urchin embryo resembles in broad outline the amphibian. Synthesis takes place during oogenesis and the dRNA is retained through gastrulation. Synthesis has been recorded during late oogenesis up to the unfertilized egg, where it decreases immediately after fertilization (Gross, 1968). The dRNA accumulates very markedly during cleavage (Emerson and Humphreys, 1971). Synthesis proceeds during the blastula stage and again rapidly rises during the gastrula stage. As in the amphibian at later stages, most of the unstable pregastular dRNA breaks down within the nucleus (Kijima and Wilt, 1969). Nemer and Infante (1967) distinguish between nuclear dRNA with 70 and 90 S peaks and a conventional mRNA with a mean 24 S value. By this criterion, synthesis of dRNA and mRNA during the early blastula stage is 4 and 73% of total RNA, respectively, indicating in general a very large proportion of dRNA. By the late pluteus (postgastrula) stage the figures are 22 and 16%, most of the remainder of RNA being rRNA. The data preclude evaluation of a possible conversion between different classes of dRNA as in the amphibian, but evidently, if it occurred, the conversion would be extremely more efficient from the earliest period. An apparent difference with the amphibian, at least in the relatively early stages investigated, is that the modal sedimentation value of dRNA increases between the blastula and pluteus stages, and even more so if calculated from the zygote stage.

Many of the echinoderm sequences made in the maturing egg and blastula stage were found to be similar (Glisin et al., 1966), suggesting a continuous addition of maternal-like sequences. From blastula to gastrula stages, however, about 40% of the sequences accumulated earlier have disappeared and been replaced by new ones. Some sequences made at the prism gastrula stage are rarely found either at earlier or later stages (Whiteley et al., 1966). As in the amphibian, these data mark the early gastrula of the echinoderm as a stage of intense transcriptive activity. Again, the course of dRNA synthesis beyond this point to the adult is little, if at all, charted.

In comparing different species of echinoderms, the homology of dRNA is considerably greater between the embryo than the adult (Whiteley et al., 1970). Most of the dRNA sequences made during oogenesis and present in the unfertilized egg of different species are very nearly identical and homologous to the same DNA sites. These sequences occur in great numbers of copies and represent the maternal-like genes which continue to be active until the blastula stage as described above. A minority of sequences in the unfertilized egg are, however, not homologous between species. After the blastula stage, when qualitatively distinct embryonic dRNA sequences appear, these show a progressive species-specific divergence in accordance with the degree of sequence divergence in the DNA. These data reveal a surprising evolutionary conservatism especially of the dRNA that functions at the very beginning of development. The full meaning of this conservatism must await understanding of the function of this early dRNA, either regulatory or informational.

It is common to interspecies hybrids in both echinoderms and amphibians to conform to a maternal type of cleavage both in morphology and rate before the paternal genome begins to assert itself for expression at gastrulation. If the cross is inviable, development stops at gastrulation. This maternal precession affects not only many morphogenetic traits such as the previous ones but also a variety of enzymes, though by no means all enzymes, involved in general metabolism. As explained already, the precession of the maternal genome may be related to an earlier programming developed in contact with its "own" cytoplasm starting during oogenesis. The phenomenon may thus be looked upon as a true extension of oogenetic activities, which also explains why it concerns mostly morphogenetic aspects of the embryo. The switchover from maternal to paternal genome, indicated by the behavior of interspecies crosses to occur at about the time of gastrulation, is again not an absolute one, since thereafter both genomes combine their expression, presumably in a Mendelian fashion. Denis and Brachet (1969) studied molecularly such an inviable echinoderm hybrid which, through chromatin elimination, contains as much as 2.5 times more maternal than paternal chromatin per embryo. Surprisingly, they found that the dRNA preferentially transcribed from the redundant genome at the blastula and gastrula stages appears to be paternal. This means that, although there is a maternal precession at the phenotypic (therefore informational) level up to these stages, there is also an ongoing paternal transcriptive activity probably anticipating the paternal phenotypic contribution at later stages.

With regard to cytoplasmic, in particular mitochondrial, gene activity during development the information is sparse. In the sea urchin, Hartmann and Comb (1969) describe that, compared to nuclear dRNA, proportionately more 6-12 S dRNA is made from these genes in the gastrula than blastula stage.

In the mouse highly polydisperse dRNA synthesis is detected at the two-cell stage (Woodland and Graham, 1969). However, autoradiographical evidence dealing with extranucleolar nuclear incorporation into RNA puts the beginning of dRNA synthesis before first cleavage (review by Davidson, 1968). Thereafter, the *in vitro* rate of synthesis rises uniformly through the blastocyst period (Ellem and Gwatkin, 1968). No molecular data are available for the early postnidation stages, no doubt partly because of the problem of accessibility in this embryo. During later organogenesis fetal liver dRNA sequences resemble adult liver sequences less, but regenerating liver sequences more, at 14 days gestation than at term (Church and McCarthy, 1967a). This indicates that the activation of the redundant genome during regeneration has a truly differentiative character.

Evidence Concerning Messenger RNA during Development

As has been explained, the principle of differential gene activation is taken as a fundamental guideline for development. This principle essentially means the programmed release of different functional mRNAs in different tissues at the right time, but it should be remembered that differentiation is also secondarily controlled at the level of translation. Moreover, although we have no evidence that it does, translation can conceivably influence the primary transcriptional program in the embryo. A general model suggested earlier in this review is that selective gene activators, produced by regulatory genes working since oogenesis, adopt a certain cytoplasmic localization which may be rearranged at fertilization. Ultimately, these activators give a cue to the genetic program for development carried in the chromosome. A characteristic which distinguishes the embryonic system from most adult systems is, therefore, the long delay before many of the (early made) activators come into action. During subsequent development the genetic program evolves further through reciprocal interactions of the nucleus and cytoplasm, whose own differentiative status is undergoing parallel changes partly dependent on the nucleus. As a result of this program new information is released gradually, obviating any need to store or realize this information before it is actually required. The molecular mechanisms in control of this release of information and its implementation during translation will be discussed in a later section on induction. This model represents a new version of old preformistic ideas on development. Equitably, the final control of the expression of this program past the level of translation validates also old ideas about the epigenesis of development. In its broadest sense, epigenesis means that much developmental cell interaction is not directly encoded in DNA but is a higher-order function of products themselves encoded in DNA.

Specific aspects of genetic programming were discussed extensively in Chapter 3. It was made clear that, first, this program necessarily involves the dynamic organization of the chromosome responding to regulatory genetic circuits. Second, although several types of proteins and RNAs are contemplated (and even must exist) in relation to this regulation, the fine mechanism responsible for conducting a program of transcription is not clarified. Third, dRNA includes labile RNA and mRNA and a possible precursorship of mRNA in labile RNA was discussed at some length. In this review this possibility was considered in connection with dRNA in the amphibian embryo. Fourth, since the mRNA precursorship remains hypothetical, an alternative hypothesis was considered, namely, that labile RNA is essentially a regulator of informational transcription. Importantly, these two hypotheses are not mutually exclusive. Hence, to the extent to which they are compatible, both hypotheses touch on the question of mRNA regulation. Fifth, subsequent to transcription, a progressive stabilization of mRNA appears to follow a characteristic cell pattern, which during development starts presumably with what was designated as unstable dRNA.

Hence, ideally, the present task would be to describe stage by stage the partition between regulatory and informational RNAs represented in embryonic dRNA, thus substantially clarifying a most important aspect of development. The evidence is that all developmental stages from oogenesis onward produce dRNA. In practice, however, because the criteria for regulatory RNAs remain subjective, this ideal partition is impossible. One can therefore only look for positive indications for mRNA in the evidence that is available and then, in a negative sort of way, hope that part, at least, of the dRNA that fails to stabilize as mRNA may have some regulatory function. (For the most part the mRNA so indicated will remain putative, yet in what follows it serves no purpose to repeat this point at every mention of mRNA. In addition, there might be some regulatory dRNA as stable as mRNA.) Lastly, equally fundamental to the concept of differential gene activation but seldom made explicit, is the opposite, namely, differential gene deactivation. Evidence for deactivation is, of course, the continuous replacement of dRNA sequences by new ones appearing throughout development.

Amphibians. A first point to note is that there is no informational amplification in the oocyte genome comparable to that for rDNA, a rule which applies to all animals. All stages of the amphibian oocyte after the lampbrush chromosome stage contain 2% of RNA with template activity as measured in a bacterial cell-free system against appropriate standards (Davidson *et al.*, 1966). As revealed by product composition the RNA would appear to be specifically translated in this system (F. R. Kramer, cited by Davidson, 1968). Nevertheless, given the heterologous nature of the system, 2% is probably a minimum estimate. At the mature stage of the oocyte an amount equal to 1.2% of unique DNA codes for RNA (Davidson and Hough, 1971). The informational complexity of this unique DNA transcript is equivalent to 2×10^4 cistrons each of 10^3 base pairs. This complexity is about 100-fold greater than that in the redundant transcript described earlier. Hence embryogenesis begins with an exceedingly large amount of information in the oocyte, suggesting that some of it may be for regulating the future developmental program (whether or not it is translated) rather than just for producing cell protein. Yet this unique transcript amounts in mass to only one-hundredth of the redundant transcript, or 0.01% of total RNA. Since the redundant transcript is also of considerably lower complexity, this means that the unique oocyte genome is much less frequently utilized. On the average, each unique cistron could be represented by 3×10^4 RNA molecules, still a respectable number, which come into service over an extended embryonic interval as described below. The premise adhered to here that these unique transcripts rather than the redundant transcripts are destined to be translated as mRNAs is discussed in Chapter 7.

A summary of the evidence for amphibian mRNA in relation to the waves of dRNA synthesis described in the previous section occupies the remainder of the section. These waves of dRNA synthesis occur at (a) oogenesis, (b) ovulation and fertilization, (c) mid-to-late blastula stages, and (d) gastrula and later stages. To begin with (a), the stored maternal dRNA just discussed in the preceding paragraph lasts from the lampbrush chromosome stage to the late blastula or even gastrula stage. It is therefore extremely stable, thus independent of immediate transcriptional control but, as will be seen, subject instead to translational control. The mRNA nature of some of this maternal dRNA is suggested by the homology with the unique genome of a small fraction which contains a great variety of sequences (see above). The informational nature is probably also shown later by the actinomycin insensitivity of early embryogenesis at a time when translations essential for embryogenesis are already occurring. Hence, early differentiation is not under direct genomic control by the embryo itself. Certain of the evidence, more direct in the echinoderm, indicates that this maternal mRNA is concerned mainly with "housekeeping" labors such as the production of proteins for the mitotic apparatus. Davidson (1968) argues however that, whereas this could be so in terms of the mass of RNA, it need not be so in terms of the number of sequences represented. As seen above, the estimated number of sequences suggests that some may act directly in regulation. Perhaps some of the sequences potentiate somehow the arrival of later mRNAs as in (c) and (d) below, and in this way regulate morphogenesis directly. Were some of the maternal dRNA to be regulatory as all this leads one to think, it would presumably have to remain attached to the chromatin or reenter the embryonic nuclei after a transient stay in the cytoplasm. It is indeed found in embryonic nuclei, but which of these two alternatives obtained is not known (Davidson and Hough, 1969). Assuming that the unique and redundant transcripts received from the oocyte persist in the same proportion in the embryo, then, from what is discussed below, they may last until not much more than a single molecule of each transcript remains per cell, at about the late blastula stage (Davidson and Hough, 1971). Some of the transcripts may last longer, but this gives an idea of their average permanence.

(b) Concerning dRNAs made during ovulation and fertilization in the amphibian, their persistence has not been adequately tested and their possible cytoplasmic origin should be kept in mind. Their function is not known. Preliminary evidence suggests that cleavage is stopped by actinomycin when the antibiotic was present in the ovary during late oogenesis (Ficg, 1964). This would indicate a lasting effect of the ovulatory dRNA and probably also that its origin is nuclear.

(c) Mid-to-late blastular dRNA persists until the early gastrula stage, whether it is retained or replenished by the same class of genes remaining at work in the embryo. The mRNA nature of this RNA is clear from the total arrest at the late blastula to early gastrula stages caused by natural lethal hybrids or the corresponding transplantation hybrids, lethal genes, or actinomycin. This mRNA is therefore obligatory for gastrulation, hence, for basic morphogenesis. Because of the need for this mRNA, the control of morphogenesis is therefore assuredly no longer maternal, as it was during early differentiation, but is now embryonic. The data on hybrids, in particular, also suggest that the foreign dRNA may by this stage no longer be compatible with the differentiating territories of embryonic cytoplasm.

(d) Although as yet not proved, there is little doubt that the massive dRNA formation from the gastrula stage onward reflects a transcriptional informational program responsible for organogenesis and accompanying cytodifferentiation. Some of these putative mRNAs become very stable during subsequent development. Well known examples (not necessarily all from the amphibian) are in lens fibrinogenesis, feather keratogenesis, and the appearance of contractile muscle proteins (review by Gross, 1968).

Other Animals. In the echinoderm, the titer of template activity remains constant at about 4% of total RNA from the egg to the blastula stage. The

DNA sites corresponding to this RNA, presumably including mRNA, remain similar during this period (Slater and Speigelman, 1968). On the basis of evidence previously cited in connection with dRNA this means that a fraction of the embryonic genome becomes repressed while, on balance, an equivalent fraction becomes activated. One should further recall here that, because of the homology between dRNA of the egg and early embryonic stages (in contrast to the amphibian), this early switch in genome is therefore only partly qualitative. Thus, in the echinoderm, genes of a maternal type remain assiduously active from oogenesis through the blastula stage. Interestingly, this very active type of genes has been conserved during evolutionary diversification of echinoderm species (Whiteley et al., 1970). As these authors point out, these ancestral or primitive genes offer a minor example of the ontogenic recapitulation of phylogeny. Evidently, it would make biological sense if the bulk of these ancestral genes were to be entrusted with the production of basic "housekeeping" proteins during the early embryonic period, as seems to be the case with the amphibian. However, because the proportion of informational genes among these echinoderm genes is not known, this remains conjecture.

In the echinoderm the unfertilized egg induced to cleave, the actinomycin-treated embryo, and the inviable hybrids all develop only up to the late cleavage stage (as in the amphibian), and therefore they last less than the maternal or maternal-like dRNA which is present until gastrulation. Hence, importantly, this presumed maternal mRNA is not sufficient for basic morphogenesis starting at gastrulation. From the data on dRNA the embryonic genome becomes active at the blastula stage in preparation for gastrulation, and then this activity increases in order to direct cytodifferentiation. Presumably, therefore, this blastular preparation includes embryonic mRNA obligatory for morphogenesis. Thus, except for the assiduous maternal-like genes, the overall pattern of mRNA synthesis and its developmental significance in the echinoderm are probably similar to the amphibian.

The pattern of mRNA synthesis that was described is reproduced in the translational activity of a cell-free system (Stavy and Gross, 1969b). In this system the activity increases rapidly during the first cleavage stages (maternal type mRNA) and again at the blastula and early gastrula stages (embryonic mRNA). As expected from the data *in vivo*, actinomycin does not alter the pattern of the system until the gastrula stage, when it causes the activity to fall to zero. On the other hand, the response to poly U is uniform throughout these stages, and the same in normal and actinomycin-treated embryos, which may be taken to indicate an equivalent potential of the translational machinery. (The initiation requirements of poly U differ from those of natural mRNAs, therefore, translation potential rather than translational capacity is measured.) In conclusion, the changes in early echinoderm translation (to be discussed later) depend upon the information that becomes available primarily by the pattern of mRNA synthesis, and, as will be seen, secondarily, upon control at the prepolysomal level of translation.

A cascade of morphogenetically meaningful transcriptions begins in fish right after fertilization and contains the totality of mRNAs for the axial structure of the embryo (Wilde and Crawford, 1968). In keeping with its premature genomic activation, the mammalian embryo probably stores much less maternal mRNA than, say, the amphibian. Actinomycin arrests development of the mouse embryo at the two- to four-cell stage, as does a Y chromosome deficiency (Woodland and Graham, 1969), from which it seems likely that the effect of the antibiotic at this stage is already partly on the synthesis of embryonic mRNA. Transcription of unique DNA in the terminal mouse embryo, presumably representing mRNA, was described in Chapter 3. These data show a much greater frequency and complexity in unique than in redundant RNA sequences. The data also show certain similarities as well as differences with the unique transcript in the amphibian oocyte, but comparison should be deferred because of the almost antipodal contrast in maturity between the two developing systems. It is a frequent occurrence in interspecies crosses in birds that early embryonic enzyme production specified by genes derived from the female precedes that by the male, virtually resulting in a maternal hemizygote. This phenomenon corresponds with the maternal informational precession discussed for the amphibian and echinoderm, and is believed to reflect an earlier maternal genetic program. If not entirely surprising, the precession of maternal mRNA revealed in enzyme production need not be the rule in intraspecies crosses in birds, in which, in fact, it can be the reverse (Ohno et al., 1968b). Nevertheless, the departure still requires a causal explanation.

There is a suggestion in ascidians and molluscs (with a mosaic type of development) that maternal templates continue to operate beyond gastrulation. The evidence is that actinomycin and hybrid crosses do not have an effect until a late embryonic event, metamorphosis (Davidson, 1968). Such persistence of maternal information would surpass that known for any organism. Hence, in various organisms embryonic transcription proper, as manifest by actinomycin sensitivity, occurs either before or after pregastrulation, which is when it occurs in the amphibian and echinoderm. The question is whether this difference indicates a different genetic programming of morphogenetic transcriptions. It need not be so in the mammal where, because of the limited supply of maternal mRNA, the early sensitivity may not correspond with the release of truly morphogenetic information. Circumstantial evidence in the mammal suggests that this information appears during the pregastrula stage as it does in the amphibian and echinoderm, but this is not certain. On the other hand, in the mollusc, presumably with an ample maternal mRNA supply (very late sensitivity), it would seem as if indeed the morphogenetic information is released past the gastrula stage: true differentiation of the mollusc actually starts also after gastrulation.

Translation during Development

In translation one observes the final outcome of developmental information and an important level of control second only to the primary transcriptional level. One ultimate way of looking at differentiation is therefore that it concerns the timing of relevant embryonic translations in order to achieve the protein endowment characteristic of each differentiated cell type, including protein to recognize its own or other cell types. Embryonic translational control must thus be as important as that which functions in the adult discussed in Chapter 5, but its understanding remains of a more general nature. It is in turn dependent upon the differentiative status of the cytoplasm, itself partly dependent on the nucleus. For example, many of the cytoplasmic factors which promote development listed in the introduction are likely to operate at the translational level. Much of what will be said here regarding translation concerns the utilization of the RNA machinery whose timely preparation for this final event was the subject of previous sections; it is in translation that all macromolecular regulation converges during development. As hinted above, it may not be possible here to do justice to this integrative aspect of development, for not enough is known of translation to match what is known about the RNAs.

Sublevels of translational control are found at the (a) prepolysomal, (b) polysomal, and (c) postpolysomal levels. These levels were examined in some detail in Chapter 5 and need be only briefly outlined here. (a) Prepolysomal control implicates foremost those aspects which regulate the half-life of mRNA. (Other aspects of this control, such as the maturation and conveyance of mRNA will not be dealt with for lack of sufficient evidence.) There are two arbitrary facets to this sublevel of control: first, the inherent chemical stability and the stability against nuclease action that presumably are not vastly different for different mRNAs and are both relatively brief; second, the biologically imposed stability that by contrast very markedly and selectively may affect the active half-life of different mRNAs, that is, the period after transcription over which a number of copies of each mRNA are released for service and including the period over which each copy remains in service. The essential phenomenon in the biological stabilization and selection of mRNA, which is very evident during development, is that of inactive but virtually preprogrammed polysomes involving participation of the informosome (Chapter 2). A good example of prepolysomal control (assuming in this case that there is no major postpolysomal control) is the different temporal pattern of transcription-translation events for several enzymes studied during slime mold aggregation (Roth *et al.,* 1968), which ranks as a true developmental process. Many other typical examples will be seen in the ordered translation of maternal mRNAs. As already pointed out, template stabilization and the differentiative process go generally together.

Modulation of embryonic translation by tRNA acting at the prepolysomal level is discussed separately later.

Several aspects of translational control are apparent at (b) the polysomal level, of which one is the loading of mRNA with ribosomes and another the overall rate of translation. These aspects have not, however, been much clarified in the early embryonic period with which we are mainly concerned. Loading of ribosomes may depend on the function of the informosome, the translation initiation complex, or initiation signals in mRNA. Because all these elements intervene in initiation, together they institute the most important regulation of translation. The rate of translation may depend on small effectors of which a classical example taken from more advanced embryonic stages is the action of heme on the translation of preexisting globin mRNA (Paul, 1968). By comparison, the effect of the differential rate of translation of individual codons (elongation) is probably small. It is at this polysomal level of translation that cytoplasmic factors, and, of course, translation factors proper, probably impinge the most. "Inactivating genes" may function at this level of control by producing a translational repressor. These genes are suggested by the superinduction effect of actinomycin on the synthesis of crystallin in the calf lens (Papaconstantinou *et al.*, 1966); their mechanism would resemble that discussed in Chapter 3 for posttranscriptional modulation. These repressors are more likely to predominate toward terminal differentiation (as in the lens), although at least one example is known in early embryonic translation. This is in the echinoderm where actinomycin suppresses the normal decline of dCMP aminohydrolase which begins with fertilization (Scarano *et al.*, 1964). In addition, these repressors may act at the next level of control. Over and above these aspects, an overall correlation is generally found in developing organs between the degree of differentiation and the polysomal content, whose increase indicates a general enhancement of translation. Unfortunately, this is not very informative in the absence of data concerning the nature of the template in the polysomes. To take an example from earlier embryos, in amphibians the gradient of polysome concentration follows on the whole the cephalocaudal and dorsoventral embryonic axes (Brachet, 1968), but nothing can be concluded except that there is a corresponding gradient of translation.

The control of (c) postpolysomal events moves into the area of epigenetics, to combine embryological with biochemical parlance (Gross, 1968). This level is out of bounds in this review concerned with basic development, not the least because, much more so than in the case of adult translation, the importance of these events during basic development can only be guessed at. A vicarious ground for the reader to exercise his own judgment is in the orderly appearance of enzymic activities during later development (reviews by Deuchar, 1965; Watts, 1968), many of which are likely to fall under this type of control. Lastly, it is clear that the size of the amino acid pool does not tightly control translation.

The pool remains fairly constant during development (as does the nucleic acid precursor pool), whether by synthesis renewal, turnover replacement, or release of free amino acid from yolk.

The general pattern of embryonic transcription would seem to disclaim an obligatory coordination between the synthesis of rRNA and mRNA. Until recently it has been considered that new mRNAs as they appear are translated by old ribosomes. However, as presented in Chapter 5, newer evidence has led to a consideration of the alternative possibility that separate classes of mRNAs may be read by separate classes of ribosomes. (A class of mRNAs would comprise functionally or developmentally related mRNAs.) On these grounds, from the general pattern of transcription, it could be argued that during development preexisting maternal information requires maternal ribosomes for expression, and that, to a similar purpose, embryonic ribosomes appear at about the same time as embryonic mRNAs. This hypothetical classification of ribosomes could be achieved by their manufacture being coordinated in time with that of the particular mRNAs they are supposed to serve, or by some intrinsic characteristic of the ribosome which permits it to recognize its mRNA irrespective of when the two were made. In the previous example of maternal and embryonic mRNAs these alternatives cannot be distinguished from each other. However, the embryonic mRNA synthesized prior to the synthesis of ribosomes (for which the echinoderm provides the best evidence) would, on this hypothesis, suggest the second alternative. The possibility of ribosomal classes has obvious implications with respect to the specificity of developmental translation, yet it remains undecided. Its mention here is a reminder that the possibility remains open.

In contrast to the previous sections on RNA species, the focus in what follows is more on synthesis than function of proteins during early development, in other words, more on what happens to the translational apparatus during development. As can be read into this approach, there is little functional characterization of early embryonic proteins (except for spindle proteins) and the discussion must in any case remain more general than it was for RNAs.

To anticipate a few of the main conclusions to follow, active translation goes on from the beginning of embryonic life, so that a sensitive regulation of translation must appear early. This initial translation is of retained maternal mRNAs as is clearly seen in the amphibian embryo which inherits large reserves, but to some extent is probably also seen in the mammalian embryo which receives very little reserves. Embryonic mRNAs appear early in development, however, and earlier in the mammal than in the amphibian where some of the maternal mRNA is long lasting. Concerning other components of the translational apparatus, in reserve-laden embryos translation begins with preformed tRNA and ribosomes, and continues from gastrulation onward with these components progressively made anew. At the other extreme, mammalian embryonic translation depends upon these components being made right from the beginning. As was to be expected, therefore, the embryonic translational machinery depends heavily sooner or later, depending on the type of embryo, on the transcriptional machinery for stable RNAs. A last, but regrettable, point to mention is the lack of analysis of the influence of translation on transcription and particularly informational transcription, for, although experimentally difficult, developmental systems (and the basic cell cycle) are the ones where any such influence must be obligatory. We begin by reviewing the echinoderm embryo, which is the best studied in respect to translation and contains abundant reserves as does the amphibian embryo.

Echinoderms

After a prodigious amount of protein synthesis during oogenesis geared to match the newly made rRNA with ribosomal protein and provide enough yolk protein, etc., synthesis comes to a low ebb with the end of oocyte maturation. The little synthesis which subsists thereafter is for maintenance turnover and is dispensable for further development (Staŷy and Gross, 1969a).

Upon fertilization protein synthesis increases sharply and consistently. Depending on the species, there has been reported a transient decrease of synthesis at metaphase of each of the first cleavage divisions. This drop in synthesis is now known only to mean impermeability to the radioactive tracers used in these studies (Fry and Gross, 1970). Concomitant with the general increase of synthesis, polysomes increase very rapidly at the expense of the preformed ribosome pool, from 5% at fertilization to 50% a little later. In the absence of marked dRNA synthesis this shows that maternal mRNA was present but not available (Humphreys, 1969), representing a clear instance of translational control over the maternal mRNA. This unavailability of maternal mRNA appears to be the specific lesion in the translational machinery of the mature egg which has to be rectified by fertilization. As far as is known, the corrective factor is not brought into the egg by the spermatozoon (see below). Such lesion is further revealed by the translation potential of the ribosomes (measured with poly U) remaining identical through fertilization, which additionally suggests that, if any fault lies with the ribosomes, this can only be in the initiation mechanism (Staŷy and Gross, 1969a). The evidence which follows argues indirectly against faulty ribosomes. Nevertheless, fertilization may increase the translational ability of the ribosome, perhaps by affecting its ability to dissociate and recycle into the polysome (Vittorelli *et al.*, 1969) or making available critical translation factors.

Precisely this unavailability of mRNA in early embryos gave origin to the "informosome" concept, that is, a masking particle consisting of nonribosomal protein and mRNA (Infante and Nemer, 1968) described in full in Chapter 2. A possible visualization of informosome formation (in the amphibian) is the wrapping of dRNA with protein as it is made in the loops of the lampbrush chromosome. Echinoderm mRNA is thought to be unmasked from the informosome by a protease activity elicited within the egg by fertilization. Moreover, since fertilization markedly affects the cortical ooplasm it is possible that the protease activation depends on changes in this plasm. Retracing a few steps, MacKintosh and Bell (1969) proposed that the lesion of the unfertilized egg is not with the masking of mRNA but with the initiation mechanism. On the strength of the evidence, this has to be taken as a minority viewpoint.

From the two-cell stage onward all classes of polysomes based on size become active in translation (Kedes and Gross, 1969). Only a small fraction at most can remain inactive. The mRNA made after fertilization with sequences resembling the maternal is distributed to all classes of polysomes and begins its translation soon after; the duration of translation was not measured in these experiments. This mRNA is 4-50 S with a distinct 9 S peak during cleavage (the largest values suggest some contaminant cytoplasmic polydisperse RNA which is not mRNA, or, alternatively, the presence of immature mRNAs). While these observations emphasize the importance of primary transcriptional control, since translation is not delayed with respect to transcription, it should be clear that they cannot and do not exclude an equally important translational regulation, examples of which will follow.

There has been some initial controversy regarding the size-activity relationships in these early polysomes, however. First, Nemer (1967) considered that large polysomes are very active and use maternal mRNA whereas small polysomes are less active and use embryonic mRNA. As development proceeds the small polysomes become more sizeable and active. It is now believed that the small polysomes make histone using embryonic 9 S mRNA (Nemer and Lindsay, 1969) and, corresponding with the polysome changes, the spectrum of the histones made changes qualitatively until gastrulation (Spiegel *et al.*, 1970; Vorobyev *et al.*, 1969). Remarkably enough, the appearance of some histones depends on actual gastrulation taking place, as they fail to appear in exogastrulae where the process is abortive (this was also found in the amphibian). It follows that the early large polysomes are engaged in synthesis of nonhistone protein, presumably using maternal or maternal-like mRNAs.

There are three main phases of early translation corresponding with the evidence described earlier for the various phases at which mRNA becomes available. (a) **Beginning** with fertilization and continuing through cleavage states, translation of maternal and possibly maternal-like embryonic templates would appear to supply "housekeeping" and histone proteins. These preexisting maternal templates, which guarantee survival to actinomycin until the blastula, appear to last up to the gastrula stage. (Puromycin, of course, arrests the embryo at any stage.) By the same token, the blastular arrest caused by actinomycin reveals that part of the template active at this phase is not maternal-like in character, and, in addition, is obligatory for gastrulation. Hence embryonic translation starts very early, and, not surprisingly, is developmentally significant. (b) By the mesenchyme blastula stage newly made embryonic templates begin translation, in all likelihood for enzymes and other products characteristic of late morphogenesis. From this time on the embryo becomes sensitive to actinomycin. In species with early rRNA transcription, ribosomal proteins must be formed also at this time, because maternal ribosomes are consumed and replaced by half with embryonic ribosomes by the next phase of translation. (c) At the gastrula stage a new generation of mRNAs, which are presumably required for early cytodifferentiation, enter translation. After this phase, the mRNA population changes progressively to meet the translational requirements for subsequent differentiation. Because of its length and the variety of organogenesis involved, this last phase includes most embryonic proteins to be made. Toward the end of embryonic life, however, many of the proteins resemble adult proteins.

Quantitatively, these three phases (a-c) overlap with one another and the overall rate of translation climbs to a maximum during the gastrula stage. Qualitatively, the temporal pattern of translation during these phases reproduces that of transcription but with a certain lag (from the actinomycin results), which is clearly indicative of a measure of translational regulation. The present conclusions apply to the average protein population and more definite conclusions must await characterization of at least some of the proteins. Doubtless some of these proteins perform intriguing morphogenetic tasks well worth our knowing, but until such time arrives translational events around the period of fertilization remain the most interesting.

Thus many species of embryonic protein, although not yet characterized, are made starting from fertilization (Gross, 1968). The chromatographical profile of pulse-labeled soluble protein is radically different between cleavage and gastrula stages. The newcomer protein, however, hardly perturbs the optical density profile of preexisting bulk protein. Even in the presence of actinomycin, the pattern of translation of maternal mRNA changes during the blastula stage, again indicative of a programmed regulation of this inherited mRNA at the translational level. By the gastrula stages, on the other hand, actinomycin prevents the coincidence of the radioactive and optical density profiles obtained by electrophoresis, which is as expected from more embryonic templates becoming active. These data reveal real changes in both the rate and quality of protein synthesis as could only be anticipated, but they are too crude to justify further elaboration.

Amphibians

Induction of maturation in the amphibian oocyte requires stimulation by progesterone. This hormonal action results in protein synthesis which is obligatory for maturation but is kept under check by the conditions in the uterine environment (Smith and Ecker, 1970). This protein synthesis is also observable when oocyte maturation is induced *in vitro*, and subsists even after enucleation (Ecker and Smith, 1971), the latter indicating a translational regulation. As mentioned, in addition these authors described a factor which is released at the time of maturation from the germinal vesicle and is necessary later for cleavage, yet not for protein synthesis before fertilization. Perhaps in relation to this factor, the protein made during maturation assumes a nuclear localization during early embryonic development. There is a parallel here to the behavior of some dRNA made in the oocyte, which perhaps suggests a joint regulatory involvement during early development. Lastly, at odds with the echinoderm, protein synthesis continues to rise in the ovulated amphibian egg irrespective of fertilization, which is believed to be a trivial consequence of the change in environment.

As seen in the echinoderm, treatment of the amphibian embryo with actinomycin reveals, first, that protein is initially made on maternal mRNA, but, second, that blastular mRNAs must be translated eventually in order for gastrulation to ensue. It would be interesting to know if this translation has any direct influence on early gastrular informational transcription, a type of influence about which nothing is known. Stored maternal mRNA then becomes gradually and selectively available to the polysome as the result of a programmed prepolysomal regulation (as the echinoderm), and most of it is translated prior to degradation (Crippa and Gross, 1969). Embryonic mRNA appears early in development and is translated concurrently with the maternal mRNA. By the late blastula stage both types of mRNA contribute equally to the polysome, therefore, probably to the protein population. By this stage embryonic mRNA represents 4% of the genome or little more than maternal mRNA during oogenesis. Thereafter, total translation increases sharply with gastrulation and depends progressively more on the embryonic supply of mRNA. This increase results in the appearance of the greatest number of proteins so far, which are to be used mainly for organogenesis. Thus, as was to be expected from the similar pattern of release of information, in gross terms the developmental pattern of amphibian translation resembles that of the echinoderm.

Whatever the meaning of embryonic nuclear protein synthesis, the uptake of methionine-^{35}S measured by autoradiography in the newt is greater in the nucleus than cytoplasm. Beginning with the gastrula stage, the pattern of nuclear activity follows the basic dorsoaxial gradient of embryonic organization and the incorporation is now most prominent in the blastopore region (Sirlin, 1955); the preponderance of nuclear RNA synthesis also in this region was mentioned earlier. In later stages, as cytodifferentiation gets under way, cytoplasmic protein synthesis becomes progressively greater than the nuclear synthesis (Waddington and Sirlin, 1959), approaching the pattern in mature cells.

Other Animals

In the mealworm *Tenebrio* the late pupal stage is marked by the appearance of a tyrosine-rich adult cuticle protein. The mRNA responsible for this protein was produced early in pupal life but remained masked until the late stage. Correspondingly, in a cell-free system cuticle translation is possible using young pupal ribosomes only if late pupal tRNA and ribosome-bound protein are supplied (Ilan, 1968). Ecdyson (moulting hormone) applied to the early pupa prevents these promoting components from appearing (Ilan et al., 1970). Thus, although cuticle protein synthesis is obviously under translational control, it is not clear which active component is responsible for this control, i.e., tRNA, synthetases, initiation or propagation factors, or even less specific components. Translational control is also implicated in *Drosophila* where the cell supernate from a mutant with abnormal tergites will promote more amino acid acylation and transfer to protein than the wild-type supernate (Rose and Hillman, 1969). In contrast to embryos derived from large yolky eggs, translation of embryonic mammalian mRNA starts almost from fertilization (Woodland and Graham, 1969).

Bacterial Spores and Plant Seeds

Some justification for including these objects under the topic of developmental translation is that certain embryos, such as the brine shrimp during gastrulation (Clegg and Golub, 1969), enter a period as a dormant cyst that bears a considerable analogy. In its own right, spore germination is the bacterial equivalent of development (review by Keynan, 1969).

In *Bacillus cereus* a large portion of the genome, perhaps a quarter, codes for stable mRNAs specific for sporulation (Sterlini and Mandelstam, 1969) and germination (review by Davidson, 1968). In sporulating cells of *B. subtilis* the RNA polymerase shows a changed template specificity (Losick et al., 1970). An altered initiation factor, but more likely a modified polymerase moiety, was suggested in this system in order to explain transcription of sporulation but not vegetative genes, the latter restarting transcription only after germination. In the *Azotobacter* spore the two rRNA species and DNA, but not tRNA, show an altered chromatographical elution behavior suggestive of a modified H bonding pattern which may reflect the special functional environment within the spore. The higher order structure of the molecules converts back to the vegetative form after germination (Olson and Wyss, 1969). Transcription of stable RNAs begins soon after spore germination, possibly concurrently with translation which is on preformed polysomes (DuPraw, 1968). Both of these processes precede DNA replication starting with the renewal of cell growth (Armstrong and Sueoka, 1968).

Only wheat embryo mRNA that was produced during the period of seed maturation is translated during the first day of development. Embryonic mRNA of sequences similar to the early mRNA then appears during the second day; about one-half each of the same type of mRNA and new types of mRNA are transcribed during the third day (Chen et al., 1968). Prior to these transcriptions (day 1) the embryo is, of course, insensitive to actinomycin. During transcription (days 2 and 3) the fraction of genome involved remains constant, indicating a balanced qualitative change in transcription very similar to that found in the early echinoderm embryo. Presumably, these progressive transcriptional changes are reflected sometime later in the proteins that are produced. Also the dormancy lesion in the plant embryo basically resembles that in the unfertilized echinoderm egg, in that the translational machinery is arrested rather than any of its components being lacking. Ungerminated plant polysomes are in effect activated by ATP (Marcus and Feely, 1966), suggesting a translational control at this level. As mentioned at the beginning, some of these findings hopefully apply to the animal cyst.

Transfer RNA and Related Enzymes in the Modulation of Developmental Translation

The rationale for the participation of tRNA and related enzymes in translational regulation was given in Chapter 5, together with some technical considerations on the nature of the evidence which is involved. Because it serves to clarify the purpose of this section, the reader is urged to reconsult this chapter. The previous discussion centered on examples taken from mature cells and is extended in this section with examples from embryonic cells. The subheadings for discussion remain as before, namely, tRNAs, synthetases, and methylases.

Transfer RNA

Of 18 tRNA species examined by RPC chromatography in the brine shrimp embryo from the encysted gastrula to the nauplius larval stage, only nine species show quantitative changes (Bayshaw et al., 1970). In the unfertilized echinoderm egg of two main activity peaks on MAK chromatography for Lys-tRNA, one species (peak II) which comprises 62% of Lys-tRNA is found exclusively in the cell supernate and another species (peak I) predominates in uncharacterized cytoplasmic particles. At the two-cell stage, species II moves into the particles, where species I still remains, to reappear in the supernate by the pluteus stage (Yang and Comb, 1968). This displacement of tRNA from cytosol to particle might reflect a more active participation in translation, if indeed the particles turn out to include cytoplasmic ribosomes. (Authentic displacement would exclude mitochondrial tRNA.) In the same embryo, Zeikus et al. (1969) examined eight amino acids by MAK chromatography and reported quantitative and qualitative shifts of activity in Leu-, Ser- and Lys-tRNAs between the egg and blastula stage; a new Lys-tRNA appears in the mesenchyme blastula. Possible variations in tRNA (and related enzymes) during insect development were mentioned already; in the mealworm ecdyson may specify a tRNA (and synthetase) which is necessary for pupal cuticle synthesis (Ilan et al., 1970). This system might serve to prove a meaningful translational regulation by tRNA.

Several tissues of the bullfrog tadpole show using chromatography on methylated albumin–silicic acid (MASA) two main species of Leu-tRNA, of which one decreases and the other increases during metamorphosis (Tonoue et al., 1969). In *Xenopus* the codon response of 12 tRNA species is the same from the neurula stage to adult liver (Marshall and Nirenberg, 1969), suggesting that what change may occur in the molecules must be elsewhere than in the anticodon. In the early chicken embryo there is evidence for the need of some new specific tRNA for turning on synthesis of hemoglobin at the six to seven somite stage (Wainwright and Wainwright, 1967). Of several tRNAs examined in red cells of this embryo, Met-tRNA (and perhaps Leu-tRNA) shows changes in activity profiles on MAK and freon RPC columns from the 4-day embryo to the adult (Lee and Ingram, 1967). In embryos older than the 10-day mouse studied by gel electrophoresis, Leu-tRNA shows quantitative temporal changes paralleling those in the corresponding synthetase described later (Rennert, 1969). Of several activities studied in the mouse placenta, an actively differentiating organ, only 2 minor His-tRNA species (out of a total 5) were found to change at 13-day gestation (De León et al., 1972). Rat liver regeneration is accompanied by an appreciable increase in tRNA, but the amino acid acceptance of 14 tRNA species and the MAK elution profile of seven species examined remains unchanged (M. K. Agarwal et al., 1970).

Light induces several new tRNA species of *Euglena* in relation to chloroplastal development (Barnett et al., 1969). Of several tRNAs examined in the wheat embryo, Ser-, Lys-, and Pro-tRNAs differ quantitatively in MAK profile from the ungerminated embryo to the 48-hour seedling (Vold and Sypherd, 1968); a reservation concerning this work is, however, that mitochondrial or chloroplastal tRNA species could have contaminated the samples studied.

This listing of changes in tRNA activity in developing cells is briefer than the listing in Chapter 5 dealing with changes in adult cells and, in particular, briefer than the listing of neoplastic cell changes given there. Since from the many changes in tRNA in neoplastic cells one would have expected them also to occur during development, it is difficult to tell whether the current deficit (supposing it reflects interest) is significant or is due to sampling. According to a recent appraisal by Zamecnik (1971), a greater number of isoaccepting tRNAs may after all be functional in the embryo than in the adult, which, if true, might mean a codon restriction in the latter. It remains to be seen how many of the changes observed in adult cell tRNA can be traced back to early development, for until this is done it is impossible to evaluate the role that tRNA plays in the specific regulation of embryonic translation.

Synthetases

In the echinoderm embryo the total synthetase activity measured against yeast tRNA shows two distinct MAK peaks. The main activity peak doubles in size from the unfertilized egg to the early blastula stage and decreases to less than the initial size by the pluteus stage (Ceccarini et al., 1967). These changes in activity indicate that synthetases are already active before fertilization and are further activated after cleavage. When considering that embryonic tRNA is already functional before fertilization, and that it is not limiting during subsequent development, this phenomenon has clear implications for developmental translation. Moreover, synthetase activities for single amino acids show changes during the period studied (Ceccarini and Maggio, 1969). During amphibian liver metamorphosis the activation of several amino acids does not vary as measured by ATP-PP$_i$ exchange (Degroot and Cohen, 1962), although individual synonym synthetase activities would not be resolved by this test. One of two Leu synthetases is lost during murine development (Rennert, 1969).

Thirteen species of synthetase examined in fungal spores (against yeast tRNA) are low or absent prior to germination, and increase rapidly during germination as they become less firmly bound intracellularly (Van Etten and Brambl, 1968); this last point is interesting as it may indicate something about the functional topography of translational enzymes. The cotyledon of the soy bean embryo contains six Leu-tRNAs detectable with homologous synthetase when analyzed by freon RPC column. The hypocotyl shows only four of these Leu-tRNAs, but traces of the other two are recognized by cotyledon synthetase. Hypocotyl enzyme does not charge the synonym cotyledon tRNAs (Anderson and Cherry, 1969). Thus hypocotyl is poor in the minor Leu-tRNA species and also lacks the synthetase to recognize them. In addition, administration of cytokinin increases the level of the minor species in the hypocotyl without affecting the enzyme; the possible precursorship of this plant growth factor for tRNA modification was explained in Chapter 5. In conclusion, as with tRNA the developmental significance of synthetase variation awaits such time when thorough comparison with adult cell synthetases has been worked out.

Methylases

Although only summarily reviewed here, the methylases are the most studied embryonic enzymes related to tRNA. In slime molds the rate and extent of methylation of tRNA decrease both for total and base-specific activities from early plasmodial aggregation to fructification, suggesting the presence or activation of specific enzyme inhibitors (Pillinger and Borek, 1969). In the mealworm the two parameters above change simultaneously from day to day of pupal life and decrease toward the end (Baliga et al., 1965). The extent of methylation of tRNA in several mammalian organs is lower in the neonatal animal than in the fetus, and still lower in the adult (Simon et al., 1967; Hancock, 1967). Thus, methylases from the early mouse embryo do not act on their homologous tRNA but act on late embryonic tRNA and even more so on adult liver tRNA, as revealed by gel electrophoretic mobility, which is influenced by the extent of methylation (Rennert, 1970). This pattern reveals more tRNA sites able to be methylated during the progress of life, which agrees with the previous pattern of methylase activity. Bovine lens methylase activity disappears during terminal fiber differentiation (Kerr and Dische, 1968). No activity changes were observed during liver regeneration (Rodeh et al., 1967).

This section cannot disclose the great contrast which is observed between embryonic and adult methylase activities (Borek, 1969), the latter described in Chapter 5. Whereas the activity is essentially constant in mature cells, in differentiating cells the activity suffers both quantitative and qualitative variations. These variations reflect not merely the amount of methylases but also complex changes of specificity with regard to their substrates. In most developmental systems studied an early increase in methylase activity is followed by a decrease during later differentiation, part of which is believed to be due to the intrinsic aspects just mentioned and the remainder to the buildup of methylase inhibitors. Hence methylase activity is a marker for early cell differentiation, with the implication of functional changes in tRNA during this process. In this respect, the high methylase activity of the neoplastic cell is taken to reflect derepression of the embryonic genome silenced with the end of normal differentiation.

The general evaluation of these sections on tRNA and allied enzymes as participants in the modulation of embryonic translation is no different from that in Chapter 5 with respect to adult translation. In short, the evidence for participation of these molecules in modulation is, in part, highly promising, but for the most part remains circumstantial rather than causally meaningful. It is possible that the rather limited number of functional changes observed reflects the small number of embryonic systems studied (as well as technical difficulties), but the missing causal evidence has also to do with the lack of rigorous experimental design to test it. It would appear however that, as in adult cells, any number of tRNA species newly made to intervene in embryonic modulation cannot be great; in the main they are minor species of the population. By contrast, modification of embryonic tRNA species seems to be more pronounced as revealed in particular by the behavior of methylases. In general, any possible regulatory function by tRNA is likely to become more important during early cytodifferentiation than it was during the previous phase of basic embryonic differentiation. The postgastrular phases of development may thus be the ground of choice for further exploration of this function. This exploration is worth pursuing because, were selective tRNA synthesis for a modulatory purpose to be established, this would represent the first clear demonstration of a developmental programming of the facultative genome for any type of regulation.

Embryonic Induction

If the molecular biology of the eukaryote may be epitomized in the process of development as this is the most comprehensive deployment of genomic potentialities, in turn development may one day be summarized in the process of induction. Induction is traceable back to the earliest embryonic program and thus involves the fundamentals of development. It plays a central role in development because all organogenesis is completely dependent upon it. Although no essential difference need apply, we shall limit our consideration of induction to the vertebrates, in which much of the process takes place between interacting cell populations, compared with the invertebrates in which the process occurs largely within cytoplasmic territories of the egg. Induction can be defined as a developmentally significant interaction between tissues (never individual cells) of dissimilar origin leading to a new pattern of differentiation in one or both of the interacting tissues. Inductions rank from the primary or body axial organizer type down to the lower-rank inductions, and tend to be more unidirectional and restrictive the lower their rank. Examples of the lower rank are kidney and pancreas inductions which are considered to be tertiary inductions generally involving epitheliomesenchymal interaction. The inductive sequence is however a hierarchical continuum in the sense that, if the primary induction fails, all others fail too—because of this any ranking is to a certain extent arbitrary. Induction clearly depends on the space-time organization of the embryo in order to bring the specifically reacting tissues together at the proper time. As mentioned, these tissues (except for those involved in primary induction) are frequently themselves the outcome of the preceding inductive sequence.

The overall process of differentiation can be resolved into four phases. First, there is (a) the acquisition of competence by undifferentiated tissue to differentiate along a certain pathway. This first phase may be called protodifferentiation. Noteworthy, this phase may sometimes advance to a stage of virtual self-differentiation (Wessells, 1968) sufficiently to obscure the following phase. (b) The competent tissue is then taken by an inductive event to a point of no return, in other words, to a determination to follow that differentiative pathway, in what finally amounts to an inheritable covert differentiation. This means that there is a stability of the determined state, which indeed may be the main reason for the near irreversibility of many differentiative processes. Occasionally, however, determination alters course by transdetermination, that is, changes in the determined state which can be revealed experimentally (Hadorn, 1966). Covert differentiation normally marks the beginning of self-differentiation. The phases until now are brief compared with those that follow. (c) Next comes the phase of activation leading to overt differentiation (Waddington, 1956), and finally (d) modulation of differentiated cell function during the mature functional phase (Rutter *et al.,* 1968), since tissue function is not stereotyped forever. Point (d) may be interpreted to mean that developmental processes do not cease in the embryo but continue right into the adult; in other words, an adult tissue represents the final equilibration of tendencies operating since the earliest formation of that tissue. The informational regulation of mature cells (discussed in Chapter 3) thus becomes the terminal stage of regulation. To correlate the previous phases with the types of proteins that are found in each: phase (a) is characterized by proteins obligatory for the cell and derived from the original embryonic program, with perhaps a few semispecific proteins as well; phases (b) and (c) involve the forming and perfecting of tissue-specific proteins, e.g., insulin and amylase. From the viewpoint of the cell, but certainly not that of the organism, the latter proteins are entirely expendable and can be considered a luxury.

What follows in this and the succeeding paragraphs is a deliberate but admittedly crude attempt to bring these basic inductive events (as a subsystem of organogenesis) conceptually in line with the molecular aspects and principles discussed previously for development in general. This requires extricating from one another a series of events, the failure to do which has clouded the issue in the past, while keeping in mind the overall directionality of the series, which can be lost amidst many nonspecific (experimental) reactions. In addition, we shall focus on the induced rather than inducing cells, although it was pointed out that they form a unitary system and effects of induced on inducing cells occur as well. The initial period during which a tissue acquires the competence to differentiate can be looked upon as one during which the chromosomal apparatus undergoes a self-programming for differentiation in reciprocal accord with the internal status of cytoplasm, itself changing according to the embryonic region where it happens to be. As mentioned, the program potential can be quite far-reaching. It can be brought experimentally to express itself by such uninformative an agent as NaCl (Barth and Barth, 1967). The less the specificity of the artificial provoking agent, though not quite as low as kitchen salt, the more apt is the term "evocator." In the embryo, however, "inducer" substances (discussed below) are produced by neighboring cells and facilitate protodifferentiation specifically, that is, help establish a certain pattern of differentiation within the context of the available program. This means that the differentiative program has a certain range within which subprograms become organized hierarchically and coextensively rather than simultaneously in time. Basically, this also means that the developmental program and induction are codependent. What is important in this inductive phase is that the genes which the tissue is to use begin to be singled out, while concurrently all other genes become more permanently silenced; the genetic potentiality is progressively restricted in the direction of the stable genetic pattern characteristic of the mature tissue. In the next postinductive phase of determination, during which genes are not as yet expressed, the predifferentiative status quo attained in the previous phase is maintained. This phase is unlikely to be static, however, and probably involves some sort of program consolidation dependent upon cell division. Until this point differentiation is foremost chemical in nature. Finally, genes are expressed in a typical pattern of cytodifferentiation and the appropriate morphological cell parameters become manifest.

What do these inductive phases in the reacting tissue represent in somewhat more concrete molecular terms, particularly with regard to gradual derepression of the tissue-specific genes? (We need not concern ourselves with the concurrent repression of other genes.) It should be remembered that any large-scale developmental change must involve a vast number of genes. It appears that, by the time induction occurs, a relevant transcription is already derepressed with the result that there is now a regulatory mechanism available to effect changes in the current embryonic program. This phase of competence requires a compatible status of the cytoplasm on which it depends. The intense regulatory (transcriptional) activity eventually leads to a new program, which precedes the tissue-specific informational transcription that the inducing agent is to elicit from this program by the end of this phase. The general point was made elsewhere (Sirlin, 1968) that different phases of induction may have a counterpart in the different levels of genomic derepression discussed in Chapter 3. A similar point is made by Davidson and Britten (1971) in an adaptation of their model illustrated in Fig. 13 (p. 155) to explain induction in mosaic and regulative eggs. Moreover, these authors propose that proteins binding to the sensor gene are the receivers of the inductive signal, whereupon these proteins adjust the regulation of the program to produce the specific informational response of this phase. These sensor gene proteins could then be one of the prior cytoplasmic products permissive of the inductive event.

As determination sets in, the genome becomes more and more committed to the program, that is, the now functional genes are further derepressed and new constellations of both regulatory and informational genes necessary for the program are engaged. Davidson and Britten (1971) considered the locking of the determined state (after withdrawal of inducer) as being achieved by means of self-sustaining positive feedbacks on the genome. Transdetermination, or the shifting of the program to a different program, may be seen as the rare breaking of this lock. Concurrent with the changes in transcription, the status of cytoplasm is adjusted and one immediate consequence is that the translational machinery is expanded to implement the program. For example, polysome rosettes appear early during induction of pancreatic acinar cells by mesenchyme (Grobstein, 1967). In keeping with the expanding machinery, the production of stable RNAs is increased. Visible changes are now beginning to show in the cytoplasm. The inductive process, until here brief, enters its longest phase. At this time a marked stimulation of dRNA probably represents more regulatory and informational transcripts. This can be seen during *in vitro* kidney tubule formation in the metanephrogenic mesenchyme (Saxén *et al.*, 1968) and during *in vivo* Wolffian lens regeneration (Reese *et al.*, 1969). Because of the stability of determination one would expect that stabilization of mRNA, that is, masking, becomes preponderant during this phase.

To this point in determination the inductive process is actinomycin sensitive. However, actinomycin sensitivity can extend well beyond this point as shown in the developing insect eye (Waddington and Robertson, 1969), indicating, as these authors cautiously point out, that more than stabilization of mRNA is at work. What probably still goes on as covert differentiation, depending on the embryonic system, is a good deal of reprogramming involving the activity of the regulatory genome to specify the final informational transcription for the system. Activation, lastly, must be a phase where the block on translation is lifted, at least for those proteins (probably the semispecific) whose pretranslational regulation is completed. This is probably accompanied by the production of late mRNAs after which a massive translation of specific proteins follows. The functional control of these proteins such as by hormone and adaptive modulation could have started during their translation, but will last the life of the cell.

Regardless of interpretation, four points require mentioning. First, acquisition of competence involves a sequence of covert genomic and cytoplasmic events that are permissive with respect to the inducer (Holtzer, 1968), to a point when this becomes essentially supportive. In this sense, induction, like developmental derepression at large, is essentially noninstructional. Second, cell division is necessary and even critical for determination to set in, the possible significance of which in relation to programming was considered earlier. After determination, however, cell division abates and sometimes even ceases altogether. Third, a certain critical mass—a recurrent biological phenomenon —of both inducer and induced tissue is obligatory, stressing the point that tissues rather than cells are involved. Fourth, and fairly obvious, the coming together of interacting tissues depends upon a range of cell displacements from discrete migrations to impressively sweeping morphogenetic movements.

What is the nature of inducers? The tortuous history of this search which began in the 1930's and which continues to this day is recounted by Needham

(1968). To start with, induction by the primary organizer may differ in character from lower rank inductions. The primary induction system probably uniquely requires direct cell-to-cell contact (tight junctions) and no responsible substance has been identified so far. Hence a justified speculation is that there may be no primary inducer, but instead, specific cell interactions at the level of membranes somehow help establish or stabilize an undecided developmental program in the competent cell. Based on this view, primary induction is not a true induction but essentially the self-assertion of what must be a basic developmental program. Lower rank inducers, or factors as they are also called, are known and, because there is no cell-to-cell contact in these systems, the inducers must traverse the extracellular space (at least the basal lamina) to reach their target cells. It is not impossible, however, that certain lower rank "inducers" arise inside the reacting cells (which would create an antinomy, for, as implied, intracellular substances are generally not regarded as inducers.) A few examples of these inducers follow. Analysis of the amphibian embryonic axis (Tiedemann, 1968) has revealed that a neural factor is present in a crude RNP fraction. A mesodermal factor is a protein of about 28,000 daltons, which is normally inhibited by another high molecular weight factor(s). The structural mRNA for mesodermal factor will have appeared in the embryo much earlier than the axis itself (Wilde and Crawford, 1968). However, at least two subfactors have been postulated to explain the multiple axial structures, and an additional postulate is that they interact along dorsoventral and anterior-posterior gradients (Yamada, 1967). It would appear, therefore, that the lower the rank of an inductive process the less likely it is that one is dealing with single substances. To give another example, several factors are known to operate in the induction of cartilage in somites, of which the purest identified is a glycopeptide of 3000 daltons (Zilliken, 1967). It has been contended that RNA plays an inductive role in systems such as kidney (Niu, 1959), but this would more likely be as a moeity of the effective substance. To summarize the answer to the initial question asked, in the case of lower rank inducers the evidence is that most are protein, sometimes conjugated to nonprotein moieties. The general mechanism of action of these substances is not known. They could be functioning in the manner of typical hormone (e.g., on sensor genes) or as typical factors encountered in macromolecular syntheses, yet unique regulatory functions cannot be excluded.

Some comments bearing on the difficulty of separating these inducers from hormones on an operational basis are as follows: (a) The range of action is typically shorter for inducers; and (b) the effect of inducers (especially those with a higher rank) is permanent compared with the transient effect of hormones (except for developmental hormones) after either is withdrawn. Both act finally on all RNA species, although clearly the action on some species must be indirect. Both are assumed to activate selectively some kind of infor-

Embryonic Development

mational transcription. Both may also act on translation, but the balance between this and the transcriptional effect is difficult to ascertain even for the best studied hormones. Moreover, these translational effects are no more than a possibility in the case of inducers, perhaps greater in those with the lowest rank. Both have morphogenetic effects and sometimes these are equally pronounced. This functional analogy of inducer to hormone serves to emphasize the capacity for self-organization of the genome acting as their common target; it reacts permissively to both.

An important difference appears to exist between the primary and lower rank inductions, since thus far only the latter have molecular inducers which have been identified. Cell membrane interactions were tentatively suggested, by default as it were, to mediate the primary induction, which is a system apparently requiring intimate cell contact. With reference to the categories of phenomena discussed under General Aspects of Development, the indication might perhaps be that the primary induction depends predominantly on supracellular or supramolecular ("cellular") signals, whereas the lower rank inductions, which do not require cell contact, rely mostly on "molecular" or macromolecular signals. We may now consider the relationship between these signals and their effects on the reacting cell; we shall deal exclusively with effects on transcription and assume, for simplicity, that they are direct. Since some emphasis was placed on the self-programming of the inductive response, it looks as if the two types of signal could intervene at different stages of the program responsible for this response. Based on the categorization above, the cellular signals appear to commit developmentally a very basic or immature program (primary induction). The molecular signals seem to act on more advanced or mature programs (lower rank inductions). What else can we say of the signals in relation to these stages of program?—very little that is definitive in the case of cellular signals, since, on the one hand, we can only guess at what a change of basic program represents in terms of the chromosome, and, on the other hand, we do not know the nature of the signal or its means of internalizing to the nucleus. Conceivably, however, only derepression of a "master" regulatory gene by a simple signal might be involved, the dramatic inductive response merely reflecting what a program has to specify at a fundamental stage. Molecular signals can instead be regarded as influencing the more terminal regulation of the program. These inducers would then resemble the derepressive signals, in particular hormones, encountered in the process of selective transcription (Chapter 3); we know the nature of these signals and something about how they internalize, but much less about their doings in the chromosome. Perhaps the one striking fact so far concerning these inducers is that most are protien. At present, the impression is that they need not be more instructional or positive in their transcriptional action than cellular signals, but because of their nature, are probably coded for more directly in the DNA of the donor cell. If these long arguments are correct, what knowledge we have seems to lead inescapably to the character of the inductive response depending on the developmental program.

Based on his *in vitro* work on mesenchymoepithelial associations, Grobstein (1967) expressed the view that transcellular macromolecular complexing at the level of the epithelial basal lamina is an important factor in the behavior of inductive interfaces. His concept is that this complexing represents a two-way interaction between inducing and induced tissue which ends by creating a microenvironment conducive to synthesis of specific proteins for the interface. Speaking for the current view, Wessels (1968) thinks that these *in vitro* observations may not truly reflect the (tertiary) inductive phenomenon defined earlier, which would have already occurred, but rather tissue interactions as lateral phenomena. Whatever their nature, these phenomena include a major supramolecular component.

Contrary to an earlier emphasis on outside influences on cells undergoing induction, a considerable degree of self-programming is now believed to be at the heart of induction in respect to both inducing and induced cells. Recognition of this self-organizing character behind many central problems of development is quite important, for it permits one to conceptualize the molecular basis of their regulation in ways which are acceptable to most cell biologists. At present, in the general case, induction is regarded as a stimulative or promoting influence on the induced cells—at times a very strong one—rather than a critical transfer of genetic information. Thus to summarize induction, it ought to explain in the end which developmental circumstances tell a group of cells when to evolve a program of their own in response to their cellular and noncellular surroundings.

A Cell's Life: An Overview

Having explored to this point many aspects of the biology of cells, let us briefly trace (again at the biological level) the whole life of an hypothetical vertebrate cell beginning with the ancestral lineage.

A group of cells, first part of a certain germinal layer in the embryo, is reshuffled or actively migrates to a given, if not necessarily final, location. By then the genome of the cells is primed to conduct a given differentiative pathway, which in the normal course of embryonic events will be gradually followed. In doing so, these cells will be responding in concert to physicochemical and cellular influences with which they have or will come in contact, first transiently and later more permanently. Many of these influences affecting the cells in an important way will be elaborated upon or, as the case may be, metabolized by their own cytoplasm before reaching the nucleus. Internal influences arise from their cytoplasm by now differentiatively attuned with the programmed genome. From now on, this nucleocytoplasmic interaction becomes responsible for the differentiative future of the cells.

We now focus on one of the cells in the group that until now have been multiplying in numbers. This cell leaves the proliferative pool and begins to differentiate overtly, say, together with the rest in the group as part of a fixed organ. That is, this cell belongs to a community with identity of purpose, that influences each other's differentiation, and whose mass and contribution as a whole to the organism is what is regulated within a certain equilibrium range. Our cell begins to acquire the tissular connections, the extracellular environment, the molecular population, and, lastly, the cytoarchitecture appropriate to its function in that organ. These aspects involve, first, recognition and reciprocal movements between cells which depend on the activity of external membranes, and, second, particularly in secreting cells, the formation of internal membrane systems. The cell is becoming a recognizable type of cell.

The final position in the organ will influence the cell's phenotype, which essentially means which and how much protein the cell will produce. The cytoplasm has thus added to its regulatory capacity a functional capacity in the sense of being able to produce cell-specific product. Not all the product will, however, be specialized and indeed most will be obligatory and common to cells in most organs. The obligatory product will, however, be regulated within the normal range for this organ. The previous chiefly nuclear (transcriptional) commitment has turned into a (translational) commitment of the cell.

Beginning prenatally the cell may or may not have been exposed to and activated by hormone, which can affect the genome and/or the cytoplasmic synthetic machinery, which in turn may affect the genome. Depending on which type of cell this is, it might receive nervous stimulation. Circulating maternal metabolic substrates (if our cell is mammalian) will either induce or inhibit certain enzymes and affect the stability of others, which can also be caused by metabolites endogenous to the cell. These early embryonic cell activities may undergo stage-dependent fluctuations and may well be necessary to the economy of the embryo as a whole. Although it need not exercise in full its mature function as yet, the young cell in the embryo is, or is almost, functionally competent.

With birth our cell may or may not acquire new synthetic capacities. It may change function cyclically well after birth and its architecture change accordingly. Let us say that this cell does not undergo a drastic functional change, in which case its function will still be intensified with respect to what it was in the embryo. That is, the facultative genome is responding faster and comes closer than before to full expression under the influence of the diurnal rhythm of light, work, and diet all acting indirectly through factors and substrates, hormones, and other signals. Many of these influences act again first on the cytoplasm, and from our viewpoint particularly on translation, before reaching the nucleus.

With mature function the cytoplasm becomes the most active part of the cell, sufficient to a large extent in short-term self-regulation and in controlling the residual (but vital) nuclear function. Functional adaptation within physiological limits, as this cell will undergo, is a measure of self-regulation. The mechanisms called upon to act are not different from the prenatal since late embryonic differentiation. During the entire life-span of this cell the macromolecular machinery will require continuous refurbishing, e.g., enzymes and structural proteins, stable RNAs, and certainly unstable mRNAs; stability of these mRNAs is partly endogenously and partly exogenously (hormonally) controlled. Sooner or later, and we do not really know the relative importance of endogenous and systemic causes, the cell reaches senescence and may die.

The purpose of this overview was to bring out the intricate temporal interplay of the various biological levels of regulation, the tracking of which may have been lost in a text preoccupied with macromolecules. Any student of cells may add a few more biological levels of regulation from his own experience. The main point is that not many of these levels of regulation can be explained in terms of molecular or the complex cellular phenomena discussed under development.

CHAPTER 7

Evolution of RNA: An Evaluation and Epilogue

General comments on evolutionary aspects of RNA have appeared at many points in this text. The attempt in this chapter is to gain, in a more systematic way, some useful perspective on these aspects of RNA which are so obviously important from the standpoint of evolutionary theory. Yet, because this attempt focuses on the biological aspects, and organic evolution prior to the modern prokaryote is not discussed, it forfeits any claim of being comprehensive. To allow tracing the evolution of the eukaryotic genetic system back to its source, the discussion will dwell extensively on DNA before dealing with the major RNA species. As shall be seen, however, much of evolutionary interest is found in DNA which, on present knowledge or at times in principle, is not correlatable with the transcription products; while many of the changes found in DNA may be duly regarded as evolutionary, it is not often that something firm can be said as to their biological advantage at the RNA level. Because in their nature most evolutionary considerations remain subjective, it will be in this chapter more than in any other that, on occasion, reader and writer must beg to differ. To suit the partly evaluatory purpose of the narrative, the information presented in previous chapters (referenced by use of the appropriate chapter number) will be only briefly recapitulated, whereas the most relevant or new information (cited directly) will be more fully described. This evaluation, finally, stresses the conceptual aspects of function that can be derived from the physical evolution of the molecules.

THE FIRST GENETIC SUBSTANCE

The first question to discuss is the order of appearance in the biosphere of the macromolecules involved in genetic function. Prior evolution leading to these macromolecules will not be discussed because its predominantly chemical character is outside the scope of this book. Starting from that arbitrary point, as a genetic code developed primitive organisms faced the problem of storing the accruing genetic information. It is pertinent to ask which was the first genetic substance capable of doing it, DNA or RNA. It has been suggested that stable polyarabinonucleotides, intermediate in character between RNA and DNA, in turn could have served as a common ancestor to these two types of molecules (Schramm and Ulmer-Schürnbrand, 1967).

Today, the simplest known genomes are those in certain RNA phages with three genes. It is thus possible and even probable that an RNA, perhaps carrying out direct translation along the lines suggested in Chapter 1, managed to perpetuate itself as the incipient genetic substance. Obviating a transcriptional step, inaccurate at this stage, could have been an advantage. According to these views, present day RNA viruses are relics of a "universe before it became DNA dominated" (Spiegelman and Haruna, 1966). Based on present evidence, only definitely in these obligatory parasites do RNA helices occur and is RNA synthesis self-directed. However, during subsequent evolution the need would have arisen for a separate genetic substance in helical DNA, partly for controlling (the now existing) transcription as distinct from its own perpetuation and partly also for metabolic stability. One view in this respect is that the single $3'$-OH in deoxyribose enables DNA both to replicate and transcribe. Conceivably, the semiconservative replication of the helix, although mechanistically complex, allows a more effective selection to operate on progeny molecules. The fact is that the genetic substance as represented in viruses, RNA or DNA, whose information is transferred by a variety of equivalent mechanisms (Chapter 1), gave way finally—regardless of the actual order in time—to DNA-transcription as the mainstay of higher evolution.

Nevertheless, it is likely that during evolution RNA and DNA appeared after polypeptides (Munro, 1969; Orgel, 1968), in particular, the primitive noncoded polypeptides described in Chapter 5. This does not mean that translation, as we know it, is more primitive than transcription; or, by definition, could it precede the genetic code. Rather, the emergence of the code depended on accurate replication and transcription, and as these processes evolved they required a progressively more faithful enzyme synthesis. Hence emergence of the genetic code meant replacing primitive by coded polypeptides, their reliable translation gradually becoming critical for the effective evolution of the code. As a result a transcription-dependent translation evolved somewhat ahead of, but in parallel with, the code. For a view of the evolutionary ancestry of translation as a process the reader should consult Lipmann (1971).

A fundamental question is, why a code based on nucleic acid rather than protein? The answer to this is based on the principle of complementarity which underlies the function of macromolecular templates (Chapter 1). By this principle, a nucleic acid is the obligatory precursor of another nucleic acid or a protein, in other words, the central dogma. These topological transforms are enormously more accurate and faster than any alternative one, i.e., from protein to protein. At best, this protein-to-protein system could operate via poorly templated enzymic activities capable of making only unsophisticated protein, in all a poor competitor against a nucleic acid system. Significant in this respect, whereas the apparatus for transcription is possibly the most complex in terms of recognitions, the apparatus for the assembly of protein is mechanistically the most complex of the nucleic acid system. The reciprocal question is, why is protein the main evolutionary endpoint as far as macromolecules are concerned? The answer is, of course, their superior functional potential as opposed to their poor genetic capacity outlined above.

THE DAWN OF THE CELL NUCLEUS

Before embarking onto evolutionary aspects of the eukaryotic chromosome at large, it is as well to recall that the origin of the more primitive bacterial chromosome is not established (Luria, 1969; Subak-Sharpe, 1969). The origin of this chromosome could be monophyletic by progressive differentiation of an ancestral chromosome or polyphyletic by integration of probacterial plasmids. For obvious reasons, discussion of the evolution of the eukaryotic chromosome from a bacterial-like chromosome can hardly be entertained seriously at this time.

Even if chromosomal evolution cannot be retraced, it is nevertheless instructive to examine first the salient characteristics of the present eukaryotic chromosome. Then, reasonably assuming that these characteristics represent the contemporary evolutionary trend, one can inquire into their functional meaning in relation to the characteristics of the present bacterial chromosome, which is presumably the closest to the ancestral one. With these exercises in hand, one can tentatively speculate in a retroactive way on what may have taken place during the transition from the bacterial to the lower eukaryote chromosome, that is, the dawn of the cell nucleus. This acknowledges that Nature was concurrently experimenting on how to achieve not just a chromosome but a nucleus, yet these other experiments are beyond our scope. Thus this discussion will center on the chromosome which, if not the only nuclear constituent, is sufficiently representative of nuclear evolution to justify the broader title of the section.

The foremost characteristic of the eukaryotic chromosome is precisely that it is a chromosome as opposed to an almost naked DNA helix in the bacterium. (It is somewhat unfortunate that the bacterial equivalent is customarily also called a chromosome.) The eukaryotic chromosome contains up to twice as much protein as it does DNA, namely, histones and acidic proteins, and these proteins clearly are eukaryotic innovations. In lesser proportion the eukaryotic chromosome also contains RNAs, lipids, and metals. Many of these components which are of special interest to us, namely, the proteins and RNAs, are directly associated with the DNA. In addition, the functional significance of polyamines (present also in the prokaryotic chromosome) is unclear. The conclusion in Chapter 3 with respect to the multimolecular organization of the eukaryotic chromosome in terms of proteins and RNAs was that, although little understood, this organization is compatible with supporting a functional continuum through which the complexly integrated expression of genetic function can take place. Because of this, the term chromosomal apparatus was preferred, and it was implied that in respect to function this apparatus means more than the sum of its parts. The current effort is to understand the functional organization of this apparatus. Yet what we do understand of it provides some clues as to what the deployment of genetic potential of the cell may depend on, namely, regulatory programming.

In essence, in order to be conducted a program of this type requires a chromosomal apparatus, since DNA cannot regulate itself even though it carries all the information for the program. The meaning of this apparatus, in contrast to the primitive prokaryotic chromosome, is that of a genetic organelle capable of internal (perhaps to some extent intrinsic) regulation. A model indicating how genetic circuits relying on the organization of the apparatus can be responsible for enacting such a regulatory program was sketched in Chapter 3. An important corollary was that a bacterial cell cannot support a chromosomal program of this kind, and whatever program this cell has must, instead, be in relation to the general regulation of metabolism (Chapter 3). Consider that something which was absent in the bacterial cell had to be created to permit the enormous diversity of eukaryotic evolution, in fact, the fastest tract of organic evolution. A view is that what was absent, more than sequestration of the chromosome within the nucleus, was proteination of the chromosome in order to allow programming, and this irrespective of the likelihood that the two events occurred interdependently. In this light, the uncoupling of transcription and translation, although physically caused by the presence of the nucleus, is also a consequence of chromosomal proteination.

Concurrent with the molecular complication of the eukaryotic chromosome, significant changes have occurred in the genome: (1) the size of the genome has increased, which in part reflects the appearance and increase of molecularly redundant DNA; (2) this increase of genome is accompanied by a disproportionate increase of regulatory genome relative to informational genome; (3) the single genomic replicon has become a multireplicon; (4) the functional units of the informational genome have scattered relative to their previously frequent operonic arrangement, yet without each unit becoming multiple in the process; (5) conversely, the cistronic units of part of the auxiliary genome (rDNA) have multiplied and consolidated to form a patent organelle within the chromosomal organelle, the nucleolus; (6) however, intermediate between points (4) and (5), the cistronic units of another part of the auxiliary genome (tDNA) have multiplied but some have also become disposed in scattered clusters; and (7) distinct nuclear polymerases have appeared to separately read the informational and auxiliary genomes, in the latter, possibly rDNA and tDNA also separately. Points (2) and (4) reflect the appearance of a regulation of informational transcription qualitatively different from the bacterial regulation. To a certain extent this is also true of the regulation of tRNA transcription (point 6). In the evolution of this new strategy of regulation, it is likely that the presence of (1) redundant DNA played an important role. Since the prokaryote has very little of this redundant DNA (most of it coding for stable RNAs), it is tempting to think that this novel structure of DNA sequences has been possible to some extent because of the proteination of the chromosome. Such a causal order is suggested by the greater proteination than redundant DNA content of the lower eukaryote chromosome. Hence, directly or indirectly, proteination of the chromosome is involved in the higher regulatory capabilities of the eukaryote.

The implications of some of the points made in the preceding paragraph are worth considering further, beginning with point (4). In the bacterium the physical linkage of functionally related cistrons within operons allows proteins to be made as required, independently of proteins that are not required. Also, blocks of physically separate operons (in particular, constitutive operons) can be jointly regulated, e.g., by a single promoter. This, on the whole, inflexible linkage accounts for the typically stereotyped response of the bacterium to its environment, but it probably also prevents separation of related cistrons during linear chromosome transfer. Evolutionary selection in the bacterium works therefore essentially at the level of individual cistrons or operons. On the other hand, in the eukaryote it is envisaged that the informational cistrons have scattered, but at the same time become linked within regulatory genetic circuits which are not physically linked. These circuits allow an integrated but, at the same time, discretionary production of proteins as required, for example, during cell differentiation. Evidently, this flexible, functional rather than physical, linkage in the eukaryote would suit its correspondingly flexible developmental and physiological response. More important, selection may now work at the level of circuits as much as on the cistrons themselves, with the potential for establishing an entirely new regulatory integration of cistrons. (Mechanistic aspects involved in creating this integration are discussed later in connection with regulatory RNA.) Point (5) in the preceding paragraph means the presence of a joint rRNA precursor with an attendant physical coregulation of the major rRNA species. In the bacterium a physical versus regulatory coordination of these species is not decided, since it is uncertain whether the common precursor occurs. Solving this uncertainty is therefore of great evolutionary interest. In the case of eukaryotic rRNA a physical coordination differs from what occurs in informational RNA. In other respects as well, the rRNA cistrons of the eukaryote remain the most prokaryotelike in character (Chapter 4). Point (6) regarding tRNA reveals an interesting situation halfway in between mRNA and rRNA. The multiplied cistrons have scattered (like informational DNA) but part of them remain in clusters which possibly are transcribed serially (like rDNA), yet the transcripts are separate. It also seems probable that certain tRNA cistrons are included in regulatory circuits very much like informational cistrons (Chapter 4). Point (7) indicates more evolved transcriptional mechanisms operating on the more evolved genome. Discussion of these evolutionary innovations of the eukaryote is the main subject in the remainder of this chapter.

The closest look one can take at what may have happened during transition of the genetic system from the prokaryote to the higher eukaryote is by examining the present day lower (unicellular) eukaryotes, such as amoebas and fungi. The decidedly eukaryotic chromosomes of these creatures have histones (Leighton et al., 1971) yet, significantly, not all these organisms appear to have them. In addition, whether these chromosomes have all the other classes of molecules and, in particular, all the macromolecules present in the higher eukaryote chromosome remains to be seen. Thus, with respect to the innovations listed above for the higher eukaryote, the situation in the lower eukaryote seems to be: (1) the genome is larger than the bacterial genome; (2) a small proportion of redundant DNA, but larger than in the bacterium, appears to be present; (3) the genome represents a multireplicon; (4) the majority of informational cistrons are scattered, but some are still linked in polycistronic operons; (5) at least some consolidation of rDNA has occurred and many lower eukaryotes possess nucleoli; and (6) distinct classes of nuclear polymerases are probably present in some cases (Horgen and Griffin, 1971), but perhaps not all (Dezeleé et al., 1970). There are no data in the lower eukaryote concerning the disposition of tDNA. Points (1) and (4) suggest a regulation of informational transcription resembling that found in the higher eukaryote, but probably simpler and, for certain cistrons, still resembling the bacterial operon. It remains an open question whether (2) the redundant DNA is transcribed (Prescott et al., 1971) at least in the size and rate at which it is transcribed in the higher eukaryote. The maximum size of the redundant transcript is about 30 S compared with up to 100 S in the higher eukaryote. To mention briefly, satellite DNA has appeared in the lower eukaryote already in connection with the centromere as in the higher eukaryote, but not in connection with any other constitutive heterochromatin. Concerning point (5), the mature rRNA species are definitely larger than the bacterial species (Chapter 2), and decidedly transcribed in a common precursor. However, this precursor (about 34 S) may not be much larger than the combined mature species (as in bacteria). It follows that the rRNA pathway may be more conservative than in the higher eukaryote, which could also explain the extremely fast maturation of the rRNA (about 3 minutes). The equivalence of (6) the polymerases with the higher eukaryote counterparts is not known.

In conclusion, perhaps not unexpectedly, in regard to all the parameters examined there does not seem to be a sharp but rather a gradual transition between the prokaryote and lower eukaryote, or between this and the higher eukaryote. The overall impression is that the early transitional steps in the direction of the eukaryote have frozen at different stages represented today by different lower eukaryotes, which incidentally, would be indicative of a polyphyletic divergence. However, it is difficult to be objective because of the scanty and often contradictory nature of the molecular data. Given this state of affairs, an area of evolutionary interest as great as this one represented by the lower eukaryote certainly deserves some vigorous exploration. After all, creating the first functional nucleus ranks equal in the scale of evolutionary achievement to completing the first prokaryotic cell cycle.

The advent of the eukaryote marks, in general, the evolutionary introduction of macromolecular degeneracy on a massive scale, that is, the occurrence of many different molecular species in the same molecular class. [It should be clear that molecular degeneracy does not imply molecular redundancy in the DNA (see Glossary), although they increase together.] The molecular degeneracy of the eukaryote is exemplified in several RNAs (labile, chromosomal, and stable), histones, hormone receptor and perhaps other binding proteins, ribosomal proteins, and perhaps synthetases. Of these, only ribosomal degeneracy is not exclusive to the eukaryote, but is already present in the prokaryote. Reasoning by extension, degeneracy is inherent in the constitution of the eukaryotic chromosomal apparatus, containing a much larger number of molecular species than the prokaryotic counterpart, even if we consider the molecules transiently associated with bacterial DNA as part of the chromosome. To grasp the value of molecular degeneracy in biological systems, perhaps an analogy with computer systems will be helpful; in these systems degenerate circuits (in the present sense) are used to stabilize decision-making functions. From this analogy, the value of molecular degeneracy should be essentially for regulation. Therefore, returning to the eukaryotic chromosome, this means a more complicated and self-regulated apparatus than in the prokaryote. An exceptional aspect of degeneracy, in this case molecular only in a special sense, which did not materially alter is codon degeneracy. The simple reason is that by the time of the advent of the eukaryote the genetic code had already completed most of its evolution. Only because most of the evolution of the code preceded eukaryotic evolution, can the code today be considered semiuniversal.

DNA

Considered as a molecule, the first striking point about eukaryotic DNA is its greater complexity in an informational, but not a chemical sense. This is not to restate the variety of sequences to be expected in the genome, but has to do with the structure of its sequences. As it turns out, this complexity could hardly have been anticipated before the advent of the procedures for molecular hybridization. To begin with the established picture of the geneticist, the DNA of present day creatures is the outcome of (a) point mutations, in which the small infidelity of DNA polymerase possibly is a major force, along with (b) gene duplications by unequal crossover and tandem duplications, e.g., Keyl's geometrical series (Chapter 3). The duplications under (b) result in partial repetition on the genome. To this level the evolutionary process is best summed up by the conclusions drawn by E. L. Smith (1968) looking at it from the endpoint of protein: amino acid substitutions representing single, double, or triple base changes; chain shortening by elimination of N and C termini; deletion and probably insertion of amino acid residues within a chain; chain extension by joining of cistrons or unequal cross-over; and gene duplication leading to chain divergence. The next evolutionary level to consider is that of (c) doublings of the entire genome capable of resulting in total repetition. Successive ploidizations in vertebrates and, of course, in plants have been known for some time, and a sequence of ploidization from primitive to higher vertebrates up to amphibian and fishes was proposed by Ohno *et al.* (1968a). Representative of the geneticist's most advanced outlook, but by now almost certainly influenced by the thinking outside his own field, stands Nei's assertion (1969) that "a great number of nonfunctional genes are present in mammals without a deleterious effect." This conclusion comes close to that obtained from the molecular approach whose discussion follows. The important difference that stems from this approach will be that these "nonfunctional" genes not only are innocuous but also are an essential ingredient of the evolving genetic system.

Structural Complexity of Eukaryotic DNA and Some of Its Biological Implications

From molecular and particularly hybridization analyses, it is known that, as DNA increases in amount from the prokaryote to the eukaryote by a factor of

10^3, it acquires a complexity in terms of sequence structure hitherto unsuspected. This structural complexity is not present in prokaryotic DNA. For reasons explained later, it is also significant that the evolutionary increase in the amount of DNA is greater than that which can be accounted for by ploidy. In addition, although postreplicative modification of DNA does occur, viz., vertebrate DNA is the most methylated (Vanyushin et al., 1970) and this modification was briefly discussed in Chapter 6 in connection with development, from what is known, modification plays no major part in the complication of sequence structure.

As presented in Chapter 3 the mammalian genome can be divided into, at one end, a large fraction of unique sequences that are transcribed, but until recently remained unexplored by transcriptional analysis. At the other end, there is a variable, but generally small, fraction of sequences which are extremely redundant moleculary, and these include distinct DNA satellites. The extreme satellites are not transcribed, which is understandable in terms of their redundancy and sometimes eccentric base composition, for such a transcription would be informationally of little meaning. In the mouse, for example, the function of satellite DNA is centromeric. In other species, however, the satellites have very different characteristics from the mouse, or are less conspicuous and their sequences represent an extremely redundant form of the following fraction of sequences, yet some of the sequences must also be centromeric. To the extent that the sequences share a common function, their marked divergence between organisms reveals an extraordinary plasticity which is evidently compatible with the noninformational nature of the function. In between the two previous fractions, unique and satellite, there is a large fraction of more typical redundant sequences whose transcription is readily observed, even though it cannot be ascribed to any one sequence unambiguously. This intermediate fraction is organized in families of related, but not identical, sequences which differ even between closely related taxonomic species. By several criteria this fraction is rapidly and dynamically evolving. As a result of their evolutionary history, the families are of different size and age, therefore different copy frequency and divergent sequence structure, each characteristic of a different species. Because viral genomes (except for their termini) and bacterial genomes have discouraged accumulation of this fraction, its steady increase in the eukaryote must have a strong functional reason. The unique and typically redundant genomes are the main concern of this chapter.

To repeat, save for the most extremely redundant sequences and in particular satellites, which all tend to predominate in constitutive heterochromatin, the other DNA fractions as a whole are transcriptively active. Hence, the less extreme their redundancy the more likely the sequences are transcribed. It was made clear, however, that even if as the sequences become more redundant and their informational content becomes less distinct from that of related sequences, correspondingly they cease to be transcribed, this does not mean that these sequences serve no functional purpose. Several other possible important functions, also obviously absent in the prokaryote, were mentioned in Chapter 3. The point is that we cannot assume that any DNA fraction which is never transcribed has no function (other than to ensure molecular continuity), even if we do not know what that function might be.

The unique genome may be taken to include informational cistrons. Although evidence to support this assignment is offered later, it must be stressed that it may not be an absolute one. For example, in the eukaryote there are some duplicated informational cistrons that subsequently may have become molecularly redundant, such as those for immunoglobulins and certain enzymes. The number of cistrons for these proteins is, in any case, small compared with that of the typical redundant families of sequences. A more important exception to the uniqueness of informational cistrons are the cistrons for histones which have a 400-fold multiplicity (Kedes and Birnstiel, 1971), even though the cistronic multiplicity of each histone species could be much less than this. A very important consideration is that the common cistron divergence by a few point mutations does not per se constitute molecular redundancy as defined by criteria presented in Chapter 3. Accepting therefore an informational assignment for the unique genome as valid for the most part, this makes the redundant genome the one most pressing to understand in terms of function.

Two explicit but somewhat opposite views concerning this redundant genome are as follow. It represents either (a) the organization assumed by the genome in order to conform with the character of eukaryotic regulation (Goodwin, 1966b; but see for criticism the discussion to his paper), or (b) the organization of the genome that has evolved but not principally determined by the character of the regulation (e.g., Scherrer and Marcaud, 1968). For the sake of contrast, these interpretations were presented here in a more extreme version than perhaps was intended by their authors. By this liberty, the first and conformist interpretation is more deterministic in respect to regulation and stresses an intrinsic redundancy of regulation. The second and romantic interpretation is more stochastic in that respect and depends on a certain gratuitousness of the genetic material. Although their basic arguments are substantial, and not as exclusive of each other as it might seem, these two interpretations, however, do not seem to explain enough the redundant genome and do not really approach a functional definition of this genome.

A view more dynamic and closer to a definition of the redundant genome is that by Britten and Davidson (1969), to the effect that the expendable genetic material represented in redundant sequences has been captured during evolution for regulatory function. This view is of main concern to us in this and the following sections. The central tenet is that, because redundant DNA suggests itself for regulation, its transcript may have a regulatory function. Comparisons of evolutionary interest between the implication of this concept for informational regulation and the well-established facts of prokaryotic regulation were included in the conclusions on eukaryotic mRNA in Chapter 3. An obvious conclusion was that, because it lacks a redundant genome, the prokaryote cannot have this type of regulation. Even if the concept has not advanced our understanding of regulation beyond the hypothetical level, the significance is that the functional emphasis is now as much on the redundant as on the unique genome. On this basis, a facultative eukaryotic genome consists of (1) regulatory sequences most, but not all, of which are redundant and (2) informational sequences most, but not all, of which are unique. A proviso to this categorization of the genome is that the redundant genome includes the stable RNA cistrons which are not regulatory but auxiliary cistrons. Also, as will become clear, on this particular view much of the growth of the eukaryotic genome during evolution depends on the redundant genome.

Two views have been offered on the emergence of families of typically redundant sequences. The first view, explained in Chapter 3, is by Walker (1969) who calculated that these sequences can stem from unique sequences if in any taxonomic species the use of a particular codon for each amino acid, rather than any of the several codons generally available, were to be enforced. This proposal formally solves the requirements for converting the sequence structure from unique to redundant, yet its biological reality depends on the actual degree of codon restriction which is not known. In addition, this view also depends on the selective pressure between synonym codons being less than between codons for different amino acids. This is not to say, as King and Jukes (1969) do, that all the synonym informational mutations which are implied are necessarily neutral in respect to evolution (see Clarke, 1970), even though neutral mutations are known to occur widely (Kimura, 1969), including unique DNA in particular (Laird et al., 1969). (If the present mutations were neutral they would come under the so-called nonDarwinian evolution because no selection at the level of protein is involved.) The second view on the emergence of redundant sequences by Britten and Kohne (1969) is again of a more dynamic evolutionary character. They considered that each family of redundant sequences derives originally from the sudden multiplication of a unique sequence (see saltation below). As the family becomes established, however, a reversal of the process may occur restrictedly and a few unique sequences emerge from each family. These propositions notwithstanding, it is clearly not known whether during evolution redundant sequences breed only redundant sequences, and, similarly, unique sequences breed only their kind. We shall return to these aspects later.

Typical redundant sequences are evolving at a faster rate than protein. A good example would be the occasional increase by saltation in numbers envisaged for these sequences by Britten and Kohne (1969). The present evidence is that these families of sequences appear suddenly, in terms of evolutionary time, rather than progressively. Probably, therefore, many redundant sequences represent noncoding sequences (Zuckerkandl, 1968), and probably also many of these sequences make up most of the balance of the increased eukaryotic genome which is not accountable by ploidy. Likewise, unique mammalian sequences seem to be evolving at a faster rate than protein (Laird et al., 1969). As a consequence of the fast evolution of the two types of sequences the rodent genome, as a whole, is evolving 15 times faster than protein (Walker, 1968). Together these considerations bring up the question that, irrespective of sequence structure, an unknown but possibly extensive fraction of the eukaryotic genome (not the small fraction dedicated to stable RNAs) does not code for protein. An implied inference here is that in particular the redundant genome is in great part not translated and, as we shall see, a part of this genome may function only at the RNA level. The reciprocal inference is that the unique genome is mostly informational. However, for this genome there is also some evidence to the effect that not all of it may be utilized for an informational purpose (Chapter 3); in addition, there is evidence that a portion of the informational transcript is redundant and most likely noncoding (Chapter 5). Hence, part of the unique genome perhaps may make RNA that serves as such (as in the case of the redundant genome) and another part perhaps may be making nothing at all. In conclusion, the possibility that much of eukaryotic DNA does not code for protein is indeed very real.

These considerations imply that the eukaryote differs from the prokaryote in that, whereas its DNA-like RNA is not all informational, probably most of this RNA in the prokaryote is informational and translated. Evidently, this difference is independent of the fact that in both types of organism part of the translatable information is ultimately regulatory by virtue of the function of some of the protein. As will become clear, this difference presumably related to the presence and absence of a redundant genome, respectively, has important evolutionary consequences. The main one indicated so far is that which pertains to the character of the regulation. Only a fraction, often a different fraction, of the eukaryotic redundant genome is transcribed during any cell cycle, even during the period of greatest cell differentiation. There is no absolute contrast in this regard between the informational genomes, for, to a varying degree, these are not all transcribed during every cell cycle in either type of organism. Truly the difference seems to reside in qualitatively distinct eukaryotic features with which the redundant genome is believed to be concerned. The emerging principle is that the redundant genome represents a progressive complication of the genome in charge of progressively new genetic requirements.

In view of all these biological aspects introduced by the structural complexity of the genome, one should hope that the more traditional concern of the geneticist with certain aspects of genetic units (Chapter 3) is now shared with a concern for the meaning of this complexity, which cannot but be the functional organization of the eukaryotic genome. What is needed is a greater awareness of the problems of molecular evolution of DNA vis-à-vis its evolution as a genetic molecule, that is, idiosyncracies in the molecule that may not be revealed genetically. Taking as common ground that these two evolutionary aspects of eukaryotic DNA must be closely interrelated, they represent a single course of evolution of the molecule looked at from two or even more different angles. It is quite clear that without the classical studies during the periods of the so-called formal, physiological, and developmental genetics there would have been no basic genetic knowledge of the eukaryote to go by. It is equally clear now that without an appreciation of the genomic complexity of the eukaryote there can be no real understanding or furtherance of this knowledge. This appreciation takes the form of a new approach—the molecular genetics of the eukaryote—which conveniently reminds us that genetic law is ultimately expressed in terms of molecules and their interaction. It is appropriate to recall here that the function of the prokaryotic genome would not have been understood without this same approach in recent years, even though, alas, this genome is structurally simple. For a fact, however, the genetic and molecular vistas on DNA redundancy are less at variance with each other than they would seem from the contents of this paragraph, which can be seen in what follows.

Current views on the evolution of DNA are that it must be concerned with at least the following aspects: (a) regulation of the genome, (b) phylogenetic divergence of proteins, and (c) emergence of new proteins. As explained, point (a) probably involves in part the function of the redundant genome and may be associated with the continuous replenishment and diversification of sequences during evolution of this genome (Chapter 3). A general argument for point (b) is that, given the premise that informational genes consist for the most part of unique monocistronic sequences, then it is impossible for a vital gene to change significantly and at the same time ensure phyletic continuity. (Significant change would mean one resulting in loss of a vital function.) Indeed, many enzymes have evolved very conservatively. Once a gene is duplicated, however, the second copy is free to evolve, provided—and quite important—that it escapes the original regulation. This evolutionary strategy has resulted in proteins which are clearly different from one another yet have a recognizable common ancestry, e.g., hemoglobin and myoglobin. In short, as stated by Ohno (1970) who is much responsible for the present argument, "genetic redundancy created by gene duplication rather than allelic mutations at preexisting loci has been the major force of evolution." Stated differently, natural selection is a rather conservative guardian of the genome which it preserves or modifies slowly, whereas redundancy acts more like a creator of new genome.

Ohno (1970) goes on to argue that of two likely mechanisms for gene duplication, tandem duplication and ploidization, the latter one principally may have been encouraged by evolution. Although tandem duplications are known to have occurred, and perhaps to a considerable extent, his argument against them in terms of selective advantage is threefold. First, tandem duplication invites subsequent unequal cross-over with potentially disastrous informational consequences. This possibility was realized early for the Bar locus in *Drosophila* and is now known for rDNA in general (however, duplication of rDNA is for the most part innocuous owing to its multiplicity). Second, tandem duplication disrupts the established gene dosage ratio of unlinked informational genes which are the general rule in the eukaryote. Third, if the informational gene but not the generally also unlinked regulatory gene is duplicated, then the informational gene does not become truly autonomous to evolve; this may be the case with rDNA which as a whole evolves very conservatively. In effect, many ancestral duplicated genes, e.g., the closely linked immunoglobulins, have not diversified much during evolution. Hence, according to this line of thinking, if ploidization does become the evolutionary mechanism of choice because it has less of the previous limitations, it nevertheless has its own special requirements. First, ploidization eventually requires diploidization for the correct segregation of linkage groups without which further evolution is not feasible. The cytogenetic evidence described by Ohno (1970) indicates that diploidization has taken place at several opportunities during vertebrate evolution. Second, and most important, ploidization is forbidden above the level of amphibians and fishes because of the advent of chromosomal sex determination. Thus, above this phylogenetic level other sources of new genetic material must of necessity take over. From considerations in this section and Chapter 3, a likely candidate for this alternative source is the redundant genome which has been increasing disproportionately in size ever since the beginning of eukaryotic evolution. Inescapably, any increase in total genome such as has occurred during this evolution has to involve some kind of duplication besides ploidy. In the case of the redundant genome, however, this is not tandem but scattered duplication and, so to speak, according to the peculiar molecular rules of this genome. It should be noted that point (c) mentioned above has been reserved for discussion later.

Three lines of circumstantial evidence which were in support of the premise of uniqueness for most informational sequences were as follows. It is irrelevant to this assumption whether (a) any redundant sequence is also present in mRNAs, provided it is noncoding; or (b) whether unique DNA can in addition have noninformational functions as considered earlier. One line of evidence was the extremely high structural homogeneity which is found in enzyme populations in the cell (Chapter 1), for, in general, this is unlikely to be compatible with multiply repeated informational cistrons. Differently stated, if unambiguous templates are an advantage—which we may call the principle of "template hygienics" —then at any evolutionary equilibrium level repeated cistrons by offering a larger target for mutation would conspire against this principle, the more so in proportion to their degree of repetition. Moreover, because transcription and translation can both amplify the protein product by virtue of their intrinsically repetitive character, what then is the need to amplify cistrons and thereby risk having unhygienic templates? A second line of evidence was the general lack of selective pressure to retain intact the occasional gene duplication. The histone cistrons (Kedes and Birnstiel, 1971) clearly show that this selection does indeed occur when beneficial, for, much like rDNA, these multiple cistrons appear to have the little inhomogeneity predicted from the evolutionary conservatism of histones; perhaps this means further that, as in the case of rRNA, a little structural inhomogeneity is not detrimental to the primitive function of histone. (Significant for the central assumption here, nonhistone mRNAs also analyzed by these authors consist mostly of unique sequences.) An exceptional use of cistronic repetition would be when a certain informational redundancy (as opposed to the general uniqueness of informational cistrons or the apparently near identical multiplicity of histone cistrons) is, in fact, beneficial, as perhaps is the case with the immunoglobulins and isozymes cited earlier. A third line of evidence, not direct but important, was that in mouse and *Drosophila* most informational sequences must definitely be unique because the redundant sequences in these genomes cannot code for more than a fraction of the total protein (Walker and Hennig, 1970; Hennig *et al.*, 1970). This argument was leveled in Chapter 3 against Callan's genetic hypothesis based on a generalized multiple cistronic repetition. Last, but not least, it was difficult to accommodate the fundamentals of Mendelian genetics with multiple determinants for each unit phenotypic trait. Collectively, therefore, the circumstantial evidence in favor of unique informational cistrons was strong. This evidence was validated recently by experimental evidence on the hemoglobin cistrons (Bishop *et al.*, 1972). We can now therefore conclude that most informational cistrons are unique, without excluding that a proportion may be present at low multiplicities.

One corollary of these arguments could be that uniqueness of informational sequence actually forces a greater diversification of protein. Let us consider the case of a unique informational sequence which becomes duplicated and let us follow the possible evolutionary course of the second or duplicate sequence as it becomes a different unique informational sequence. On the basis that the first or original sequence remains unaffected and functional, it does not further concern us directly. Recalling the companion assumption that redundant sequences are mostly noninformational, one possibility is that, during the period it takes the duplicate sequence to evolve into a different unique sequence, the sequence becoming molecularly redundant and therefore no longer translatable into protein. (Redundancy with respect to the original sequence is seen here as being the product of a large number of mutations, prior to further mutations producing eventually a different unique sequence from the duplicate sequence.) Thus, at the level of protein, evolution of the duplicate sequence may be envisaged as follows: initially it withdraws from the translational pool (where the original sequence remains) and ultimately reenters the pool as a new translatable sequence.

For simplicity, we shall first consider the last phase, namely, reentry into the translational pool starting from the period of redundancy of the sequence. This would be the period during which the duplicate sequence is freer to evolve and perhaps be tested at the RNA level. The principle being implied, namely, the economy of transcriptional as opposed to translational testing, was presented in Chapter 3. According to this principle, the redundant sequence would undergo transcriptional testing, which might also serve to ensure survival of the sequence, until such time as it reacquires informational value. Yet the key concept of "testing" must be left undefined, for it is not obvious how a prospective informational RNA can be effectively tested at its own level; all that is implied is that the coding sequence is evolving, either by selection or drift. The assumption is then that by the time the sequence reacquires informational value, evidently different from the original one, it has become again unique and its transcript is no longer discarded but tested translationally. Ultimate fixation of the sequence probably has to conform with constraints in the DNA composition (to be discussed later). The sequence may also require transportational or translational elements in the form of additional noncoding sequence. In any event, provided the transcript can reach the cytoplasm and be translated, it is not difficult to visualize how this sequence may be selected for and eventually fixed. Once translated, the actual selective process is, in principle, not radically different from that for fixation of a valuable mutant (extant) protein. The crucial question to understand is rather by what criterion does the sequence begin to be translated. Alternately stated, assuming that not all transcripts but only those that stand a fair chance of being translated arrive in the cytoplasm, what makes the sequence arrive there for the first time and undergo its acid translational test? There is no answer to this question, which implies a controlled transit of transcripts (Chapter 3). However, it could be the concern of the nuclear envelope (or its pores), the informosome, special noncoding sequences in the transcript (vide supra) or its inclusion in a putative precursor containing these sequences, the higher order structure of transcript or precursor, etc. Two general possibilities deserve comment. One possibility is that the special noncoding sequences are redundant and this redundancy is somehow responsible for the selective transit. The other and really intriguing possibility, however difficult to explain mechanistically, is that, on the contrary, it is the degree of uniqueness of the informational transcript (or its lack of redundancy) which the cell is looking at, whether at the level of the nuclear envelope as the first logical choice or perhaps at the polysomal level. These possibilities therefore touch on the question of how translational testing begins.

We may now consider the first phase of evolution of the duplicate sequence, that is, withdrawal from the translational pool as the sequence is becoming redundant and nontranslatable. In this phase the real interest is in what happens to the duplicate transcript while it can still reach the cytoplasm, which may eventually cease by a reversed control of the transit (or other reasons mentioned later). First, it is possible that adverse selection speeds up this reverse control and/or favors evolution of the sequence toward greater redundancy, in either case barring the transcript from translation. For driving this selective force it can be assumed that a duplicate protein will, in general, be subfunctional relative to the protein which is still being made from the original sequence. Selection against the transcript would therefore mean enforcing the principle of template hygienics. Second, selection probably also operates to eliminate the duplicate protein for as long as this can be made, a process which may also concern subfunctional protein, such as mutant protein or indeed the newly appearing protein in the previous phase. (In the more extreme case, the duplicate protein may bring about elimination of the cell first, that is, an evolutionary dead end.) This elimination of unwanted or "defective" protein implies an active proteolytic mechanism as is believed to operate in the bacterial cell (Goldberg, 1972), but for which there are no firm indications in the eukaryotic cell (unless the protein fails to stabilize, etc.). Such a mechanism would represent Darwinian selection at the protein level. An alternative non-Darwinian selection would be if the duplicate protein were to be rejected during translation, perhaps by a scavenger function attached to the polysome. Either of these mechanisms implies that the cell is looking again at redundancy but this time not in RNA but in protein. Whatever the mechanism, judging from the homogeneity of cell enzyme populations, the net outcome is that not much wasteful protein is made or is left to accumulate in the cell. Lastly, given enough diversification of its proximal regulatory genes, the duplicate sequence is severed from the original regulation and transcription could cease altogether. Had the severance occurred physically during duplication, this could have caused an immediate withdrawal from the translational pool, thus obviating all selection against the protein. (Escape from the original regulation is essential if the sequence is to evolve autonomously. From previous arguments, however, to continue evolving the duplicate sequence has to remain in or reenter the transcriptional pool.) In summary, we have speculated about how a duplicated informational sequence may evolve toward molecular redundancy and at some point become nontranslatable. The sequence may emerge as a new unique sequence as it undergoes transcriptional testing and eventually be translated again. The sequential evolutionary pathway envisaged here is thus twofold: the protein made from the divergent sequence is selected against until it sooner or later ceases to be made; the new sequence is selected for, first, at the RNA and, finally, at the protein level. As stated in the initial corollary, this evolutionary detour of protein would have been forced by the need to preserve the uniqueness of the original informational sequence but, in turn, results in a new unique informational sequence. A further biological implication of this detour is considered below.

The question was asked previously whether or not during evolution redundant DNA sequences give rise only to other redundant sequences, and, similarly, unique sequences give rise only to other unique sequences. From the preceding paragraph the answer is no. That is, a unique sequence may evolve into a redundant sequence perhaps making only RNA, and, vice versa, a redundant sequence could evolve into a unique sequence eventually bearing a protein product. In addition, making only RNA gives a sequence a potential to assume some kind of regulatory function which ensures its survival. These points bring us back to an aspect mentioned several paragraphs above: which was that the evolutionary replenishment of redundant sequences is a conceivable source for phylogenetically novel protein, presumably after these sequences have gone through a stage of producing only RNA. Two pathways of protein evolution that we may have to consider are illustrated in Fig. 23: (A) from a unique sequence to a family of redundant sequences and again to a unique sequence (Britten and Kohne, 1969) for the creation of phylogenetically novel protein; and (B) from a unique sequence to a single redundant sequence (not a family) and again to a unique sequence for the creation of new but phylogenetically related protein, yet sufficiently different to be considered a new type of protein. Pathway (A) involves initial saltation or the occasional sudden birth of a redundant family, presumably involving extensive chromosomal duplication. Pathway (B) is the one proposed here and elaborated upon in the preceding paragraphs. In regard to it, it is perhaps significant that small redundant families become extremely rare, and abruptly so, above the two-member sequence level (Britten and Kohne, 1969). This pathway necessitates some sort of initial gene duplication and is the one that Ohno (1970) has in mind for kindred reasons at the level of chromosomal evolution. In conclusion, according to the present views, the redundant genome constitutes a pool where part of the protein ultimately originates and probably part of the regulation originates. By itself this duplicity tells us that, although we have discussed evolution of protein, we have yet to discuss evolution of the regulation and integration of the genome. At this point the implication is that the redundant genome can function at the RNA level, a consideration that emerges from several lines of argument in this book, but does not exclude function at the DNA level. It will not escape notice that, if this implication is correct, the benefit of a redundant genome with all its possibilities for regulation and transcriptional testing is denied the prokaryote. Absence of this testing leaves, therefore, Darwinian selection as the mainstay of bacterial informational evolution.

This discussion of the structural complexity of eukaryotic DNA is possible because of the work on molecular redundancy pioneered by Britten and Kohne (1968, 1969). The eukaryotic concept of DNA redundancy resulting from this

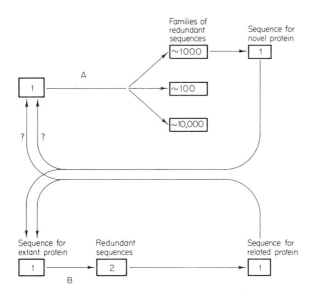

Fig. 23. Postulated pathways of evolution of eukaryotic informational sequences. Pathways (A) and (B) are drawn to begin with a unique informational sequence of unknown previous evolutionary history. The figures inside blocks indicate the number of cistrons involved at each step in the evolution from protein to protein. Pathway (A) is the saltation pathway proposed by Britten and Kohne (1969), which eventually results in a phylogenetically novel protein. Pathway (B) is proposed here to result in a different but phylogenetically related protein. The main difference between the two pathways is the greater diversification of the original sequence in A while traversing the larger selective pool represented by families of redundant sequences. Saltation in (A) presumably involves extensive chromosomal duplication. In (B) a single sequence separates from the original sequence via some form of gene duplication as envisaged by Ohno (1970). A second possible difference between the pathways is that in (A) the original sequence may or may not function informationally, whereas in (B) the original sequence is informationally functional and remains so during divergence of the duplicate sequence. The divergent sequence in (B) is expected to be tested transcriptionally: the preceding withdrawal from the translational pool and the subsequent reentry into this pool are discussed in the text. Whether or not they were previously translated, some of the divergent sequences in (A) may also be tested transcriptionally. The reverse arrows feeding back into pathway (B) indicate a repetition of the pathway leading to a second related protein. Whether also pathway (A) can be repeated (as indicated with question marks) is not known because of the uncertainty surrounding the translational status required of the original sequence.

work is almost certainly one of the most important to have appeared in recent years, and the underlying biological facts must be taken into account in any functional explanation of the genome. Although many of them remain conjecture, these facts have already opened as many valuable vistas on the structure of the genome as the initial elucidation of the DNA helix opened on the function of the molecule. The point is that, from a functional angle, understanding the structure of the genome is every bit as important for the eukaryote as understanding the function of the helix. One of the evolutionary paradoxes of the eukaryote has been the so-called C paradox (discussed later) which refers to the "excess" of DNA relative to what is presumed to be the informational content. Previously, this paradox had been rationalized by several ad hoc genetical hypotheses (Chapter 3). At present, the concept of redundancy permits one to begin describing this paradox in terms of sequence structure, as this varies from one taxonomic class to another, and eventually to understand it as befits any evolutionary DNA parameter. As implied above, however, the concept encompasses much more than this paradox. For example, recent hypotheses on the regulatory organization of the genome are based on the concept (Chapter 3), and more refined hypotheses along these lines are sure to appear in time. In conclusion, DNA redundancy throws light on the evolution of the eukaryotic genome by showing it to be the problem of evolution of a supermolecule, including coding and noncoding functions as equally relevant aspects. Such an integral view of this intricate genome as is now possible signifies an important departure from the more simplistic views to which, thus far, geneticists and evolutionists alike have inclined.

Evolutionary Parameters of DNA

Together with the qualitative complexities just described, quantitative vicissitudes attend DNA during the life cycle of the eukaryotic cell, although certainly not all to be described occurring in the same cell type or even organism. These quantitative aspects are discussed below.

(1) Amplification, which represents an increase in specific genome presumably by faithful copy of preexistent sequences, themselves sometimes molecularly redundant. Its raison d'etre is believed to be to augment the amount of template available. Yet, because the template becomes dose regulated, its utilization may not be proportionally augmented. The most common type of amplification, the germinal amplification of rDNA, can be visualized in certain, if perhaps not all, oocyte nucleoli (Chapters 4 and 6). In those oocytes where it most clearly occurs, it is a regular phenomenon resulting in a multinucleolate cell. The rDNA amplification in these free germinal nucleoli is in addition to the basic somatic cistronic multiplicity carried in the lampbrush chromosome. As mentioned in Chapter 4, the amplification could originate from rRNA derived from the chromosomal cistrons, by means of a reverse transcriptase. In contrast to the basic chromosomal multiplicity, which is called linear because it is a covalent part of chromosomal DNA, the germinal nucleolar amplification is, therefore, extrachromosomal and is called lateral. (It should be noted that the multiplicity of rDNA in ordinary somatic chromosomes is also contained within the nucleolus,

but it is described as chromosomal because in this case the nucleolus, contrary to those oocyte nucleoli which are detached, is part of the chromosome. Even in those oocyte nucleoli which present a low level of germinal amplification and remain attached to the chromosome, this amplification is presumed to be lateral. As far as is known, the somatic multiplicity is very rarely amplified physiologically. One possible exception is in the ovarian nurse cells of insects, where, however, any amplification might be more properly considered as being paragerminal. However, the somatic multiplicity can be amplified for reparative purposes as described under point 2.) An entirely different type of amplification is the apparently linear amplification of nonribosomal DNA which is observed only in certain DNA puffs of insect polytene chromosomes. The mechanism is by tandem duplication followed by misreplication. These chromosomes belong in a somatic cell lineage but this type of nonribosomal DNA amplification is alleged to occur also in germinal lineages, and to be of possible widespread but still selective occurrence in ordinary somatic chromosomes (Pavan and Da Cunha, 1969). This latter, broad generalization lacks supporting evidence and remains to be proved. At present, nonribosomal amplification in somatic chromosomes has been shown only for the histone cistrons (Kedes and Birnstiel, 1971), where it is a stable amplification.

(2) Magnification is a special somatic amplification for adaptive repair of an rDNA deficiency. Magnification seems to be a more variable phenomenon than germinal amplification and its molecular basis is not known. One wonders, however, whether this basis could also be RNA priming, as indicated for rDNA amplification. The indications are that magnification also involves lateral increase of rDNA, but an open question remains whether under some circumstances the magnified rDNA can integrate in the chromosome and become a linear increase. Amplification and magnification may be more fully compared with each other from their detailed description in Chapter 4. They seem to share more than a passing resemblance, particularly when it is considered that amplification, although germinal, is, in fact, postsynaptic: both may be lateral and to a different degree both are impermanent. Thus, to a different degree, both amplification and magnification reveal similar replicative idiosyncracies of rDNA, but, at present, they are best regarded as different yet related phenomena. As in the case of amplification, magnification has been suggested as a general possibility for nonribosomal DNA (Ritossa and Scala, 1969) but without supporting evidence. The possibility that informational templates may be selectively amplified according to the needs of the cell seems therefore to be in more than one mind, and perhaps it does occur on a very selective genomic scale as it does (stably) in the case of histone. Amplification on a grand scale as proposed by Callan (1967) was however argued against in Chapter 3 and the preceding section.

(3) Underreplication of DNA is found in chromocentric constitutive heterochromatin during polytenization of larval insect salivary chromosomes (Rudkin,

1969). In these chromosomes the chromocenter includes centromeric DNA (the so-called α-heterochromatin) and noncentromeric DNA (β-heterochromatin), both of which contain highly redundant DNA. Underreplication involves most of centromeric DNA but not, or less, the other DNA (Gall et al., 1971). In addition, constitutive heterochromatin intercalated in euchromatin also contains highly redundant DNA and is not or is little underreplicated. Thus this phenomenon seems to be confined to certain fractions of redundant DNA in larval cells (i.e., centromeric DNA), which it reduces in amount presumably because this DNA is no longer required in these nondividing cells.

(4) Elimination of either entire chromosomes or parts of them, also known as diminution, was described in Chapter 3.

(5) In contrast to the previous ones, two other forms of quantitative variation can and frequently do affect all DNA in the cell in a systemic fashion: these are polyteny (or endoreduplication of DNA in permanent register) and polyploidy. Of the two, polyteny is incompatible with mitosis and only occurs in terminal somatic cell lineages. Polyploidy can and often is compatible with mitosis, particularly in plants, and thus affects either cell or phyletic lineages. Of all forms of quantitative variation under (1) to (5), amplification of rDNA and principally polyploidy have a major evolutionary significance, and hence figure prominently in this chapter. In addition, as far as one can tell, all of the major evolutionary increase of DNA to be discussed, except for amplification, is by linear integration with the preexisting genome and therefore represents a true expansion of the genome. In particular, the increase does not represent polynemy, which means an increase in the number of copies of each DNA molecule in the chromosome.

During evolution from the lower eukaryote to vertebrates the amount of nuclear DNA increases progressively (Mirsky, 1966; Britten and Davidson, 1969). For example, the haploid amount of DNA in a sponge and mammal is 0.06 and 3 pg, respectively. However, by the time evolution reaches the vertebrate stage, the momentum of the increase has slowed down. This general increase refers to the minimum amount of DNA typical of each evolutionary level, which is really the instructive one, for, as we shall see, there are vast departures from this minimum at many levels below reptiles. Within vertebrates the amount of nuclear DNA is roughly correlated with cell size. A minority of nuclei within an organism may increase physiologically in their level of ploidy with a corresponding adjustment in cell size, very clearly so in the case of mammalian liver (Epstein, 1967). However, Mirsky's argument (1966) that the vertebrate correlation with cell size is an adjustment on the part of DNA to the amount of cytoplasm is not persuasive and the converse may prove to be true. Thus the correlation between ploidy and cell size in invertebrates (mealy bug; S. W. Brown, 1969), plants, and vertebrates (haploid *Xenopus*; Brown, 1967) suggests that the primary change is in ploidy rather than amount of cytoplasm. The important point is that this derivative correlation cannot offer an explanation for the "excess" vertebrate DNA, which, therefore, must come from intrinsic evolutionary tendencies of the genome. (Derivatively also, the correlation between cell size and DNA applies as well between cell size and RNA or protein turnover in both nucleus and cytoplasm; see Cell Cycle.)

The evolutionary tendency of eukaryotic DNA seems to be toward the expansion of both the unique and redundant genomes. Of these genomes, the redundant expands relatively much more than the unique, that is, from a very small proportion of redundant genome in the lower eukaryote to about half or more of the total genome in the higher vertebrate (Chapter 3 and preceding section). Thus the redundant genome grows in absolute size considerably more than the unique genome. As it increases, part of the redundant genome it is firmly believed diverges into unique genome. The evolutionary increase of total DNA can, therefore, be explained by an increase of the basic haploid genome presumably caused by chromosomal duplications (as far as we know, affecting any part of the genome) and independent increases of the redundant genome. In addition, the basic genome at certain levels would be increased by a series of ploidizations (discussed earlier), but by this means only up to the level of fishes and amphibians. However, in many classes up to amphibians there are erratic increases of DNA superimposed onto the basic trend of increase in an almost unpredictable fashion. For example, certain amphibians contain much more DNA than the mammal, whereas other amphibians contain less than the mammal. This erratic increase is believed to be caused by extemporary regional duplications involving both unique and redundant DNA alike, but principally by sudden increases of redundant DNA; perhaps, there is also extemporary ploidy. In crossing the level of reptiles the amount of DNA has stabilized at a haploid value of 1.5 pg in birds, double that value in mammals, with reptiles somewhere in between. Given the prior trend of increase of the genome, the stabilization starting with the bird or reptile could be due to a leveling of this trend or to a compensatory loss of genome (albeit on a lower scale than in plants; see below). On this basis, in mammals changing rates of increase and loss of genome, not necessarily in phase with each other, could account for the minor differences in the total size of genome and redundant fraction thereof, which are found. Nevertheless, the near constancy of DNA which is seen in mammals is the more remarkable in that, for example, in placental mammals the known diploid chromosome numbers range from a high of 84 in the black rhinoceros to a low of 17 in certain rodents; hence chromosomal speciation is independent of genome size. Whatever the cause, it is clear that the stabilization of genome size does not depend on the evolutionary onset of homeothermy. Rather, the implication is that, after this onset, mutation may begin to take a greater share than before in the evolutionary process, for the genome is growing progressively less. Of interest, speciation in plants may differ from that in animals. In plants the generation of redundant DNA, which to begin with tends to be relatively more abundant than in animals, is characteristically followed by an appreciable loss of DNA occurring concurrently with the diversification of redundant into unique DNA (Bendich and McCarthy, 1970).

The overall trend of increase of eukaryotic DNA, resulting at one point or another in amounts of DNA presumed to be excessive with respect to informational requirements, is known from classical cytogenetics as the C paradox (from C, the basic amount of DNA per organism). Of course, it is virtually impossible to know just how much DNA may be required in a noninformational capacity at any point. Yet the paradox remains valid whether or not the presumption of "excess" of any kind of DNA is correct, in particular for the mammal, for it must be correct for other vertebrates (viz., amphibians) with more of both unique and redundant DNA than the mammal. As can be gathered from this, the paradox has nothing to do with a massive amplification of the informational genome as had been thought by Callan (1967), or with a massive amplification of regulatory genome for that matter, but instead it probably has to do with the extemporary tendencies of the genome. They suggest, as far as their carrier is concerned, that part of the DNA is not functional, although, it should be remembered, this argument for triviality is only conjecture. These organisms with more DNA than their congeners with the basic amount corresponding to their evolutionary level tend to preserve archaic features, and, biologically speaking, are less successful. This phenomenon is observed within insects (Bier and Müller, 1969), urodeles versus anurans, and primitive versus modern fishes (Hinegardner, 1968). Extra DNA, therefore, slows down evolution. A borderline situation was discussed in Chapter 3 in respect to the supernumerary B chromosomes which are detrimental to the individual which carries them, yet are preserved in the population because of their recombinational benefits. The lesson is that, at any given evolutionary level, to carry too much DNA, perhaps meaning to carry nonfunctional DNA, becomes a liability.

Evolution from the prokaryote to the lowest eukaryote involved a tenfold increase in haploid DNA, and from the prokaryote to the mammal a 1000-fold increase. As discussed presently, this evolution has been accompanied by a narrowing of the compositional variability of DNA. It is revealing that despite the difference of two orders of magnitude in the DNA increase, evolution up to the lower eukaryote took 3×10^9 years, but only 10^9 years from this to the highest eukaryote. Hence early evolution has been much slower. It is tempting to suggest that the faster evolutionary tempo in terms of DNA amounts beginning with the lower eukaryote has been enhanced by the acquisition of new DNA by means of molecular redundancy. More generally, however, since the introduction of redundancy, in turn, probably depended on the proteination of DNA, the newly created cell nucleus should really be viewed as the primary factor in determining the potential for DNA increase. It was mentioned earlier that, to the contrary, cell size does not determine but instead reflects the nuclear DNA amount. It is not suggested that the rapid qualitative diversification of the eukaryote is explainable by the increasing amount of DNA alone, for, to cite an example, meiotic recombination is another principal factor. The point remains that an increase of redundant DNA in relation to control of the genetic apparatus and leading to a greater regulatory capability of the genome could have been an important factor for the diversification. Indeed, according to Walker (1971), highly redundant DNA itself would have helped consolidate the meiotic process.

A molecular parameter of interest in the evolution of DNA is composition. In bacteria the composition varies in the range of 25-75% GC, and the amino acid composition of protein is slightly but definitely correlated with the GC content (Sueoka, 1965). This joint variation in composition has been interpreted to mean that, although the genetic code is semiuniversal (Chapter 1), different codons are preferentially used by different bacteria. According to this interpretation, the variation is therefore contingent on the code degeneracy. There have been reports of sudden major changes in DNA composition in bacterial strains the significance of which is very obscure (discussed by Ycas, 1969), particularly in view of the fact that most bacterial DNA is a unique sequence. DNA composition remains greatly variable in protozoa, fungi, and algae, but less so in plants and invertebrates. In vertebrates, however, it fluctuates within narrow limits around 42% GC. No doubt, bacterial compositional drift is facilitated by short generation times. One explanation proposed for the bacterial drift toward high GC values is in order to avoid dimerization of thymine by natural ultraviolet radiation (Singer and Ames, 1970). The possibility that thermophilic bacteria undergo a similar directional drift in order to stabilize DNA against high ambient temperature may also be true, but leaves unexplained why the same drift occurs in nonthermophilic bacteria. These two explanations are inferential and their validity is uncertain, however. In vertebrates it is not clear whether or not the fixed composition reflects the ancestral composition combined with longer generation times, which would have given it less opportunity to drift, or is somehow more subtly related to intrinsic constraints in the genome. These two possibilities are not mutually exclusive and both suggest an inherent compositional stability as opposed to a convergent compositional variation, either one typically eukaryotic. In keeping with the notion of stability, the fixation in vertebrate DNA composition is more rigid than the stabilization of the amount of DNA past the reptilian stage. To consider briefly the nature of intrinsic compositional constraints, one immediate possibility that comes to mind is the pattern of preferential codon usage as in bacteria; this is the indication from the doublet pattern to be mentioned below. However, since much of the DNA may be noncoding now, this makes one think also of some explanation in the DNA molecules themselves, particularly in view of their much larger size (Chapter 1) and more complex sequence structure than in bacteria. What in the molecules may explain how they fix their composition remains obscure, but one observation (De Voe and Tinoco, 1962) suggests that intramolecular electrostatic forces in DNA may somewhat influence the sequence. Thus it was proposed that regions of high sequence instability could result in mutational hotspots such as are known in phage (Benzer, 1961) at the single base level. Perhaps effects the opposite to this type are prevalent in vertebrate DNA molecules. In summary, toward the upper reaches of evolution the DNA has become very restricted with respect to both amount per nucleus and composition, but the full meaning in regard to the earlier flexibility of either of these parameters is not solved.

The evolution of DNA composition gains an interesting thermodynamical perspective from communication theory analysis (Gatlin, 1968), at the same time stressing the importance of preferential codon usage. Considering a sequence of symbols, the maximum entropy which characterizes a completely random sequence may be reduced either through (a) divergence from equiprobability of symbols, D_1; or (b) divergence from independence of symbols, D_2. These parameters, D_1 and D_2, can be calculated from nearest neighbor data on DNA sequences (Chapter 1). The analysis shows that vertebrate DNA evolves toward its final compositional monotony by increasing D_2 while holding D_1 constant. D_2 thus becomes a quantitative evolutionary index which, moreover, now suggests that a decline in entropy is at the root of DNA evolution. Presumably, a lower entropy is to ensure a more reliable informational system. Gatlin (1968) further shows that, at the level of the protein sequence, this lowering of entropy is a function of the number of codons as well as the particular codons used within each degenerate set. From this conclusion, evolution of DNA in the higher eukaryote would aim not only at a more diversified protein synthesis but also at a more accurate protein synthesis, a dual concept whose internal logic is its best recommendation.

Despite compositional differences over the evolutionary continuum, within any organism the total DNA composition is homogeneous (unimodal) in contrast to the heterogeneous protein composition. For the most part the greater homogeneity reflects the four-monomer composition of DNA versus the 20-monomer composition of protein. Mathematical models have been proposed to further explain this essential DNA homogeneity (Freese, 1962). There are exceptions to it, however, such as the extreme DNA satellites which differ compositionally from main DNA, perhaps the most extreme being crab poly dAT. It is obvious, moreover, that the exceptions will mount simply as the single gene level is approached, for instance, note the rDNA cistrons. Not surprisingly, therefore, in looking at DNA fragments larger than 10^6 daltons the standard deviation from the mean GC value is about 2-3% in bacteria and twice this in mammals, both figures being larger than expected on a random base sequence. A rhythmical compositional periodicity of DNA has been seen in yeast but not in the mammal (Chapter 1). On the other hand, the frequency of small isostich fragments (Chargaff, 1968), which is random in bacteria and yeast, is nonrandom in mammals and indicates a greater directional replacement between purines and pyrimidines. A nonrandom sequence in the mammal, as is clear from the data, is in accordance with the nonrandom pattern of base doublets which is found to be similar in all vertebrates (Subak-Sharpe, 1969), evidently in agreement with the compositional stability of their DNA. This doublet pattern reveals interestingly that mammalian DNA has too few GC doublets to permit equal coding by all possible GC-containing codons, that is, a definite restriction with respect to these codons. Concerning the residual inhomogeneity of mammalian DNA that is also disclosed by the data, no explanation is at hand but it may reflect a moderate phyletic divergence overriding the essential homogeneity. Finer analysis could conceivably show that these inhomogeneities have a specific functional meaning. For example, it was seen in Chapter 1 that (a) structural links may occur along mammalian DNA molecules; (b) these molecules may require structural links between strands for reasons connected with replication; and (c) during evolution of protein-nucleic acid interactions, such as occurring in promoter and initiation sequences, certain sequences could be favored over others. These situations could create certain compositional constraints (vide supra) showing up as minor DNA heterogeneities, especially between taxonomic groups, and contributing to the randomness typical of the large genome, whatever other reasons there may be for it.

Homology of sequence between organisms is related to DNA composition. Bacteria with their wide compositional range and almost exclusively unique sequences can share as little as 5% of their sequences in common (De Ley, 1968). Mammals have a much narrower compositional range and have unique and redundant sequences in roughly equal proportions. Considering unique sequences first, the mouse and guinea pig share up to 77% of their sequences, the cow and pig 83% (Laird et al., 1969). These figures convey an idea of the considerable degree of homology between related unique genomes, which of course is expected to decrease between distant genomes. Considering redundant sequences, mammals share 20% of these sequences, but within primates as much as 95%. This partial homology is likely to include sequences with a greatly variable complexity (frequency distribution) in the typical and extremely redundant families alike. Irrespective of their age, within any of these families the homology of sequences remains greater, however, than between sequences of related taxonomic species. By the criteria of hybridization employed, between vertebrates the correspondence in the redundant genome decays exponentially over evolutionary time with a half-life of 10^8 years (Hoyer and Roberts, 1967). Superimposed on this decay is the rapid replenishment of the redundant genome (Chapter 3). For example, that part of the redundant genome which codes for RNA restricted to the nucleus seems to be of more recent evolutionary origin than the genome which codes for cytoplasmic RNA including mRNA. In addition, part of the replenished redundant genome diverges, perhaps not so gradually as steadily, into unique genome. Many of the redundant sequences are known to be evolving at a faster rate than protein. Given their general homology, an interesting question that remains to be answered is to what extent recombination or other forms of exchange occur preferentially within each family of these sequences. It is also known that the unique sequences are evolving at an even faster rate. For example, despite the above mentioned sharing of sequences, in artiodactyls (cow and pig) and rodents the rate of divergence of unique DNA is three and thirty times, respectively, faster than the rate of amino acid substitutions calculated in hemoglobin and insulin (Laird et al., 1969). Part at least of the divergence must therefore be neutral with respect to evolution.

On the premise that the unique genome is the one which codes for mRNA, the following overall picture seems to emerge. The redundant genome is replenished more frequently yet evolves more slowly than the unique genome, and this, in turn, evolves faster while assimilating portions of what was previously redundant genome. If this appraisal is correct, the resulting implication is that there should occur a continuous buildup of redundant genome, which would agree well with its faster evolutionary growth in size. A manifold paradox confronts us here: (a) is the mammalian genome still expanding as it has expanded during its early evolutionary history; or (b) how important is a concomitant loss of redundant genome to balance a buildup (as is found in plants); and (c) what proportion of redundant genome and at what rate is contributing to the unique genome? Some of the biological implications of these data in relation to the function of the unique and redundant genomes have been considered in the preceding section.

Evolution of DNA as a genetic code was examined in Chapter 1. To recapitulate briefly on the early beginnings of the code, initially the DNA could have had two instead of four bases, i.e., adenine and guanine or hypoxanthine (Crick, 1968), a possibility which has been entertained more in coding than structural considerations. With his "principle of continuity" Crick proposes that, if there is to be evolution, no step in the process can invalidate any previous step but can only adjust it. On this basis, the initial codewords must have been triplets, even if perhaps read also only in two bases. Although not the only possible explanation, codon size may have been determined by the helical diameter of the adaptor tRNA, for this optimizes the fitting together of two adaptor molecules on the mRNA. If correct, this view implies a formative influence of tRNA on the code, for codon size is one of its basic parameters. A more ancestral and hypothetical reason for codon size would be that it was determined by the unit dimensions of polynucleotides originally acting as direct templates for amino acids (Chapter 1). These two views resemble each other in that what determines codon size is not in the DNA but in decoding.

However, the general point was made in the previous section that evolution of the code is not dissociable from evolution of DNA as a molecule; it is just conceivable that DNA dictated some of the code parameters, perhaps this one.

Introduction of degeneracy in the form of the wobble pairing has significantly added to the plasticity of the code, this representing an important aspect of the late code evolution. Additional functional possibilities contemplated for the wobble pairing, all derived from its primary coding function, are: (a) to permit evolution of DNA composition without altering protein composition as much (Sueoka, 1965); (b) conversely, to propitiate non-Darwinian evolution of protein (King and Jukes, 1969); (c) to absorb the physical stress in codon-anticodon pairing (Fuller and Hodgson, 1967), (d) to modulate translation (Chapter 5); (e) to prevent mRNA autocomplementarity (Rich, 1968); (f) to the contrary, to maintain the primary structure of protein in the face of mutational changes affecting the higher order structure of (RNA phage) mRNA (Adams *et al.,* 1969); and (g) to create conditions for the emergence of families of redundant sequences (Walker, 1969). Not all of these views need be true, or, conversely, exclusive of one another; yet there are still other views. From this, the wobble begins to look as a most cunning device of the genetic code not just for economizing tRNAs—which it does not do certainly in higher organisms. Being that this functional plasticity of the wobble matches, in particular, that of tRNA, itself a cunning molecule, their evolutions may be just as ancient. Among all RNA species, especially tRNA retains many of the physical and functional markings of an ancestral molecule (Chapter 2 and this Chapter). Certain facets of interaction between tRNA and the genetic code, including the wobble, during a common evolutionary history were pointed out in discussion in Chapter 1. This interaction could correspond with the stage during which the code was still moulding its plasticity.

Macromolecular Associations of Eukaryotic DNA

In their permanence the macromolecular associations of DNA are so unlike one another in the prokaryote and eukaryote that they unquestionably represent a fundamental difference in organization of the genetic system in these organisms. Bacterial DNA is essentially naked and freely accessible to regulatory molecules, except for a possible but still undecided structural association with polyamines. These regulatory molecules are proteins (repressors) and small compounds (inducers or corepressors) which act as effectors on the repressors and thereby elicit or inhibit, respectively, the informational response (Chapter 3). This response can affect any part of the facultative genome, can be as fast as it is, and can be coupled to translation, all because the bacterial DNA is free, that is to say, all macromolecules which associate with the DNA do so impermanently. These conditions do not apply to DNA replication which, by contrast, is a continuous process (assisted by a complex of replicatory proteins) operating in the proximity of the bacterial cell membrane. Basically, the expression of genetic information is negotiated by two proteins, repressor and polymerase. However, a host of other proteins interact specifically with the polymerase or the DNA, a distinction not always visible: a σ initiation factor which is a freely exchangeable moiety of the polymerase; a ρ termination factor (most studied in bacteriophage); a stimulatory protein activated by cAMP; and, in the case of rDNA, but perhaps interacting with informational DNA as well, a stimulatory ψ protein controlled by GT_4X (Chapter 4). The interactions mediated by these specific regulatory proteins are essential to the metabolic homeostasis of the bacterial cell, and, again, their immediacy is possible because DNA is free.

Proteination of eukaryotic DNA entails a tremendous rise in multimolecular complexity resulting now in a recognizable chromosomal apparatus. In reality, this apparatus functions as a dynamic supramolecular complex of which DNA is only a part, from what was said, with more perhaps than a coding function. The fact is, although the basic molecular attributes of DNA have not changed since the prokaryote, to consider DNA by itself in the eukaryote from the functional standpoint becomes now a biochemist's abstraction. Evidently, this condition both limits the availability of DNA to any external effector and demands specific mechanisms to bring about their interaction, which is precisely one of the functions of the chromosome. As will become clear, however, the present interest in the chromosome is only matched by the ignorance (or, what is equal, some knowledge at a rather superficial level) concerning its functional properties. First of all, the chromosome functions within a very controlled obligatory environment, the nucleus. Second, as may already have been gathered, it represents the combination of DNA with macromolecules which administer its expression from within the chromosome, even though some of these molecules may be in a state of constant flux with respect to the chromosome, in which case the nuclear environment plays an even more important role. As far as is known, most histones and certain residual (acidic) proteins never, or rarely, leave the chromosome; this reveals that we now have a group of proteins whose function is exclusively in the chromosome, although, of course, they continue to be produced in the cytoplasm. Other acidic proteins and perhaps chromosomal RNAs certainly do leave the chromosome, and, in addition, some acidic proteins most probably establish an exchange with the cytoplasm. The functional relationship between these molecules, and between them and DNA, is today more a matter of speculation than informed debate. Nevertheless, from discussion of these molecules in Chapter 3, the implication is that their function is colligative: histones function as general repressors but not as the exclusive repressors; acidic proteins and chromosomal RNAs are believed to intervene in depression but not to the exclusion of histones.

The physical arrangement underlying the function of these molecules in the chromosome is also not understood. It is clear, however, that this arrangement represents an elaborate set of interrecognitions, namely, protein-protein, protein-DNA, and DNA-RNA. This set of recognitions, which must themselves have evolved concurrently with the chromosomal apparatus, may be qualitatively one of the most complex in the eukaryote. That part of histone bonded ionically to phosphate groups in DNA has the simplest pattern of recognition and has evolved the least. Indeed, evolution of the histone molecule has been remarkably conservative, to the point where it reminds one of the evolution of DNA; this, no doubt, attests to the essential functions of the whole histone molecule in relation to the DNA. Contrary to histones, which are relatively few, acidic proteins and chromosomal RNAs represent highly heterogeneous molecular populations. Acidic proteins, in particular, consist of many classes of proteins, none of them too well characterized with respect to their function, which thus far has permitted only their tentative assignment to a derepressive function. The participation of chromosomal RNAs in derepression is not established either, and more about this will be said later in connection with regulatory RNA. Although the nature of the evidence on the RNA lends itself more readily to a design of experiments than that on acidic proteins, recently there has been some serious questioning of this evidence (Chapter 3). These proteins and RNAs are partly tissue and even function-specific, hence are more differentiated than histones. On this basis, therefore, they have evolved more recently or more dynamically than histones, which are largely not tissue or function-specific. Perhaps partly a reflection of this is that the coordination with DNA replication is much less for acidic protein than for histone synthesis.

As far as one can tell, these eukaryotic molecular classes (histone, acidic chromosomal protein, chromosomal RNA) have no directly recognizable prokaryotic ancestry, yet this remains an interesting question to settle. From a functional viewpoint, however, some of the molecules may be the analogous counterpart of the bacterial proteins freely interacting with DNA, except that they are now a more integral part of the chromosome. By comparison to the prokaryote, the general import of this discussion is that any interaction from outside the eukaryotic chromosome intended to mobilize the information in DNA is directed as much or more to these macromolecules than to DNA itself, hence, among other consequences, the less immediate informational response of the eukaryote. As elaborated upon in Chapter 3, the basic mechanism of directed transcription would thus lie within this DNP complex which actually carries the template specificity of chromatin. As a working model it was proposed that as new informational requirements arise, for example, during a developmental situation, the chromosomal apparatus undergoes a series of preordained organizational changes in its regulatory components, including exchange with the nuclear environment, resulting in the appropriate release of information. As further discussed below, the model goes on to propose that these interactions within the apparatus are needed for progressively implementing a genetic program that is intrinsic to the chromosome but capable of adjusting to the cell environment. It is suggestive in this respect that cytodifferentiation (and later hormonal effects) frequently depends on chromosomal replication, since this provides a good opportunity for any reorganization necessary for a change in program to occur. In the prokaryote, by contrast, any equivalent of a program can only result from free interaction of DNA with the cell, that is, at a general metabolic level.

According to this model, the eukaryotic program is always in relation to genetic circuits involving the function of a regulatory genome and whose orderly realization is accompanied by changes in the state of organization of the chromosome. Basic to the model is that the regulatory genome must be highly responsive to the status of the cytoplasm in order to generate the signal for the next release of information with its attendant organizational state. This release of information in turn may modify the status of cytoplasm, and so on. These regulatory signals would therefore act on the chromosome in relation to progressive phases or cycles of intracellular nucleocytoplasmic interaction, an aspect which is again most clear during embryonic differentiation. (The present search for these signals and their possible nature were described in Chapter 3.) Even during differentiation the chromosome is already responding to extracellular influences (Chapter 6), which will become more pronounced as the cell approaches maturation in the postemoryonic organism. As the period of intense differentiation finally comes to an end, in the mature functional cell most of the obligatory programming as defined above would have been achieved. As the result of the facultative genome prevailing at this stage, the informational response will follow upon activation of the regulatory genome, for example, by external hormone, more immediately than before. A less extensive regulation of response is expected at all stages in the case of the auxiliary genome for rRNA. On the other hand, particularly at the mature functional stage, the response of part of the tRNA genome may be as regulated as that of the informational genome, in other words, part of the tRNA genome may function as a facultative rather than auxiliary genome. To be sure, as clarified above, these generalizations are about an intricate hierarchy of chromosomal specificities involving genes and the macromolecules associated with them, the nature of which we fail as yet to understand. The main point here is that these specificities represent the emergence of a genetic apparatus qualitatively different from the ancestral prokaryotic apparatus. By prokaryotic standards, the genetic apparatus of the eukaryote is much further evolved than the translational apparatus.

At some time the eukaryotic system must respond to extracellular signals, chief among which are hormones and embryonic inducers. In all cases these signals seem to evoke a chromosomal response noninstructionally, hence permissively on the part of the preexisting program. In short, external hormones and inducers cannot evoke any response from a cell unless this cell is programmed for that response. It follows that this noninstructionality of signal is generally in relation to a limited aspect of the genomic potential, which is essentially that represented in the facultative genome. It also follows that this particular genome is mostly responsible for the sort of compromise which results between the intracellular and extracellular regulation of the cell. Although not the place to dwell on the broader implications of noninstructionality, there is one further point which deserves mention here. The great profusion of hormonal signals would appear to demand an adjustment in the molecular mechanisms mediating the chromosomal response such that, taken collectively, these mechanisms might seem to be rather uneconomical compared to those mediating the response to less complicated (nonhormonal) signals. Above all, this adjustment is absorbed in the molecular degeneracy of the chromosomal apparatus outlined earlier, that is, the excess number of molecules in each molecular class. In the last analysis, however, the adjustment calls for a permissive organization of the genetic system more than it does for any costly circuits in the system. Noninstructional signals acting on permissive systems are a widespread property of biological systems. As an example, the reader may consult Jerne (1966) in order to see how this property applies also to antibody formation.

Eukaryotic Cytoplasmic DNA

Eukaryotic cytoplasmic DNA is present mainly in organelles, mitochondria and chloroplasts, which appear to have a bacterial-like symbiotic ancestry (Chapter 4). (An additional DNA-containing organelle, the centriole, was succinctly considered in connection with the cell cycle in Chapter 6.) Nevertheless, these eukaryotic organelles have come to differ in their stable permanence in the cytoplasm from viral episomes or plasmids whose sojourn in the bacterial cytoplasm depends on their genetic constitution (Signer, 1969), among other factors. Attardi and Attardi (1969) have put forward a fundamental hypothesis for the origin of the mitochondrion. These authors draw attention to the similarity in structural organization between the mesosome of gram-positive bacteria and the primitive mitochondria of protozoa. The mesosome is a most important organelle in bacterial replication, which, moreover, because of its content in respiratory enzymes and membranelike function, qualifies as the nearest equivalent to a bacterial mitochondrion. The Attardi's then go on to propose that the true mitochondrion originated from such a mesosome which "escaped" from an earlier organism carrying with itself the origin of the bacterial chromosome. This region of the bacterial chromosome is known to contain information for DNA replication, the translational apparatus, and, perhaps, certain membranous structures. Correspondingly, at least these categories of genes have been preserved in the mitochondrion up to the higher eukaryote.

Mitochondrial DNA is a unique sequence except in the lower eukaryote for its redundant rDNA portion. These properties of mitochondrial DNA agree with a putative prokaryotic ancestry. By contrast, chloroplastal DNA is considerably sequence redundant, which is presumably indicative of divergence from its primitive ancestry.

During evolution mitochondrial DNA has been subject to composition drift. For example, yeast mitochondrial DNA is not homologous to *Xenopus* or chick mitochondrial DNAs, which are homologous to each other (Dawid and Wolstenholme, 1968). As with nuclear DNA, the mean GC composition of mitochondrial DNA varies widely in the lower eukaryote but not in higher plants or vertebrates. In the lower organisms the variation in composition is independent of nuclear DNA. In the higher organisms the two DNAs resemble each other, presumably indicating that constraints similar to those mentioned earlier for nuclear DNA are in operation for mitochondrial DNA. This might also indicate a progressively greater functional dependence of the mitochondrion on the nucleus (see below), perhaps witnessed further by the great reduction in size of the mitochondrial genome in higher organisms. This reduced genome is a rare instance of evolutionary simplification.

The organelles maintain a separate genetic system within the cytoplasm, yet they function semiautonomously, that is, depend on nuclear DNA for part of the genetic information they need to function. Much of this nuclear information is converted into protein for organelle function within the cytoplasm, although some may be converted within the organelle (Chapter 4). Organelles and cytoplasm form, therefore, an integrated metabolic system that is specified by two genetic systems. Whether or not the comparison is accurate, the obligatory dependence of certain DNA-containing particles in the cytoplasm of *Paramecium* on the so-called metagons (a form of masked mRNA made by nuclear genes) serves as a fruitful model for the nuclear dependence of the eukaryotic organelles. These particles in *Paramecium* are designated K, μ, λ, and σ, and all resemble gram-negative bacteria (Gibson and Beale, 1963). In view of the dependence of the organelles, and since their functional integration with the nucleus increases during evolution, the permanent lack of physical integration of the organelle genome with the nuclear genome remains somewhat of an intriguing problem. It has been argued, therefore (Attardi and Attardi, 1969), that the organelles fulfill certain functions of general value to the cell, which cannot be delegated to the cytoplasm. (These functions considered in Chapter 4 do not exclude other well-accepted functions concerned with respiration and self-maintenance of the organelle.) Lastly, in a similar vein it was proposed (M.M.K. Nass, 1969) that the nuclear genome may have retained an ancestral blueprint of the prokaryotic DNA sequences later appearing in the organelle, so that the sharing of sequences would obviate the need for integration. However, apart from lacking direct support, carried to the extreme this argument obviates the need even for the mitochondrion.

The informational content of mitochondrial DNA is larger in the lower than higher eukaryote (Chapter 4). Indeed, the latter seems to have approached a minimum of informational content. A first possible explanation for this loss of information is that the predicated capture of the prokaryotic ancestor occurred only once in the eukaryotic realm. It can be imagined in this case that the capture was followed by a gradual simplification of the organelle genome concurrent with a progressive loss of function to the nucleus, as evolution proceeded toward the higher eukaryote. Since the number of generations elapsed from the ancestor to the present organelles have been much less in the case of the higher than lower eukaryote, one must then argue that the informational simplification was most conditional on the greater informational store accruing simultaneously in the higher eukaryotic nucleus. In addition, if the rapidly dividing cells of the lower eukaryote required more supporting organelle genome, this would have tended to preserve or even encourage a further complication of this genome. This explanation contends that the lower eukaryote mitochondrion is closer in its characteristics to the ancestral form. The hypothesis of Attardi and Attardi (1969) described above probably belongs in this category of explanation. A second possible explanation for the loss of information is that the organelles in the lower and higher eukaryote originated as separate events and from separate ancestors (Raven, 1970). Each ancestor, for all we know, could have had a different genome size and a different subsequent evolution. This explanation contends that the informational loss may be more ostensible than real. Obviously too, this explanation carries no prediction as to what determined evolution of the organelle.

RNA

One useful way of looking at the evolution of RNA species is that it pertains to the broader question of the evolution of the genetic code and the translational apparatus. Because this way of looking at the evolution of RNAs ought to provide the completest insight, it has been adopted for the remainder of the chapter. Inspection of the headings should reveal that when appropriate the code and translation are considered together with the RNAs rather than as separate subjects.

A convenient distinction drawn by Woese (1970b) is between essential and nonessential or trivial evolution of an RNA species. Essential evolution concerns those aspects affecting basic functional properties of the molecules. Trivial evolution does not affect these basic properties but allows a certain amount of incidental variation in molecular structure to occur. Taking this distinction to the level of the genetic code, one implication of a code which is universal in all its fundamental traits is that all organisms using such a code shared a common ancestor possessing an already fully evolved code, even though this ancestor need not have been the only primeval organism. On the other hand, a genetic code which is not universal in its fundamental traits implies either a nonfully evolved code at the time of divergence of these organisms from the common ancestor, or a fully evolved code at that time but which subsequently underwent divergent evolution. Hence any significant degree of universality as in the present code of the prokaryote and eukaryote is inevitably presumptive of some common ancestry in these organisms. This much is obvious enough, but the point that matters here is that it remains inconclusive (Chapter 1) to what extent this universality represents essential or trivial determinants of the code evolution. This means that the degree of universality with respect to fundamental traits of the code is not known, and, therefore, how much these aspects of the code had evolved by the time of divergence cannot be known.

In the discussion of RNAs and other translational molecules that follows we will be confronted with a general universality spiced with substantial refinements particular to either the prokaryote or the eukaryote. The essential versus trivial nature of these refinements is again difficult to assess. The same therefore applies to the extent to which the evolution of the RNAs and translational molecules was shared. As an example, in the usual case it will be impossible to decide whether a certain molecular conformation in the eukaryote is actually necessary (essential), or functionally better, or just occurs (trivial). The remarkably few exceptions to the general trend of conformational change which is found in any molecular type—almost invariably toward complication—suggest nevertheless that much of the change is essential. In practice, once the present limitation is understood, it becomes less critical than it might appear to be, since any evaluation of evolutionary relevance in the molecular data of necessity remains conjecture. However, because of this limitation, the intention in what follows is to present a general evolution-

TABLE 6
Species of RNA in the Prokaryote and Higher Eukaryote[a]

Species	No. of cistrons	Precursor (S)	Mature (S)	No. of sequences
rRNA		A. Prokaryote		
Major	10-40 (poly-cistronic?)	17	16	> 1?
		24	23	> 1?
Minor	5-10	> 5	5	1-2
mRNA	~ 4×10^3 (many polycistronic)	—[b]	Varies with each mRNA	Each species 1
tRNA				
(~60 species)	Total > 60	> 4	4	Each species 1
Other RNAs				
7 S	?	?	7	1?
Poly A	—	—	~ 6	1?

TABLE 6 (Continued)

Species	No. of cistrons	Nuclear species (S)	Cytoplasmic species (S)	No. of sequences
rRNA		B. Higher Eukaryote		
Major	10^2-10^3 (polycistronic)	~45	18 28 and 5.5 (H bonded)	Each species 1-2
Minor	?	8	–	Several
	10^3-10^4	>5	5	1-2
DNA-like RNA				
mRNA[c]	10^5?	>Cytoplasmic?[f]	Varies with each mRNA[g]	Each species mostly 1
Labile RNA[d]	10^6?	up to 100	up to 90	High redundancy
Chromosomal RNA	High	>3	>3?	High redundancy
tRNA (>80 species)	Total 10^2-10^3	>4	4	Each species 1
Other RNAs				
4-7 S (~25 species)	?	4-7	4-6?	Each species 1 or >
Poly A[e]	–	up to 45 or >	–	1?

[a]Excluding the higher eukaryotic cytoplasmic organelles. The data are discussed in full in Chapters 2, 3, and 4. Number of cistrons in the eukaryote are rounded to nearest order of magnitude. Number of sequences disregards the possibility of a few point mutations. S means approximate sedimentation value.

[b]The informational sequence is transcribed in its final size but may carry some noncoding sequence derived from the proximal regulatory genes.

[c]The number of informational cistrons in the mammalian genome (complexity 3×10^9 nucleotide pairs) was calculated by assuming a hypothetical 50% of unique genome and a complexity of 3×10^3 nucleotide pairs for each cistron: this number, representing a maximum estimate, is 5×10^5. By definition, the informational genome represents a unique DNA sequence.

[d]The number of sequences in the mammalian redundant genome was calculated using the same assumptions as in footnote (c), except for an arbitrary complexity of 10^3 nucleotide pairs for each sequence: this number is 1.5×10^6. It is not known whether all these sequences would be transcribed. The relationship between labile RNA and the possible precursor of mRNA is discussed in Chapters 3 and 7.

[e]Possibly not a separate species but a part of DNA-like RNAs.

[f]This RNA under question stands for a possible precursor of mRNA.

[g]This RNA may arrive in the cytoplasm in the form of an immature precursor. Probably contains extensive noncoding sequence additional to the informational sequence.

ary appraisal of RNA species in regard only to their structure and function, as far as the data permit and without presuming to deal with first evolutionary causes. This appraisal will assume for directness a sequential evolution of the species, although the phyletic evolution to which it relates has been a parallel one. The biogenesis and regulation of these molecules were treated partly from this evolutionary angle in Chapters 2 through 5. These aspects will not be covered again except insofar as they relate clearly to structure or function of the molecules.

A summary of RNA species is presented in Table 6 to serve as a frame of reference for the sections that follow. The table shows the greater quantitative complexity of eukaryotic RNA as judged by the number of cistrons. More significantly, it shows the qualitatively greater molecular complexity of the eukaryote in the form of new RNA species and subspecies in three out of the four major categories tabulated. Because some of these categories are conventional in the case of the eukaryote, the true complexity is, in reality, even greater. The exceptional category is tRNA where only the number but not the character of species has changed. However, in common with rRNA, eukaryotic tRNA is more sophisticatedly substituted, which probably means a correspondingly more sophisticated conformation of the molecule. Evolution of the structural characteristics of each RNA category in relation to function will be discussed. As a conceptual introduction to these categories, Table 7 indicates their relationship with the types of sequences according to which the eukaryotic genome is organized. The principal functional domain(s) of each type of sequence is also shown.

A generalization is that eukaryotic RNAs are more highly controlled, and hence, subject to more complex regulatory mechanisms than their prokaryotic counterparts. This is reflected at the level of the precursors of the RNAs whose function is to provide a scaffolding for the shaping of the mature molecules, and which are larger relative to the mature molecules in the eukaryote. A larger discarded portion of the precursor as this implies is most evident in rRNA, probable but not established in mRNA, and possibly least evident in tRNA. One could ask whether each precursor has a more similar conformation than the resulting mature molecule in the two types of organism, as predicted on grounds of evolutionary recapitulation.

Transfer RNA

As must be fairly obvious, without the adaptor function of tRNA in conjunction with the synthetase function, the present genetic code has no meaning. In this section we shall contend that this obligatory need for tRNA in translation is related to the ancestry of the molecule, whose origin can be traced back to a prototRNA (Dayhoff, 1969). In fact, the evolution of DNA sequences is nowhere more directly traceable at present than in the evolution of this RNA molecule.

Primitive tRNA may have functioned in various capacities in macromolecular anabolism other than in its present adaptor role, a few of which may still remain as atavistic functions (Chapters 1 and 2). Later, as tRNA achieved the adaptor role, a multiplication and diversification of the primitive structural gene occurred until it accommodated all late-appearing amino acids. The result is the present basic number of tRNA species, which is the same in all living beings.

TABLE 7
Categories of Eukaryotic Transcripts in Relation to the Structural Organization of the Genome[a]

Structure of sequences	Prevalent type of function	First-order function	Second-order function	RNA species	Max. order of intrinsic functional RNA structure	Functional half-life of species
Unique	Informational	Template for translation	Ultimately regulatory	mRNA	Secondary or possibly lower tertiary	Wide range
Redundant	Noninformational	Possible regulator and structural component of informational transcription	Partly informational	Labile RNA	Probably secondary	Mostly brief
		Auxiliary for translation	Partly regulator of translation (tRNA)	rRNA, tRNA	Higher tertiary (highly modified)	Relatively stable

[a] Among redundant sequences only the typically redundant are considered. Not all RNA species entered in Table 6 are included.

On the other hand, it is almost certain that the total number of species above this basic number has increased during subsequent evolution from the prokaryote to the eukaryote. The cistronic multiplicity of each species certainly has increased. Notably, however, despite vast differences in genome size, eukaryotic tDNA has ultimately attained an essentially constant absolute amount, e.g., between insects and man, indicating that the cistronic multiplicity has stabilized during evolution. The ancestry of the tRNA molecule is clearly shown in Table 6 where, apart from the previous quantitative aspects by and large, the tRNA category is the one with the least qualitative innovation—indeed, reminiscent of DNA and histone evolution.

The common origin of tRNAs is further attested to by the overwhelming preservation of size (slightly larger in the eukaryote) as well as primary and secondary structure. Gross composition of total tRNA has remained constant at about 60% GC from bacteria to mammals despite the widely different DNA composition between these organisms. Obviously, this constancy will not be expected to apply at the level of each individual tRNA species, although it holds the most for those tRNAs which handle the same amino acid. There have been occasional insertions as well as deletions in the primary structure resulting in conservation of base numbers only in the paired or helical regions of the molecule, in this case, with effects on the secondary structure. Frequently, however, there have been reciprocal base changes in the paired strands, again indicative of a pressure to conserve these regions. Presumably for reasons of functional tertiary structure, some of the bases in the looped or unpaired regions were also conserved. All this has suggested that as a whole the structure of tRNA is strongly refractory to mutational drift and from this general similarities in tertiary structure can be expected, despite a larger eukaryotic molecule. This prediction seems to be well in agreement with present evidence (Chapter 2). A well-defined and conservative tertiary structure is, of course, a prerequisite for the common basic functions of the molecule: the conservatism has therefore to do with the function of the molecule rather than, as could be argued, with the fact that it is a final gene product.

Even if a strong conservatism is disclosed in the preceding paragraph, a certain phyletic divergence of tRNA and synthetase is only to be expected and is probably found. An incipient difference may be present in loop II with mostly purines and cytosine in the prokaryote, but purines and dihydrouracil in the eukaryote (Woese, 1970b). Another difference seems to be the preferential use by the prokaryote of the 2 + 2 wobble pattern (shown in Table 3), as opposed to the 3 + 1 pattern by the eukaryote. That eukaryotic tRNA works sufficiently well *in vitro* with the prokaryotic ribosome should not, therefore, be construed to indicate a perfect functional harmony. Likewise, a slight functional distinction between prokaryotic synthetases and synthetases of eukaryotic organelles (Chapter 4) points to a latter day evolutionary divergence. Further aspects of refined divergence are illustrated below.

In the remote past, primitive tRNA may have participated significantly in the early evolution of the genetic code, if not more generally in the early evolution of translation. Although realistic when considering all the evidence, this view—implying that the present translational role stems from an initial deterministic influence of tRNA on the evolution of these systems rather than from a late incidental appearance of the molecule—is not held unanimously (Chapter 1). Evidently, because of their conjectural nature, neither of these views can be proved or disproved. Be that as it may, through unknown intervening stages tRNA eventually settled in its translational role. Thereafter, most of the protracted divergence leading to tRNA species for different amino acids appears to have occurred prior to the time when *E. coli* branched off the evolutionary mainstream (Dayhoff, 1969). This constitutes one of the few instances where the semiuniversality of the code and an RNA species can be followed in parallel. However, according to genealogical analysis by computer, the appearance of bacterial initiator tRNA possibly occurred after the branching of *E. coli*. If that is the case, the presence of this initiator tRNA in eukaryotic organelles can only be explained by a symbiotic capture later than the branching. For any tRNA there is a greater distance in terms of structural differences between the prokaryote and eukaryote than between eukaryotic kingdoms (McLaughlin and Dayhoff, 1970). This wider structural gap between the prokaryote and eukaryote is now reflected in the barrier against cross-acylation. By contrast, the lesser structural gap between eukaryotes has resulted in no such clear-cut barrier (Sueoka, 1965). Nevertheless, the overall rate of evolution of the molecule has been similar in the prokaryote and eukaryote, and, despite their effects, the structural differences are relatively minor. These differences must therefore be proportional to the longer span of evolutionary time elapsed between the prokaryote and lower eukaryote than between the lower and higher eukaryote; the first span is approximately three times longer. Using this argument, the greater diversification which is characteristic of higher eukaryote tRNA would not be primarily a function of time but instead a true physiological necessity.

In many of its functional traits tRNA behaves like a protein (Chapter 2), or, to use Crick's aphorism (1968), "like Nature's attempt to make RNA do the job of a protein." Hence tRNA not only helps to translate from nucleic acid to protein but also behaves in between nucleic acid and a protein, which must mean that the function of tRNA strongly depends on tertiary structure. In agreement with this, the higher order structure of tRNA is one of the most (if not, indeed, the most) elaborate of any nucleic acid, and, like that of a complex protein, behaves allosterically during function. Catalysis having been the business of proteins since their manufacture was perfected, the question arises as to why they did not capture the tRNA function. One reason may be the ancestral coadaptive value of tRNA with regard to the code that was mentioned and which is clear on general evolutionary grounds (Chapter 1) and, particularly, in the case of the wobble; suggestive in respect to the code is the possible mutual influence between codon composition and total tRNA composition, hence, tRNA function. A second and perhaps more cogent reason is efficiency; recognition of mRNA by tRNA takes less room (a triplet) than by protein (Crick, 1968). That tRNA is, all told, an economical molecule can be seen from a minimal calculation: recognition sites for CCA transferase, amino acid, synthetase, codon, (possibly) 5 S RNA, ribosome, translocase, plus at least two translation factors, divided into 80 nucleotides gives an average of about ten nucleotides per site. Even if there must be overlap between sites (many of which are put together by tertiary structure), this figure should be compared with the vast amount of supportive structure relative to the active center which is found in any enzyme. These aspects—coadaptation and efficient translation of the code—however, could have been interrelated at some time.

Because of its high degree of posttranscriptional modification, each tRNA, in actuality, requires the support of a host of informational cistrons to specify the appropriate modificatory enzymes. In their structural effects these modifications are the equivalent of the proteination which, in the eukaryote, is common for most if not all RNA species with the exception of tRNA. (Another exception is perhaps the low molecular weight nuclear RNAs described in Chapter 2, but this is not certain.) Accordingly, modification has more evolutionary momentum in tRNA than in any other nucleic acid. The range stretches from apparently a single modified base per molecule in the probably ancestral-like tRNAs exclusively involved in bacterial cell wall synthesis to roughly between 10 and 20 bases per molecule in the more evolved mammalian tRNAs. However, while it is evident that the increased molecular sophistication introduced by modification must serve a purpose related to functional maturation, it is not clear what precisely this is (Chapter 2). Because of their nature, these modifications relate the unfinished tRNA molecule to many intermediary metabolic pathways as well as to hormones (Chapter 5). In fact, this relationship of tRNA to hormone in respect to structural specification strengthens the qualifications of tRNA as a frequent facultative product of the genome (Chapter 4).

A protean variety of tRNA function was listed in Chapter 2. In all probability this functional variety has strong evolutionary connotations, or, stated differently, it is difficult for one to imagine such variety without some very deep-rooted motivation. It is clear that the functional differentiation of tRNA species basically subserves the need for multiple amino acids in translation. However, tRNA is also specifically involved in the translation initiation complex (Chapter 5), a most important point of translational regulation. The additional possibility that the differentiation of the molecule is exploited for modulation of translation was explored in Chapter 5. Transfer RNA also participates in non-translational regulatory roles in the biosynthesis of amino acids (Chapter 4), which means it participates directly in intermediary metabolism. In some of these syntheses a reciprocal relationship is established with respect to the synthetase, whereby a moderate regulation of this enzyme ensues. Multiple connections with intermediary metabolism for the modification of the tRNA molecule were noted in the preceding paragraph. Still several other nontranslational roles were pointed out in Chapters 2 and 5, of which one particularly suggestive of a primitive function is the intervention in nontemplated peptidoglycan syntheses for the bacterial cell wall. All these functions considered, tRNA begins to look like an ubiquitous auxiliary of translation which also has many central regulatory and metabolic roles; some of these nontranslational roles could be another reason for the progressive complication and degeneration of the molecule. Such uncommon versatility of tRNA, quite likely in some cases assisted by a similar versatility of the synthetase, is typical of an ancestral molecule present since early times at the crossways of many important cell functions. As mentioned, many of the nontranslational roles indicated above may qualify as atavistic. In conclusion, it is difficult not to think of tRNA as an ancestral participant in the total chemical evolution of the cell rather than just as an adaptor for a translational job (discussed later). Partly because of these reasons, tRNA is now being extensively used to probe meaningful aspects of functional structure. However, practical reasons for this probing are also important, such as its small and regular but complex structure and the ease of the functional (acylation) test. Indeed, size by size, there is probably no other macromolecule with so much information to offer the faithful inquirer.

Given that polymerization of amino acid was an early macromolecular function in evolution, tRNA may have acted as its own primitive enzyme (Crick, 1968). This may be another answer to the question asked above as to why protein has not taken over the tRNA function. Eventually, the synthetase appeared in its present function (for it could have had earlier nontranslational functions in relation or not to tRNA). In the prokaryote there is little or no cistronic multiplicity of individual tRNA species, and, from the unique sequence in the genome, the synthetases also probably have single cistrons. Since ancestral cistronic singularity is not binding, as is revealed by the greater cistronic multiplicity of eukaryotic tRNAs, it cannot be predicted whether or not the eukaryotic synthetase cistrons have remained unique and the enzyme amplification has remained at the transcriptional-translational level. Related to this question, it should be kept in mind that the functional degeneracy of the synthetase has apparently decreased during evolution to the eukaryote (Chapter 5). Because functionally less degenerate enzymes demand more enzymes of their class, this decreased degeneracy of eukaryotic synthetase immediately suggests a greater molecular degeneracy, that is, a greater variety (not multiplicity) of cistrons. Since eukaryotic tRNA is also more degenerate molecularly, both enzyme and RNA seem to comply with the general eukaryotic trend in this direction. As in the case of tRNA, moreover, this increased degeneracy seems to be the major difference with the prokaryotic enzyme.

The lack of cross-reaction between tRNA and synthetase from prokaryotic and eukaryotic sources stands in contrast to the cross-reaction between these molecules from distant eukaryotes. The probable interpretation is that present decoding was still incomplete at the stage of the common ancestor of the prokaryote and eukaryote, but it was well under way before the onset of eukaryotic diversification. Considering the time scale of these landmarks, and although most of the evolution of decoding had already occurred prior to the diversification, its subsequent evolution suggests that the general faster tempo of eukaryotic evolution applies also to decoding. As anticipated in Chapter 1, the unavoidable implication is that contemporary decoding is less universal than coding.

Present day synthetases carry perhaps the most dramatic recognition function in the transfer of genetic information, for they alone (not tRNA) translate between the nucleic acid and protein languages. For this task, logically, there can be no template and only an intrinsic recognition function can be responsible. The coordinate function of tRNA and synthetase—the adaptor and the translator—remains essential, however, for decoding to the point where this function institutes "a code within the code." By way of recapitulating the meaning of the function (Chapter 1), it serves to displace the recognition between codon and amino acid to the stereochemically more demanding recognition between tRNA and synthetase. These considerations reveal that an important, unknown force exists to preserve the congruency of synthetase recognition with the semiuniversal dictates of the code.

To say a word about evolutionary aspects of translation, it is clear from Chapter 1, speaking only in terms of basic mechanism, that the all inclusive evolution of modern translation is more intricate than the evolution of replication or transcription. This is true even though the peptidization reaction itself is as ancient or more so than nucleic acid polymerization. The bacterial translational apparatus involves at least 80 macromolecules other than tRNA (Lengyel and Söll, 1969). The corresponding figure for the higher eukaryote has not been estimated but, judging from the greater number of ribosomal proteins and tRNAs that are involved, the translational apparatus is bound to be significantly larger. As suggested by certain properties of ribosomes (Chapter 5), the lower eukaryote may occupy an intermediate position and retain a simpler, more bacterial-like translational apparatus. To what extent these differences in the apparatus between prokaryote and eukaryote reflect mechanistic or regulatory differences in translation is not clear, but the latter has a far greater possibility. This would mean that the evolutionary phase of translation of which a living record is available (i.e., the final tract of the inclusive evolution) has concerned mainly the regulatory aspects. These considerations should not obscure one important point, however, which is that in every aspect bacterial translation resembles eukaryotic translation much more than bacterial transcription resembles eukaryotic transcription. In other words, translation has evolved faster to a more terminal stage in bacteria than transcription.

Looking now at the translational function more generally, it is not easy for this writer to see how the apparatus for it could have evolved not in concert,

RNA

if not actually built around, the adaptor function of tRNA. Based on this view, the adaptor would not be an ad hoc addition to translation but instead coresponsible for its evolution from an early stage, as was intimated above. This particular stage would depend on whether or not tRNA acted initially as its own enzyme. Several specific arguments pertaining to the need for this coevolution have been made on different occasions in this text. Suffice it to add, first, that the need is most clear when the interaction of tRNA with the templated ribosome is considered mechanistically, since such an intimate interaction seems too much to expect of a molecule new to the translational system. Hence, second, coevolution could depend in part on the higher structures of tRNA and ribosome being coresponsible not only for their mutual interaction but also for controlling codon-anticodon binding.

Moving on to a different aspect of translation, it is puzzling why GTP has become established as a key translational molecule in preference to ATP. Perhaps it may have been to delineate the translational domain sharply against the metabolic and energetic pathways where ATP dominates. In contrast, the equal standing of the two nucleotides in respect to replication and transcription is obviously because both are equally essential precursors. Yet, within its own domain, ATP function comes as close to translation as acylation, thereby providing the energy for the peptide bond. It is because of this that, differently from replication and transcription (Chapter 1), in translation precursors, energy and information enter separately the anabolic pathway. In turn, the translational involvement of the ubiquitous cAMP is distinct from that of GTP and ATP and clearly delimited from their energetic pathways. This distinction is almost certainly related to the ancestral kinase-activator function of the cyclic nucleotide already operative in bacteria and now being exploited as a secondary hormonal signal. When the complex effects on the cell are considered, this signal is as economical as the hormonal.

Ribosomal RNA

In contrast to tRNA whose manifold roles partly depend on sequence diversification, the bulk of auxiliary RNA function in the cell is carried out by rRNA with a relatively minor diversification of sequence. To begin with composition, in both major rRNA species this has changed during evolution from the prokaryote to the eukaryote (Chapter 2), but the change is considerably greater in the large species. Because the two bacterial RNA species are rather similar in composition, a consequence of this differential change is that the two mammalian species are markedly dissimilar. This change of composition is in contrast to tRNA composition which as a whole (not the individual species) has changed less. In bacteria with widely divergent GC content in DNA (25-75%) the rRNA remains stable at around 54% GC. Obviously, the constant composition of rDNA has no effect on total DNA composition because it constitutes a minor fraction. In vertebrates the GC content is stable in both DNA (42%) and rRNA and, in the latter, is higher than in bacteria, i.e., 64 and 56% for the large and small species, respectively, in mammals; these ratios show how the composition of the two species diverged progressively along the evolutionary scale (Amaldi, 1969). The increase in GC for the larger species has resulted in a more developed higher order structure of the RNA perhaps second only to that of tRNA, although the extent to which the general design of the rRNA molecules has been maintained (as in the case of tRNA) is not clear, particularly in view of the drastic changes in sequence to be described later. Insects show a strikingly reversed trend (Wyatt, 1968) with the GC content in rRNA decreasing from the primitive (orthopterans) to the most specialized insects (dipterans; approximately 43% GC); it was mentioned that there is also a tendency toward the minimum genome size during this evolutionary specialization of the insect. (Of interest, this reversed evolutionary trend is one of the very few documented for any nonviral RNA molecule). Although generally less than in tRNA, the relative substitution (mainly methylation) of rRNA molecules has increased during evolution from the prokaryote to the eukaryote even more than the GC content (Chapter 2). The size of the molecules, particularly in the large species, has also increased but much more than the GC content (and relatively more than the increase in size of tRNA). How this drastic change in size has come about is not known, but it could have been by a series of partial repeats similar to those which are mentioned later for bacterial genes. Within the eukaryote, however, the increase in size applies only to the large rRNA species and does not occur at all in the plant kingdom. In close proportion to this increase, in the eukaryote (except plants) the large ribosomal subunit has progressively grown in size (Kokileva *et al.,* 1971). Thus, comparing the evolution of rRNA and tRNA in respect to composition and size, this has been less conservative in the case of the large, but not the small, rRNA species.

One of the smallest organisms known, *Mycoplasma,* has only one rRNA cistron and 40 tRNA cistrons. This number of cistrons approaches the minimum stable RNA genome and must be close to the smallest genome of its kind in any organism other than a virus. In itself, this strongly suggests that the first true cistronic multiplicity on living record, i.e., the 10-40 rRNA cistrons in bacteria, is the result of duplications of an ancestral cistron. It follows that any subsequent increase of multiplicity during evolution could have arisen from duplications of the multiplicity in the ancestral lineage. The absolute multiplicity of cistrons is greater in the higher eukaryote than the prokaryote, although the proportion of rDNA to total DNA becomes less as a consequence of the disproportionately larger growth of the genome (Chapter 4). The multiplicity is even larger in plants with generally larger genomes. Thus animal rDNA shows, not a rigid conservatism, but at least a certain conservative tendency, except for the occasional discontinuity between related organisms. This general situation parallels, therefore, that for tDNA. From these comparisons, the indication may be for a certain fundamental tendency aimed at preserving the functional ratio between stable RNA molecules. According to present data, in the eukaryote the number of cistrons for each stable RNA is about one order of magnitude above the number in the prokaryote, but this difference is still less than the difference in size of the respective genomes. Another aspect is that, given that the stable RNA cistrons belong in the redundant DNA, their higher multiplicity held in common by all eukaryotes agrees with the general evolutionary trend toward greater molecular redundancy Quite clearly, on the other hand, the evolutionary increase in molecular degeneracy has been much greater in tRNA.

The previous indication at the cistron level for a functional homeostasis between the stable RNAs should be taken to be only suggestive, if nothing else, because over most organisms the cistronic ratio of total tRNA to rRNA varies between 2.5 and 10 (Attardi and Amaldi, 1970). (The total amount of tDNA remains always lower than that of rDNA.) By itself the excess of tRNA cistrons is not surprising in view of the 10- to 15-fold molar excess of the RNA in the cell. The variability of the cistronic ratio is, however, more disquieting, although it may partly obey experimental fluctuation that is not uncommon in the hybridization data on which the ratios are based. Perhaps at present it may be more realistic to consider that the cistronic ratio is no more important than the individual dosage regulatory capacity of each organism to maintain the molar ratio of the stable RNAs, which is what is relatively constant in all organisms irrespective of the amount of cistronic DNA. Clearly, because of the larger cell size, the total number of stable RNA molecules per cell is nevertheless about two orders of magnitude higher in the eukaryote than the prokaryote. By reference to the cistronic multiplicity, which is only about one order of magnitude higher in the eukaryote, the additional order of magnitude in molecular numbers of the eukaryote is probably accounted for by the longer cell generation times rather than the synthetic rates, which are not vastly different.

Each major rRNA species has a moderate compositional heterogeneity which possibly is not functionally significant. The assumption here is that a certain heterogeneity of sequence in this auxiliary RNA is not incompatible with a similarly basic function, contrasting with either mRNA or tRNA in which heterogeneity is actually required by a diversified function. This, of course, would be the explanation why a massive amplification of cistrons is apparently exclusive to rRNA. Yet, given the large cistronic multiplicity and the amplification, the little internal heterogeneity in all rRNA species (including 5 S RNA) in one organism is, indeed, remarkable. To begin with the eukaryote, all rRNA molecules in a taxonomic species belong in one or at the most a few families of sequences. These sequences are more similar to one another than to sequences of any related taxonomic species. Between these related species the molecules show slight differences in electrophoretic mobility, e.g., between primates (Elicieri and Green, 1970), which is indicative of a certain compositional rather than conformational change. However, a certain conservatism of sequence applies even between eukaryotic phyla (Chapter 2), suggesting a relatively slow evolution of rRNA. A general view is that within the eukaryote the degree of sequence homology seems to depend on the number of cell generations elapsed from the ancestral sequence (Bendich and McCarthy, 1970), but in reaching the higher eukaryote the sequence tends to stabilize (Sinclair and Brown, 1971). In keeping with this view, between the prokaryote and eukaryote the sequences are completely dissimilar. Concerning the prokaryote, according to Britten and Kohne (1969), it is possible that also not all rRNA molecules belong in a single family of sequences, which would mean an appreciable diversification during multiplication of the ancestral cistron. A general conservatism of the molecules, however, still holds between closely related taxonomic species of bacteria, for instance, much more so than for mRNA. To a varying degree, this conservatism extends to other principal components of the bacterial translational apparatus such as tRNA and ribosomal protein. In conclusion, as with composition and size, with respect to sequence the overall evolution of rRNA has been less conservative than that of tRNA, and again this may apply especially to the large species.

A partial conservatism of rRNA serves two related purposes. At the evolutionary level, within the prokaryote or eukaryote, the conservatism of the molecule evidently represents the preservation of a well tested and efficient structure. One is justified in using this term "structure" because it is not rRNA alone which is preserved, but its ability to associate meaningfully with a large ensemble of ribosomal proteins. As mentioned, however, from the prokaryote to eukaryote, it is not clear how much of the basic design of the rRNA molecules has been preserved in the face of an increasing higher order structure. Since the function is unchanged, clearly at least part of the design has been maintained, despite the changes in sequence and, in the homeotherm, the much larger precursor and mature molecular size. (If that is the case, the changes in sequence that occurred would be as directional as the reciprocal changes in the helical regions of tRNA: a directional change of sequence in the stable RNAs constitutes a form of noninformational degeneracy.) At the organismal level, the homogeneity of molecule is clearly in order to ensure a uniform backbone for the ribosomes in all cells, which is the same argument as before applied now at this other level. One suspects that in the eukaryote a small-scale mechanism, similar to the related two by which the cistronic complement is so readily reconditioned on a larger scale (namely, amplification and magnification), is physiologically responsible at the cistron level for safeguarding the sequence homogeneity or even the cistronic multiplicity within the organism. (Assuming that magnification represents a general phenomenon and not one peculiar to the insect where it has been recognized thus far.) A possible mechanical safeguard mentioned earlier was the constitutive heterochromatin, rich in highly redundant DNA, which is frequently disposed in the proximity of the nucleolus organizers. It is also conceivable that a cistronic heterogeneity above a certain tolerable range can be prevented by some other means from being reproduced at the RNA level. Conversely, under selective pressure, the previous ostensibly conservative mechanism could help secure the rapid propagation of whatever cistronic changes occur during evolution, featured among which might be a cistron consolidation to form a joint precursor. Perhaps relevant in general to these points, several characteristics of the eukaryotic rRNA cistrons were pointed out in Chapter 4, which set them aside from the informational cistrons. In the case of the prokaryotic organism the mechanism by which the sequence homogeneity is largely preserved is not clear. The lower cistronic multiplicity by itself sets a limit to the possible degree of heterogeneity, and, by extension, to the degree of evolutionary diversification. Nevertheless, in contrast to the eukaryote, it is likely that the prokaryotic sequence is conserved for the most part passively through natural selection. In conclusion, different mechanisms may be operative for the preservation of sequence according to the organism: an active cellular mechanism in the eukaryote and a passive selective mechanism in the prokaryote.

A first obligatory corollary from the previous considerations is that, in contrast to tRNA, the rRNA molecules are not functionally differentiated and all serve the same function in the organism. Hence, in principle, RNA regulation need be mainly modular or quantitative (Chapter 4). With respect to function this corollary therefore says that the molecular redundancy which is present in rDNA is incidental, and can be seen as a mere deviation from accurate cistronic multiplicity; this is completely different from the fundamental role which is postulated for the redundancy (irrespective of origin) in relation to the function of regulatory RNAs. The redundancy characterizing rDNA remains important with respect to evolution, however, for it is one source of rRNA diversification. A second, perhaps less obvious corollary is that the rRNA molecule need be less regionally differentiated than tRNA as well, since, contrary to this, rRNA serves mainly as a binder of proteins which themselves carry much, and probably most, of the specificity of ribosomal function. All this means that rRNA is a functionally less substantive molecule but a truer auxiliary molecule than tRNA; it is the closest to a "mechanical" molecule among all well-known RNAs, although surely not an inflexible one. In addition, because of its size, the total amount of tertiary structure in rRNA is the largest in any RNA. Where the two stable RNAs resemble each other is in having a precursor with a larger and simpler (less modified) primary structure than the mature molecule but already with a considerable higher order structure. This structure of the precursor provides a scaffolding for enzymes to achieve the final higher order structure. Thus as the precursor matures it becomes shorter and more complex in its substitutions, which is a general procedure for molecular maturation followed by many complex proteins. That the precursor is instrumental only to achieve the final structure is clear from the fact that, once this structure is achieved, it remains highly stable after purification from ribosomal protein. An exception that confirms the rule is insect rRNA with poor stability, because of its low GC content.

An interesting lack of conservatism which stands in contrast to the partial molecular and cistronic conservatism is found at the level of the eukaryotic precursor rRNA representing the actual transcriptional unit. In plants, invertebrates and vertebrates up to reptiles, the precursor is about 2.7×10^6 daltons and the mature 18 and 28 S RNAs are about 1.5 and 0.65×10^6 daltons, respectively. In birds and mammals the corresponding figures are 4.0, 1.7, and 0.65×10^6. Thus in the homeotherm both precursor and mature rRNAs are larger but the precursor is disproportionately larger (Perry et al., 1970). These data further reveal that nonmammalian (multicellular) eukaryotes conserve 80% of their precursor as mature molecules and mammals only about half, the rest of the RNA being discarded. [By contrast the more primitive (unicellular) lower eukaryote may conserve near to 100%.] A simple, though perhaps not final, interpretation is that the mammalian precursor has grown at the expense of the nontranscribed DNA, which is intercalated in the rDNA (Chapter 4). A first implication of this interpretation is that the difference in the mammal may be more in the length transcribed than in the actual length of DNA in the cistrons. The prediction is, therefore, that there should be less intercalated DNA in the mammal, which as yet remains unidentified. The envisaged evolutionary variation in the transcript length may be achieved by shifting the initiation or termination sites on the DNA. A second implication is that the discarded RNA sequences evolve more rapidly than the rRNA itself. Indeed, these sequences seem to vary considerably between related taxonomic species and even within the same species (Tiollais et al., 1971). It would seem then that all eukaryotes must share a common minimum precursor to form the immediate precursors (20 and 32 S RNAs) of the mature molecules (Rogers et al., 1970). For example, on this basis, the first precursor in the HeLa cell is 45 S RNA and the minimum precursor is 41 S RNA. The excess over this minimum in the first precursor, i.e., from zero in plants to 1.0×10^6 daltons in mammals, represents most of the RNA that is discarded. All this suggests that in terms of evolution the eukaryotic rRNA pathway described in Chapter 4 is generally relaxed with respect to transcription but becomes progressively more stringent with respect to processing of the molecule, while concurrently resulting in mature molecules of progressively larger size. Whether or not coincidental, the formal analogy of this evolutionary trend with that of the physiological regulation of the pathway (except, of course, for final molecular size) is fairly obvious. The present interpretation regarding the evolutionary lengthening of the precursor strongly points to a function for the discarded RNA, despite its great instability. This function is as yet unknown but the present indications are for its greater importance in the homeotherm. Indeed, given the phyletic divergence of mammals and birds, the joint appearance of homeothermy and the excessive first precursor might represent a convergent evolution, itself indicative of a functional value for the discarded RNA. There remains a strange compositional correlation between this RNA and the mature rRNAs whose functional meaning is equally unknown; when the rRNA is high GC, as in most animals, the discarded RNA is even higher GC, and conversely, when the rRNA is low GC, as in insects, the discarded RNA is even lower.

Superficially, 5 S RNA has a gross composition resembling that of total tRNA but the degree of substitution, if any, is much less. The nature of its evident higher order structure and the function of 5 S RNA are not known, except that the locale of function is the ribosome and the function is universal to all known types of ribosomes (with one possible exception given below). The structure-function relationships of the two RNAs remain therefore of great interest to compare, the more so in view of the evolutionary similarities that will be pointed out. To this comparative purpose a provisional model of their interaction in the ribosome was described in Chapter 2. Only as a possibility suggested by this model, 5 S RNA came as a second-order permanent adaptor, namely, between the first-order transient adaptor (tRNA) and the ribosome. If an adaptor function were really involved, the suspected absence of mitochondrial 5 S RNA, or alternatively, its inclusion in the major rRNA species would present an interesting problem in the mitochondrial ribosome. This possibility of an adaptor need not, however, exclude other specific functions of 5 S RNA.

Microheterogeneity of 5 S RNA has been demonstrated in *E. coli* and man (Chapter 2) to the extent that, although they belong in one family of sequences, the molecules differ in a few bases. As for the large rRNAs, it is doubtful that this kind of heterogeneity has any functional significance or leads to molecular differentiation as it probably does in the long run with tRNA. Moreover, the bacterial and human sequences are closely related and this almost certainly amounts to common ancestry. In other words, the evolution of 5 S RNA has been astonishingly conservative, in fact much more so than that of the major rRNAs, in which again it resembles tRNA.

Major structural repeats present both in 5 S RNA and large rRNAs of the bacterium were discussed in Chapter 2 together with their possible, very basic evolutionary explanations. These repeats indeed offer a unique record, not available for tRNA, of how the structure of the ancestral stable RNA genes first emerged. These repeats could condition the regularly observed structural fragility of the molecules (particularly the large rRNAs) which would then be a reflection of their past evolution. In addition, a common ancestry across the same evolutionary span indicated for 5 S RNA is suspected also for the large rRNAs, but in their case the scanty evidence on sequences is hardly compelling. If such a view is correct, a clear idea of the processes (e.g., duplication, lengthening) by which these bacterial RNAs have become the much longer eukaryotic molecules must await further evidence on sequences.

The cistronic multiplicity of 5 S RNA is of the same order as that of the large rRNAs in the prokaryote, but of a higher order in the eukaryote. How this latter difference relates to the homeostasis between these stable RNAs is not obvious (as it is likewise not obvious between tRNA and rRNA), but it shows that the eukaryotic 5 S RNA cistrons are less efficiently utilized. Perhaps, however, the most interesting comparative evolutionary aspect of 5 S RNA is that the physical dispersal of cistrons with respect to the prokaryotic arrangement stands the very reverse of the consolidation of the rRNA cistrons (Chapter 6). As in the case of informational cistrons, in general, a functional coordination (in this case with rRNA) appears to have become more important than a physical proximity of the cistrons. The contrast would be even greater if bacterial 5 S RNA were indeed transcribed covalently with 23 S RNA, for in this case not only proximity of cistrons but also their physical coordination as in informational operons would be involved. Since a dispersal of cistrons has occurred in the case of tRNA as well, the conclusion is that there have been considerable similarities not only in the molecular evolution but also in the regulatory evolution of 5 S RNA and tRNA. Perhaps when the translational function of 5 S RNA is resolved, the trivial or fundamental nature of these similarities can be decided.

The 5.5 S RNA is the only RNA known definitely to be derived from another, 28 S RNA, after which they remain functionally and physically associated, presumably for a structural reason. Thus far, this phenomenon is authenticated only in the eukaryotic realm (but another possible candidate remains mitochonrial 5 S RNA). Also eukaryotic 8 S RNA is physically and presumably functionally associated with 28 S RNA, but only while this is present in the nucleolus. A common derivation of the two RNAs in a single precursor is not certain in this case. It needs no insight to realize that these molecular additions reflect an evolutionary complication of the. ribosomal pathway, an aspect to which we shall return.

The question arises as to why there is so much noninformational, mainly ribosomal, RNA in the cell (Crick, 1968). The answer may simply be because rRNA was present in the primitive translational machinery or because of economy. With regard to economy, replacing the RNA by protein would commit a considerable portion of translation just to serve itself and in addition to the considerable portion already committed to ribosomal proteins, translational enzymes, and factors. In fact, because rRNA forms a huge backbone for ribosomal function and assembly, its replacement, if as efficient, would require not one but many proteins. This suggests another answer to the question put initially: perhaps the job cannot be done by protein alone.

In the eukaryote the assembly of rRNAs into the ribosomal subparticles takes place in the nucleolus. This is an organelle which has arisen as a subapparatus of the chromosomal apparatus and probably in relation to the evolutionary trend to multiply and mass together the rRNA cistrons in one or a few stretches of the genome (Chapter 6). It is conceivable that the first steps in the direction of the nucleolar organelle were encouraged by the incipient multimolecularization and, in particular, proteination of the eukaryotic chromosome, for as a part of the chromosome the organelle is also multimolecular.

Although the integrity of the nucleolar structure depends on ribosomal function, it is premature, however, to equate the organelle fully with this function, especially for no other reason than it is the only one vindicated thus far. For in principle, there is no reason why nonribosomal nucleolar functions related to other aspects of nuclear function may not occur as well. For example, nonribosomal RNAs are present in the nucleolus and some of them, like the small stable RNA species (Chapter 2) and perhaps a few tRNA species (Chapter 4), may even be made in the nucleolus. Whether these nonribosomal RNAs participate in the ribosomal function or in some unknown nonribosomal function remains to be seen. A second example is the presence of dynamic mineral pools (Chapter 6) which perhaps suggest that other kinds of nonribosomal pools may occur, including the metabolic machinery to maintain them.
In this respect it is not known how these mineral pools, or for that matter any of the ribosomal pools listed below, are maintained. At present, in conclusion, nonribosomal nucleolar functions remain a possibility. Still another area of ignorance is the exchange of nonribosomal RNA between the nucleolus and chromosome (5 S RNA excluded), which seemingly applies to the small stable RNAs. The situation is the more regrettable because a direct exchange of this nature, perhaps involving RNAs other than these, might be a means by which a genomic control over nucleolar function is exercised. Knowledge of this kind could permit one to understand the observed dominance relationship between multiple nucleolus organizers which appears to have evolved in different directions according to the particular organism.

Returning to the ribosomal function, in order to minimize rationalization it will be useful to find a reason why a nucleolus is required for it, in the first place. After all, the production of the bacterial ribosome is basically similar and does not require any organelle. As far as one can tell, the most sensible postulate seems to be that the nucleolus is needed for the regulation of the ribosomal pathway whose intricacies were discussed in Chapter 4. Based on this view, the massing of rRNA cistrons at the organizer locus, which was noted above to correlate with the nucleolar presence, would basically subserve this regulation. [The current molecular model of dosage regulation of these cistrons (Perry and Kelley, 1970) demands, in effect, that they be massed.] In further support of that view, the eukaryotic prerogative of a far more elaborate processing and proteination of rRNA starting with a definite common precursor also correlates with the presence of the organelle and—this is the postulate—may causally depend on it. Yet, while more complex, processing of the precursor is not unique to the eukaryote, since it also occurs for the relatively much smaller prokaryotic precursor(s). It could then be that the eukaryotic processing requires a discrete topology of the transcription-modification-proteination-cleavage series such as can only be provided by a structured organelle. This, moreover, could have provided the pathway with machinery still to be defined, but including a distinct polymerase, a specific repressor, and compartmentalized pools of phosphorus and other minerals, ribosomal proteins, 5 S RNA and perhaps nonribosomal RNAs, all of which are conditions found exclusively in the eukaryote.

According to this view, the nucleolus would be responsible for the more involved regulation of the eukaryotic ribosomal pathway that has evolved. Compared with the prokaryotic regulation, the salient characteristics of this regulation include: (a) a more flexible dosage of cistrons whose number far exceeds that which is utilized during most or all of the time; (b) the possible integration of cistrons into genetic circuits bringing the regulation partly under hormonal control; (c) an equal importance of the regulation of both synthesis and maturation of rRNA; (d) the large localized pools of subparticles, protein, RNA, and minerals described above, which may be a requirement for (c). In addition, there is (e) the independent regulation of the two ribosomal subpathways also typical of the eukaryote. In regard to (d) and (e), for example, it is particularly difficult to see how the enormous concentration of inorganic phosphorus which behaves as if held in a closed pool, or the regulatory block at the level of 32 S RNA, could occur without the presence of the organelle. The markedly different allocation of cistrons and character of the pathway regulation have suggested to Woese (1970b) that the final cistronic multiplicity of the eukaryote has arisen independently of the prokaryote and from a common monocistronic ancestor. This conjectural view is debatable but serves finally to emphasize the large extent to which the evolution of the pathway may have surpassed the evolution of the RNA molecules themselves, the contention here being, of course, that the former evolution implicates the nucleolar organelle. In conclusion, the strict causal relationship ruling between the evolution of pathway and molecules, on the one hand, and the evolution of the nucleolar organelle, on the other hand, has not been resolved.

Having gone through the possible reasons for a nucleolus, let us now as an exercise ask the question: why did not the bacterium develop one? One reason why the bacterium could not form a nucleolus is that it lacks a multimolecular chromosome from which to form it. Another reason, although probably not as causal, may be that whereas the bacterial rRNA cistrons are clustered, they are not solidly massed together. These clusters include most of the ribosomal protein cistrons whose translation is coupled to their transcription and coordinated with that of rDNA; 5 S DNA is also in the clusters. Hence these molecules are made right next to the nascent rRNAs for their orderly assembly to ensue, which, in addition, explains the small pools of them that are found. In all, we may say that the bacterium did not form the nucleolar organelle because the overall cistronic arrangement sufficed the needs of the regulation, which remains virtually at the transcriptional level. By contrast, although the orderly assembly of the ribosomal components is equally necessary, the complex eukaryotic system had to rely on a more dynamic regulation of the assembly, and because this regulation is more at the pathway level it became dependent on the presence of the organelle. Because of the difficulties inherent in structural-functional analyses, we still know more about the function than about the genesis or organization of this organelle.

Compared with the prokaryotic ribosome, the eukaryotic ribosome has almost doubled in size and protein complexity. Little can be said about the arrangement or function of the individual proteins in either organism, but the eukaryotic ribosome has acquired some distinct functional characteristics (Chapter 5). The increase in complexity is real, since most eukaryotic ribosomal proteins appear to be single copy. Consequently, it also represents a gain in molecular degeneracy. Yet this remains only an increase in the extensive molecular degeneracy already present in the prokaryotic ribosome, which, as pointed out, is considerably evolved from a mechanistic viewpoint. Since the basic ribosomal function has not changed it is nevertheless difficult to escape the conclusion that the increased eukaryotic complexity was a requirement for further evolution, prominent in which was the concurrent evolution of the vacuome. This complication of the ribosome goes hand in hand with the considerably greater self-regulation of translation, as a whole, of which the eukaryote is capable. Clearly, it is the general properties of the regulation that have changed rather than the primary change being the complication of the ribosome. On this basis, although closely interrelated, the mechanistic complication of the ribosome has followed the change in regulation. (The situation parallels that in the eukaryotic chromosome where, however, the mechanistic complication is also greatly increased, as indicated by the first appearance of chromosomal molecular degeneracy). Since during evolution of the eukaryote only the large ribosomal subunit has increased in size, and this was accompanied by the most substantial changes to occur in either size or sequence of any rRNA, two alternative possibilities are further suggested: the function of the small subunit has been refractory to the changes in the regulation or, conversely, the function of the large subunit has attracted them the most. Noteworthy, therefore, is the fact that the large rRNA species of plants has remained of protozoan size, for example, smaller than in mammals. Concurring with the regulatory complication of translation, Loening (1968a) has asked whether the difference reflects, somehow, the totipotency for differentiation of the plant as compared with the generally more irreversible differentiation of the animal.

Informational RNA

Because of the uniquely informational nature of mRNA, more than for any other RNA, its evolution is expected to be mostly at the level of primary structure, that is, sequence. This expectation for the evolution of sequence is based exclusively on the informational function of the molecule and not, for example, on the fact that it is only an intermediate gene product; however, the degeneracy of the code in regard to this function, which is of course manifest in the molecule, has most probably been an important factor in evolution. Despite these considerations, and depending on whether there is some higher order structure in mRNA (particularly eukaryotic mRNA) perhaps involving noninformational aspects of translation, evolution of this structure also has to be considered. In any event, it needs pointing out that the discussion that follows of a highly heterogeneous RNA population to correspond with a protein population of vastly different molecular weights, and dealing with RNA sequences partly examined in less than a handful of cases, must of necessity be at a more general level than it was for the preceding stable RNAs. The discussion will concentrate on what appear to be certain salient functional aspects of mRNA.

Prokaryotic informational sequences which are situated near the stable RNA cistrons have been partly conserved during evolution (Chilton and McCarthy, 1969). This may either passively reflect the conservatism of the stable RNA cistrons because of physical linkage, or be more significant in respect to the informational cistrons themselves and depend on their function, e.g., the cistron cluster near the chromosome origin (where part of the stable RNA cistrons are) includes cistrons for ribosomal proteins. Nevertheless, the majority of informational sequences are not near the stable RNA cistrons and have evolved much faster than stable RNA; there is little homology between mRNAs of distantly related bacteria. Needless to say, similar data on RNA are not available in the eukaryote. [Data on eukaryotic proteins (Dayhoff, 1969) are not directly comparable because of codon degeneracy, but they certainly suggest that the homology of sequence decreases with taxonomic distance, as in bacteria.] It was mentioned that bacterial protein composition is slightly but definitely correlated with DNA composition. Given the wider compositional divergence of prokaryotic than eukaryotic DNA, a pertinent question to ask is therefore whether, on the average, the compositional variation of prokaryotic protein is correspondingly wider. The answer to this question is not known because of the lack of data on total eukaryotic protein, but any correlation would be affected by the almost certain possibility that a much greater proportion of eukaryotic than prokaryotic DNA is noncoding. In particular, the presence of the eukaryotic redundant genome with a most doubtful translational status would completely obscure any comparison.

What follows is a series of guideline inferences concerning the informational and regulatory content of the uncharted eukaryotic genome. As seen earlier in this chapter, the unique and redundant genomes are evolving in sequence faster than the average protein sequence. However, there is a conservative obligatory genome for enzymes of intermediary metabolism which have remained practically immutable with respect to the ancestral prokaryotic enzymes and this genome is excluded from the present considerations. For similar reasons, other conservative genes such as those active during early embryonic development (Chapter 6) are also excluded. These exclusions do serve to remind us that, as far as one can tell, part of the eukaryotic evolution seems to be truly neutral with respect to protein, which is a form of the so-called non-Darwinian evolution. It is assumed in this case that due to the code degeneracy the informational sequence could have evolved without its protein product evolving to anywhere near the same extent. Of the two genomes, the unique genome seems to be evolving even faster with respect to protein than the redundant genome. Part of the redundant genome appears to be replenished with new sequences faster than the unique genome. Taken together, these conditions mean that the redundant genome evolves more slowly yet it is partly of more recent evolutionary origin, therefore, it builds up. In addition, part of the redundant genome diverges progressively into the unique genome, acting as a partial outlet for the buildup; a concomitant loss of redundant genome also in part counteracts the buildup. In turn, this means that the unique genome evolves faster while at the same time incorporating part of what was previously redundant genome. Because, therefore, the basic amount of the two genomes is increasing and particularly the redundant genome is increasing faster, the amount of total genome has increased markedly during evolution. On occasion, this combined growth of the two genomes has resulted in amounts of eukaryotic DNA presumed to be excessive with regard to the total information required. This is the so-called C paradox which clearly does not represent an increase of useful information. The large-scale informational amplification, which earlier models of regulation based on cytogenetical data (Chapter 3) postulated to account for this paradox, has no molecular basis.

Without going into all the possible implications of the previous patterns, some of which were already considered in some detail earlier in this chapter, the essential points are as follows. A central premise will be (1) that most of the unique genome in use (not necessarily all) represents informational genome, and (2) most of the redundant genome (not all) includes regulatory genome, more particularly the genome believed to act in regulation at the RNA level. Based on this premise, four closely related conclusions from the previous patterns are: (a) a significant proportion of the genome may be noncoding. (b) The expanding redundant genome contributes to the informational genome as an important source of new genetic material in addition, but possibly in some kind of relationship, to any regulatory function that it may have. (c) The informational content in a mammal is far greater than in the lower eukaryote and potentially of the order of half the total genome (Table 6), yet how much of this information is actually realized is not known (Chapter 3). (d) The functional genetic unit is supposed to consist of both informational and regulatory genes, but what the unit is, is not known. The evolutionary circuits by which it is proposed that sequences in the genome recycle or diverge into related or entirely novel informational sequences passing through a stage of varying degrees of molecular redundancy is shown in Fig. 23. The possible relationship of this hypothetical informational evolution with certain probable facts of gene and chromosomal evolution was also considered in connection with these circuits. The argument was made from this molecular approach that much of the evolving eukaryotic information is tested at the transcriptional level, which would represent a non-Darwinian form of evolution. According to this argument, an enormous amount of potential information would be transcriptionally tested in the nucleus, of which only a portion is realized. By this type of testing genes could be created which are useful in the future without reference to immediate use, that is, in a nondirectional fashion anticipating evolutionary needs. This type of testing would therefore have much more evolutionary potential than that which the prokaryote, lacking redundant DNA, has to manage entirely at the translational level representing Darwinian evolution. Because the prokaryote lacks this genetic reserve, most of its evolution (except from new genetic material acquired by transduction, etc.) is instead by divergence of already existent informational sequence. In conclusion, these considerations on eukaryotic evolution present us with several important differences with respect to prokaryotic evolution such as a faster tempo, new mechanisms, and an intensified regulation; what seem to be the most coherent explanations at a conceptual level were attempted for them on several occasions.

Framed by these lengthy considerations, we begin now to discuss informational RNAs. Both prokaryotic and eukaryotic stable RNAs have larger precursors whose primary function is to provide a working model for securing the final higher order structure of the mature molecules, on which alone function depends. Because the precursor itself has a well developed higher order structure, the interesting concept deriving from this procedure of molecular maturation is that the spatial complexity of the RNA is, in essence, refined rather than augmented during maturation. Concerning bacterial mRNA, as far as one is able to tell, this may also have a precursor but only by virtue of the fact that sequences transcribed as part of the mRNA from the short contiguous regulatory genes (promoter and operator) may be eliminated or not translated. As we shall see, however, such relatively small bacterial precursor, derived essentially from no longer functional regulatory genes, would or would not contribute to the higher order structure, hence function, of mRNA. On the other hand, derivation of eukaryotic mRNA from a large precursor is at present a matter of much dispute which, because of technical difficulties, is still more at the dialectic than factual level. For one thing, the vicissitudes which attend the mRNA molecule in the nucleus are not clarified; for another, the significance of these processing stages of the molecule are not understood. Nevertheless, on balance, the evidence seems to indicate that there is a mRNA precursor in the nucleus. This dispute brings forth two related aspects of nuclear function in which a mRNA precursor may play a role: (a) the basic regulation of informational transcription or selective derepression; (b) the ensuing highly involved posttranscriptional selection or specification of functional mRNA, which is believed to be concurrent with processing of the molecule. According to point (b), the eukaryotic mRNA precursor would be subject to considerably more regulation than even the stable RNA precursors are and certainly much more regulation than the prokaryotic mRNA precursor which practically is not regulated at all and results almost automatically in functional mRNA. Although these aspects are poorly understood, it is nevertheless clear that they have a great conceptual importance for understanding how the regulated expression of the eukaryotic informational genome works out, and, for this reason, these aspects were considered in that light in Chapter 3.

Two models have been proposed for the derivation of eukaryotic mRNA based on the recent knowledge of the molecular biology of DNA. According to one model by Georgiev (1969b), the derivation of mRNA is polycistronic (as is frequently the prokaryotic) and the polycistronic transcript would be a precursor mainly because of the several individual mRNAs it contains. Since functional eukaryotic mRNA is monocistronic, the model goes on to postulate that the individual mRNAs are eventually released from the common transcript. According to this model, but possibly occurring physiologically irrespective of any model, there is a qualitative waste of informational transcript in the sense that only some of these mRNAs will be used at any time. Such a waste implies that the character of the regulation may have to be so flexible as to result in the making of more mRNAs than are required. By contrast, somewhat antithetically, the inflexible prokaryotic regulation is economical because it results only in those mRNAs that are required and which, in fact, are directly specified by the regulation (Chapter 3). According to another model by Britten and Davidson (1969), the regulation of mRNA has evolved toward monocistrony, which would be a departure from the prokaryotic trend. This direct monocistronic transcription denies the occurrence of a mRNA precursor. A reasonable compromise between this model and the previous model predicting a precursor was proposed in Chapter 3, wherein the reasons for compromising were explained. According to this compromise model, an informational monocistrony is accompanied by a greatly complex regulatory polycistrony, but only at the level of an initial precursor; later during processing the regulatory component is removed. Some regulatory complexity is again most likely to occur irrespective of any model, and additional evolutionary implications of all three models were considered together with the conclusions on eukaryotic mRNA in Chapter 3. These models are not to be considered final, since, for instance, the mechanism of specification of the mRNA molecule may not be explained. They are, nevertheless, the only ones accounting for coordinate regulation.

The possibility also of the terminal maturation of a eukaryotic mRNA precursor in the cytoplasm may be suggested by some recent evidence (Aronson, 1972). It is this precursor still present in the cytoplasm, although much less certain than in the nucleus, which commands special interest here. As we come later to discuss higher order structure in the polysome it will become apparent that the need for a precursor could be for actual translational pruposes as much as it was needed in the nucleus for preparing the molecule for its mature structure. Hence, from nucleus to cytoplasm, the mRNA precursor would resemble the stable RNA precursor in having a considerable extra sequence and a complex maturation (without modification), but differ from it in its functional significance. In this regard, by contrast to the putative bacterial mRNA precursor, the extra sequence of the eukaryotic mRNA precursor would discharge a more substantive translational function.

At this writing the sole mRNA which is available in sufficient quantity for direct biochemical analysis is polycistronic RNA phage mRNA, which conveniently contains only three informational cistrons (Chapters 3 and 5). The occurrence of secondary structure in the form of partial helicoidality in this mRNA has been revealed from the partial sequences so far established. The function of this higher order structure is probably in the control of internal translation, an important aspect which is just beginning to be understood. The basic fact is that this mRNA contains an extensive proportion of noncoding sequence whose function is to serve as specific binding sites for the ribosome, one site being attached to each of the three cistrons. Supposedly, these binding sites are not normally available but become available to the ribosome as required during translation because of the functional plasticity of the secondary structure. This structural mechanism would serve to prevent an undesirable operonlike linear translation of the phage mRNA. That is, since the RNA phage lacks a transcription discretely separate from replication, the secondary structure of mRNA would act as a regulatory translational device in lieu of bacterial-like regulatory genes with effects on translation, but, in reality, acting during transcription. In addition, it seems possible that, by suitable use of the wobble ambiguity, the phage mRNA may change its secondary structure (as the result of mutation) without necessarily changing its informational message (Adams et al., 1969). Historically, Rich (1968) had previously speculated that one reason for the wobble ambiguity was to prevent detrimental autocomplementarity in mRNA, in fact the contrary of the wobble accommodation of useful autocomplementarity which appears to be occurring.

It would seem unlikely that these findings in phage mRNA can be extrapolated to bacterial mRNA where there is no evidence for any extensive secondary structure. Whereas this lack of evidence for secondary structure can certainly not disprove it, it is nevertheless clear that the coupling of translation to informational transcription, initially by the initiation complex, obviates any major autonomous regulation of translation in which a secondary structure could serve as in the case of phage. Yet it cannot be excluded that some translational regulation occurs and requires a structural device, e.g., at internal reinitiation. As another hypothetical example, a transcript of the operator gene that remained covalently linked to mRNA (the virtual mRNA precursor considered earlier) could have a noncoding translational function dependent on secondary structure. In this example such structure would be small because of the shortness of the operator gene. (This translational function is deliberately left undefined, but it is not likely to correspond to that of a ribosomal binding site in phage mRNA, as explained in Chapter 5.) An even more hypothetical example suggested by the coupling of translation to transcription would be that the resulting proximity of the DNA to functional mRNA configures some sort of higher order structure with local regulatory effects on translation. As stated earlier, however, there is no clear conclusion regarding the questions of precursor or higher order structure in bacterial mRNA.

There is likewise no conclusive evidence for a higher order structure in eukaryotic mRNA, but, in this case, the possibility is not negligible. Thus partial helicoidality (as in viral mRNA) is suggested by a 50% base pairing in globin mRNA (Williamson et al., 1971); recent data on protein sequences (White et al., 1972) are generally compatible with helicoidality. In the first place, there is a truly autonomous regulation of translation in which, as suggested above, a higher order structure could serve. In the second place, the mRNA cannot derive any of this superstructure from a coupling with transcription (as is possible in the bacterium) which does not exist. Although its nature must also be left undefined at this time, a regulation of the polysome which depended on the higher order structure of mRNA could therefore be conceptually important and is worth the exploration. These structural aspects are really of greater interest to us from the point of view of evolution of the functional mRNA molecule than the sequence of the molecule whose interest is mainly with respect to the evolution of protein.

The only possible evidence concerning a higher order structure of eukaryotic mRNA has to do with certain observations presented in Chapter 5. A basic observation is that an extensive additional sequence is found in mRNAs such as globin mRNA (Gaskill and Kabat, 1971). This suggests that there are sequences other than for punctuation which are not translated, hence are noncoding. These mRNAs appear to contain an extensive poly A run at the 3' terminus (Burr and Lingrel, 1971), probably as the result of a covalent posttranscriptional insertion (Darnell et al., 1971). For those mRNAs which contain the homopolymer (histone mRNAs are one exception), it could be an obligatory portion for their function in the cytoplasm. Moreover, HeLa cell mRNAs (but again not histone mRNAs) appear also to contain a typically redundant sequence close to the poly A. Although many potential functions exist, one possibility is that these nontranslatable sequences are involved in securing a partial higher order structure. It is not at all obvious how this could happen, but it may be recalled that poly A has a distinct ability to stack, and, perhaps related to this, polysomes frequently adopt a closed-circle disposition. Another aspect to consider is that eukaryotic mRNA is believed to remain associated with informosomal proteins down to the time when it is translated (Chapter 2). The function of informosomal protein during translation is not known, but certain aspects of prepolysomal regulation which could possibly be mediated by this protein were considered in Chapter 5. Interesting among these aspects, since we are exploring a physical basis for higher order structure in mRNA, is one whereby the protein might serve to impose this structure on the molecule, perhaps acting in collaboration with the noncoding sequences just mentioned. Of further interest in respect to this collaboration is that both these proteins and sequences (especially poly A) are assumed to participate in the selective transport of mRNA to the cytoplasm. Hence the collaboration could start within the nucleus and structure and transport represent two intimately related aspects. As stated already, however, there is no firm evidence in favor of a higher order structure in eukaryotic mRNA, and the present purpose was solely to identify observations which might pertain to this structure, if it occurs at all. Evidently, any observation that showed the helicoidality suggested above would confirm the structure. What is clear is that, irrespective of function, such a structure would have to be regionally canceled during translation, so that it would be even more flexible than it is in the case of stable RNAs.

To recapitulate these considerations, eukaryotic mRNA seems to consist of a coding unique portion for translation and a noncoding and sometimes redundant portion for other purposes, some of which we have considered. We may now turn our attention to other aspects of translation. We must bear in mind that, because of the monocistronic translation, typical ribosome binding sequences (as occur in phage mRNA) are even less likely to be required than in the case of bacterial mRNA. Addressed to the control of initiation, the "ticket theory" of Sussman (1970) proposes that a sequence of detachable 5' terminal nucleotides in mRNA would bind and count the ribosomes that circulate on the mRNA. By such means the mRNA would be capable of dosing the amount of protein to be made, and, through additional elements, perhaps even dictate which protein is made at which time. By this means the mRNA could also control its own functional half-life whose range is enormously greater than in bacteria. Thus, for a speculative theory, this one can explain many things. However, despite the ingenuity of this theory, this writer conceives eukaryotic regulation to be of a much more dynamic character than the cell "counting with its fingers" that the theory implies. (For comparison, the mechanism by which the amount of bacterial protein is dosed is not clarified either, but the stoichiometry of transcription-translation and the instability of mRNA make the mechanism, in this case, conceptually less pressing to understand.) Experimental observation has led, on the other hand, to the proposals that 5' terminal sequences of mRNA as such (Sabatini and Blobel, 1970) or in association with the translation initiation complex (Baglioni et al., 1971) help specify the free versus attached nature of the polysome, which still remains a most important specification to explain. Contrary to the ticket theory, the first of these two proposals requires translation of the terminal sequences, as was explained in Chapter 5. A necessary caution is, however, that none of the 5' terminal sequences in these proposals, all different from initiation codons, has been demonstrated. To mention briefly termination of translation last, it is conceivable that the function of the 3' terminal poly A runs also involves this function, but then it must be remembered that not all mRNAs carry terminal homopolymer.

In summary, the hypothetical purpose of noncoding sequences in eukaryotic mRNA would be as (a) signals for specifying certain facets of translation such as initiation, dosage, and destination of the protein, or (b) elements for specifying a higher order structure of the mRNA, which might be required for its transport or regulation during translation. These sequences would all have a general regulatory purpose and some might derive from redundant DNA; however, none of their functions has been established. A further speculation could be that the functions under (a) and (b), each regarded as a class, need not be exclusive of each other, and either of them could be assisted by informosomal protein acting in a manner essentially as yet unknown. It should be clear now that these noncoding sequences could bear on the question of a mRNA precursor considered earlier, particularly if this were to be tailored in the cytoplasm according to any one of these functions. It follows, therefore, that an important phase in the evolution of eukaryotic mRNA starting from the ancestral prokaryotic mRNA could have consisted of the acquisition of this precursor, which in the cytoplasm stands very putative, and this extensive noncoding sequence, which, whatever its function, occurs for certain.

This section has stressed the translational value of a precursor, whereas other values considered in Chapter 3 were from a transcriptional angle and for the task of extricating mRNA out of the nucleus—two processes which demand the outmost specification. Moreover, with the likely participation of informosomal protein in polysomal function, the eukaryotic polysome becomes an even more formidable quinternary arrangement than past knowledge (extrapolated from the prokaryote) had led one to believe, yet an arrangement which one must be realistically prepared to accept in view of the forthcoming evidence. A sure prediction is that, when all translational aspects are understood, little will remain of the traditional "mRNA tape" metaphor that still applies to eukaryotic mRNAs.

As befits the direct informational role of mRNA (Fig. 1, p. 10), the macromolecular aspects related to its function involve the largest series in the cell ranging from transcription to translation. This section has dealt with some of these mechanisms in the eukaryote, which listed by categories are: (a) mobilization of regulatory transcription in relation to (b) selective derepression of the genome; (c) informational transcription proper; (d) proteination and perhaps maturation and selection of functional mRNA in the nucleus followed by active transport to the cytoplasm, all of these highly discriminating steps with the possible intervention of both regulatory RNA and protein; (e) possible completion of maturation in the cytoplasm; (f) switch-off of hypothetical translation inactivator genes; and (g) translational events prior to, during, and after the polysomal stage. Aspects (a) to (f) were discussed in Chapter 3 and then evaluated jointly with (g) in Chapter 5. It was pointed out that this integral series of events frequently creates a more asymmetrical response than in the prokaryote (Fig. 20). It was also noted that the functional organization of the ensemble under (g) engages multiple facets of the cell architecture without which the coordination of this ensemble may be impossible to understand. This coordination has to explain not only a local autonomous regulation of translation but also some kind of intracellular feedback from this regulation on informational transcription to create an internal cell regulation. Such a feedback plainly cannot be visualized as readily as it can, for example, for inducible bacterial transcription.

Taking a grand view, the fine molecular basis of translational regulation in the eukaryote is as little known as is that of transcription. Although eukaryotic translation resembles prokaryotic translation mechanistically very closely—which is equivalent to saying that the translational mechanism "froze" at an early stage of evolution—its regulation is however much more complex. As just mentioned, this regulation depends on and, in turn, contributes to a much more organized cytoplasm. The greater complexity notwithstanding, the universal mechanistic similarity of translation leaves little doubt as to its common ancestry. Precisely because of this mechanistic similarity, the regulation of translation is the main course open for further evolution. In this similarity, translation also stands in sharp contrast with informational transcription; here the mechanism itself, let alone the regulation, has evolved so significantly in qualitative terms with respect to the bacterial system that one might even question a common ancestry, viz., the bacterial chromosome has none of the molecular degeneracy which is typical of the animal chromosome. A profound consequence of this difference at the very basis of the information transfer system stressed throughout is that, despite the many resemblances that subsist, the entire system must have diverged substantially between the two types of organism.

Eukaryotic Regulatory RNA

Although the RNAs to be discussed in this section occur in the cell, in contrast to the previously discussed RNAs, their regulatory function directed toward the genome and transcription, in particular, is not unambiguous. From the beginning it should be understood that this function rests on postulate and circumstantial evidence, and in no single instance has it been established. It follows, therefore, that the contents of this section are speculative. However, it would be surprising if all the noninformational RNA that has appeared in the eukaryotic cell, including highly unstable RNA in the nucleus, were to have only a mechanical function. Some regulatory function of RNA is thus likely to occur, and, despite all of the incertitude, its conceptual importance is too obvious to ignore.

The notion was put forward in Chapters 3, 5, and 6 that much of the regulation of the transcriptional program carried in the eukaryotic chromosome is achieved by means of regulatory RNA, presumably interacting in some way with regulatory protein. This noninformational RNA would act in the capacity of a direct regulator of informational transcription. As indicated in Chapter 4, stable RNAs and, in particular, tRNA may in part also be subject to this regulation. Evidence for possible candidates for this regulatory role was described and evaluated in Chapters 2 and 3: these are labile RNA, chromosomal (derepressor)

RNA and perhaps, although this is even less clear, the small stable RNAs present in the nucleus; all these RNAs seem to be represented by more numerous species in the higher than the lower eukaryote. Evidently, however, the two stable RNAs would regulate in a manner very different from that of labile RNA. Whereas it serves little purpose to reevaluate the evidence for regulation here, which, as intimated above, is far from unequivocal, it does serve a purpose to explore from a general standpoint the possible implications of an informational regulation mediated by RNA. This is the intention in what follows on the implicit assumption that the evidence as a whole lends some support to such a regulatory function. Mainly discrete transcriptional signals will be considered, but briefly mentioned are signals for other purposes, possibly conjoint with the mRNA molecule. The narrative has a partly historical character to help trace the evolution of the concepts devoted to this regulation.

Many informational RNAs must intervene in regulation via their protein products, and these may indeed be the only regulatory RNAs in the prokaryote, e.g., repressor mRNAs. [Another possible regulatory bacterial RNA might be 7 S RNA, whose function remains unassigned. If that is the case, this function would have to depend on the one or few sequences of which the RNA appears to consist. The evolutionary relationship of 7 S RNA to the more numerous small stable RNAs in the eukaryotic nucleus is somewhat of a possibility to keep in mind.] Noninformational regulatory RNAs would therefore be exclusive to the eukaryote and, as such, one of the major qualitative innovations of evolution; these RNAs belong in the category of nonmessenger DNA-like RNAs in Table 6. An underlying postulate in this section will be that these RNAs have been responsible for permitting the inflexible regulation of the virus or bacterium to change into the dynamic regulation of the nucleated organism. These RNAs deserve at present a functional consideration without much emphasis on their structural parameters. Except perhaps for chromosomal RNAs, these parameters are relatively uninteresting to the present consideration, and, in point of fact, as described in Chapter 2, poorly known. What is interesting is that labile and chromosomal RNAs are both generated from the exclusively eukaryotic redundant genome. In consequence, the discussion will again have to touch on the functional meaning of DNA redundancy, but particularly that which can be transcribed (overlooking any aspects that do not depend on this activity). The possible involvement of redundancy in the evolution of the chromosomal apparatus, in regulatory function, and in the generation of new genetic material were already pointed out in this chapter. These remarks amount, therefore, to a renewed suggestion that the appearance of redundant DNA is at the root of eukaryotic evolution.

Among the first to postulate a regulatory RNA that would function in derepression of the eukaryotic genome was Frenster (1965). Sometime later a large family of small RNAs was described under the name of chromosomal (derepressor) RNA and proposed to function in such a regulatory capacity (Chapters 2 and 3). [According to Heyden and Zachau (1971), chromosomal RNA represents mostly a fractionation artifact, in particular, degraded tRNA. It is premature to reject the RNA on the basis of this report (see Bonner, 1971), and it will be regarded in this section as a bona fide species, but its validity requires reinvestigation. Whatever the outcome, this will not affect the general import of the section which, as it stands, does not demand this species in particular.] With this proposal, the ideas on eukaryotic regulation had to await the discovery of redundant DNA for a new and major conceptual advance. Thereafter, insofar as being mediated by noninformational RNA is concerned, regulation began to be increasingly viewed in relation to this DNA. For example, it became clear that chromosomal RNA was coded by redundant DNA, and, since this RNA was proposed to be regulatory, it was only logical to further propose that a function of redundant DNA was to regulate the bulk of the remainder (unique) DNA which was taken to be informational (Sivolap and Bonner, 1971). On this premise and knowing the proportion of redundant DNA in different genomes, Bonner (1969) calculated the average numerical relationship between the hypothetical regulatory (redundant) units and informational (unique) units. This turned out to be one in yeast, four in *Drosophila*, and 20 in higher vertebrates. If one now considers that in the bacterium all the regulatory genes contiguous to the informational cistrons constitute a single regulatory unit, then in the lower eukaryote the numerical relationship remains of the same order as before but in the higher eukaryote it is 20 times higher. On this conceptual basis, there is a progressive evolutionary complication of the regulatory genome represented in redundant DNA. Qualitatively, as will be seen, the same concept could be derived from labile RNA which is also coded by a progressively larger fraction of redundant DNA along the evolutionary scale.

The last concept figures prominently in the most elaborate model proposed for eukaryotic regulation (Britten and Davidson, 1969), which was discussed in detail in Chapter 3 (Fig. 13). According to this model, the total number of regulatory genes surpasses by far the number of informational genes (although there can still be a greater total amount of informational than regulatory DNA), so that the relative proportion of regulatory genes becomes much higher than in the bacterial system. Briefly, the model proposes that RNA acts as a transcriptional signal between regulatory genes (integrators), which are discontiguous to the informational cistrons, and other regulatory genes (receptors) which are contiguous. Both classes of regulatory genes would consist of sequences belonging in one family of redundant DNA and the RNA signal generated by one integrator gene would recognize all receptor genes in that family. Thus a regulatory ensemble is created whereby different combinations of integrator and receptor genes result in the activation of informational cistrons to produce either a multivalent or pleiotropic response in a typically eukaryotic pattern, for example, as occurs during hormonal stimulation. This accounting of the expected informational response is certainly a strong quality of the model. If other ensembles, in turn, respond to products direct or indirect of the first ensemble, a more complex genetic circuit is created with a correspondingly more complex pattern of response. These authors then go on to propose that labile RNA could be such a signal which on recognizing a receptor sequence would have a positive regulatory effect on transcription, much as had been proposed earlier for chromosomal RNA. Interestingly, the first evolutionary appearance of labile RNA seems to be at the level of the unicellular lower eukaryote (Prescott et al., 1971) and more or less together with the first appearance of redundant DNA. As described already, this earliest labile RNA may not be very typical of evolved labile RNA. In any case, this RNA does not become really prominent before the advent of the multicellular eukaryote, and then at an equal rate with an enlarging redundant genome. Chromosomal and nuclear stable RNAs are also less represented in the lower eukaryote.

The concept of a regulatory RNA signal between genes has several important implications. First, it has more immediate effects than a translation-mediated signal such as protein, proposed by Georgiev (1969b). Second, if the RNA signal derives from redundant DNA and thereafter functions by recognizing this type of DNA as is postulated, then, because of the molecular nature of redundancy, this RNA-DNA recognition can work even under conditions of different degrees of sequence complementarity. By definition of redundancy, all that is needed for the two polynucleotides to recognize each other is that they belong in the same redundant family; contrast this relaxed form of recognition with the more stringent and discretely localized recognition expected between a protein signal and DNA. It follows that the degree of complementarity can affect the efficiency of recognition, or signal strength, which thus becomes available for a fine regulation of informational transcription. (The molecular mechanism whereby signal strength is converted into a differential rate of transcription as implied here is not difficult to imagine, but is best not to attempt to define it for the present.) To clarify this point, it is not mandatory for recognition that redundant DNA sequences be at both ends of an inbetween RNA signal, since sheer repetition of unique sequences at the ends would suffice, but a redundant system has a modulatory advantage. This general advantage of a redundant RNA-DNA recognition at the functional level extends to the evolutionary level, and here the implications become particularly important because the resulting evolutionary flexibility of the genetic system is great. For example, extant redundant sequences can diverge gradually and thus change the relative efficiency of one of the regulatory genomic circuits just described or ultimately change the connections of the circuit and thus create an entirely new one, all this without any risk of sudden discontinuity. Newly appearing redundant sequences likewise may serve to create new circuits responding to different external effectors. Clearly, this flexibility of recognition between sequences can be a means for reshuffling the regulative integration of informational genes under the surveillance of natural selection, which may be compounded with the physical reshuffling brought about by meiotic recombination. In eukaryotic evolution these changes in regulation are probably of as much consequence, or more, than the creation or divergence of the informational sequences themselves. In this respect, third, a regulation based on molecular redundancy can absorb a much greater mutational load than is tolerable in informational sequences (Crow and Kimura, 1970). The implications listed so far give a regulatory RNA signal a functional and evolutionary potential superior to that of a protein signal. Fourth, this type of regulation involves a certain gratuitousness of non-informational RNA synthesis partly because of the numerically degenerate nature of the regulation, which is intrinsically much more wasteful than in bacteria. As pointed out in the previous section, such a gratuitousness may extend to mRNA. For example, in quiescent cells (Chapter 3) much more mRNA is made than appears to find its way to the cytoplasm. Fifth, if, related to the regulation part of the noninformational RNA also represents an evolving fraction as considered in preceding sections, then at any evolutionary point there will be a lot of RNA with no immediate functional purpose. A suggestion for this is the great divergence of redundant transcript between related species.

It will not escape notice that the type of regulation being considered gives DNA itself, particularly the redundant DNA, most of the regulatory function of the eukaryotic genetic system. Nevertheless, this is not to say that eukaryotic DNA works in a regulatory vacuum, more specifically, without the participation of a multimolecular chromosomal apparatus. On either of these reciprocal premises, it follows that the prokaryote without redundant DNA or a true chromosomal apparatus cannot have a RNA-mediated regulation of its informational DNA, but only a more metabolic regulation. However, alongside of all these attractive implications, at present regulatory RNA poses many major unsolved problems for particularized discussion of which the reader is referred to Chapter 3. To single out one of these problems revealing our precarious knowledge, this is the contemplated conversion of the enormous labile RNA molecules eventually to signals of more manageable dimensions such as the chromosomal RNA molecules. It is, of course, entirely possible that these two classes of molecules function independently of one another and are not filially related, but this would do little to simplify the complexity of the regulatory system as it is. To complicate matters even further, it is necessary to remember that labile RNA is not only visualized as a signal but, on increasing evidence, also as a precursor to mRNA. Whether these two propositions are mutually compatible remains to be decided, since again, there could be separate molecular classes. A simple compromise however would be, for example, that the discarded part of the precursor serves as an internal feedback to the nucleus (as suggested at the close of the preceding section) to signal the impending exit of mRNA to the cytoplasm. Any evidence in favor of an excessively large pre-

RNA

cursor certainly adds to the likelihood of a conjoint signal of some kind. It bears repeating at this point that the regulatory assignment being evaluated for RNA is speculative and in absolute need of proof. Moreover, as considered in Chapter 3, a regulatory RNA function does not exclude a similar function for protein. Consideration of regulatory protein was omitted here because, even though they are as undocumented as regulatory RNA, the valid bacterial precedent of repressor protein gives them a certain immediate credibility, hence a lesser conceptual importance.

Germane to the evolutionary potential of regulatory RNA, the recent origin of much of the redundant DNA implicated in the regulation is borne out by a comparison of sequences with those presumably coding for informational RNA. Between taxonomic species the degree of divergence of redundant families of sequences transcribing the nucleus-restricted labile RNA is greater than that of the sequences whose transcripts are destined for the cytoplasm. Within species the pattern is reversed and the sequences transcribing the nucleus-restricted RNA are less divergent, and also have less informational content, than those transcribing cytoplasmic RNA (Shearer and McCarthy, 1970b). These differences signify a faster evolutionary renewal of sequences coding for the exclusive nuclear transcript, which is the one most likely to include any regulatory RNA. Such a renewal is predicted by the principle of transcriptional testing. It is also interesting that, in mammals, for example, the previous evolutionary pattern is the opposite of that for rRNAs. Thus rRNA sequences tend to be preserved between taxonomic species but diverge slightly within any of them. A conclusion is that all mammals share the same ancestor rDNA but their redundant DNAs are to a considerable extent unrelated occurrences. Evidently, this contrast between the DNAs is understandable in terms of their fundamentally different functional significance.

The possibility exists for regulatory RNA acting also at the translational level, namely, responsible for the inactivator function described in Chapter 3 under the heading of posttranscriptional modulation of mRNA. The evidence pertaining to this function need not be repeated here, but it is indirect. Were this evidence to be substantiated, such RNA would be exclusive to the eukaryote, much as the regulatory RNA for transcription considered thus far. Last, but not least, in dealing with the cytoplasm it should not be forgotten that cytoplasmic polydisperse RNA (Chapter 2) apparently has no informational function. What its function is remains a question even difficult to guess at, except loosely in the sense that it may involve some aspect of polysomal regulation.

Control of the bacterial regulator gene offers an analogy, although a most primitive one, to an eukaryotic transcriptional program. To orient the reader, this program was defined in Chapter 3, considered further in Chapter 4 in the light of additional information, discussed later in connection with cell differentiation in Chapter 6, and then reconsidered from several evolutionary angles throughout this chapter. What follows now is frankly free speculation into the emergence of such a program starting with the bacterial regulator gene: of necessity, the approach will be general enough to forego consideration of any relevant change in the biosphere, metabolism, etc. The mechanism by which the bacterial regulator gene is controlled is not known but appears to be geared, perhaps indirectly, to the periodicity of the cell cycle of growth (Chapter 3). What is probably involved in this control are endogenous signals ultimately traceable to some undefined metabolic or macromolecular event, itself geared to the cell cycle. In any case, a primitive transcriptional program is instituted which is essentially metabolic in nature rather than intrinsic to the chromosome. Converting these endogenous signals the regulator gene is responding to signals emanating from vicinal cells as in a colonial microorganism, would seem to be the next logical step along the evolutionary ladder. For example, individual ameboid cells in the slime mold (a lower eukaryote) respond to chemotactic acrasin produced by a few cells, first by aggregating around them and then by undergoing collective differentiation into a fructifying organ. Acrasin is now known to be cAMP (Konijn et al., 1968), which is a bacterial molecule, but almost certainly more complex secondary signals are responsible for the later phase of mold organ differentiation. Also perhaps molecular patterns external to each aggregated cell and positional signals involving cell membrane phenomena (Chapter 6) have begun to play an organizational role at this early stage. Without wishing to press the issue, the order of evolutionary appearance of biological phenomena could well have been that listed under development, namely, cellular, supramolecular, and supracellular, in the order of compliance with principles of macromolecular syntheses.

We have arrived, rather precipitously it must be admitted, at the crucial era of the dawn of the nucleus considered in detail earlier in this chapter. In terms of evolutionary time this period is the late occurring transition period between the prokaryote and the unicellular lower eukaryote. We are not concerned, however, with how the cell nucleus originated. What concerns us is that the basic innovation of the eukaryote, a true chromosome, is now present in all its complexity, even if not in all its final complexity. With it has come the possibility of incorporating gradually into the chromosome a transcriptional program, which until now was nonchromosomal. The bacterial-like regulator gene has become a regulatory gene responding to external effectors such as molecules from other cells, including hormones, which are already present in a primitive form at this stage. Effects of this kind are probably happening during the slime mold differentiation mentioned above. (Steroid hormone may have been used first for surface to surface recognition, and later internalized to reach the nucleus; it appears as sex hormones in fungi. Protein hormone is a later evolutionary appearance.) This eukaryotic regulatory gene would roughly correspond to the sensor gene of Britten and Davidson (1969) seen in Fig. 13. Noninformational RNA, i.e., not producing a bacterial-like repressor protein, would have begun functioning with the ability to start a rudimentary but nevertheless recognizably eukaryotic transcriptional program. This onset of a chromosomal program must have also required the creation of many other regulatory genes, either earlier or later than the sensorlike gene, but ultimately capable of establishing connections with it. Early as well, histones would have become available as general informational repressors (a possibility being that they originated as miscoded ribosomal protein). The complement of histones soon froze, however, as witnessed by their present remarkable conservatism (Chapter 3) and primitive traits (Chapter 6). One reason for this conservatism may have been to preserve intact the interaction with the DNA backbone, itself also conserved. Conservatism also means that the chromosomal machinery since then has evolved without a concomitant need for histone evolution, and, as far as evolution is concerned, histones are among the fastest typical eukaryotic macromolecules. Further evolution of the chromosome, including nonhistone proteins and RNAs, then required a numerical degeneracy of all these macromolecules mainly for the purpose of accurate regulatory specification. An open question is to what extent bacterial-like specific (nonhistone) informational repressors remain present, or, alternatively, have been replaced by this multimolecular specification in the chromosome. This question of single versus multiple specification applies equally to the informational activators of the genome. Last but not least in considering proteination, the appearance of mRNA-transporting (informosomal) proteins is perhaps best regarded as an extension of the proteination of the chromosome.

All previous changes during the early period of the nucleus, essentially represented in a richly proteinated chromosome capable of accommodating a growing regulatory capacity, may have depended on the appearance of redundant DNA on the evolutionary scene. A correlation is clear as follows. The first noninformational (DNA-like) RNA depended on the appearance of this DNA, for it is coded by this DNA and neither is present in the prokaryote. To the extent that this RNA represents potential regulation, some of the regulatory capacity must have come from redundant DNA even if (as is probable) this first required the presence of chromosomal protein to combine with and become established. In accord with this capacity, during subsequent evolution the concurrent need seems to have been for the more rapid expansion of the redundant genome relative to the generally expanding total genome. As it expanded the redundant genome would assume another function of a genetic reservoir for generation and diversification (Fig. 23) of the unique genome, the informational genome. However, as implied above, not all the unique genome expanded in this way through the agency of the redundant genome. This information generative function of the redundant genome would probably be correlated to its regulatory genetic function: it was explained how this noncoding function could allow a convenient testing of prospective informational function at the transcriptional level, thereby leading to fixation of this function. Within the total context of nuclear organization, both these functions of the redundant genome may have been important in creating the faster tempo of organic evolution as this approached the stage of the multicellular organism. By this stage the increased genomic complexity was probably more the result of the growth and consolidation of the regulatory component than of the actual increase of informational content, although the latter has been increasing considerably to this stage. The net result was a growth of genetic circuitry more than gene numbers as reflected in the complexity of genetic patterns in multicellular organisms.

Part of this circuitry must have also been concerned with maintaining a rapport between genome and cytoplasm. Indeed, up to here, the progressively higher differentiative capacity of the cell (in fact, the cytoplasm) has been both cause and effect with respect to the elaboration of genetic programs; cytoplasmic expression remains the end-product as well as the rationale for ongoing evolution. Specifically, this implied a dependence of the program on dynamic nucleocytoplasmic molecular interactions, for which, moreover, a parallel evolution of the regulation rather than the (already evolved) mechanism of the ancestral translational apparatus (not considered here) was essential. As discussed in connection with development, these nucleocytoplasmic interactions reached a point where it is justified to speak of two accruing programs, the genetic program in the chromosome and the epigenetic program in the cytoplasm. In addition, a supracellular organizational component in every organism must have had its demands on the genome, even if, as discussed again in connection with development, what these demands are is not at all clear. This component probably dates back to the first colonial organism and has since been growing during evolution.

On the one hand, higher eukaryote cells represent a conservative evolution of part of the genome. This is the obligatory genome whose transcriptional program still remains entrained to the cell cycle (Chapter 3) as envisaged at the start of this speculative account for the bacterial regulator gene. Even if cell character has drastically changed by the end of the account, this obligatory genome is basically an extension of the prokaryotic constitutive genome and in quite a few metabolic aspects it has not changed much with respect to the ancestral genome. On the other hand, higher eukaryote cells have evolved their distinctive facultative genomes to provide each cell type with the discretionary information required to meet specialized function within a multicellular organism. Bearing in mind our initial resolve to track the emergence of a transcriptional program, it is these highly refined facultative genomes which, as a class, tell one what the informational evolution and the underlying, so little known, regulatory evolution of the eukaryote have been all about.

Appendix

RESUME OF TECHNIQUES

The techniques included here are standard for the study of RNA. There is, however, an ever increasing number of refinements of these techniques as well as an arsenal of ancillary procedures which are not included. The purpose of this resume is to provide only that reader who has no prior technical expertise with an introductory list of references in which the basic procedures are described. Original rather than more recent references have received preference when they offered more complete descriptions of principle and procedure.

Biochemical work begins either with whole cells or tissue, or with cells fractionated into subcellular components by means of differential centrifugation (Allfrey, 1959; Schneider and Kuff, 1964; for nuclei, Siebert, 1967; for nucleoli, Muramatsu and Busch, 1967) followed by ultracentrifugation (Anderson, 1966). The RNA is then extracted by deproteination with several procedures involving detergents, phenol, etc. (Kirby, 1964). A recent mild procedure employs a column of silicic acid to avoid RNA denaturation (Sueoka and Hardy, 1968).

Analysis of base composition after RNA hydrolysis to base derivatives is carried out by paper chromatography and electrophoresis (I. Smith, 1968, 1969), microelectrophoresis (or microphoresis; Edström, 1964), ion-exchange chromatography (Dowex; Cohn, 1957), and nearest-neighbor analysis of RNA-^{32}P (Kornberg, 1961). The amount of bases is measured by ultraviolet spectrophotometry or microbiological assay (Davidson, 1969). Oligonucleotide sequences for structural analysis can be elegantly recognized by ^{32}P-fingerprinting (Sanger and Brownlee, 1967) combined with specific nuclease cleavage (Laskowski, 1968); the procedure has been made more sensitive by chemical labeling of the RNA *in vitro*.

Preparative separation of polynucleotides is achieved with sucrose density gradient sedimentation (Britten and Roberts, 1960); chromatography on ion-exchange substituted cellulose (ECTEOLA, Bradley and Rich, 1956; DEAE, Tomlinson and Tener, 1962) and adsorbed methylated albumin (MAK, Mandell and Hershey, 1960; MASA, Stern and Littauer, 1968); partition (Everett *et al.*, 1960) and reversed-phase (RPC; Kelmers *et al.*, 1965) column chromatography; polyacrylamide gel electrophoresis (Loening, 1965); Sephadex-dextran gel filtration (Boman and Hjerten, 1962); and countercurrent distribution (CCD; Holley and Merrill, 1959). These techniques rely on different molecular parameters and, therefore, vary in the resolution and degree of purification of the molecules (review by Steele and Busch, 1967). Some of these techniques can be used semianalytically for determination of weight and overall shape of the molecule.

A method for determination of molecular weight is by sedimentation velocity in the analytical ultracentrifuge (Schachman, 1959), sometimes using density gradients. However, sedimentation velocity depends on molecular shape. Less dependent on shape is the buoyant density of DNA, DNA-RNA hybrids (Meselson *et al.*, 1957), and RNA (Szybalski, 1968), measured by CsCl density equilibrium sedimentation. The technique permits also the estimation of the degree of H bonding in the tertiary structure, but this can be more precisely analyzed by melting temperature profile (Marmur *et al.*, 1963). Other methods for molecular weight determination involve the diffusion constant, intrinsic viscosity, light scattering (Spirin, 1963), and end-group analysis (Midgley, 1965).

Electron microscopy permits direct measurement of molecular length and correlation with molecular weight (Kleinschmidt, 1967), in the case of DNA assuming a linear density of 196 daltons per Å in the B form. An important advantage is, of course, the direct visualization of molecular contour.

Codon binding can be examined in an *in vitro* ribosomal system (Nirenberg and Leder, 1964). The capacity of mRNA to direct protein synthesis *in vitro* can be tested in the original cell-free ribosomal system of Nirenberg and Matthei (1961).

The degree of sequence complementarity is studied by molecular hybridization of DNA (Schildkraut *et al.*, 1961) and DNA-RNA (Gillespie and Spiegelman, 1965). As the reader of the text may have gathered, this technique, although it requires precise interpretation (Kennell and Kotoulas, 1968; McCarthy and Church, 1970), is extremely useful because it enables one to compare sequences without actually knowing their base composition. One of the reacting polynucleotides may also be labeled chemically (Smith *et al.,* 1967). The hybridization technique has been adapted cytologically for some RNA species (Gall and Pardue, 1969). Analysis of an aspect of hybridization, reannealing kinetics, in conjunction with biochemical analysis permits estimation of the genome complexity (Wetmur and Davidson, 1968); this technique is evaluated in footnotes 11 and 12.

In addition to biochemical methods, the specific pharmacological approach using many analogs and antibiotics, for example, is invaluable in the analysis of molecular structure and function. References to it are widespread in the literature. Physical measurements applicable to the study of RNA structure such as chromicity, optical rotary dispersion, circular dichroism, and nuclear magnetic resonance, are described in textbooks.

Lastly, fluorescent antibodies (Coons and Peters, 1972), cytochemical techniques (Smetana, 1967), and autoradiography (Baserga, 1967)–the last two in conjunction with the electron microscope–can be used to supplement the biochemical approach with direct intracellular localization.

LIST OF ABBREVIATIONS

The following abbreviations are used in the text: DNP, deoxyribonucleoprotein; dRNA, DNA-like RNA, used generically in contrast to stable RNA; mRNA, messenger RNA; RNP, ribonucleoprotein; rRNA, ribosomal RNA; tRNA, transfer RNA.

In addition, (a) stable RNA, refers conventionally to both rRNA and tRNA; (b) rDNA, tDNA, 5 S DNA, etc., mean DNA possessing sequence homology with rRNA, tRNA, 5 S RNA, etc., respectively; (c) abbreviated names of techniques are explained in the preceding section, Resume of Techniques; (d) standard abbreviations are used for amino acids.

GLOSSARY

The following is a list of words requiring general definition or used in a specific sense in the text. Words within quotation marks mean that they are referred to elsewhere in the glossary.

Amplification: When the basic "multiplicity" of "redundant DNA" is transiently increased as in oocyte rDNA.
Complexity: The weight in daltons of a "unique DNA" sequence or, derivatively, of a "unique RNA" sequence. If the sequence is in double-stranded DNA the complexity is equivalent to the number of nucleotide pairs. If the sequence is in single-stranded RNA the equivalence is to number of nucleotides.

Constitutive function: Regulator- or operator-negative function in the prokaryote which escapes "control" by these regulatory genes.

Constitutive heterochromatin: That chromatin which is never transcribed and is generally rich in highly "redundant" or "satellite" DNA. Compare with "facultative heterochromatin."

Control: As used here means control *by* something, e.g., transcriptional control means control *by* transcription. Compare with "regulation."

Darwinian selection: That which occurs at the level of protein or subsequent levels. See "non-Darwinian selection."

Degeneracy: The occurrence of many distinct molecular species of related but not necessarily identical function in the same molecular class, but not the multiple representation of any particular molecular species in that class. Examples are histones and ribosomal proteins. By definition, molecular degeneracy does not apply to mRNA. By extension, also used for codons and anticodons (see "degenerate tRNA"), though here degeneracy implies identical function. In order to contrast with "redundancy," which concerns the sequence structure of DNA irrespective of whether it codes or not, "degeneracy" refers to the functional products of DNA (except when it refers to codons and anticodons) and therefore implies "structural DNA."

Degenerate tRNA: "Synonym" species of tRNA differing in their anticodon. Compare with "redundant tRNA."

Euchromatin: Used here in a biochemical sense to mean chromatin that is transcribed.

Excess DNA: Used trivially for the amount of eukaryotic DNA assumed to be greater than required for encoding protein and stable RNA, but which may code for other RNA.

Factor: Differs from an enzyme in that it mediates a recognitory reaction which does not essentially alter the reactants, as opposed to a synthetic or degrative reaction which does; it acts therefore stoichiometrically rather than catalytically. Factors are active in all major macromolecular functions such as replication, transcription, and translation. Their properties make factors capable of regulating these functions very directly.

Facultative function: A function encoded in the genome of an eukaryotic cell but exogenously regulated in whole or in part, e.g., by hormone. Can be a hypertrophied "obligatory" function or a nonobligatory (truly facultative) function necessary to maintain a certain differentiated state of that cell but also of other cells. Compare with "luxury" function.

Facultative heterochromatin: Used here in a biochemical sense to mean chromatin that, although it can be transcribed, is repressed at the time; hence it represents a nonfunctional state of "euchromatin." Compare with "constitutive heterochromatin."

Gene dosage regulation: The expression of a "structural" gene for RNA or protein when present in multiple copies. The lower the "multiplicity," the more rigid is the regulation of transcription. As the multiplicity increases, transcription becomes progressively more flexibly subregulated, e.g., the eukaryotic rRNA genes.

Higher order structure of RNA: See definition in footnote 5.

Informational DNA: Codes for mRNA which, in turn, codes for protein. For further explanation, see Informational Macromolecules in Chapter 1. Compare with "noninformational DNA."

Informational transcription: That of "informational DNA" resulting in mRNA.

Isoaccepting tRNA: One in a set of "synonym" tRNAs; interchangeable with synonym tRNA.

Luxury function: A nonobligatory "facultative function" in an eukaryotic cell that is largely unnecessary to that particular cell, e.g., synthesis of collagen.

Multiplicity: As applied to DNA, numerical repetition of sequences which generally, but not necessarily, implies "redundant DNA." In "noninformational DNA" the multiple cistrons may be redundant to a greater or lesser extent. In "informational DNA," by definition there is no multiplicity of "unique DNA." However, originally unique DNA may be present at a low multiplicity as the result, e.g., of duplications; there also appears to be a more multiple informational DNA, e.g., the histone genes, possibly involving some degree of redundancy.

Noncoding DNA: One which is not transcribed but may have mechanically structural or other functions. Compare with "structural DNA."

Non-Darwinian selection: That operating at levels other than of finished protein, hence more neutral in respect to protein than "Darwinian selection."

Noninformational DNA: Codes for RNA other than mRNA.

Noninformational transcription: That of "noninformational DNA," e.g., stable RNA.

Obligatory function: A function encoded in the genome of an eukaryotic cell, endogenously regulated, and absolutely required for the cell's reproduction. The term is preferred to "constitutive function" to avoid a genetic connotation as in bacteria, for which there is no experimental basis. (Constitutive heterochromatin is, however, used for permanent eukaryotic heterochromatin as sanctioned by usage.) Compare with "facultative function."

Polarity mutant: A bacterial mutant resulting in progressively decreased translation of enzymes whose cistrons (all in one operon) are progressively operator-distal to the mutant site. The related concepts of transcriptional and translational polarity are discussed in Chapter 3.

Redundancy: Refers to molecular redundancy of DNA. See "redundant DNA" for explanation.

Redundant DNA: Synonymous with "reiterated DNA" but not with "excess DNA." Refers to molecular redundancy of DNA, i.e., the presence of related but not identical sequences. Thus, molecular redundancy implies multiplicity but not exact repetition in a trivial sense. Eukaryotic genomes possess several discrete fractions of redundant sequences each characterized by a different copy frequency. The extremely repetitive types of redundant DNA sometimes band as density "satellites." Assumed to represent mostly "noninformational DNA."

Redundant RNA: The transcript of "redundant DNA."

Redundant tRNA: "Synonym" species of tRNA differing in their sequence other than the anticodon. Compare with "degenerate tRNA."

Regulation: As used here means regulation *of* something, e.g., transcriptional regulation means regulation *of* transcription. Compare with "control."

Reiterated DNA: Synonymous with "redundant DNA."

Satellite DNA: Conventionally used for the extremely repetitive (highest copy frequency) forms of "redundant DNA" sometimes with aberrant composition, therefore different buoyant density from main DNA. Depending on variation in these characteristics, different taxonomic species have different satellites in different amounts. Preferably, the term should not be used for rDNA which can also band as a satellite.

Structural DNA: A coding, hence transcribable, DNA. A connotation which alludes to the structure of the product RNA or protein, not that of the DNA itself. Compare with "noncoding DNA."

Structural transcription: That of "structural DNA." Structural transcription is either "informational" or "noninformational."

Synonym: Refers to different codons, and, by extension, different anticodons or more generally tRNAs for the same amino acid.

Unique DNA: A sequence occurring once per haploid genome. By definition, therefore, unique DNA cannot be multiple. Includes most of the "informational DNA."

Unique RNA: The transcript of "unique DNA."

Bibliography

Aaij̇, C., and Borst, P. (1970). *Biochim. Biophys. Acta* **217**, 560.
AB, G., and Malt, R. A. (1970). *J. Cell. Biol.* **46**, 362.
Abd-el-Wahab, A., and Sirlin, J. L. (1959). *Exp. Cell Res.* **18**, 3011.
Abelson, J., Barnett, L., Brenner, S., Gefter, M., Landy, A., Russell, R., and Smith, J. D. (1969). *FEBS Lett.* **3**, 1.
Ackermann, W. W., Cox, D. C., and Dinka, S. (1965). *Biochem. Biophys. Res. Commun.* **19**, 745.
Adams, A., Lindhal, R., and Fresco, J. R. (1967). *Proc. Nat. Acad. Sci. U. S.* **57**, 1684.
Adams, J. M., Jeppesen, P. G., Sanger, F., and Barrell, B. G. (1969). *Nature (London)* **223**, 1009.
Adesnik, M., and Levinthal, C. (1969). *J. Mol. Biol.* **46**, 281.
Agarwal, K. L., Büchi, H., Caruthers, M. H., Gupta, N., Khorana, H. G., Kleppe, K., Kumar, A., Ohtsuka, E., RajBhandary, U. L., Van De Sande, J. H., Sgaramella, V., Weber, H., and Yamada, T. (1970). *Nature (London)* **227**, 27.
Agarwal, M. K., and Weinstein, I. B. (1970). *Biochemistry* **9**, 503.
Agarwal, M. K., Hanoune, J., Yu, F. L., Weinstein, I. B., and Feigelson, P. (1969). *Biochemistry* **8**, 4806.
Agarwal, M. K., Hanoune, J., and Weinstein, I. B. (1970). *Biochim. Biophys. Acta* **224**, 259.
Agranoff, B. W. (1971). *In* "Handbook of Neurochemistry" (A. Lajtha, ed.), Vol. 6, p. 203. Academic Press, New York.
Alberts, B. M., and Frey, L. (1970). *Nature (London)* **277**, 1313.
Alff-Steinberger, C. (1969). *Proc. Nat. Acad. Sci. U. S.* **64**, 584.
Allen, E. H., and Schweet, R. S. (1960). *Biochim. Biophys. Acta* **39**, 185.
Allen, R. E., Ramis, P. L., and Reger, D. M. (1969). *Biochim. Biophys. Acta* **190**, 323.
Allende, C. C., Chaimovich, H., Gatica, M., and Allende, J. E. (1970). *J. Biol. Chem.* **245**, 93.

Allfrey, V. G. (1959). *In* "The Cell" (J. Brachet and A. E. Mirsky, eds.), Vol. 1, p. 193. Academic Press, New York.
Allfrey, V. G., Pogo, B. G. T., Pogo, A. O., Kleinsmith, L. J., and Mirsky, A. E. (1966). *In* "Histones" (A. V. S. de Reuck and J. Knight, eds.), p. 42. Little, Brown, Boston, Massachusetts.
Aloni, Y., and Attardi, G. (1971). *J. Mol. Biol.* **55**, 271.
Altman, S. (1971). *Nature (New Biol.)* **229**, 19.
Amaldi, F. (1969). *Nature (London)* **221**, 95.
Amaldi, F., and Attardi, G. (1968). *J. Mol. Biol.* **33**, 737.
Amaldi, F., and Buongiorno-Nardelli, M. (1971). *Exp. Cell Res.* **65**, 329.
Amano, M., Leblond, C. P., and Nadler, N. J. (1965). *Exp. Cell Res.* **38**, 314.
Ames, B. N., and Hartman, P. E. (1963). *Cold Spring Harbor Symp. Quant. Biol.* **28**, 349.
Ames, B. N., and Martin, R. G. (1964). *Annu. Rev. Biochem.* **33**, 235.
Amos, H. (1967). *Nat. Cancer Inst. Monogr.* **26**, 23.
Anderson, M. B., and Cherry, J. H. (1969). *Proc. Nat. Acad. Sci. U. S.* **62**, 202.
Anderson, N. G. (1966). *Science* **154**, 103.
Anderson, W. F. (1969a). *Biochemistry* **8**, 3687.
Anderson, W. F. (1969b). *Proc. Nat. Acad. Sci. U. S.* **62**, 566.
Anderson, W. F., and Gilbert, J. M. (1969). *Biochem. Biophys. Res. Commun.* **36**, 456.
Andoh, T., and Ozaki, H. (1968). *Proc. Nat. Acad. Sci. U. S.* **59**, 792.
Arceneaux, J. L., and Sueoka, N. (1969). *J. Biol. Chem.* **244**, 5959.
Arion, V. J., Mantieva, V. L., and Georgiev, G. P. (1967). *Biochim. Biophys. Acta* **138**, 436.
Arms, K. (1968). *J. Embryol. Exp. Morphol.* **20**, 367.
Armstrong, D. J., Burrows, W. J., Skoog, F., Roy, K. L., and Söll, D. (1969). *Proc. Nat. Acad. Sci. U. S.* **63**, 834.
Armstrong, R. L., and Sueoka, N. (1968). *Proc. Nat. Acad. Sci. U. S.* **59**, 153.
Arndt, D. J., and Berg, P. (1970). *J. Biol. Chem.* **245**, 665.
Arnott, S., Wilkins, M. H. F., Hamilton, L. D., and Langridge, R. J. (1965). *J. Mol. Biol.* **11**, 391.
Arnott, S., Wilkins, M. H. F., Fuller, W., and Langridge, R. J. (1967). *J. Mol. Biol.* **27**, 535.
Arnott, S., Fuller, W., Hodgson, A., and Prutton, I. (1968). *Nature (London)* **220**, 561.
Arnstein, H. R. V., and Rahamimoff, H. (1968). *Nature (London)* **219**, 942.
Aronson, A. I. (1963). *Biochim. Biophys. Acta* **72**, 176.
Aronson, A. I. (1972). *Nature (New Biol.)* **235**, 40.
Ascione, R., and Woude, G. F. V. (1969). *J. Virol.* **4**, 727.
Attardi, G., and Amaldi, F. (1970). *Annu. Rev. Biochem.* **39**, 183.
Attardi, G., and Attardi, B. (1967). *Proc. Nat. Acad. Sci. U. S.* **58**, 1051.
Attardi, G., and Attardi, B. (1968). *Proc. Nat. Acad. Sci. U. S.* **61**, 281.
Attardi, G., and Attardi, B. (1969). *In* "Problems in Biology: RNA in Development" (E. W. Hanly, ed.), p. 245. University of Utah Press, Salt Lake City, Utah.
Attardi, G., Huang, P., and Kabat, S. (1965a). *Proc. Nat. Acad. Sci. U. S.* **53**, 1490.
Attardi, G., Huang, P., and Kabat, S. (1965b). *Proc. Nat. Acad. Sci. U. S.* **54**, 185.
Attardi, G., Parnas, H., Hwang, M. I. H., and Attardi, B. (1966). *J. Mol. Biol.* **20**, 145.
Attardi, G., Parnas, H., and Attardi, B. (1970). *Exp. Cell Res.* **62**, 11.
Aubert, M., Scott, J. F., Reynier, M., and Monier, R. (1968). *Proc. Nat. Acad. Sci. U. S.* **61**, 292.
Auer, G., Zetterberg, A., and Foley, G. E. (1970). *J. Cell. Phys.* **76**, 357.
August, J. T. (1969). *Nature (London)* **222**, 121.

Bibliography

Avery, O. T., MacLeod, C. M., and McCarty, M. (1944). *J. Exp. Med.* **79**, 137.
Axel, R., Weinstein, I. B., and Farber, E. (1967). *Proc. Nat. Acad. Sci. U. S.* **58**, 1255.
Bachrach, U. (1970). *Annu. Rev. Microbiol.* **24**, 109.
Bachvarova, R., and Davidson, E. H. (1966). *J. Exp. Zool.* **163**, 285.
Bär, H. P., and Hechter, O. (1969). *Proc. Nat. Acad. Sci. U. S.* **63**, 350.
Baglioni, C., Bleiberg, I., and Zauderer, M. (1971). *Nature (New Biol.)* **232**, 8.
Baguley, B. C., and Staehelin, M. (1969). *Biochemistry* **8**, 259.
Baker, R. F., and Yanofsky, C. (1968). *Nature (London)* **219**, 26.
Baker, W. K. (1968). *Advan. Genet.* **14**, 133.
Baldwin, A. N., and Berg, P. (1966). *J. Biol. Chem.* **241**, 839.
Baliga, B. S., Srinivasan, P. R., and Borek, E. (1965). *Nature (London)* **208**, 555.
Baliga, B. S., Borek, E., Weinstein, I. B., and Srinivasan, P. R. (1969). *Proc. Nat. Acad. Sci. U. S.* **62**, 899.
Balinsky, B. I. (1965). "An Introduction to Embryology," 2nd Ed. Saunders, Philadelphia, Pennsylvania.
Baltimore, D. (1971). *Bacteriol. Rev.* **35**, 235.
Bandyopadhyay, A. K., and Deutscher, M. P. (1971). *J. Mol. Biol.* **60**, 113.
Banerjee, A. K., Rensing, M., and August, J. T. (1969). *J. Mol. Biol.* **45**, 181.
Barbieri, M., Pettazoni, P., Bersani, F., and Maraldi, N. M. (1970). *J. Mol. Biol.* **54**, 121.
Barden, N., and Korner, A. (1969). *Biochem. J.* **114**, 30 P.
Barnett, W. E., Brown, D. H., and Epler, J. L. (1967). *Proc. Nat. Acad. Sci. U. S.* **57**, 1775.
Barnett, W. E., Pennington, C. J., and Fairfield, S. A. (1969). *Proc. Nat. Acad. Sci. U. S.* **63**, 1261.
Barth, L. G., and Barth, L. J. (1967). *Biol. Bull.* **133**, 495.
Baserga, R. (1967). *Methods Cancer Res.* **1**, 45.
Bassel, B. A. (1968). *Proc. Nat. Acad. Sci. U. S.* **60**, 321.
Baudisch, W., and Panitz, R. (1968). *Exp. Cell Res.* **49**, 470.
Bautz, E. K. F. (1967). *Mol. Genet.* **2**, 21.
Bautz, E. K. F. (1970). *Rep. 7th Pathobiol. Conf.* Aspen, Colorado.
Bautz, E. K. F. and Bautz, F. A. (1970). *Nature (London)* **226**, 1219.
Bautz, E. K. F., and Heding, L. (1964). *Biochemistry* **3**, 1010.
Bautz, E. K. F., and Reilly, E. (1966). *Science* **151**, 328.
Bayev, A. A., Venkstern, T. V., Mirzabekov, A. D., Krutilina, A. I., Axelrod, V. D., Li, L., Fodor, L., Kasarinova, L. Y., and Engelhardt, V. A. (1968). *In* "Structure and Function of Transfer RNA and 5 S-RNA" (L. O. Fröholm and S. G. Laland, eds.), p. 17, Academic Press, New York.
Bayshaw, J. C., Finamore, F. J., and Novelli, G. D. (1970). *Dev. Biol.* **23**, 23.
Beale, G. H. (1969). *In* "The Biological Basis of Medicine" (E. E. Bittar and N. Bittar, eds.), Vol. 4, p. 81. Academic Press, New York.
Beadle, G. W., and Tatum, E. C. (1941). *Proc. Nat. Acad. Sci. U. S.* **27**, 499.
Beardsley, K., and Cantor, C. R. (1970). *Proc. Nat. Acad. Sci. U. S.* **65**, 39.
Beatty, B. G., and Wong, J. T-F. (1970). *Biochem. Biophys. Res. Commun.* **41**, 99.
Beck, G., Hentzen, D., and Ebel, J-P. (1970). *Biochim. Biophys. Acta* **213**, 55.
Becker, F. F. (1969). *Annu. Rev. Med.* **20**, 243.
Beermann, W. (1960). *Chromosoma* **11**, 263.
Beermann, W. (1967). *In* "Heritage from Mendel" (R. A. Brink and A. D. Styles, eds.), p. 179, Univ. of Wisconsin Press, Madison, Wisconsin.
Bekhor, I., Bonner, J., and Dahmus, G. K. (1969a). *Proc. Nat. Acad. Sci. U. S.* **62**, 271.
Bekhor, I., Kung, G. M., and Bonner, J. (1969b). *J. Mol. Biol.* **39**, 351.
Belitsina, N. V., and Spirin, A. S. (1970). *J. Mol. Biol.* **52**, 45.

Beljanski, M., and Beljanski, M. (1963). *Biochim. Biophys. Acta* **72**, 585.
Bell, E. (1969). *Nature (London)* **224**, 326.
Bellamy, A. R. (1966). *Biochim. Biophys. Acta* **123**, 102.
Bello, L. J. (1968). *Biochim. Biophys. Acta* **157**, 8.
Belozersky, A. N., and Spirin, A. S. (1960). *In* "The Nucleic Acids" (E. Chargaff and J. N. Davidson, eds.), Vol. 3, p. 147. Academic Press, New York.
Bendich, A. J., and McCarthy, B. J. (1970a). *Proc. Nat. Acad. Sci. U. S.* **65**, 349.
Bendich, A. J., and McCarthy, B. J. (1970b). *Genetics* **65**, 545.
Benzer, S. (1959). *Proc. Nat. Acad. Sci..U. S.* **45**, 1607.
Benzer, S. (1961). *Proc. Nat. Acad. Sci. U. S.* **47**, 403.
Berendes, H. D. (1968). *Chromosoma* **24**, 418.
Berendes, H. D., and Boyd, J. B. (1969). *J. Cell Biol.* **41**, 591.
Bergman, S., Rodman, T. C., and Sirlin, J. L. (1972). *Humangenetik*, in press.
Berlowitz, L., and Birnstiel, M. L. (1967). *Science* **156**, 78.
Bernhard, W., and Granboulan, N. (1968). *In* "Ultrastructure in Biological Systems, Vol. 3: The Nucleus" (A. J. Dalton and F. Haguenau, eds.), p. 81. Academic Press, New York.
Bernhardt, D., and Darnell, J. E. (1969). *J. Mol. Biol.* **42**, 43.
Bier, K., and Müller, W. (1969). *Biol. Zentralbl.* **88**, 425.
Billeter, M. A., Dahlberg, J. E., Goodman, H. M., Hindley, J., and Weissmann, C. (1969). *Nature (London)* **224**, 1083.
Billing, R. J., Barbiroli, B., and Smellie, R. M. S. (1969a). *Biochem. J.* **112**, 563.
Billing, R. J., Barbiroli, B., and Smellie, R. M. S. (1969b). *Biochem. J.* **114**, 37 P.
Birnbaum, L. S. and Kaplan, S. (1971). *Proc: Nat. Acad. Sci. U. S.* **68**, 925.
Birnstiel, M. L., Speirs, J., Purdom, I., Jones, K. W., and Loening, U. E. (1968) *Nature, (London)* **219**, 454.
Birnstiel, M. L., Grunstein, M., Speirs, J., and Hennig, W. (1969). *Nature (London)* **223**, 1265.
Birnstiel, M. L., Chipchase, M., and Speirs, J. (1971). *Progr. Nucleic Acid Res. Mol. Biol.* **11**, 351.
Bishop, J. O., Leahy, J., and Schweet, R. (1960). *Proc. Nat. Acad. Sci. U. S.* **46**, 1030.
Bishop, J. O., Pemberton, R., and Baglioni, C. (1972). *Nature (New Biol.)* **235**, 231.
Blake, R. D., Fresco, J. R., and Langridge, R. (1970). *Nature (London)* **225**, 32.
Bleyman, M., Kondo, M., Hecht, N., and Woese, C. R. (1969). *J. Bacteriol.* **99**, 535.
Bock, R. M. (1969). *In* "Exploitable Molecular Mechanisms and Neoplasia" (R. B. Hurlbert, ed.) p. 191. Williams & Wilkins, Baltimore, Maryland.
Boedtker, H., and Kelling, D. G. (1967). *Biochem. Biophys. Res. Commun.* **29**, 758.
Bolle, A., Epstein, R. H., Salser, W., and Geiduschek, E. P. (1968). *J. Mol. Biol.* **31**, 325.
Boman, H. G., and Hjerten, S. (1962). *Arch. Biochem. Biophys. Suppl.* **1**, 276.
Bond, H. E., Cooper, J. A., Courington, D. P., and Wood, J. S. (1969). *Science* **165**, 705.
Bonner, J. (1969). Presented in seminar course at Rockefeller Univ., New York, Nov., 1969.
Bonner, J. (1971). *Nature* **231**, 543.
Bonner, J., Dahmus, M. D., Fambrough, D., Huang, R. C., Marushige, K., and Tuan, D. Y. H. (1968). *Science* **159**, 47.
Bonner, J., Griffin, J., Shih, T., and Sadgopal, A. (1969). *Biophys. J.* **9**, A-87.
Borek, E. (1969). *In* "Exploitable Molecular Mechanisms and Neoplasia" (R. B. Hurlbert, ed.) p. 163. Williams & Wilkins, Baltimore, Maryland.
Borek, E. (ed.) (1971). *Cancer Res.* **31**, 591.
Borst, P., and Grivell, L. A. (1971). *FEBS Lett.* **13**, 73.

Brachet, J. (1942). *Arch. Biol.* **53**, 207.
Brachet, J. (1957). "Biochemical Cytology," Academic Press, New York.
Brachet, J. (1968). *Biol. Rev. Cambridge Phil. Soc.* **43**, 1.
Brachet, J., and Jeener, R. (1944). *Enzymologia* **11**, 196.
Brachet, J., Hulin, N., and Guermant, J. (1968). *Exp. Cell Res.* **51**, 509.
Bradley, D. F., and Rich, A. (1956). *J. Amer. Chem. Soc.* **78**, 5898.
Bramwell, M. E., and Harris, H. (1967). *Biochem. J.* **103**, 816.
Breier, B., and Holley, R. W. (1970). *Biochim. Biophys. Acta* **213**, 365.
Bremer, H., and Yuan, D. (1968). *J. Mol. Biol.* **38**, 163.
Brenner, S., Jacob, F., and Meselson, M. (1961). *Nature (London)* **190**, 576.
Brenner, S., Kaplan, S., and Stretton, A. O. W. (1966). *J. Mol. Biol.* **19**, 574.
Brentani, R., and Brentani, M. (1969). *Genetics* **61**, Suppl. 1, 391.
Bretscher, M. S. (1968a). *Nature (London)* **217**, 509.
Bretscher, M. S. (1968b). *Nature (London)* **220**, 1088.
Bretscher, M. S. (1968c). *J. Mol. Biol.* **34**, 131.
Briggs, R., and Justus, J. T. (1968). *J. Exp. Zool.* **167**, 105.
Brink, A., Styles, E. D., and Axtell, J. D. (1968). *Science* **159**, 161.
Britten, R. J., and Davidson, E. H. (1969). *Science* **165**, 349.
Britten, R. J., and Kohne, D. E. (1968). *Science* **161**, 529.
Britten, R. J., and Kohne, D. E. (1969). *In* "Handbook of Molecular Cytology" (A. Lima-de-Faria, ed.), pp. 21 and 37. North-Holland Publ., Amsterdam.
Britten, R. J., and Roberts, R. B. (1960). *Science* **131**, 32.
Brown, D. D. (1965). *In* "Development and Metabolic Control Mechanisms and Neoplasia" (D. N. Ward, ed.), p. 219. Williams & Wilkins, Baltimore, Maryland.
Brown, D. D. (1967). *In* "Current Topics in Developmental Biology" (A. A. Moscona, ed.) Vol. 2, p. 48. Academic Press, New York.
Brown, D. D., and Dawid, I. B. (1968). *Science* **160**, 272.
Brown, D. D., and Dawid, I. B. (1969). *Annu. Rev. Genet.* **3**, 127.
Brown, D. D., and Gurdon, J. B. (1964). *Proc. Nat. Acad. Sci. U. S.* **51**, 139.
Brown, D. D., and Gurdon, J. B. (1966). *J. Mol. Biol.* **19**, 399.
Brown, D. D., and Littna, E. (1966a). *J. Mol. Biol.* **20**, 81.
Brown, D. D., and Littna, E. (1966b). *J. Mol. Biol.* **20**, 95.
Brown, D. D., and Weber, C. S. (1968a). *J. Mol. Biol.* **34**, 661.
Brown, D. D., and Weber, C. S. (1968b). *J. Mol. Biol.* **34**, 681.
Brown, G. M., and Attardi, G. (1965). *Biochem. Biophys. Res. Commun.* **20**, 298.
Brown, I. R., and Church, R. B. (1971). *Biochem. Biophys. Res. Commun.* **42**, 850.
Brown, S. W. (1966). *Science* **151**, 417.
Brown, S. W. (1969). *Genetics* **61**, Suppl. 1, 191.
Brownlee, G. G. (1971). *Nature (New Biol.)* **229**, 147.
Brownlee, G. G., Sanger, F., and Barrell, B. G. (1967). *Nature (London)* **215**, 735.
Brownlee, G. G., Sanger, F., and Barrell, B. G. (1968). *J. Mol. Biol.* **34**, 379.
Brunschede, H. and Bremer, H. (1971). *J. Mol. Biol.* **57**, 35.
Bucher, N. L. R. (1963). *Int. Rev. Cytol.* **15**, 245.
Bullough, W. S. (1968). *In* "The Biological Basis of Medicine" (E. E. Bittar and N. Bittar, eds.), Vol. 1, p. 311. Academic Press, New York.
Bulova, S. I., and Burka, E. R. (1970). *J. Biol. Chem.* **245**, 4907.
Bultmann, H., and Clever, U. (1969). *Chromosoma* **28**, 120.
Burdon, R. H., and Clason, A. E. (1969). *J. Mol. Biol.* **39**, 113.
Burgess, R. R., Travers, A. A., Dunn, J. J., and Bautz, E. K. F. (1969). *Nature (London)* **221**, 43.

Burr, H., and Lingrel, J. B. (1971). *Nature (New Biol.)* **233**, 41.
Butler, J. A. V., Johns, E. W., and Phillips, D. M. P. (1968). *Progr. Biophys. Mol. Biol.* **18**, 209.
Byars, N., and Kidson, D. (1970). *Nature (London)* **226**, 648.
Cairns, J. (1963). *Cold Spring Harbor Symp. Quant. Biol.* **28**, 43.
Cairns, J., Stent, G. S., and Watson, J. D., eds. (1966). "Phage and the Origins of Molecular Biology," Cold Spring Harbor Lab. Quant. Biol., Cold Spring Harbor, New York.
Callan, H. G. (1963). *Int. Rev. Cytol.* **15**, 1.
Callan, H. G. (1967). *J. Cell Sci.* **2**, 1.
Callan, H. G. (1970). *Rep. 7th Annu. Pathobiol. Conf.*, Aspen, Colorado.
Campbell, A. (1967). *Mol. Genet.* **2**, 323.
Campbell, R. J. (1967). *J. Theoret. Biol.* **16**, 321.
Candelas, G. C., and Gelfant, S. (1970). *Abstr. 10th Annu. Meeting. Amer. Soc. Cell Biol., San Diego, California*, p. 30a.
Capecchi, M. R., and Gussin, G. N. (1965). *Science* **149**, 417.
Capra, J. D., and Peterkofsky, A. (1968). *J. Mol. Biol.* **33**, 591.
Carriere, R. (1969). *Int. Rev. Cytol.* **25**, 201.
Cashel, M., and Gallant, J. (1969). *Nature (London)* **221**, 838.
Caskey, C. T., Beaudet, A., and Nirenberg, M. (1968). *J. Mol. Biol.* **37**, 99.
Caspersson, T. (1941). *Naturwissenschaften* **29**, 33.
Cattanach, B. M., Perez, J. N., and Pollard, C. E. (1970). *Genet. Res.* **15**, 183.
Cavalieri, L. F., and Carroll, E. (1968). *Proc. Nat. Acad. Sci. U. S.* **59**, 951.
Cavalieri, L. F., and Nemchin, R. G. (1968). *Biochim. Biophys. Acta* **166**, 722.
Ceccarini, C., and Maggio, R. (1969). *Biochim. Biophys. Acta* **190**, 556.
Ceccarini, C., Maggio, R., and Barbata, G. (1967). *Proc. Nat. Acad. Sci. U. S.* **58**, 2235.
Chamberlin, M., McGrath, J., and Waskell, L. (1970). *Nature (London)* **228**, 227.
Chambon, P., Weill, J. D., Doly, J., Strosser, M. T., and Mandel, P. (1966). *Biochem. Biophys. Res. Commun.* **25**, 638.
Chantrenne, H., Burny, A., and Marbaix, G. (1967). *Progr. Nucleic Acid Res. Mol. Biol.* **7**, 173.
Chargaff, E. (1968). *Progr. Nucleic Acid Res. Mol. Biol.* **8**, 297.
Chargaff, E., and Davidson, J. N., eds. (1955). "The Nucleic Acids," Vol. 1 and 2. Academic Press, New York.
Chargaff, E., and Davidson, J. N., eds. (1960). "The Nucleic Acids," Vol. 3. Academic Press, New York.
Chargaff, E., Lipschitz, R., Green, C., and Hodes, M. D. (1951). *J. Biol. Chem.* **192**, 223.
Chen, D., Sarid, S., and Katchalski, E. (1968). *Proc. Nat. Acad. Sci. U. S.* **60**, 902.
Chilton, M.-D., and McCarthy, B. J. (1969). *Genetics* **62**, 697.
Choi, Y. C., and Busch, H. (1969). *Biochim. Biophys. Acta* **174**, 766.
Church, R. B., and McCarthy, B. J. (1967a). *J. Mol. Biol.* **23**, 459.
Church, R. B., and McCarthy, B. J. (1967b). *Proc. Nat. Acad. Sci. U. S.* **58**, 1548.
Church, R. B., and McCarthy, B. J. (1970). *Biochim. Biophys. Acta* **199**, 103.
Church, R. B., Luther, S. W., and McCarthy, B. J. (1969). *Biochim. Biophys. Acta* **190**, 30.
Ciferri, O., and Parisi, B. (1970). *Progr. Nucleic Acid Res. Mol. Biol.* **10**, 121.
Clark, B. F. C., and Marcker, K. A. (1966). *J. Mol. Biol.* **17**, 394.
Clark, B. F. C., and Marcker, K. A. (1968). *Sci. Amer.* **218**, 36.
Clark, B. F. C., Doctor, B. P., Holmes, A. C., Klug, A., Marcker, K. A., Morris, S. J., and Paradies, H. H. (1968). *Nature (London)* **219**, 1222.
Clark, M. F., Matthews, R. E., and Ralph, R. K. (1964). *Biochim. Biophys. Acta* **91**, 289.
Clarke, B. (1970). *Nature (London)* **228**, 159.

Clason, A. E., and Burdon, R. H. (1969). *Nature, (London)* **223**, 1063.
Clegg, J. S., and Golub, A. L. (1969). *Develop. Biol.* **19**, 178.
Clever, U. (1964). *Science* **146**, 794.
Clever, U. (1968). *Annu. Rev. Genet.* **2**, 11.
Clever, U. (1969). *Exp. Cell Res.* **55**, 317.
Clever, U., and Ellgaard, E. G. (1970). *Science* **169**, 373.
Cocucci, S,.M., and Sussman, M. (1970). *J. Cell Biol.* **45**, 399.
Cohen, P., Chin, R-C., and Kidson, C. (1969). *Biochemistry* **8**, 3603.
Cohen, S. S., Morgan, S., and Streibel, E. (1969). *Proc. Nat. Acad. Sci. U. S.* **64**, 669.
Cohn, W. E. (1957). *Methods Enzymol.* **3**, 724.
Colby, C., and Duesberg, P. H. (1969). *Nature (London)* **222**, 940.
Cold Spring Harbor Symp. Quant. Biol. (1963). "Synthesis and Structure of Macromolecules," Vol. 28.
Cold Spring Harbor Symp. Quant. Biol. (1966). "The Genetic Code," Vol. 31.
Cold Spring Harbor Symp. Quant. Biol. (1968). "The Replication of DNA in Microorganisms," Vol. 33.
Cold Spring Harbor Symp. Quant. Biol. (1969). "The Mechanism of Protein Synthesis," Vol. 34.
Cold Spring Harbor Symp. Quant. Biol. (1970). "Transcription of Genetic Material," Vol. 35.
Colli, W., Smith, I., and Oishi, M. (1971) *J. Mol. Biol.* **56**, 117.
Colombo, B., Vesco, C., and Baglioni, C. (1968). *Proc. Nat. Acad. Sci. U. S.* **61**, 651.
Comb, D. G., and Katz, S. (1964). *J. Mol. Biol.* **8**, 790.
Comb, D. G., and Silver, D. J. (1966). *Nat. Cancer Inst. Monogr.* **23**, 325.
Comb, D. G., and Zehavi-Willner, T. (1967). *J. Mol. Biol.* **23**, 441.
Comings, D. E. (1968). *Amer. J. Hum. Genet.* **20**, 440.
Comings, D. E. (1970). *Chromosoma* **29**, 434.
Comings, D. E., and Kakefuda, T. (1968). *J. Mol. Biol.* **33**, 225.
Comings, D. E., and Mattoccia, E. (1969). *J. Cell Biol.* **43**, 25a.
Comings, D. E., and Okada, T. A. (1970). *Nature (London)* **227**, 451.
Connors, P. G. and Beeman, W. W. (1970). *Fed. Proc. Fed. Amer. Soc. Exp. Biol.* **29**, 672.
Coon, H., and Cahn, R. D. (1966). *Science* **153**, 1116.
Coons, A. H., and Peters, J. H. (1972). *In* "Methods in Immunology and Immunochemistry" (C. A. Williams and M. W. Chase, eds.), Vol. 4. Academic Press, New York (in press).
Cooper, D. W. (1971). *Nature (London)* **230**, 292.
Cooper, H. L. (1969). *In* "Biochemistry of Cell Division" (R. Baserga, ed.), p. 91. Thomas, Illinois.
Cooper, H. L. (1970). *Nature (London)* **227**, 1105.
Cory, S., and Marcker, K. A. (1970). *Eur. J. Biochem.* **12**, 177.
Corbin, J. D., and Krebs, E. G. (1969). *Biochem. Biophys. Res. Commun.* **36**, 328.
Cotter, R. I., McPhie, P., and Gratzer, W. G. (1967). *Nature (London)* **216**, 864.
Cox, R. A., and Bonanou, S. A. (1969). *Biochem. J.* **114**, 769.
Craddock, V. M. (1970). *Nature (London)* **228**, 1264.
Craig, N. C. (1970). *Abstr. 10th Annu. Meeting Amer. Soc. Cell. Biol., San Diego, California*, p. 43a.
Craig, N. C., and Perry, R. P. (1970). *J. Cell Biol.* **45**, 554.
Cramer, F. (1971). *Progr. Nucleic Acid Res. Mol. Biol.* **11**, 391.
Cramer, F., Doepner, H., von der Haar, F., Schlimme, E., and Seidel, H. (1968). *Proc. Nat. Acad. Sci. U. S.* **61**, 1384.
Cramer, F., von der Haar, F., Holmes, K. C., Saenger, W., Schlimme, E., and Schultz, G. E. (1970). *J. Mol. Biol.* **51**, 523.

Crick, F. H. C. (1957). *Biochem. Soc. Symp.* **14**, 25.
Crick, F. H. C. (1958). *Symp. Soc. Exp. Biol.* **12**, 138.
Crick, F. H. C. (1963). *Science* **139**, 461.
Crick, F. H. C. (1966a). *Cold Spring Harbor Symp. Quant. Biol.* **31**, 3.
Crick, F. H. C. (1966b). *J. Mol. Biol.* **19**, 548.
Crick, F. H. C. (1967). *Proc. Roy. Soc., Ser. B* **167**, 331.
Crick, F. H. C. (1968). *J. Mol. Biol.* **38**, 367.
Crick, F. H. C. (1970). *Nature, (London)* **227**, 561.
Crick, F. H. C., Barnett, L., Brenner, S., and Watts-Tobin, R. J. (1961). *Nature (London)* **192**, 1227.
Crippa, M. (1966). *Exp. Cell Res.* **42**, 371.
Crippa, M. (1970). *Nature (London)* **227**, 1138.
Crippa, M., and Gross, P. R. (1969). *Proc. Nat. Acad. Sci. U. S.* **62**, 120.
Crippa, M., and Tocchini-Valentini, G. P. (1970). *Nature (London)* **226**, 1243.
Crippa, M., and Tocchini-Valentini, G. P. (1971). *Proc. Nat. Acad. Sci. U. S.* **68**, 2769.
Crippa, M., Davidson, E. H., and Mirsky, A. E. (1967). *Proc. Nat. Acad. Sci. U. S.* **57**, 885.
Crouse, H. V., and Keyl, H. G. (1968). *Chromosoma* **25**, 357.
Crow, F., and Kimura, M. (1970). "An Introduction to Population Genetics Theory." Harper, New York.
Culp, W., McKeehan, W., and Hardesty, B. (1969). *Proc. Nat. Acad. Sci. U. S.* **64**, 388.
Curgo, C., Apirion, D., and Schlessinger, D. (1969). *J. Mol. Biol.* **45**, 205.
Curtis, A. S. G. (1965). *Arch. Biol.* **76**, 353.
Cutler, R. G., and Evans, J. E. (1967). *J. Mol. Biol.* **26**, 91.
Dahlberg, J. E. (1968). *Nature (London)* **220**, 548.
Dahmus, M. E., and McConnell, D. J. (1969). *Biochemistry* **8**, 1524.
Daneholt, B. (1970). *J. Mol. Biol.* **49**, 381.
Daneholt, B., and Edström, J-E. (1967). *Cytogenetics* **6**, 350.
Daneholt, B., and Edström, J-E. (1969). *J. Cell Biol.* **41**, 62.
Daneholt, B., Edström, J-E., Egyhazi, E., Lambert, B., and Ringborg, V. (1969). *Chromosoma* **28**, 399.
Daniel, V., Sarid, S., and Littauer, U. Z. (1970). *Science* **167**, 1682.
Darnell, J. E. (1968). *Bacteriol. Rev.* **32**, 262.
Darnell, J. E. (1971). Personal communication.
Darnell, J. E., Philipson, L., Wall, R., and Adesnik, M. (1971). *Science* **174**, 507.
Davern, C. I., and Meselson, M. (1960). *J. Mol. Biol.* **2**, 153.
Davey, P. J., Yu, R., and Linnane, A. W. (1969). *Biochem. Biophys. Res. Commun.* **36**, 30.
Davidson, E. H. (1968). "Gene Activity in Early Development," Academic Press, New York.
Davidson, E. H., and Britten, R. J. (1971). *J. Theoret. Biol.* **32**, 123.
Davidson, E. H., and Hough, B. R. (1969). *J. Exp. Zool.* **172**, 25.
Davidson, E. H., and Hough, B. R. (1971). *J. Mol. Biol.* **56**, 491.
Davidson, E. H., Crippa, M., Kramer, F. R., and Mirsky, A. E. (1966). *Proc. Nat. Acad. Sci. U. S.* **56**, 856.
Davidson, E. H., Crippa, M., and Mirsky, A. E. (1968). *Proc. Nat. Acad. Sci. U. S.* **60**, 152.
Davidson, J. N. (1969). "The Biochemistry of the Nucleic Acids," 6th Ed. Methuen, London.

Davidson, J. N., and Cohn, W. E., eds. (1963-1972). *Progr. Nucleic Acid Res. Mol. Biol.* **1-12**.
Davies, H. G., and Small, J. V. (1968). *Nature (London)* **217**, 1122.
Davies, J., and Davis, B. D. (1968). *J. Biol. Chem.* **243**, 3312.
Davies, L. M., Priest, J. H., and Priest, R. E. (1968). *Science* **159**, 91.
Dawid, I. B. (1969). *Fed. Proc. Fed. Amer. Soc. Exp. Biol.* **28**, 349.
Dawid, I. B., and Wolstenholme, D. R. (1968). *Biophys. J.* **8**, 65.
Dayhoff, M. O. (1969). "Atlas of Protein Sequences and Structure, 1969," Vol. 4. Nat. Biomed. Res. Found., Silver Springs, Maryland.
De Angelo, A. B., and Gorski, J. (1970). *Proc. Nat. Acad. Sci. U. S.* **66**, 693.
Degroot, N., and Cohen, P. P. (1962). *Biochim. Biophys. Acta* **59**, 595.
De Lange, R. J., Fambrough, D. M., Smith, E. L., and Bonner, J. (1969). *J. Biol. Chem.* **244**, 319.
De León, V., Yang, W.-K., and Sirlin, J. L. (1972). In press.
De Ley, J. (1968). *Evol. Biol.* **2**, 103.
de Lucia, P., and Cairns, J. (1969). *Nature (London)* **224**, 1164.
Denis, H. (1968). *Advan. Morphog.* **7**, 115.
Denis, H., and Brachet, J. (1969). *Proc. Nat. Acad. Sci. U. S.* **62**, 438.
de Terra, N. (1969). *Int. Rev. Cytol.* **25**, 1.
Deuchar, E. M. (1965). In "The Biochemistry of Animal Development" (R. Weber, ed.), Vol. 1, p. 245. Academic Press, New York.
De Voe, H., and Tinoco, I. (1962). *J. Mol. Biol.* **4**, 500.
De Wachter, R., and Fiers, W. (1969). *Nature (London)* **221**, 233.
Dezeleé, S., Sentenac, A., and Fromageot, P. (1970). *FEBS Lett.* **7**, 220.
Dingman, C. W., Aronow, A., Bunting, S. L., Peacock, A. C., and O'Malley, B. W. (1969). *Biochemistry* **8**, 849.
Doctor, B. P., Fuller, W., and Webb, N. L. W. (1969). *Nature (London)* **221**, 58.
Doi, R. H., Kaneko, I., and Igarashi, R. T. (1968). *J. Biol. Chem.* **243**, 945.
Donachie, W., and Masters, M. (1969). In "The Cell Cycle: Gene-Enzyme Interactions" (G. M. Padilla, I. L. Cameron, and G. L. Whitson, eds.), p. 37. Academic Press, New York.
Doolittle, N. F., and Pace, N. R. (1970). *Nature (London)* **228**, 125.
Doty, P., and Lundberg, R. D. (1957). *Proc. Nat. Acad. Sci. U. S.* **43**, 213.
Dounce, A. L. (1952). *Enzymologia* **12**, 251.
Doyle, D., and Laufer, H. (1969). *J. Cell Biol.* **40**, 61.
Dresden, M. H. (1969). *Develop. Biol.* **19**, 311.
Drews, J. (1969). *Eur. J. Biochem.* **7**, 200.
Dube, S. K., Marcker, K. A., and Yudelevich, A. (1970). *FEBS Lett.* **9**, 168.
Dubin, D. T., and Günlap, A. (1967). *Biochim. Biophys. Acta* **134**, 106.
Duda, E., Staub, M., Venetianer, P., and Denes, G. (1968). *Biochem. Biophys. Res. Commun.* **32**, 992.
Dudock, B., DiPeri, C., Scileppi, K., and Reszelbach, R. (1971). *Proc. Nat. Acad. Sci. U. S.* **68**, 681.
DuPraw, E. J. (1968). "Cell and Molecular Biology." Academic Press, New York.
Ecker, R. E., and Smith, L. D. (1971). *Dev. Biol.* **24**, 559.
Edelman, G. M. (1970). *Sci. Amer.* **223**, 34.
Edelman, M., Verma, I. M., and Littauer, U. Z. (1970). *J. Mol. Biol.* **49**, 67.
Edlin, G., and Broda, P. (1968). *Bacteriol. Rev.* **32**, 206.
Edlin, G., and Stent, G. S. (1969). *Proc. Nat. Acad. Sci. U. S.* **62**, 475.
Edlin, G., Stent, G. S., Baker, R. F., and Yanofsky, C. (1968). *J. Mol. Biol.* **37**, 257.
Edmonds, M., and Caramela, M. G. (1969). *J. Biol. Chem.* **244**, 1314.
Edmonds, M., and Kopp, D. W. (1970). *Biochem. Biophys. Res. Commun.* **41**, 1531.

Edmonds, M., and Vaughan, M. H. (1970). *Fed. Proc. Fed. Amer. Soc. Exp. Biol.* **29**, 672.
Edström, J-E. (1964). *Methods Cell Physiol.* **1**, 417.
Edström, J-E. (1968). *Nature (London)* **220**, 1196.
Edström, J-E., and Beermann, W. (1962). *J. Cell Biol.* **1**, 371.
Edström, J-E., and Daneholt, B. (1967). *J. Mol. Biol.* **28**, 331.
Edström, J-E., Daneholt, B., Egyhazi, E., Lambert, B., and Ringborg, U. (1969). *Biochem. J.* **114**, 51P.
Eggen, F., and Nathans, D. (1969). *J. Mol. Biol.* **39**, 293.
Egyhazi, E., Daneholt, B., Edström, J-E., Lambert, B., and Ringborg, U. (1969). *J. Mol. Biol.* **44**, 517.
Elgin, S. C. R., and Bonner, J. (1970). *Biochemistry* **9**, 4440.
Eliceiri, G. L., and Green, H. (1969). *J. Mol. Biol* **41**, 253.
Elicieri, G. L., and Green, H. (1970). *Biochim. Biophys. Acta* **199**, 543.
Ellem, K. A. O. (1966). *J. Mol. Biol.* **20**, 283.
Ellem, K. A. O., and Gwatkin, R. B. L. (1968). *Develop. Biol.* **18**, 311.
Emerson, C. P. (1971). *Nature (New Biol.)* **232**, 101.
Emerson, C. P., and Humphreys, T. (1971). *Science* **171**, 898.
Endy, W. W., and Dobrogosz, W. J. (1967). *J. Cell Biol.* **35**, 37A.
Engelhardt, D. L., Robertson, H. D., and Zinder, N. D. (1968). *Proc. Nat. Acad. Sci. U. S.* **59**, 972.
Enger, M. D., and Tobey, R. A. (1969). *J. Cell Biol.* **42**, 308.
Enger, M. D., and Walters, R. A. (1970). *Biochemistry* **9**, 3551.
Englesberg, E., Squires, C., and Merouk, F. (1969). *Proc. Nat. Acad. Sci. U. S.* **62**, 1100.
Epler, J. L. (1969). *Biochemistry* **8**, 2285.
Epler, J. L., and Barnett, W. E. (1967). *Biochem. Biophys. Res. Commun.* **28**, 328.
Epstein, C. J. (1967). *Proc. Nat. Acad. Sci. U. S.* **57**, 327.
Epstein, W., and Beckwith, J. R. (1968). *Annu. Rev. Biochem.* **37**, 411.
Erbe, R. W., Nau, M. M., and Leder, P. (1969). *J. Mol. Biol.* **39**, 441.
Erikson, R. L. (1968). *Annu. Rev. Microbiol.* **22**, 305.
Evans, D., and Birnstiel, M. L. (1968). *Biochim. Biophys. Acta* **166**, 274.
Everett, G. A., Merrill, S. H., and Holley, R. W. (1960). *J. Amer. Chem. Soc.* **82**, 5757.
Fahnestock, S., Neumann, H., Shashoua, V., and Rich, A. (1970). *Biochemistry* **9**, 2477.
Faiferman, I., Hamilton, M. G., and Pogo, A. O. (1970). *Biochim. Biophys. Acta* **204**, 550.
Fambrough, D. M. (1969). *In* "Handbook of Molecular Cytology" (A. Lima-de-Faria, ed.), p. 437. North-Holland Publ., Amsterdam.
Fan, H., and Penman, S. (1971). *J. Mol. Biol.* **59**, 27.
Fan, D. P., Higa, A., and Levinthal, C. (1964). *J. Mol. Biol* **8**, 210.
Feldmann, H., and Zachau, H. G. (1964). *Biochem. Biophys. Res. Commun.* **15**, 13.
Fell, H., and Mellanby, E. (1953). *J. Physiol.* **119**, 470.
Fellner, P. (1969). *Eur. J. Biochem.* **11**, 12.
Fellner, P., and Sanger, F. (1968). *Nature (London)* **219**, 236.
Ficq, A. (1964). *Exp. Cell Res.* **34**, 5811.
Fiers, W. F. (1966). *Nature (London)* **212**, 822.
Filner, P., Wray, J. L., and Varner, J. E. (1969). *Science* **165**, 358.
Fishman, B., Wurtman, R. J., and Munro, H. N. (1969). *Proc. Nat. Acad. Sci. U. S.* **64**, 677.
Fittler, F., and Hall, R. H. (1966). *Biochem. Biophys. Res. Commun.* **25**, 441.

Flaherty, L., Bennett, D., and Graef, S. (1972). *Exp. Cell Res.* **70**, 13.
Flamm, W. G., Walker, P. M. B., and McCallum, M. (1969). *J. Mol. Biol.* **40**, 423.
Flickinger, R. A., Greene, R., Kohl, D. M., and Miyagi, M. (1966). *Proc. Nat. Acad. Sci. U. S.* **56**, 1712.
Flickinger, R. A., Daniel, J. C., and Greene, R. F. (1970a). *Nature (London)* **228**, 557.
Flickinger, R. A., Kohl, D. M., Lauth, M. R., and Stambrook, P. J. (1970b). *Biochim. Biophys. Acta* **209**, 260.
Forchhammer, J., and Kjeldgaard, N. O. (1968). *J. Mol. Biol.* **37**, 245.
Forget, B. G., and Reynier, M. (1970). *Biochem. Biophys. Res. Commun.* **39**, 114.
Forget, B. G., and Varricchio, F. (1970). *J. Mol. Biol.* **48**, 409.
Forget, B. G., and Weissman, S. M. (1967). *Nature (London)* **213**, 878.
Forget, B. C., and Weissman, S. M. (1969). *J. Biol. Chem.* **244**, 3148.
Fournier, M. J., Doctor, B. P., and Peterkofsky, A. (1970). *Fed. Proc. Fed. Amer. Soc. Exp. Biol.* **29**, 468.
Fox, A. S., Yoon, S. B., and Gelbart, W. M. (1971). *Proc. Nat. Acad. Sci. U. S.* **68**, 342.
Franze-Fernández, M. T., and Pogo, A. O. (1971). *Proc. Nat. Acad. Sci. U. S.* **68**, 3040.
Freese, E. (1962). *J. Theoret. Biol.* **3**, 82.
Freese, E. (1963). *Mol. Genet.* **1**, 207.
Frenster, J. H. (1965). *Nature (London)* **206**, 1269.
Frenster, J. H. (1966). In "The Cell Nucleus-Metabolism and RAdiosensitivity (H. M. Klouwen, ed.), p. 26. Taylor & Francis, London.
Frenster, J. H. (1969). In "Handbook of Molecular Cytology" (A. Lima-de-Faria, ed.), p. 251. North-Holland Publ., Amsterdam.
Fresco, J. R., Blake, R. D., and Langridge, R. (1968). *Nature (London)* **220**, 1285.
Fridlender, B. R., and Wettstein, F. O. (1970). *Biochem. Biophys. Res. Commun.* **39**, 247.
Friesen, J. D. (1969). *J. Mol. Biol.* **46**, 349.
Fromson, D., and Nemer, M. (1970). *Science* **168**, 266.
Fry, B. J., and Gross, P. R. (1970). *Develop. Biol.* **21**, 105.
Fujinaga, K., Mak, S., and Green, M. (1968). *Proc. Nat. Acad. Sci. U. S.* **60**, 959.
Fuller, W., and Hodgson, A. (1967). *Nature (London)* **215**, 817.
Fuller, W., Arnott, S., and Creek, J. (1969). *Biochem. J.* **114**, 26P.
Gagné, R., Tanguay, R., and Laberge, C. (1971). *Nature (New Biol.)* **232**, 29.
Galizzi, A. (1969). *Eur. J. Biochem.* **10**, 561.
Gall, J. G. (1968). *Proc. Nat. Acad. Sci. U. S.* **60**, 553.
Gall, J. G. (1969). *Genetics* **61**, Suppl. 1, 121.
Gall, J. G. (1970). *Abstr. 10th Annu. Meeting Amer. Soc. Cell. Biol., San Diego, California,* p. 68a.
Gall, J. G., and Callan, H. G. (1962). *Proc. Nat. Acad. Sci. U. S.* **48**, 562.
Gall, J. G., and Pardue, M. L. (1969). *Proc. Nat. Acad. Sci. U. S.* **63**, 378.
Gall, J. G., Cohen, E. H., and Polan, M. L. (1971). *Chromosoma* **33**, 319.
Gallo, R. C., Longmore, J. L., and Adamson, R. H. (1970). *Nature (London)* **227**, 1134.
Gallwitz, D., and Mueller, G. C. (1969a). *Eur. J. Biochem.* **9**, 431.
Gallwitz, D., and Mueller, G. C. (1969b). *J. Biol. Chem.* **244**, 5947.
Galper, J. B., and Darnell, J. E. (1969). *Biochem. Biophys. Res. Commun.* **34**, 205.
Galston, A. W., and Davies, P. J. (1969). *Science* **163**, 1288.
Gamow, G. (1954). *Nature (London)* **173**, 318.
Ganoza, M. C., and Williams, C. A. (1969). *Proc. Nat. Acad. Sci. U. S.* **63**, 1370.
Garel, J. P., Mandel, P., Chavancy, G., and Daillie, J. (1970). *FEBS Lett.* **7**, 327.
Garen, A. (1968). *Science* **160**, 149.

Gartland, W. J., Ishida, T., Sueoka, N., and Nirenberg, M. W. (1969). *J. Mol. Biol.* **44**, 403.
Gaskill, P., and Kabat, D. (1971). *Proc. Nat. Acad. Sci. U. S.* **68**, 72.
Gatlin, L. L. (1968). *J. Theoret. Biol.* **18**, 181.
Gefter, M. L., and Russell, R. L. (1969). *J. Mol. Biol.* **39**, 145.
Geiduschek, E. P., and Grau, O. (1970). *In* "RNA Polymerase and Transcription" (L. Silverstri, ed.), p. 190. North-Holland Publ., Amsterdam.
Geiduschek, E. P., and Haselkorn, R. (1969). *Annu. Rev. Biochem.* **38**, 647.
Geiduschek, E. P., and Sklar, J. (1969). *Nature (London)* **221**, 833.
Gelderman, A. H., Rake, A. V., and Britten, R. J. (1968). *Carnegie Inst. Yearb.* **67**, 320.
Gelderman, A. H., Rake, A. V., and Britten, R. J. (1971). *Proc. Nat. Acad. Sci. U. S.* **68**, 172.
Georgiev, G. P. (1969a). *Annu. Rev. Genet.* **3**, 155.
Georgiev, G. P. (1969b). *J. Theoret. Biol.* **25**, 473.
Gerbi, S. A. (1971). *J. Mol. Biol.* **58**, 499.
Gevers, W., Kleinkauf, H., and Lipmann, F. (1968). *Proc. Nat. Acad. Sci. U. S.* **60**, 269.
Giacomoni, D., and Spiegelman, S. (1962). *Science* **138**, 1328.
Gibson, I., and Beale, G. H. (1963). *Genet. Res.* **4**, 42.
Gibson, I., and Hewitt, G. (1970). *Nature (London)* **225**, 67.
Giege, R., and Ebel, J-P. (1968). *Biochim. Biophys. Acta* **161**, 125.
Gierer, A., and Schramm, G. (1956). *Nature (London)* **177**, 702.
Gilbert, W., and Dressler, D. (1968). *Cold Spring Harbor Symp. Quant. Biol.* **33**, 473.
Gillespie, D., and Spiegelman, S. (1965). *J. Mol. Biol.* **12**, 829.
Ginsburg, A., and Stadtman, E. F. (1970). *Annu. Rev. Biochem.* **39**, 429.
Girard, M., and Baltimore, D. (1966). *Proc. Nat. Acad. Sci. U. S.* **56**, 999.
Glassman, E. (1969). *Annu. Rev. Biochem.* **38**, 605.
Glick, M. C., and Warren, L. (1969). *Proc. Nat. Acad. Sci. U. S.* **63**, 563.
Glisin, V. R., Glisin, M. V., and Doty, P. (1966). *Proc. Nat. Acad. Sci. U. S.* **56**, 285.
Glücksmann, A. (1951). *Biol. Rev. Cambridge Phil. Soc.* **26**, 59.
Goehler, B., Kaneko, I., and Doi, R. H. (1966). *Biochem. Biophys. Res. Commun.* **24**, 466.
Goldberg, A. L. (1972). *Proc. Nat. Acad. Sci. U. S.* **69**, 422.
Goldman, M., Johnston, W. M., and Griffin, A. C. (1969). *Cancer Res.* **29**, 1051.
Goodman, D., Manor, H., and Rombauts, W. (1969). *J. Mol. Biol.* **40**, 247.
Goodman, H. M., Abelson, J., Landy, A., Brenner, S., and Smith, J. D. (1968). *Nature (London)* **217**, 1019.
Goodwin, B. C. (1966a). *Nature (London)* **209**, 479.
Goodwin, B. C. (1966b). *In* "The Histones" (A. V. S. de Reuck and J. Knight, eds.), p. 68. Little, Brown, Boston, Massachusetts.
Goodwin, B. C., and Cohen, M. H. (1969). *J. Theoret. Biol.* **25**, 49.
Gordon, J. (1970). *Biochemistry* **9**, 912.
Gorini, L. (1970). *Annu. Rev. Genet.* **4**, 107.
Gould, R. M., Thornton, M. P., Liepkalns, V., and Lennarz, W. J. (1968). *J. Biol. Chem.* **243**, 3096.
Goulian, M. (1971). *Annu. Rev. Biochem.* **40**, 855.
Goulian, M., Kornberg, A., and Sinsheimer, R. (1967). *Proc. Nat. Acad. Sci. U. S.* **58**, 2321.
Granboulan, N., and Scherrer, K. (1969). *Eur. J. Biochem.* **9**, 1.
Grandi, M., and Küntzel, H. (1970). *FEBS Lett.* **10**, 25.

Granholm, N. H., and Baker, J. R. (1970). *Develop. Biol.* **23**, 563.
Griffin, M. J., and Ber, R. (1968). *J. Cell Biol.* **40**, 297.
Griffin, M. J., and Cox, R. P. (1967). *J. Cell Sci.* **2**, 545.
Griffiths, E., and Bayley, S. T. (1969). *Biochemistry* **8**, 541.
Grobstein, C. (1967). *Nat. Cancer Inst. Monogr.* **26**, 279.
Grodzicker, T., and Zipser, D. (1968). *J. Mol. Biol.* **38**, 305.
Groner, Y., Herzberg, M., and Revel, M. (1970). *FEBS Lett.* **6**, 315.
Gros, F., Hiatt, H., Gilbert, W., Kurland, C. G., Risebrough, R. W., and Watson, J. D. (1961). *Nature (London)* **190**, 581.
Gross, P. R. (1968). *Annu. Rev. Biochem.* **37**, 631.
Gross, S. R., McCoy, M. T., and Gilmore, E. B. (1968). *Proc. Nat. Acad. Sci. U. S.* **61**, 253.
Grossbach, U. (1969). *Chromosoma* **28**, 136.
Grüneberg, H. (1969). *J. Embryol. Exp. Morphol.* **22**, 145.
Gurdon, J. B. (1967). *In* "Heritage from Mendel" (R. A. Brink and A. D. Styles, eds.), p. 203. Univ. of Wisconsin Press, Madison, Wisconsin.
Gurdon, J. B. (1968). *In* "Essays in Biochemistry" (P. N. Campbell and G. D. Greville, eds.), Vol. 4, p. 25. Academic Press, New York.
Gurdon, J. B., and Woodland, H. R. (1968). *Biol. Rev. Cambridge Phil. Soc.* **43**, 233.
Gurdon, J. B., and Woodland, H. R. (1969). *Proc. Roy. Soc., Ser. B.* **173**, 99.
Gurley, L. R., and Hardin, J. M. (1968). *Arch. Biochem. Biophys.* **128**, 285.
Gustafson, T., and Toneby, M. I. (1971). *Amer. Nat.* **59**, 452.
Hadjiolov, A. A. (1966). *Biochim. Biophys. Acta* **119**, 547.
Hadjiolov, A. A. (1967). *Progr. Nucleic Acid Res. Mol. Biol.* **7**, 196.
Hadorn, E. (1966). *Develop. Biol.* **13**, 424.
Hämmerling, J. (1953). *Int. Rev. Cytol.* **2**, 475.
Halkka, L., and Halkka, O. (1968). *Science* **162**, 803.
Hall, B. D., and Spiegelman, S. (1961). *Proc. Nat. Acad. Sci. U. S.* **47**, 137.
Hall, R. H. (1970a). "Modified Nucleosides in Nucleic Acids." Columbia Univ. Press, New York.
Hall, R. H. (1970b). *Progr. Nucleic Acid Res. Mol. Biol.* **10**, 57.
Hallberg, R. L., and Brown, D. D. (1969). *J. Mol. Biol.* **46**, 393.
Hallick, L., Boyce, R. P., and Echols, H. (1969). *Nature (London)* **223**, 1239.
Hatlen, L. E., Amaldi, F., and Attardi, G. (1969). *Biochemistry* **8**, 4989.
Hamilton, L. D. (1968). *Nature (London)* **218**, 633.
Hamilton, T. H. (1968). *Science* **161**, 649.
Hamilton, T. H., Teng, C. S., and Means, A. R. (1968). *Proc. Nat. Acad. Sci. U. S.* **59**, 1265.
Hancock, R. L. (1967). *Can. J. Biochem.* **45**, 1513.
Hanoune, J., and Agarwal, M. K. (1970). *FEBS Lett.* **11**, 78.
Harbers, E. (1968). "Introduction to Nucleic Acids." Reinhold, New York.
Harewood, K., and Goldstein, J. (1970). *Fed. Proc. Fed. Amer. Soc. Exp. Biol.* **29**, 885.
Harney, C. E., and Nakada, D. (1970). *Biochim. Biophys. Acta* **213**, 529.
Harris, H. (1959). *Biochem. J.* **73**, 362.
Harris, H. (1963). *Progr. Nucleic Acid Res. Mol. Biol.* **2**, 20.
Harris, H. (1964). *Nature (London)* **201**, 863.
Harris, H. (1968). "Nucleus and Cytoplasm." Oxford Univ. Press (Clarendon), London and New York.
Harris, H., Sidebottom, E., Grace, D. M., and Bramwell, M. E. (1969). *J. Cell Sci.* **4**, 499.
Hartmann, J. F., and Comb, D. G. (1969). *J. Mol. Biol.* **41**, 155.

Harvey, R. J. (1970). *J. Bacteriol.* **101**, 574.
Haschemeyer, A. E. U. (1969). *Proc. Nat. Acad. Sci. U. S.* **62**, 128.
Hatfield, G. W., and Burns, R. O. (1970). *Proc. Nat. Acad. Sci. U. S.* **66**, 1027.
Hatfield, B., and Caicuts, M. (1969). *Fed. Proc. Fed. Amer. Soc. Exp. Biol.* **28**, 349.
Haussler, M. R., and Norman, A. W. (1969). *Proc. Nat. Acad. Sci. U. S.* **62**, 155.
Hay, E. D. (1968a). *In* "Ultrastructure in Biological Systems, Vol. 3: The Nucleus" (A. J. Dalton and F. Haguenau, eds.), p. 2. Academic Press, New York.
Hay, E. D. (1968b). *In* "Epithelial-Mesenchymal Interactions" (R. Fleischmajer and R. E. Billingham, eds.), p. 31, Williams & Wilkins, Baltimore, Maryland.
Hay, E. D., and Gurdon, J. B. (1967). *J. Cell Sci.* **2**, 151.
Hayashi, H., Fisher, H., and Söll, D. (1969). *Biochemistry* **8**, 3680.
Hayashi, Y., Osawa, S., and Miura, K. (1966). *Biochim. Biophys. Acta* **129**, 519.
Hayes, D. H., Hayes, F., and Guerin, M. F. (1966). *J. Mol. Biol.* **18**, 499.
Hayward, W. S., and Green, M. H. (1969). *Proc. Nat. Acad. Sci. U. S.* **64**, 962.
Haywood, A. M. (1971). *Proc. Nat. Acad. Sci. U. S.* **68**, 435.
Hecht, N. B., and Woese, C. R. (1968). *J. Bacteriol.* **95**, 986.
Hecht, N. B., Bleyman, M., and Woese, C. R. (1968). *Proc. Nat. Acad. Sci. U. S.* **59**, 1278.
Hegelson, J. P. (1968). *Science* **161**, 974.
Heinemann, S. F., and Spiegelman, W. G. (1970). *Proc. Nat. Acad. Sci. U. S.* **67**, 1122.
Helinski, D. R., and Yanofsky, C. (1962). *Proc. Nat. Acad. Sci. U. S.* **48**, 173.
Hendler, R. N. (1965). *Nature (London)* **207**, 1053.
Hennig, W. (1968). *J. Mol. Biol.* **38**, 227.
Hennig, W., Hennig, I., and Stein, H. (1970). *Chromosoma* **32**, 31.
Henshaw, E. C. (1968). *J. Mol. Biol.* **36**, 401.
Henshaw, E. C., and Loewenstein, J. (1970). *Biochim. Biophys. Acta* **199**, 405.
Herrington, M. D., and Hawtrey, A. O. (1971). *Biochem. J.* **121**, 279.
Hershey, A. D., and Chase, M. (1952). *J. Gen. Physiol.* **36**, 39.
Herz, S. J. (1970). *Biochemistry* **9**, 690.
Herzberg, M., Lelong, J. L., and Revel, M. (1969). *J. Mol. Biol.* **44**, 297.
Hess, E. (1970). *Science* **168**, 664.
Hess, E. L., Herranen, A. M., and Lagg, S. E. (1961). *J. Biol. Chem.* **236**, 3020.
Hess, H., and Meyer, G. F. (1968). *Advan. Genet.* **14**, 171.
Heyden, H. W. v., and Zachau, H. G. (1971). *Biochim. Biophys. Acta* **232**, 651.
Heyman, T., Seror, S., Desseaux, B., and Legault-Demare, J. (1967). *Biochim. Biophys. Acta* **145**, 596.
Heywood, S. M. (1970). *Nature (London)* **225**, 696.
Heywood, S. M., and Nwagwu, M. (1969). *Biochemistry* **8**, 3839.
Heywood, S. M., and Rich, A. (1968). *Proc. Nat. Acad. Sci. U. S.* **59**, 590.
Hill, C. W., Squires, C., and Carbon, J. (1970). *J. Mol. Biol.* **52**, 557.
Hindley, J. (1967). *J. Mol. Biol.* **30**, 125.
Hinegardner, R. (1968). *Amer. Nat.* **101**, 357.
Hirsch, C. A. (1966). *Biochim. Biophys. Acta* **123**, 246.
Hnilica, L. S. (1967). *Progr. Nucleic Acid Res. Mol. Biol.* **7**, 25.
Hoagland, M. D. (1969). Lecture at Symposium on Polypeptides, organized by Miles Laboratories, New York, June.
Hoagland, M. B., Zamecnik, P. C., and Stephenson, M. L. (1957). *Biochim. Biophys. Acta* **24**, 215.
Hoagland, M. B., Wilson, S. H., and Quincey, R. V. (1968). *In* "Regulatory Mechanisms for Protein Synthesis in Mammalian Cells" (A. San Pietro, M. R. Lamborg, and F. T. Kenney, eds.), p. 179. Academic Press, New York.

Bibliography

Hodge, L. D., Robbins, E., and Scharff, M. D. (1969). *J. Cell Biol.* **40**, 497.
Hodnett, J. L., and Busch, H. (1968). *J. Biol. Chem.* **243**, 6334.
Hogness, D. S. (1966). *In* "Macromolecular Metabolism" (J. Hurwitz, ed.), p. 29. Little, Brown, Boston, Massachusetts.
Holland, J. J., and Kiehn, E. D. (1968). *Proc. Nat. Acad. Sci. U. S.* **60**, 1015.
Holley, R. W., and Merrill, S. H. (1959). *Fed. Proc. Fed. Amer. Soc. Exp. Biol.* **18**, 249.
Holley, R. W., Apgar, J., Everett, G. A., Madison, J. T., Marquisse, M., Merrill, S. H. Penswick, J. R., and Zamir, R. (1965). *Science* **147**, 1462.
Holliday, R. (1969). *Nature (London)* **221**, 1224.
Holoubek, U., and Crocker, T. T. (1968). *Biochim. Biophys. Acta* **157**, 352.
Holtzer, H. (1968). *In* "Epithelial-Mesenchymal Interactions" (R. Fleischmajer and R. E. Billingham, eds.), p. 152. Williams & Wilkins, Baltimore, Maryland.
Hoober, J. K. (1970). *J. Biol. Chem.* **245**, 4327.
Hoogsteen, K. (1963). *Acta Crystallogr.* **16**, 907.
Horgen, P. A., and Griffin, D. H. (1971). *Proc. Nat. Acad. Sci. U. S.* **68**, 338.
Hotta, Y., and Bassel, A. (1965). *Proc. Nat. Acad. Sci. U. S.* **53**, 356.
Howard, E. F., and Plaut, W. (1968). *J. Cell Biol.* **39**, 415.
Howard-Flanders, P. (1968). *Annu. Rev. Biochem.* **37**, 175.
Hoyer, B. H., and Roberts, R. B. (1967). *Mol. Genet.* **2**, 425.
Hsu, T. C., Brinkley, B. R., and Arrighi, F. R. (1967). *Chromosoma* **23**, 137.
Hsu, W-T., and Weiss, S. B. (1969). *Proc. Nat. Acad. Sci. U. S.* **64**, 345.
Huberman, J. A. (1969). *In* "Exploitable Molecular Mechanisms and Neoplasia" (R. B. Hurlbert, ed.), p. 337. Williams & Wilkins, Baltimore, Maryland.
Huberman, J. A., and Attardi, G. (1967). *J. Mol. Biol.* **29**, 487.
Hultin, T. (1956). *Exp. Cell Res.* **11**, 222.
Humphrey, R. R. (1961). *Amer. Zool.* **1**, 361.
Humphreys, T. (1969). *Develop. Biol.* **20**, 435.
Hung, P. P., and Overby, L. R. (1968). *J. Biol. Chem.* **243**, 5525.
Hurwitz, J., and August, J. T. (1963). *Progr. Nucleic Acid Res. Mol. Biol.* **1**, 59.
Hurwitz, J., Furth, J. J., Anders, M., and Evans, A. (1962). *J. Biol. Chem.* **237**, 3752.
Ilan, J. (1968). *J. Biol. Chem.* **243**, 5859.
Ilan, J., Ilan, J., and Patel, N. (1970). *J. Biol. Chem.* **245**, 1275.
Imamoto, F. (1970). *Nature (London)* **228**, 232.
Infante, A. A., and Nemer, M. (1968). *J. Mol. Biol.* **32**, 543.
Ingram, V. M. (1957). *Nature (London)* **180**, 326.
Ingram, V. M. (1964). *Ann. N.Y. Acad. Sci.* **119**, 485.
Ingram, V. M. (1966). "The Biosynthesis of Macromolecules." Benjamin, New York.
Irr, J., and Gallant, J. (1969). *J. Biol. Chem.* **244**, 2233.
Ishida, T., and Sueoka, N. (1967). *Proc. Nat. Acad. Sci U. S.* **58**, 1080.
Ishida, T., and Sueoka, N. (1968). *J. Mol. Biol.* **37**, 313.
Ishihama, A., and Hurwitz, J. (1969). *J. Biol. Chem.* **244**, 6680.
Itano, H. (1964). Cited in Stent (1964).
Jacob, F., and Monod, J. (1961). *J. Mol. Biol.* **3**, 318.
Jacob, F., and Monod, J. (1963). *In* "Cytodifferentiation and Macromolecular Synthesis" (Locke, M., Ed.), p. 30. Academic Press, New York.
Jacob, F., Brenner, S., and Cuzin, F. (1963). *Cold Spring Harbor Symp. Quant. Biol.* **28**, 329.
Jacob, J. (1969). *Exp. Cell Res.* **54**, 281.
Jacob, J., and Sirlin, J. L. (1959). *Chromosoma* **10**, 210.
Jacob, J., and Sirlin, J. L. (1963). *J. Cell Biol.* **17**, 153.

Jacob, J., and Sirlin, J. L. (1964). *J. Ultrastruct. Res.* **11**, 315.
Jacob, S. T., Sajdel, E. M., and Munro, H. N. (1968). *Biochim. Biophys. Acta* **157**, 421.
Jacobson, K. B. (1971a). *Progr. Nucleic Acid Res. Mol. Biol.* **11**, 461.
Jacobson, K. B. (1971b). *Nature (New Biol.)* **231**, 17.
Jacobson, M. F., and Baltimore, D. (1968). *Proc. Nat. Acad. Sci. U. S.* **61**, 77.
Jamakosmanović, A., and Loewenstein, W. R. (1968). *J. Cell Biol.* **38**, 556.
Jaworska, H., and Lima-de-Faria, A. (1969). *Chromosoma* **28**, 309.
Jeanteur, P., Amaldi, F., and Attardi, G. (1968). *J. Mol. Biol.* **33**, 757.
Jelinek, W. R. (1969). *J. Cell Biol.* **43**, 59a.
Jensen, E. V., Suzuki, T., Numata, M., Smith, S., and De Sombre, E. R. (1969). *Steroids* **13**, 417.
Jeppesen, P. G. N., Steitz, J. A., Gesteland, R. F., and Spahr, P. F. (1970). *Nature (London)* **226**, 230.
Jerne, N. K. (1966). In "Phage and the Origins of Molecular Biology" (J. Cairns, G. S. Stent, and J. D. Watson, eds.), p. 301. Cold Spring Harbor Lab. Quant. Biol., Cold Spring Harbor, New York.
Jerne, N. K. (1967). *Cold Spring Harbor Symp. Quant. Biol.* **32**, 591.
Johnson, J. M. (1969). *J. Cell Biol.* **43**, 197.
Jones, K. W. (1965). *J. Ultrastruct. Res.* **13**, 257.
Jones, K. W. (1970). *Nature (London)* **225**, 912.
Jordan, B. R., and Monier, R. (1971). *J. Mol. Biol.* **59**, 219.
Jordan, B. R., Feunteun, J., and Monier, R. (1970). *J. Mol. Biol.* **50**, 605.
Jukes, T. H. (1965). *Biochem. Biophys. Res. Commun.* **19**, 391.
Kabat, D. (1970). *Biochemistry* **9**, 4160.
Kabat, D., and Rich, A. (1969). *Biochemistry* **8**, 3742.
Kaempfer, R. (1969). *Nature (London)* **222**, 951.
Kaji, H., and Tanaka, Y. (1967). *Biochim. Biophys. Acta* **138**, 642.
Kamen, R. (1969). *Nature (London)* **221**, 321.
Kan, J., Kano-Sueoka, T., and Sueoka, N. (1968). *J. Biol. Chem.* **243**, 5584.
Kan, J., Nirenberg, M. W., and Sueoka, N. (1970). *J. Mol. Biol.* **52**, 179.
Kano-Sueoka, T., and Sueoka, N. (1966). *J. Mol. Biol.* **20**, 183.
Kano-Sueoka, T., and Sueoka, N. (1968). *J. Mol. Biol.* **37**, 475.
Kano-Sueoka, T., and Sueoka, N. (1969). *Proc. Nat. Acad. Sci. U. S.* **2**, 1229.
Kano-Sueoka, T., Nirenberg, M., and Sueoka, N. (1968). *J. Mol. Biol.* **35**, 1.
Karasaki, S. (1965). *J. Cell Biol.* **26**, 937.
Karasaki, S. (1968). *Exp. Cell Res.* **52**, 13.
Karkas, J. D., Rudner, R., and Chargaff, E. (1968). *Proc. Nat. Acad. Sci. U. S.* **60**, 915.
Kasai, T., and Bautz, E. K. F. (1969). *J. Mol. Biol.* **41**, 401.
Kates, J., and Beeson, J. (1970). *J. Mol. Biol.* **50**, 19.
Kato, T., and Kurokawa, M. (1970). *Biochem. J.* **116**, 599.
Kaulenas, M. S., Foor, W. E., and Fairbairn, D. (1969). *Science* **163**, 1201.
Kawashima, K., Izawa, M., and Sato, S. (1971). *Biochim. Biophys. Acta* **232**, 192.
Kaye, A. M., and Leboy, R. S. (1968). *Biochim. Biophys. Acta* **157**, 289.
Kedes, L. H., and Birnstiel, M. L. (1971). *Nature (New Biol.)* **230**, 165.
Kedes, L. H., and Gross, P. R. (1969). *J. Mol. Biol.* **42**, 559.
Keep, E. (1962). *Can. J. Genet. Cytol.* **4**, 206.
Kelley, R. B., Atkinson, M. R., Huberman, J. A., and Kornberg, A. (1969). *Nature (London)* **224**, 495.
Kelly, R. E., and Rice, R. V. (1969). *J. Cell Biol.* **42**, 683.

Kelmers, A. D., Novelli, G. D., and Stulberg, M. P. (1965). *J. Biol. Chem.* **240**, 3979.
Kennell, D., and Kotoulas, A. (1968). *J. Mol. Biol.* **34**, 71.
Kenney, F. T. (1970). *In* "Mammalian Protein Metabolism" (H. N. Munro, ed.), Vol. 4, p. 131. Academic Press, New York.
Kenney, F. T., Reel, J. R., Hager, C. B., and Wittliff, J. L. (1968). *In* "Regulatory Mechinisms for Protein Synthesis in Mammalian Cells" (A. San Pietro, M. R. Lamborg, and F. T. Kenney, eds.), p. 119. Academic Press, New York.
Kerkof, P. R., and Tata, J. R. (1969). *Biochem. J.* **112**, 729.
Kerr, S. J., and Dische, Z. (1968). Cited in Borek (1969).
Key, J. L. (1969). *Annu. Rev. Plant. Physiol.* **20**, 449.
Keyl, H. G. (1965). *Chromosoma* **17**, 139.
Keynan, A. (1969). *In* "Current Topics in Developmental Biology" (A. A. Moscona and A. Monroy, eds.), Vol. 4, p. 2. Academic Press, New York.
Khairallah, E. A., and Pitot, H. C. (1967). *Biochem. Biophys. Res. Commun.* **29**, 269.
Khorana, H. G. (1968). *Harvey Lect.* **62**, 79.
Kidson, C., and Kirby, K. S. (1965). *Cancer Res.* **25**, 472.
Kiefer, B. I. (1968). *Proc. Nat. Acad. Sci. U. S.* **61**, 85.
Kijima, S., and Wilt, F. H. (1969). *J. Mol. Biol.* **40**, 235.
Kimura, M. (1969). *Proc. Nat. Acad. Sci. U. S.* **63**, 1181.
King, H. W. S., and Gould, H. (1970). *J. Mol. Biol.* **51**, 687.
King, J. L., and Jukes, T. H. (1969). *Science* **164**, 789.
Kirby, K. S. (1964). *Progr. Nucleic Acid Res. Mol. Biol.* **3**, 1.
Kirk, D., and Jones, R. N. (1970). *Chromosoma* **31**, 240.
Kleinfeld, R. G. (1966). *Nat. Cancer Inst. Monogr.* **23**, 369.
Kleinschmidt, A. K. (1967). *Mol. Genet.* **2**, 47,
Kline, L. K., Fittler, K., and Hall, R. H. (1969). *Biochemistry* **8**, 4361.
Klinger, H. P., Davis, J., Goldhuber, P., and Ditta, T. (1968). *Cytogenetics* **7**, 39.
Knight, E., and Darnell, J. E. (1967). *J. Mol. Biol.* **28**, 401.
Knippers, R., and Strätling, W. (1970). *Nature (London)* **226**, 713.
Knowland, J. S. (1970). *Biochim. Biophys. Acta* **204**, 416.
Kohl, D. M., Greene, R. F., and Flickinger, R. A. (1969). *Biochim. Biophys. Acta* **179**, 28.
Kohne, D. W. (1968). *Biophys. J.* **8**, 1104.
Kokileva, L., Mladenova, I., and Tsanev, R. (1971). *FEBS Lett.* **16**, 17.
Kolodny, G. M., and Gross, P. R. (1969). *Exp. Cell Res.* **56**, 117.
Kondo, M., Eggerston, G., Eisenstadt, J., and Lengyel, P. (1968). *Nature (London)* **220**, 368.
Kondo, M., Gallerani, R., and Weissman, C. (1970). *Nature (London)* **228**, 525.
Konijn, T. M., Barkley, D. S., Cheng, Y. Y., and Bonner, J. T. (1968). *Amer. Natur.* **102**, 225.
Korge, G. (1970). *Chromosoma* **30**, 430.
Kornberg, A. (1961). "The Enzymatic Synthesis of DNA." Wiley, London.
Kornberg, A. (1969). *Science* **163**, 1410.
Kovach, J. S., Phang, J. M., Blasi, F., Barton, R. W., Ballesteros-Olmo, A., and Goldberger, R. F. (1970). *J. Bacteriol.* **104**, 787.
Kroeger, H. (1968). *In* "Metamorphosis" (W. Etkin and L. I. Gilbert, eds.), p. 185. Appleton-Century-Crofts, New York.
Kroeger, H., Jacob, J., and Sirlin, J. L. (1963). *Exp. Cell Res.* **31**, 416.
Krug, R. M., and Gomatos, P. (1969). *J. Virol.* **4**, 642.
Kuempel, P. L. (1970). *Advan. Cell Biol.* **1**, 3.

Küntzel, H. (1969). *Nature (London)* **222**, 143.
Küntzel, H., and Noll, H. (1967). *Nature (London)* **215**, 1340.
Küntzel, H., and Schäfer, K. P. (1971). *Nature (New Biol.)* **231**, 265.
Kuff, E. L., and Roberts, N. E. (1967). *J. Mol. Biol.* **26**, 211.
Kull, F. J., and Jacobson, K. B. (1969). *Proc. Nat. Acad. Sci. U. S.* **62**, 1137.
Kurland, C. G. (1970). *Science* **169**, 1171.
Kushner, D. J. (1969). *Bacteriol. Rev.* **33**, 202.
Kuwano, M., and Schlessinger, D. (1970). *Proc. Nat. Acad. Sci. U. S.* **66**, 146.
Kwan, C. N., Apirion, D., and Schlessinger, D. (1968). *Biochemistry* **7**, 427.
Lacey, J. C., and Pruitt, K. M. (1969). *Nature (London)* **223**, 799.
Laico, M. T., Ruoslahti, E. I., Papermaster, D. S., and Dreyer, W. S. (1970). *Proc. Nat. Acad. Sci. U. S.* **67**, 120.
Laird, C. D., McConaughy, B. L., and McCarthy, B. J. (1969). *Nature (London)* **224**, 149.
Lake, J. A., and Beeman, W. W. (1968). *J. Mol. Biol.* **31**, 115.
Lakhotia, S. C., and Mukherjee, A. S. (1969). *Genet. Res.* **14**, 137.
Landesman, R., and Gross, P. R. (1968). *Develop. Biol.* **18**, 571.
Lane, B. G., and Tamaoki, T. (1969). *Biochim. Biophys. Acta* **179**, 332.
Lane, N. J. (1967). *J. Cell Biol.* **35**, 421.
Langan, T. A. (1969). *Proc. Nat. Acad. Sci. U. S.* **64**, 1276.
Lapidus, I. R., and Rosen, B. (1970). *J. Theoret. Biol.* **27**, 417.
Lara, F. J. S., and Hollander, F. M. (1967). *Nat. Cancer Inst. Monogr.* **27**, 235.
Lark, K. G. (1969). *Annu. Rev. Biochem.* **38**, 569.
Lark, K. G., Consigli, R., and Toliver, A. (1971). *J. Mol. Biol.* **58**, 873.
Laskowski, M. (1968). In "Handbook of Biochemistry" (H. A. Sober, ed.), p. H-20. Chem. Rubber. Publ. Co., Cleveland, Ohio.
Laycock, D. G., and Hunt, J. A. (1969). *Nature (London)* **221**, 1118.
Lazzarini, R. A., and Santangelo, E. (1967). *J. Bacteriol.* **94**, 125.
Lazzarini, R. A., Nakata, K., and Winslow, R. M. (1969). *J. Biol. Chem.* **244**, 3092.
Leder, P., Skogerson, L. E., and Nau, M. M. (1969). *Proc. Nat. Acad. Sci. U. S.* **62**, 454.
Lederberg, J. (1966). In "Current Topics in Developmental Biology" (A. A. Moscona and A. Monroy, eds.), Vol. 1, p. xi. Academic Press, New York.
Lee, J. C., and Ingram, V. M. (1967). *Science* **158**, 1330.
Lee, J. C., and Yunis, J. J. (1971). *Chromosoma* **32**, 237.
Lee, M-L., and Muench, K. H. (1969). *J. Biol. Chem.* **244**, 223.
Lehmann, H., and Huntsman, R. G. (1966). "Man's Haemoglobin." North-Holland Publ., Amsterdam.
Leibowitz, M. J., and Soffer, R. L. (1970). *J. Biol. Chem.* **245**, 2066.
Leighton, T. J., Dill, B. C., Stock, J. J., and Phillips, C. (1971). *Proc. Nat. Acad. Sci. U. S.* **68**, 677.
Leisinger, T., and Vogel, H. J. (1969). *Biochim. Biophys. Acta* **182**, 572.
Lengyel, P. (1967). *Mol. Genet.* **2**, 194.
Lengyel, P., and Söll, D. (1969). *Bacteriol. Rev.* **33**, 264.
Leonard, N. J., Iwamura, H., and Eisinger, J. (1969). *Proc. Nat. Acad. Sci. U. S.* **64**, 352.
Lesk, A. M. (1969). *J. Theoret. Biol.* **22**, 537.
Lett, J. T., Klucis, E. S., and Sun, C. (1970). *Biophys. J.* **10**, 277.
Levi-Montalcini, R., Angeletti, R. H., and Angeletti, P. U. (1972). In "The Structure and Function of Nervous Tissue" (G. H. Bourne, ed.), Vol. 5. Academic Press, New York (in press).
Levin, J. G., and Nirenberg, M. (1968). *J. Mol. Biol.* **34**, 467.

Levitt, M. (1969). *Nature (London)* **224,** 759.
Levy, H. B., and Carter, W. A. (1968). *J. Mol. Biol.* **31,** 561.
Liao, S., Leininger, K. R., Sagher, D., and Barton, R. W. (1965). *Endocrinology* **77,** 763.
Liau, M. C., and Perry, R. P. (1969). *J. Cell Biol.* **42,** 272.
Libanati, C. M., and Tandler, C. J. (1969). *J. Cell Biol.* **42,** 754.
Lietman, P. S. (1968). *J. Biol. Chem.* **243,** 2837.
Liew, C. C., Haslett, G. W., and Allfrey, V. G. (1970). *Nature (London)* **226,** 414.
Lim, L., and Canellakis, E. S. (1970). *Nature (London)* **227,** 710.
Lima-de-Faria, A., and Jaworska, H. (1968). *Nature (London)* **217,** 138.
Lima-de-Faria, A., Birnstiel, M. L., and Jaworska, H. (1969). *Genetics* **61,** Suppl. 1, 145.
Lin, M. (1955). *Chromosoma* **7,** 340.
Lipmann, F. (1941). *Advan. Enzymol.* **1,** 99.
Lipmann, F. (1963). *Progr. Nucleic Acid Res. Mol. Biol.* **1,** 135.
Lipmann, F. (1965). *In* "The Origins of Prebiological Systems" (S. W. Fox, ed.), p. 259. Academic Press, New York.
Lipmann, F. (1969). *Science* **164,** 1024.
Lipmann, F. (1971). *Science* **173,** 875.
Lizardi, P. M., and Luck, D. J. L. (1971). *Nature (New Biol.)* **229,** 140.
Lockwood, D. H., Stockdale, F. E., and Topper, Y. J. (1967). *Science* **156,** 945.
Lodish, H. F. (1968a). *Nature (London)* **220,** 345.
Lodish, H. F. (1968b). *Progr. Biophys. Mol. Biol.* **18,** 285.
Lodish, H. F. (1970). *J. Mol. Biol.* **50,** 689.
Loening, U. E. (1965). *Biochem. J.* **97,** 125.
Loening, U. E. (1968a). *J. Mol. Biol.* **38,** 355.
Loening, U. E. (1968b). *Annu. Rev. Plant Physiol.* **19,** 37.
Loening, U. E. (1970). *Symp. Soc. Gen. Microbiol.* **20,** 77.
Loening, U. E., Jones, K. W., and Birnstiel, M. L. (1969). *J. Mol. Biol.* **45,** 353.
Loewenstein, W. R. (1968). *Develop. Biol. Suppl.* **2,** 151.
Loftfield, R. B. (1963). *Biochem. J.* **89,** 82.
Loftfield, R. B., and Eigner, E. A. (1969). *J. Biol. Chem.* **244,** 1746.
Loftfield, R. B., Hecht, L. I., and Eigner, E. A. (1963). *Biochim. Biophys. Acta* **72,** 383.
Longwell, A. C., and Svihla, G. (1960). *Exp. Cell Res.* **20,** 294.
Losick, R., Shorenstein, R. G., and Sonensheim, A. L. (1970). *Nature (London)* **227,** 910.
Lu, P., and Rich, A. (1971). *J. Mol. Biol.* **58,** 513.
Lubin, M. (1968). *Proc. Nat. Acad. Sci. U. S.* **61,** 1454.
Lubin, M. (1969). *J. Mol. Biol.* **39,** 219.
Lucas-Lenard, J., and Lipmann, F. (1971). *Annu. Rev. Biochem.* **40,** 409.
Luria, S. E. (1969). *In* "Bacterial Episomes and Plasmids" (G. E. W. Wolstenholme and M. O'Connor, eds.), p. 1. Little, Brown, Boston, Massachusetts.
Lyon, M. F. (1968). *Annu. Rev. Genet.* **2,** 31.
Maaløe, O., and Kjeldgaard, N. O. (1966). "Control of Macromolecular Synthesis." Benjamin, New York.
Maas, W. K., and Clark, A. J. (1964). *J. Mol. Biol.* **8,** 365.
McCarthy, B. J., and Bolton, E. T. (1964). *J. Mol. Biol.* **8,** 184.
McCarthy, B. J., and Church, R. B. (1970). *Annu. Rev. Biochem.* **39,** 131.
McClintock, B. (1934). *Z. Zellforsch. Mikrosk. Anat.* **21,** 294.
McClintock, B. (1967). *Develop. Biol. Suppl.* **1,** 84.

McCormick, W., and Penman, S. (1969). *J. Mol. Biol.* **39**, 315.
McCorquodale, D. J., and Mueller, G. C. (1958). *J. Biol. Chem.* **232**, 31.
McFall, E., and Maas, W. K. (1967). *Mol. Genet.* **2**, 255.
MacGregor, H. C. (1968). *J. Cell Sci.* **3**, 437.
McGuire, J. L., and Lisk, R. D. (1968). *Proc. Nat. Acad. Sci. U. S.* **61**, 497.
Mach, B., Koblet, H., and Gros, D. (1967). *Cold Spring Harbor Symp. Quant. Biol.* **32**, 269.
McIndoe, W., and Munro, H. N. (1967). *Biochim. Biophys. Acta* **134**, 458.
MacKintosh, F. R., and Bell, E. (1969). *J. Mol. Biol.* **41**, 365.
McLaughlin, C. S., Dondon, J., Grunberg-Manago, M., Michelson, A. M., and Sanders, G. (1968). *J. Mol. Biol.* **32**, 521.
McLaughlin, P. J. and Dayhoff, M. O. (1970). *Science* **168**, 1469.
McLellan, W. L., and Vogel, H. J. (1970). *Proc. Nat. Acad. Sci. U. S.* **67**, 1703.
Maden, B. E. H. (1968). *Nature (London)* **219**, 685.
Maden, B. E. H. (1969). *Nature (London)* **224**, 1203.
Maden, B. E. H., Vaughan, M. H., Warner, J. R., and Darnell, J. E. (1969). *J. Mol. Biol.* **45**, 265.
Madison, J. T. (1968). *Annu. Rev. Biochem.* **37**, 131.
Maënpää, P. H., and Bernfield, M. R. (1969). *Biochemistry* **8**, 4926.
Maio, J. J., and Schildkraut, C. L. (1969). *J. Mol. Biol.* **40**, 203.
Mairy, M., and Denis, H. (1971). *Develop. Biol.* **24**, 143.
Mandell, J. D., and Hershey, A. D. (1960). *Anal. Biochem.* **1**, 66.
Mandelstam, J. (1960). *Bacteriol. Rev.* **24**, 289.
Mandelstam, J. (1968). *In* "Biochemistry of Bacterial Growth" (J. Mandelstam and K. McQuillen, eds.), p. 414. Wiley, New York.
Mangiarotti, G. (1969). *Nature (London)* **222**, 947.
Manor, H., Goodman, D., and Stent, G. S. (1969). *J. Mol. Biol.* **39**, 1.
Marcker, K. A., and Sanger, F. (1964). *J. Mol. Biol.* **8**, 835.
Marcus, A., and Feely, J. (1966). *Proc. Nat. Acad. Sci. U. S.* **56**, 1770.
Markam, B. (1961). *Exp. Cell Res.* **25**, 224.
Marks, P. A., Fantoni, A., and de la Chapelle, A. (1968). *Vitamins Hormones* **26**, 331.
Marmur, J., Round, R., and Schildkraut, C. L. (1963). *Progr. Nucleic Acid Res. Mol. Biol.* **1**, 231.
Marshall, R., and Nirenberg, M. (1969). *Develop. Biol.* **19**, 1.
Martin, D. W., Tomkins, G. M., and Bresler, M. A. (1969). *Proc. Nat. Acad. Sci. U. S.* **63**, 842.
Martin, R. G., (1969). *Annu. Rev. Genet.* **3**, 181.
Martin, R. G., Silbert, D. F., Smith, D. W. F., and Whitfield, H. J. (1966). *J. Mol. Biol.* **21**, 457.
Martin, T. E., and Wool, I. G. (1969). *J. Mol. Biol.* **43**, 151.
Marushige, K., and Ozaki, H. (1967). *Develop. Biol.* **16**, 474.
Marushige, K., Brutlag, D., and Bonner, J. (1968). *Biochemistry* **7**, 3149.
Mascarenhas, J. P., and Bell, E. (1970). *Develop. Biol.* **21**, 475.
Masters, M., and Broda, P. (1971). *Nature (New Biol.)* **232**, 137.
Masui, Y., and Markert, C. L. (1971). *J. Exp. Zool.* **177**, 129.
Matsushita, T., White, K. P., and Sueoka, N. (1971). *Nature (New Biol.)* **232**, 111.
Matthei, J. H., and Nirenberg, M. W. (1961). *Biochem. Biophys. Res. Commun.* **4**, 404.
Matthysse, A. G., and Abrams, M. (1970). *Biochim. Biophys. Acta* **199**, 511.
Matthysse, A. G., and Phillips, C. (1969). *Proc. Nat. Acad. Sci. U. S.* **63**, 897.
Mazia, D. (1966). *In* "The Cell Nucleus-Metabolism and Radiosensitivity" (H. M. Klouwen, ed.), p. 15. Taylor & Francis, London.

Mehler, A. H. (1970). *Progr. Nucleic Acid Res. Mol. Biol.* **10**, 1.
Mehrotra, B. D., and Mahler, H. R. (1968). *Arch. Biochem. Biophys.* **128**, 685.
Melcher, G. (1969). *FEBS Lett.* **3**, 185.
Melli, M., and Bishop, J. O. (1969). *J. Mol. Biol.* **40**, 117.
Mepham, R. J., and Lane, G. R. (1969). *Nature (London)* **221**, 288.
Meselson, M., and Stahl, F. W. (1958). *Proc. Nat. Acad. Sci. U. S.* **44**, 671.
Meselson, M., and Yuan, R. (1968). *Nature (London)* **217**, 1110.
Meselson, M., Stahl, F. W., and Vinograd, J. (1957). *Proc. Nat. Acad. Sci. U. S.* **43**, 581.
Metz, D. H., and Brown, G. L. (1969). *Biochem. J.* **114**, 35P.
Michelson, A. M. (1963). "The Chemistry of Nucleosides and Nucleotides." Academic Press, New York.
Midgley, J. (1965). *Biochim. Biophys. Acta* **108**, 340.
Miller, J. H., Ippen, K., Scaife, J. G., and Beckwith, J. R. (1968). *J. Mol. Biol.* **38**, 413.
Miller, L., and Brown, D. D. (1969). *Chromosoma* **28**, 430.
Miller, O. L. (1966). *Nat. Cancer Inst. Monogr.* **23**, 53.
Miller, O. L., and Beatty, B. R. (1969a). *J. Cell. Physiol.* **74**, Suppl. 1, 225.
Miller, O. L., and Beatty, B. R. (1969b). *Genetics* **61**, Suppl. 1, 133.
Miller, O. L., Hamkalo, B. A., and Thomas, C. A. (1970a). *Science* **169**, 392.
Miller, O. L., Beatty, B. R., Hamkalo, B. A., and Thomas, C. A. (1970b). *Cold Spring Harbor Symp. Quant. Biol.* **35**, 505.
Milner, G. R. (1969). *Nature (London)* **221**, 71.
Mills, D. R., Peterson, R. L., and Spiegelman, S. (1967). *Proc. Nat. Acad. Sci. U. S.* **58**, 217.
Mintz, B. (1964). *J. Exp. Zool.* **157**, 85.
Mirsky, A. E. (1966). *In* "Histones" (A. V. S. de Reuck and J. Knight, eds.), p. 78. Churchill, London.
Mirsky, A. E., Burdick, C. J., Davidson, E. H., and Littau, V. C. (1968). *Proc. Nat. Acad. Sci. U. S.* **61**, 592.
Mitchison, J. M. (1969). *Science* **165**, 657.
Mitchison, J. M., and Creanor, J. (1969). *J. Cell Sci.* **5**, 373.
Mitra, S. K., Chakraburthy, K., and Mehler, A. H. (1970). *J. Mol. Biol.* **49**, 139.
Mittermayer, C., Braun, R., and Rusch, H. P. (1964). *Biochim. Biophys. Acta* **91**, 399.
Miura, K.-I. (1967). *Progr. Nucleic Acid Res. Mol. Biol.* **6**, 39.
Mizushima, S., and Nomura, M. (1970). *Nature (London)* **226**, 1214.
Mohan, J., and Ritossa, F. M. (1970). *Develop. Biol.* **22**, 495.
Moldave, K., Ibuki, F., Rao, P., Schneir, M., Skogerson, L., and Sutter, R. P. (1968). *In* "Regulatory Mechanisms for Protein Synthesis in Mammalian Cells" (A. San Pietro, M. R. Lamborg, and F. T. Kenney, eds.), p. 191. Academic Press, New York.
Mondal, H., Mandal, R. K., and Biswas, B. B. (1970). *Biochem. Biophys. Res. Commun.* **40**, 1194.
Monesi, V. (1969). *In* "Handbook of Molecular Cytology" (A. Lima-de-Faria, ed.), p. 472. North-Holland Publ., Amsterdam.
Monneron, A., and Bernhard, W. (1969). *J. Ultrastruct. Res.* **27**, 266.
Monneron, A., Burgleu, J., and Bernhard, W. (1970). *J. Ultrastruct. Res.* **32**, 370.
Monod, J., Changeux, J., and Jacob, F. (1963). *J. Mol. Biol.* **6**, 306.
Monro, R. E., Cerna, J., and Marcker, K. A. (1968). *Proc. Nat. Acad. Sci. U. S.* **61**, 1042.
Montenecourt, B. S., Langsam, M. E., and Dubin, D. T. (1970). *J. Cell Biol.* **46**, 245.
Moore, P. B., Traut, R. R., Noller, H., Pearson, P., and Delius, H. (1968). *J. Mol. Biol.* **31**, 441.

Moore, R. L., and McCarthy, B. J. (1968). *Biochem. Genet.* **2**, 75.
Mooz, E. D., and Meister, A. (1967). *Biochemistry* **6**, 1722.
Morell, P., Smith, I., Dubnau, D., and Marmur, J. (1967). *Biochemistry* **6**, 258.
Morikawa, N., and Imamoto, F. (1969). *Nature (London)* **223**, 37.
Morris, V. L., Wagner, E. K., and Roizman, B. (1970). *J. Mol. Biol.* **52**, 247.
Morse, D. E., and Yanofsky, C. (1969a). *J. Mol. Biol.* **44**, 185.
Morse, D. E., and Yanofsky, C. (1969b). *Nature (London)* **224**, 329.
Morse, D. E., Baker, R. F., and Yanofsky, C. (1968). *Proc. Nat. Acad. Sci. U. S.* **60**, 1428.
Moses, M. J. M. (1968). *Annu. Rev. Genet.* **2**, 363.
Moses, M. J. M., and Coleman, J. R. (1964). *In* "Role of Chromosomes in Development" (M. Locke, ed.), p. 11. Academic Press, New York.
Mosteller, R. D., Culp, W. J., and Hardesty, B. (1968). *J. Biol. Chem.* **243**, 6343.
Mueller, G. C. (1969). *Fed. Proc. Fed. Amer. Soc. Exp. Biol.* **28**, 1780.
Müller-Hill, B., Crapo, L., and Gilbert, W. (1968). *Proc. Nat. Acad. Sci. U. S.* **59**, 1259.
Mueller, K., and Bremer, H. (1968). *J. Mol. Biol.* **38**, 329.
Muench, K. H., and Safille, P. A. (1968). *Biochemistry* **7**, 2799.
Munro, H. N. (1969). *In* "Mammalian Protein Metabolism" (H. N. Munro, ed.), Vol. 3, p. 3. Academic Press, New York.
Munro, H. N. (1970). *In* "Mammalian Protein Metabolism" (H. N. Munro, ed.), Vol. 4, p. 3. Academic Press, New York.
Muramatsu, M., and Busch, H. (1967). *Methods Cancer Res.* **2**, 303.
Muramatsu, M., and Fujisawa, T. (1968). *Biochim. Biophys. Acta* **157**, 476.
Muramatsu, M., Takoji, T. U., and Sugano, H. (1968). *Exp. Cell Res.* **53**, 278.
Mushinski, J. F., and Potter, M. (1969). *Biochemistry* **8**, 1684.
Muto, A. (1968). *J. Mol. Biol.* **36**, 1.
Nagl, M. (1969). *Nature (London)* **221**, 70.
Nagl, W. (1970). *Caryologia* **23**, 71.
Nakamura, T., Prestayko, A. W., and Busch, H. (1968). *J. Biol. Chem.* **243**, 1368.
Narayan, K. S., Steele, W. J., and Busch, H. (1966). *Exp. Cell Res.* **43**, 483.
Nass, M. M. K. (1969). *Science* **165**, 25.
Nass, M. M. K., and Buck, C. A. (1969). *Proc. Nat. Acad. Sci. U. S.* **62**, 506.
Nass, S. (1969). *Int. Rev. Cytol.* **25**, 55.
Needham, J. (1968). *In* "Haldane and Modern Biology" (K. R. Dronamraju, ed.), p. 277. Johns Hopkins Press, Baltimore, Maryland.
Nei, M. (1969). *Nature (London)* **221**, 40.
Neidhardt, F. C. (1963). *Biochim. Biophys. Acta* **68**, 365.
Neidhardt, F. C. (1966). *Bacteriol. Rev.* **30**, 701.
Neidhardt, F. C., Marchin, G. L., McClain, W. H., Boyd, R. F., and Earhart, C. F. (1969). *J. Cell. Physiol.* **74**, Suppl 1, 87.
Nemer, M. (1967). *Progr. Nucleic Acid Res. Mol. Biol.* **7**, 243.
Nemer, M., and Infante, A. A. (1967). *In* "The Control of Nuclear Activity" (L. Goldstein, ed.), p. 101. Prentice Hall, Englewood Cliffs, New Jersey.
Nemer, M., and Lindsay, D. T. (1969). *Biochem. Biophys. Res. Commun.* **35**, 156.
Nichols, J. L. (1970). *Nature (London)* **225**, 147.
Niessing, J., and Sekeris, C. E. (1970). *Biochim. Biophys. Acta* **209**, 484.
Ninio, J., Fàure, A., and Yaniv, M. (1969). *Nature (London)* **223**, 1333.
Nirenberg, M. W., and Leder, P. (1964). *Science* **145**, 1399.
Nirenberg, M. W., and Matthei, J. H. (1961). *Proc. Nat. Acad. Sci. U. S.* **47**, 1588.
Nishimura, S., Jones, D. S., and Khorana, H. G. (1965). *J. Mol. Biol.* **13**, 302.

Niu, M. C. (1959). *In* "Evolution of Nervous Control" (A. D. Bass, ed.), p. 7. Amer. Ass. Advan. Sci., Washington, D. C.
Nomura, M., and Erdmann, V. A. (1970). *Nautre (London)* **288**, 744.
Nomura, M., Traub, P., and Bechmann, H. (1968). *Nature (London)* **219**, 793.
Nørrevang, A. (1968). *Int. Rev. Cytol.* **23**, 113.
Nossal, N. G., and Singer, M. F. (1968). *J. Biol. Chem.* **243**, 913.
Novelli, G. D. (1967). *Annu. Rev. Biochem.* **36**, 449.
Odartchenko, N., and Pavillard, M. (1970). *Science* **167**, 1133.
Ofengand, J., and Henes, C. (1969). *Fed. Proc. Fed. Amer. Soc. Exp. Biol.* **28**, 350.
Ohno, S. (1970). "Evolution by Gene Duplication." Springer-Verlag.
Ohno, S., Wolf, M., and Atkin, N. B. (1968a). *Hereditas* **59**, 169.
Ohno, S., Stenius, C., Christian, L. C., and Harris, C. (1968b). *Biochem. Genet.* **2**, 197.
Oishi, M., and Sueoka, N. (1965). *Proc. Nat. Acad. Sci. U. S.* **54**, 483.
Okazaki, O., Sugimoto, O., Okazaki, T., Imae, Y., and Sugino, A. (1970). *Nature (London)* **228**, 223.
Olsnes, S., and Hauge, J. G. (1968). *Eur. J. Biochem.* **7**, 128.
Olson, K. E., and Wyss, O. (1969). *Biochem. Biophys. Res. Commun.* **35**, 713.
O'Malley, B. W., McGuire, W. L., Kohler, P. O., and Korenman, S. G. (1969). *Rec. Progr. Hormone Res.* **25**, 105.
Ono, Y., Skoultchi, A., Klein, A., and Lengyel, P. (1968). *Nature (London)* **220**, 1304.
Ono, Y., Skoultchi, A., Waterson, J., and Lengyel, P. (1969). *Nature (London)* **222**, 645.
Oppenheim, A. B., Neubauer, Z., and Calef, E. (1970). *Nature (London)* **226**, 31.
Orgel, L. E. (1963). *Proc. Nat. Acad. Sci. U. S.* **49**, 517.
Orgel, L. E. (1968). *J. Mol. Biol.* **38**, 381.
Ortwerth, B. J., and Novelli, G. D. (1969). *Cancer Res.* **29**, 380.
Osawa, S. (1968). *Annu. Rev. Biochem.* **37**, 109.
Osawa, S., Otaka, E., Itoh, R., and Fukui, T. (1969). *J. Mol. Biol.* **40**, 321.
Ottensmeyer, F. P. (1969). *Biophys. J.* **9**, 1144.
Owens, S. L., and Bell, F. E. (1968). *J. Mol. Biol.* **38**, 145.
Ozaki, M., Mizushima, S., and Nomura, M. (1969). *Nature (London)* **222**, 333.
Pace, B., Peterson, R. L., and Pace, N. R. (1970). *Proc. Nat. Acad. Sci. U. S.* **65**, 1097.
Pagoulatos, G. N., and Darnell, J. E. (1970). *J. Cell Biol.* **44**, 476.
Painter, R. B., and Schaefer, A. W. (1969). *J. Mol. Biol.* **45**, 467.
Panym, S., and Chalkley, R. (1969). *Biochemistry* **8**, 3972.
Papaconstantinou, J., and Julku, E. M. (1968). *J. Cell. Physiol.* **72**, Suppl. 1, 161.
Papaconstantinou, J., Stewart, J. A., and Koehn, P. V. (1966). *Biochim. Biophys. Acta* **114**, 428.
Pardee, A. B. (1965). *In* "Control of Energy Metabolism" (B. Chance, R. W. Estabrook, and J. R. Williamson, eds.), p. 329. Academic Press, New York.
Pardue, M. L., Gerbi, S. A., Eckhardt, R. A., and Gall, J. G. (1970). *Chromosoma* **29**, 268.
Parish, J. H., and Kirby, K. S. (1966). *Biochim. Biophys. Acta* **129**, 554.
Pastan, I., and Perlman, R. (1970). *Science* **169**, 339.
Paul, J. (1968). *Advan. Comp. Physiol. Biochem.* **3**, 116.
Paul, J., and Gilmour, R. S. (1968). *J. Mol. Biol.* **34**, 305.
Pauling, L., and Delbrück, M. (1950). *Science* **92**, 77.
Pavan, C., and Breuer, M. E. (1952). *J. Hered.* **43**, 151.
Pavan, C., and Da Cunha, A. B. (1969). *Annu. Rev. Genet.* **3**, 425.
Pedersen, R. A. (1971). *J. Exp. Zool.* **177**, 65.
Pederson, T., and Gelfant, S. (1970). *Exp. Cell Res.* **59**, 32.

Pederson, T., and Robbins, E. (1970a). *J. Cell Biol.* **45**, 509.
Pederson, T., and Robbins, E. (1970b). *J. Cell Biol.* **47**, 734.
Pelc, S. R. (1968). *Nature (London)* **219**, 162.
Pelling, C. (1969). *Progr. Biophys. Mol. Biol.* **19**, 237.
Pene, J. J., Knight, E., and Darnell, J. E. (1968). *J. Mol. Biol.* **33**, 609.
Penman, S. (1966). *J. Mol. Biol.* **17**, 117.
Penman, S., Scherrer, K., Becker, Y., and Darnell, J. E. (1963). *Proc. Nat. Acad. Sci. U. S.* **49**, 654.
Penman, S., Vesco, C., and Penman, M. (1968). *J. Mol. Biol.* **34**, 49.
Penman, S., Fan, H., Perlman, S., Rosbash, M., Weinberg, R., and Zylber, E. (1970). *Cold Spring Harbor Symp. Quant. Biol.* **35**, 561.
Perkowska, E., MacGregor, H. C., and Birnstiel, M. L. (1968). *Nature (London)* **217**, 649.
Perry, R. P. (1962). *Proc. Nat. Acad. Sci. U. S.* **48**, 2179.
Perry, R. P. (1967). *Progr. Nucleic Acid Res. Mol. Biol.* **6**, 220.
Perry, R. P., and Kelley, D. E. (1968a). *J. Mol. Biol.* **35**, 37.
Perry, R. P., and Kelley, D. E. (1968b). *J. Cell. Physiol.* **72**, 235.
Perry, R. P., and Kelley, D. E. (1970). *J. Cell. Physiol.* **76**, 127.
Perry, R. P., Hell, A., and Errera, M. (1961). *Biochim. Biophys. Acta* **49**, 47.
Perry, R. P., Srinivasan, P. R., and Kelley, D. E. (1964). *Science* **145**, 504.
Perry, R. P., Cheng, T-Y., Freed, J. J., Greenberg, J. R., Kelley, D. E., and Tartof, K. D. (1970). *Proc. Nat. Acad. Sci. U. S.* **65**, 609.
Peterkofsky, A. (1964). *Proc. Nat. Acad. Sci. U. S.* **52**, 1233.
Peterkofsky, A., and Jensensky, C. (1969). *Biochemistry* **8**, 3798.
Petermann, M. L., Pavlovec, A., and Weinstein, I. B. (1969). *Fed. Proc. Fed. Amer. Soc. Exp. Biol.* **28**, 725.
Peterson, P. J. (1967). *Biol. Rev.* **42**, 552.
Petska, S. (1971). *Annu. Rev. Microbiol.* **25**, 487.
Pfeiffer, S. E. (1968). *J. Cell. Physiol.* **71**, 95.
Philippsen, R., Thiebe, R., Wintermeyer, W., and Zachau, H. G. (1968). *Biochem. Biophys. Res. Commun.* **33**, 922.
Phillips, L. A., Hotham-Iglewski, B., and Franklin, R. M. (1969). *J. Mol. Biol.* **45**, 23.
Phillips, S. G., and Phillips, D. M. (1969). *J. Cell Biol.* **40**, 248.
Pikó, L. (1970). *Develop. Biol.* **21**, 257.
Pillinger, D., and Borek, E. (1969). *Proc. Nat. Acad. Sci. U. S.* **62**, 1145.
Pitot, H. C. (1967). *Mol. Genet.* **2**, 383.
Pitot, H. C. (1969). *Arch. Pathol.* **87**, 212.
Platt, J. R. (1964). *Science* **146**, 347.
Pogo, A. O. (1969). *Biochim. Biophys. Acta* **182**, 57.
Polanyi, M. (1968). *Science* **160**, 1308.
Pollak, M., and Rein, R. (1968). *J. Theoret. Biol.* **19**, 241.
Polz, G., and Kreil, G. (1970). *Biochem. Biophys. Res. Commun.* **39**, 516.
Potter, V. R. (1969). *Proc. Can. Cancer Res. Conf.* **8**, 9.
Prescott, D. M. (1964). *Progr. Nucleic Acid Res. Mol. Biol.* **3**, 35.
Prescott, D. M. (1969). *In* "Normal and Malignant Cell Growth" (R. J. M. Fry, M. L. Griem, and W. H. Kirsten, eds.), p. 79. Springer-Verlag, New York.
Prescott, D. M., and Bender, M. A. (1962). *Exp. Cell Res.* **26**, 260.
Prescott, D. M., and Goldstein, L. (1968). *J. Cell Biol.* **39**, 404.
Prescott, D. M., and Kimball, R. F. (1961). *Proc. Nat. Acad. Sci. U. S.* **47**, 686.
Prescott, D. M., Stevens, A. R., and Lauth, M. R. (1971). *Exp. Cell Res.* **64**, 145.
Prestayko, A. W., Tonato, M., and Busch, H. (1970). *J. Mol. Biol.* **47**, 505.

Price, J. M., Rotterham, J., and Evans, V. J. (1967). *J. Nat. Cancer Inst.* **39**, 529.
Printz, D. B., and Gross, S. R. (1967). *Genetics* **55**, 451.
Printz, M. P., and von Hippel, P. H. (1965). *Proc. Nat. Acad. Sci. U. S.* **53**, 363.
Ptashne, M., and Hopkins, M. (1968). *Proc. Nat. Acad. Sci. U. S.* **60**, 1282.
Pullman, B., and Pullman, A. (1969). *Progr. Nucleic Acid Res. Mol. Biol.* **9**, 328.
Purdom, I., Bishop, J. O., and Birnstiel, M. L. (1970). *Nature (London)* **227**, 239.
Quagliarotti, G., and Ritossa, F. M. (1968). *J. Mol. Biol.* **36**, 57.
Quincey, R. V., and Wilson, S. H. (1969). *Proc. Nat. Acad. Sci. U. S.* **64**, 981.
Raacke, I. D. (1968). *Biochem. Biophys. Res. Commun.* **31**, 528.
Rabinowitz, M., and Swift, H. (1970). *Physiol. Rev.* **50**, 376.
Rae, P. M. M. (1969). *J. Cell Biol.* **43**, 109a.
RajBhandary, U. L., and Kumar, A. (1970). *J. Mol. Biol.* **50**, 707.
Rake, A. V., and Graham, A. F. (1962). *J. Cell. Comp. Physiol.* **60**, 139.
Ralph, R. K. (1968a). *Biochem. Biophys. Res. Commun.* **30**, 192.
Ralph, R. K. (1968b). *Biochem. Biophys. Res. Commun.* **33**, 213.
Rao, P. N., and Johnson, R. T. (1970). *Nature (London)* **225**, 159.
Răska, K., Frohwirth, D. H., and Schlesinger, R. W. (1970). *J. Virol.* **5**, 464.
Rasmussen, H. (1970). *Science* **170**, 404.
Raven, P. H. (1970). *Science* **169**, 641.
Reeder, R. H., and Brown, D. D. (1969). *J. Cell Biol.* **43**, 114a.
Reese, D. H., Puccia, E., and Yamada, T. (1969). *J. Exp. Zool.* **170**, 259.
Reich, E., and Goldberg, I. H. (1964). *Progr. Nucleic Acid Res. Mol. Biol.* **3**, 183.
Reichmann, M. E., and Clark, J. M. (1966). *Cold Spring Harbor Symp. Quant. Biol.* **31**, 139.
Rennert, O. M. (1969). *Proc. Nat. Acad. Sci. U. S.* **63**, 878.
Rennert, O. M. (1970). *Life Sci.* **9**, 277.
Revel, M., Herzberg, M., Becarevic, A., and Gros, F. (1968a). *J. Mol. Biol.* **33**, 231.
Revel, M., Lelong, J. C., Brawerman, G., and Gros, F. (1968b). *Nature (London)* **219**, 1016.
Rich, A. (1968). *In* "Structural Chemistry and Molecular Biology" (A. Rich and N. Davidson, eds.), p. 223. Freeman, San Francisco, California.
Richardson, J. P. (1969). *Progr. Nucleic Acid Res. Mol. Biol.* **9**, 75.
Richmond, M. H. (1970). *In* "Organization and Control in Prokaryotic and Eukaryotic Cells" (H. P. Charles and B. C. J. G. Knight, Eds.), p. 249. Cambridge Univ. Press, Cambridge.
Riddick, D. H., and Gallo, R. C. (1970). *Cancer Res.* **30**, 2484.
Riley, M., Pardee, A. B., Jacob, F., and Monod, J. (1960). *J. Mol. Biol.* **2**, 216.
Rines, H. W., Case, M. E., and Giles, N. H. (1969). *Genetics* **61**, 789.
Ringborg, U., Daneholt, B., Edström, J-E., Egyhàzi, E., and Rydlander, L. (1970). *J. Mol. Biol.* **57**, 679.
Ringertz, N. R., and Bolund, L. (1969). *Exp. Cell Res.* **55**, 205.
Ris, H. (1969). *In* "Handbook of Molecular Cytology" (A. Lima-de-Faria, ed.), p. 221. North-Holland Publ., Amsterdam.
Ritossa, F. M. (1968). *Proc. Nat. Acad. Sci. U. S.* **59**, 1124.
Ritossa, F. M., and Scala, G. (1969). *Genetics* **61**, Suppl. 1, 305.
Ritossa, F. M., Atwood, K. C., Lindsley, D. L., and Spiegelman, S. (1966a). *Nat. Cancer Inst. Monogr.* **23**, 449.
Ritossa, F. M., Atwood, K. C., and Spiegelman, S. (1966b). *Genetics* **54**, 663.
Ritossa, F. M., Atwood, K. C., and Spiegelman, S. (1966c). *Genetics* **54**, 819.
Robbins, E., and Morrill, G. A. (1969). *J. Cell Biol.* **43**, 629.

Robbins, E., and Pederson, T. (1970). *Proc. Nat. Acad. Sci. U. S.* **66**, 1245.
Roberts, E. B. (1958). *In* "Microsomal Particles and Protein Synthesis" (E. B. Roberts, ed.), p. vii, Macmillan (Pergamon), New York.
Roberts, J. W. (1969). *Nature (London)* **224**, 1168.
Roberts, W. K., and Quinlivan, V. D. (1969). *Biochemistry* **8**, 288.
Robertson, H. D., Webster, R. E., and Zinder, N. D. (1968). *Nature (London)* **218**, 533.
Robison, G. A., Butcher, R. W., and Sutherland, E. W. (1968). *Annu. Rev. Biochem.* **37**, 149.
Rodeh, R., Feldman, M., and Littauer, U. Z. (1967). *Biochemistry* **6**, 451.
Rodman, T. C. (1967). *J. Cell. Phys.* **70**, 179.
Rodman, T. C. (1969). *J. Cell Biol.* **42**, 575.
Roeder, R. G., and Rutter, W. J. (1970). *Proc. Nat. Acad. Sci. U. S.* **65**, 675.
Roeder, R. G., Reeder, R. H., and Brown, D. D. (1970). *Cold Spring Harbor Symp. Quant. Biol.* **35**, 727.
Römer, R., Riesner, D., and Maass, G. (1970). *FEBS Lett.* **10**, 352.
Rogers, M. E., Loening, U. E., and Fraser, R. S. S. (1970). *J. Mol. Biol.* **49**, 681.
Rose, J. K., Mosteller, R. D., and Yanofsky, C. (1970). *J. Mol. Biol.* **51**, 541.
Rose, R., and Hillman, R. (1969). *Biochem. Biophys. Res. Commun.* **35**, 197.
Rosen, R. (1968). *Int. Rev. Cytol.* **23**, 25.
Rosenberg, B. H., Cavalieri, L. F., and Ungers, G. (1969). *Proc. Nat. Acad. Sci. U. S.* **63**, 1410.
Ross, E. J. (1968). *In* "Recent Advances in Endocrinology" (V. H. T. James, ed.), 8th Ed., p. 293. Little, Brown, Boston, Massachusetts.
Rosset, R., and Gorini, L. (1968). *J. Mol. Biol.* **39**, 95.
Rosset, R., Monier, R., and Julien, J. (1964). *Bull. Soc. Chim. Biol.* **46**, 87.
Roth, R., Ashworth, J. M., and Sussman, M. (1968). *Proc. Nat. Acad. Sci. U. S.* **59**, 1235.
Rudkin, G. T. (1969). *Genetics* **61**, Suppl. 1, 227.
Rudland, P. S., and Dube, S. K. (1969). *J. Mol. Biol.* **43**, 273.
Rudland, P. S., Whybrow, W. A., Marcker, K. A., and Clark, B. F. C. (1969). *Nature (London)* **222**, 750.
Russell, D. H. (1971). *Proc. Nat. Acad. Sci. U. S.* **68**, 523.
Russell, R. L., Abelson, J. N., Landy, A., Gefter, M. L., Brenner, S., and Smith, J. D. (1970). *J. Mol. Biol.* **47**, 1.
Rutter, W. J., Clark, W. R., Kemp, J. D., Bradshaw, W. S., Sanders, T. G., and Ball, W. D. (1968). *In* "Epithelial-Mesenchymal Interactions" (R. Fleischmajer and R. E. Billingham, eds.), p. 114. Williams & Wilkins, Baltimore, Maryland.
Ryan, J. L., and Morowitz, H. J. (1969). *Proc. Nat. Acad. Sci. U. S.* **63**, 1282.
Rymo, L., and Lagerkvist, V. (1970). *Nature (London)* **226**, 77.
Sabatini, E., and Blobel, G. (1970). *J. Cell Biol.* **45**, 146.
Sachs, H. (1969). *Advan. Enzymol.* **32**, 327.
Sachsenmaier, W. (1966). *In* "Probleme der biologischen Reduplikation" (P. Sitte, ed.), p. 329. Springer, Berlin.
Sadgopal, A. (1968). *Advan. Genet.* **14**, 325.
Sadgopal, A., and Bonner, J. (1969). *Biochim. Biophys. Acta* **186**, 349.
Salb, J. M., and Marcus, P. I. (1965). *Proc. Nat. Acad. Sci. U. S.* **54**, 1353.
Sanger, F., and Brownlee, G. G. (1967). *Methods Enzymol.* **12**, 361.
Sanger, F., and Thompson, E. O. P. (1953). *Biochem. J.* **53**, 353.
Sarabhai, A. S., Stretton, A. O. W., Brenner, S., and Bolle, A. (1964). *Nature (London)* **201**, 13.
Sarff, M. A., and Gorski, J. (1969). *J. Cell Biol.* **43**, 122a.

Sarkar, S., and Thach, R. E. (1968). *Proc. Nat. Acad. Sci. U. S.* **60**, 1479.
Saxén, L., Koskimies, O., Lahti, A., Miettinen, H., Rapola, J., and Wartiovaara, J. (1968). *Advan. Morphog.* **7**, 251.
Scarano, E., de Petrocellis, B., and Agusti-Tocco, G. (1964). *Biochim. Biophys. Acta* **87**, 174.
Scarano, E., Iaccarino, M., Grippo, R., and Parisi, E. (1967). *Proc. Nat. Acad. Sci. U. S.* **57**, 1394.
Schachman, H. K. (1959). "Ultracentrifugation in Biochemistry." Academic Press, New York.
Scharff, M. D., and Robbins, E. (1965). *Nature (London)* **208**, 464.
Scherrer, K., and Marcaud, L. (1968). *J. Cell. Physiol.* **72**, Suppl. 1, 181.
Schildkraut, C. L., Marmur, J., and Doty, P. (1961). *J. Mol. Biol.* **3**, 595.
Schimke, R. T. (1970). *In* "Mammalian Protein Metabolism" (H. N. Munro, ed.), Vol. 4, p. 178. Academic Press, New York.
Schlessinger, D. (1969). *Bacteriol. Rev.* **33**, 445.
Schlessinger, D., and Apirion, D. A. (1969). *Annu. Rev. Microbiol.* **23**, 387.
Schmidt, D. A., Mazaitis, A. J., Kasai, T., and Bautz, E. K. F. (1970). *Nature (London)* **225**, 1012.
Schneider, W. D., and Kuff, E. L. (1964). *In* "Cytology and Cell Physiology" (G. H. Bourne, ed.), 3rd Ed., p. 19. Academic Press, New York.
Schrader, F., and Leuchtenberger, C. (1950). *Exp. Cell Res.* **1**, 421.
Schramm, G. (1965). *In* "Nucleic Acids-Structure, Biosynthesis and Function" (P. M. Barghava, ed.), p. 340. Counc. of Sci. and Ind. Res., New Delhi.
Schramm, G., and Ulmer-Schürnbrand, I. (1967). *Biochim. Biophys. Acta* **145**, 7.
Schreier, M. H., and Noll, H. (1970). *Nature (London)* **227**, 128.
Schubert, M. (1969). *In* "The Biological Basis of Medicine" (E. E. Bittar and N. Bittar, eds.), Vol. 3, p. 211. Academic Press, New York.
Schulman, L. H., and Chambers, R. W. (1968). *Proc. Nat. Acad. Sci. U. S.* **61**, 308.
Schultz, J. (1965). *Brookhaven Symp. Biol.* **18**, 116.
Schultz, J., and Travaglini, E. C. (1965). *Genetics* **52**, 473.
Schultze, B., and Maurer, W. (1967). *In* "The Control of Nuclear Activity" (L. Goldstein, ed.), p. 319. Prentice-Hall, Englewood Cliffs, New Jersey.
Schweiger, A., and Hannig, K. (1970). *Biochim. Biophys. Acta* **204**, 317.
Schweizer, E., MacKechnie, C., and Halvorson, H. O. (1969). *J. Mol. Biol.* **40**, 261.
Schwimmer, S., and Bonner, J. (1965). *Biochim. Biophys. Acta* **108**, 67.
Scolnick, E., Tompkins, R., Caskey, T., and Nirenberg, M. (1968). *Proc. Nat. Acad. Sci. U. S.* **61**, 768.
Scornik, O. A., Hoagland, M. B., Pfefferkorn, L. C., and Bishop, E. A. (1967). *J. Biol. Chem.* **242**, 131.
Scott, J. F., Monier, R., Aubert, M., and Reynier, M. (1968). *Biochem. Biophys. Res. Commun.* **33**, 794.
Scott, N. S., Munns, R., and Smillie, R. M. (1970). *FEBS Lett.* **10**, 149.
Sells, B. H., and Davis, F. C. (1970). *J. Mol. Biol.* **47**, 155.
Sharma, O. K., and Borek, E. (1970). *Biochemistry* **9**, 2507.
Shatkin, A. J. (1969). *Advan. Virus Res.* **14**, 63.
Shatkin, A. J., and Sipe, A. D. (1968). *Proc. Nat. Acad. Sci. U. S.* **61**, 1462.
Shearer, R. W., and McCarthy, B. J. (1967). *Biochemistry* **6**, 283.
Shearer, R. W., and McCarthy, B. J. (1970a). *J. Cell. Phys.* **75**, 97.
Shearer, R. W., and McCarthy, B. J. (1970b). *Biochem. Genet.* **4**, 395.
Shearn, A., and Horowitz, N. H. (1969). *Biochemistry* **8**, 295.
Sherbet, G. V., and Lakshmi, M. S. (1967). *Int. Rev. Cytol.* **22**, 147.

Sherr, C. J., and Uhr, J. W. (1969). *Proc. Nat. Acad. Sci. U. S.* **64**, 381.
Shooter, E. M., and Einstein, E. R. (1971). *Annu. Rev. Biochem.* **40**, 635.
Shugart, L., Novelli, G. D., and Stulberg, M. P. (1968). *Biochim. Biophys. Acta* **157**, 83.
Shyamala, G., and Gorski, J. (1969). *J. Biol. Chem.* **244**, 1097.
Sibatani, A. (1966). *Progr. Biophys. Mol. Biol.* **16**, 15.
Siddiqui, M. A. Q., and Hosokawa, K. (1969). *Biochem. Biophys. Res. Commun.* **36**, 711.
Sidebottom, E., and Harris, H..(1969). *J. Cell Sci.* **5**, 351.
Siebert, G. (1967). *Methods Cancer Res.* **2**, 287.
Siev, M., Weinberg, R., and Penman, S. (1969). *J. Cell Biol.* **41**, 510.
Signer, E. R. (1969). *Nature (London)* **223**, 158.
Simon, L. N., Glasky, A. J., and Rejal, R. J. (1967). *Biochem. Biophys. Res. Commun.* **142**, 99.
Sinclair, J. H., and Brown, D. D. (1971). *Biochemistry* **10**, 2761.
Singer, C. E., and Ames, B. N. (1970). *Science* **170**, 822.
Singer, M. F., and Leder, P. (1966). *Annu. Rev. Biochem.* **35**, 195.
Sirlin, J. L. (1955). *Experientia* **11**, 112.
Sirlin, J. L. (1960). *In* "The Cell Nucleus" (J. S. Mitchell, ed.), p. 35. Butterworth, London and Washington, D. C.
Sirlin, J. L. (1962). *Progr. Bioph. Biophys. Chem.* **12**, 27, 319.
Sirlin, J. L. (1963). *Int. Rev. Cytol.* **15**, 35.
Sirlin, J. L. (1968). *In* "The Biological Basis of Medicine" (E. E. Bittar and N. Bittar, eds.), Vol. 1, p. 283. Academic Press, New York.
Sirlin, J. L., and Jacob, J. (1960). *Exp. Cell Res.* **20**, 283.
Sirlin, J. L., and Knight, G. R. (1958). *Chromosoma* **9**, 119.
Sirlin, J. L., and Loening. U. E. (1968). *Biochem. J.* **109**, 375.
Sirlin, J. L., and Schor, N. A. (1962a). *Exp. Cell Res.* **27**, 165.
Sirlin, J. L., and Schor, N. A. (1962b). *Exp. Cell Res.* **27**, 363.
Sivolap, Y. M., and Bonner, J. (1971). *Proc. Nat. Acad. Sci. U. S.* **68**, 387.
Skinner, D. M. (1969). *Biochemistry* **8**, 1467.
Slater, D. W., and Spiegelman, S. (1968). *Biochim. Biophys. Acta* **166**, 82.
Smellie, R. M. S. (1968). *In* "The Biological Basis of Medicine" (E. E. Bittar and N. Bittar, eds.), Vol. 1, p. 243. Academic Press, New York.
Smetana, K. (1967). *Methods Cancer Res.* **2**, 362.
Smillie, E. J., and Burdon, R. H. (1970). *Biochim. Biophys. Acta* **213**, 248.
Smith, A. E., and Marcker, K. A. (1968). *J. Mol. Biol.* **38**, 241.
Smith, A. E., and Marcker, K. A. (1970). *Nature (London)* **226**, 607.
Smith, B. J. (1970). *J. Mol. Biol.* **47**, 101.
Smith, D. W. E. (1968). *J. Biol. Chem.* **243**, 3361.
Smith, D. W. E. (1969). *J. Biol. Chem.* **244**, 896.
Smith, E. L. (1968). *Harvey Lect.* **62**, 231.
Smith, I. (ed.) (1968). "Chromatography and Electrophoresis Techniques," 2nd ed., Vol. 2. Wiley (Interscience), New York.
Smith, I. (ed.) (1969). "Chromatography and Electrophoresis Techniques," 3rd ed., Vol. 1. Wiley (Interscience), New York.
Smith, I. Dubnau, D., Morell, P., and Marmur, J. (1968). *J. Mol. Biol.* **33**, 123.
Smith, I., Colli, W., and Oishi, M. (1971). *J. Mol. Biol.* **62**, 111.
Smith, J. D., Barnett, L., Brenner, S., and Russell, R. L. (1970). *J. Mol. Biol.* **54**, 1.
Smith, K. D., Armstrong, J. L., and McCarthy, B. J. (1967). *Biochim. Biophys. Acta* **142**, 323.
Smith, L. D., and Ecker, R. E. (1969). *Develop. Biol.* **19**, 281.

Bibliography

Smith, L. D., and Ecker, R. E. (1970). *Develop. Biol.* **22**, 622.
Smulson, M. (1970). *Biochim. Biophys. Acta* **199**, 537.
Smythies, J. R. (1970). *Intern. Rev. Neurobiol.* **13**, 181.
Sneider, T. W., and Potter, V. R. (1969). *J. Mol. Biol.* **42**, 271.
Snow, M. H. L., and Callan, H. G. (1969). *J. Cell Sci.* **5**, 1.
Soeiro, R. (1968). *In* "Regulatory Mechanisms for Protein Synthesis in Mammalian Cells" (A. San Pietro, M. R. Lamborg, and F. T. Kenney, eds.), p. 49. Academic Press, New York.
Soeiro, R., and Darnell, J. E. (1969). *J. Mol. Biol.* **44**, 551.
Soeiro, R., and Darnell, J. E. (1970). *J. Cell Biol.* **44**, 467.
Soeiro, R., Vaughan, M. H., Warner, J. R., and Darnell, J. E. (1968). *J. Cell Biol.* **39**, 112.
Soll, D. (1971). *Science* **173**, 293.
Soll, D., and RajBhandary, U. L. (1967). *J. Mol. Biol.* **29**, 113.
Soll, L., and Berg, P. (1969). *Nature (London)* **223**, 1340.
Sonneborn, T. M. (1965). *In* "Evolving Genes and Proteins" (V. Bryson and H. J. Vogel, eds.), p. 377. Academic Press, New York.
Sonneborn, T. M. (1967). *In* "Heritage from Mendel" (R. A. Brink and A. D. Styles, eds.), p. 375. Univ. of Wisconsin Press, Madison, Wisconsin.
Sotelo, J. R., and Wettstein, R. (1966). *Nat. Cancer Inst. Monogr.* **23**, 77.
Southern, E. M. (1970). *Nature (London)* **227**, 794.
Sox, H. C., and Hoagland, M. B. (1966). *J. Mol. Biol.* **20**, 113.
Spadari, S., and Ritossa, F. M. (1970). *J. Mol. Biol.* **53**, 357.
Spelsberg, T. C., Tankersley, S., and Hnilica, L. S. (1969). *Proc. Nat. Acad. Sci. U. S.* **62**, 1218.
Speyer, J. F. (1965). *Biochem. Biophys. Res. Commun.* **21**, 6.
Speyer, J. F. (1967). *Mol. Genet.* **2**, 137.
Spiegel, M., Spiegel, E. S., and Meltzer, P. S. (1970). *Develop. Biol.* **21**, 73.
Spiegelman, S. (1962). *In* "Acides Ribonucleiques et Polyphosphates" (M. Grunberg-Manago and M. J. P. Ebel, eds.), p. 407. C.N.R.S., Paris.
Spiegelman, S. (1969). *In* "Exploitable Molecular Mechanisms and Neoplasia" (R. B. Hurlbert, ed.), p. 7. Williams & Wilkins, Baltimore, Maryland.
Spiegelman, S., and Haruna, I. (1966). *In* "Macromolecular Metabolism" (J. Hurwitz, ed.), p. 263. Little, Brown, Boston, Massachusetts.
Spirin, A. S. (1963). *Progr. Nucleic Acid Res. Mol. Biol.* **1**, 301.
Spirin, A. S. (1964). "Macromolecular Structure of Ribonucleic Acids." Reinhold, New York.
Spirin, A. S. (1968). *Curr. Mod. Biol.* **2**, 115.
Spirin, A. S. (1969). *Eur. J. Biochem.* **10**, 20.
Spirin, A. S., and Gavrilova, L. P. (1969). "The Ribosome," Springer-Verlag, New York.
Staehelin, T., Wettstein, F. O., Oura, H., and Noll, H. (1964). *Nature (London)* **201**, 264.
Starr, J. L., and Sells, B. H. (1969). *Physiol. Rev.* **49**, 623.
Stavis, R. L., and August, J. T. (1970). *Annu. Rev. Biochem.* **39**, 527.
Stavy, L., and Gross, P. R. (1969a). *Biochim. Biophys. Acta* **182**, 193.
Stavy, L., and Gross, P. R. (1969b). *Biochim. Biophys. Acta* **182**, 203.
Stedman, E., and Stedman, E. (1950). *Nature (London)* **166**, 780.
Steele, W. J. (1966). *In* "The Cell Nucleus–Metabolism and Radiosensitivity" (H. M. Klouwen, ed.), p. 203. Taylor & Francis, London.
Steele, W. J. (1968). *J. Biol. Chem.* **243**, 3333.
Steele, W. J., and Busch, H. (1967). *Methods Cancer Res.* **3**, 62.

Steffensen, D. M., and Wimber, D. E. (1970). *Abstr. 10th Annu. Meeting Amer. Soc. Cell Biol., San Diego, California,* p. 202a.
Stein, G., and Baserga, R. (1971). *Biochem. Biophys. Res. Commun.* **14**, 218.
Stein, G., Pegoraro, L., Borun, T., and Baserga, R. (1970). *Abstr. 10th Annu. Meeting Amer. Soc. Cell Biol., San Diego, California,* p. 202a.
Steiner, R. F., and Beers, R. F. (1961). "Polynucleotides." Elsevier Publ., Amsterdam.
Steinhart, W. L. (1971). *Biochim. Biophys. Acta* **228**, 301.
Steitz, J. A. (1969). *Nature (London)* **224**, 957.
Stellwagen, R. H., and Cole, R. D. (1969). *Annu. Rev. Biochem.* **38**, 951.
Stenram, U. (1962). *Z. Zellforsch. Mikrosk. Anat.* **58**, 107.
Stent, G. S. (1964). *Science* **144**, 816.
Stent, G. S. (1966). *Proc. Roy. Soc., Ser. B.* **164**, 181.
Stent, G. S. (1968). *Science* **160**. 390.
Stent, G. S. (1969). " The Coming of the Golden Age," p. 70. Natural History Press, New York.
Sterlini, J. M., and Mandelstam, J. (1969). *Biochem. J.* **113**, 29.
Stern, H., and Hotta, Y. (1969). *Genetics* **61**, Suppl. 1, 27.
Stern, R., and Littauer, U. Z. (1968). *Biochemistry* **7**, 3469.
Stern, R., Zutra, L. E., and Littauer, U. Z. (1969). *Biochemistry* **8**, 313.
Stern, R., Gonano, F., Fleissner, E., and Littauer, U. Z. (1970). *Biochemistry* **9**, 10.
Stevens, A. R., and Prescott, D. M. (1968). *J. Cell Biol.* **39**, 129a.
Stevens, B. J., and Swift, H. (1966). *J. Cell Biol.* **31**, 55.
Steward, D. L., Shaeffer, J. R., and Humphrey, R. M. (1968). *Science* **161**, 791.
Steward, F. C., Mapes, M. O., and Mears, K. (1958). *Amer. J. Bot.* **45**, 705.
Stewart, T. S., Roberts, R. J., and Strominger, J. L. (1971). *Nature (London)* **230**, 36.
Stocken, L. A., and Ord, M. G. (1966). *In* "The Cell Nucleus-Metabolism and Radiosensitivity" (H. M. Klouwen, ed.), p. 141. Taylor & Francis, London.
Stockert, J. D., Fernández-Gómez, M. E., Giménez-Martin, G., and Lopez-Sáez, J. F. (1970). *Protoplasma* **69**, 265.
Stöcker, E. (1964). *Z. Zellforsch. Mikrosk. Anat.* **62**, 80.
Strehler, B. L., Hendley, D. D., and Hirsch, G. P. (1967). *Proc. Nat. Acad. Sci. U. S.* **57**, 1751.
Stubblefield, E. (1966). *J. Nat. Cancer Inst.* **37**, 799.
Stubblefield, E., and Wray, W. (1971). *Chromosoma* **32**, 262.
Stubbs, J. D., and Stubbs, E. A. (1970). *J. Mol. Biol.* **51**, 717.
Stulberg, M. P., Isham, K. R., and Stevens, A. (1969). *Biochim. Biophys. Acta* **186**, 297.
Subak-Sharpe, J. H. (1968). *Symp. Soc. Gen. Microbiol.* **18**, 47.
Subak-Sharpe, J. H. (1969). *Proc. Can. Cancer Res. Conf.* **8**, 242.
Subramanian, A. R., Ron, E. Z., and Davis, B. D. (1968). *Proc. Nat. Acad. Sci. U. S.* **61**, 761.
Sueoka, N. (1965). *In* "Evolving Genes and Proteins" (V. Bryson and H. J. Vogel, eds.), p. 479. Academic Press, New York.
Sueoka, N. (1967). *Mol. Genet.* **2**, 1.
Sueoka, N., and Cheng, T. Y. (1962). *Proc. Nat. Acad. Sci. U. S.* **48**, 1851.
Sueoka, N., and Hardy, J. (1968). *Arch. Biochem. Biophys.* **125**, 558.
Sueoka, N., and Kano-Sueoka, T. (1970). *Progr. Nucleic Acid Mol. Biol.* **10**, 23.
Sugiura, M., Okamoto, T., and Takanami, M. (1970). *Nature (London)* **225**, 598.
Sullivan, D. T. (1968). *Proc. Nat. Acad. Sci. U. S.* **59**, 846.
Sundharadas, G., Katze, J. R., Söll, D., Konigsberg, W., and Lengyel, P. (1968). *Proc. Nat. Acad. Sci. U. S.* **61**, 693.
Sunshine, G. H., Williams, D. J., and Rabin, B. R. (1971). *Nature (London)* **230**, 133.

Sussman, M. (1970). *Nature (London)* **225**, 1245.
Suyama, Y. (1967). *Biochemistry* **6**, 2829.
Suzuki, T., and Garen, A. (1969). *J. Mol. Biol.* **45**, 549.
Swanson, R. F., and Dawid, I. B. (1970). *Proc. Nat. Acad. Sci. U. S.* **66**, 117.
Svensson, I., Björk, G. R., and Lundahl, P. (1969). *Eur. J. Biochem.* **9**, 216.
Swann, M. M. (1958). *Cancer Res.* **18**, 1118.
Swift, H. (1965). *In Vitro* **1**, 26.
Swift, H., and Stevens, B. J. (1966). *Nat. Cancer Inst. Monogr.* **23**, 145.
Szego, C. M., and Davis, J. S. (1967). *Proc. Nat. Acad. Sci. U. S.* **58**, 1711.
Szent-Györgi, A., Együd, L. G., and McLaughlin, J. A. (1967). *Science* **155**, 539.
Szybalski, W. (1968). *Methods Enzymol.* **12B**, 330.
Szybalski, W., Bøvre, K., Fiandt, M., Guha, A., Hradecna, Z., Kumar, S., Lozeron, H. A., Maher, V. M., Nijkamp, H. J. J., Summers, W. C., and Taylor, K. (1969). *J. Cell. Physiol.* **74**, Suppl. 1, 33.
Taber, R. L., and Vincent, W. S. (1969). *Biochem. Biophys. Res. Commun.* **34**, 488.
Takahashi, I. (1965). *J. Bacteriol.* **89**, 1065.
Takeda, Y., and Igarashi, K. (1970). *Biochim. Biophys. Acta* **204**, 406.
Talwar, G. P., Segal, S. J., Evans, A., and Davidson, O. W. (1964). *Proc. Nat. Acad. Sci. U. S.* **52**, 1059.
Tandler, C. J. (1966). *Nat. Cancer Inst. Monogr.* **23**, 181.
Tandler, C. J., and Sirlin, J. L. (1964). *Biochim. Biophys. Acta* **80**, 315.
Tandler, C. J., Libanati, C. M., and Sanchis, C. A. (1970). *J. Cell Biol.* **45**, 355.
Tartar, V. (1961). "The Biology of *Stentor*." Macmillan (Pergamon), New York.
Tartof, K. D. (1971). *Science* **171**, 294.
Tartof, K. D., and Perry, R. P. (1970). *J. Mol. Biol.* **51**, 171.
Tata, J. R. (1968). *Nature (London)* **219**, 331.
Taylor, A. L., and Trotter, C. D. (1967). *Bacteriol. Rev.* **31**, 337.
Taylor, J. H. (1963). *Mol. Genet.* **1**, 65.
Taylor, J. H., Woods, P. S., and Hughes, W. L. (1957). *Proc. Nat. Acad. Sci. U. S.* **53**, 122.
Taylor, J. H., Mego, W. A., and Evenson, D. P. (1970). In "The Neurosciences: Second Study Program (F. O. Schmitt, ed.), p. 998. Rockefeller Univ. Press, New York.
Taylor, J. M., Cohen, S., and Mitchell, W. M. (1970). *Proc. Nat. Acad. Sci. U. S.* **67**, 164.
Taylor, K., Hradecna, Z., and Szybalski, W. (1967). *Proc. Nat. Acad. Sci. U. S.* **57**, 1618.
Taylor, M. W. (1969). *Cancer Res.* **29**, 1681.
Taylor, M. W., Buck, C. A., Granger, G. A., and Holland, J. J. (1968). *J. Mol. Biol.* **33**, 809.
Temin, H. M., and Mizutani, S. (1970). *Nature (London)* **226**, 1211.
Teng, C. S., and Hamilton, T. H. (1969). *Proc. Nat. Acad. Sci. U. S.* **63**, 465.
Teng, C. S., Teng, C. T., and Allfrey, V. G. (1971). *J. Biol. Chem.* **246**, 3597.
Terao, K., Sugano, H., and Ogata, K. (1968). *J. Biochem. (Tokyo)* **64**, 407.
Terzaghi, E., Okada, Y., Streisinger, G., Emrich, J., Inouye, M., and Tsugita, A. (1966). *Proc. Nat. Acad. Sci. U. S.* **56**, 500.
Tewari, K. K., and Wildman, S. G. (1968). *Proc. Nat. Acad. Sci. U. S.* **59**, 569.
Tewari, K. K., and Wildman, S. G. (1969). *Biochim. Biophys. Acta* **186**, 358.
Thiebe, R., and Zachau, H. G. (1968). *Biochem. Biophys. Res. Commun.* **33**, 260.
Thomas, C. (1970). *Biochim. Biophys. Acta* **224**, 99.
Thomas, C. A., Jr. (1963). *Mol. Genet.* **1**, 113.
Thomas, C. A., Jr. (1966). *Progr. Nucleic Acid Res. Mol. Biol.* **5**, 315.
Thomas, C. A., Jr., and MacHattie, L. A. (1967). *Annu. Rev. Biochem.* **36**, 485.

Thomas, C. A., Jr., Hamkalo, B. A., Misra, D. N., and Lee, C. S. (1970). *J. Mol. Biol.* **51**, 621.
Thompson, L. R., and McCarthy, B. J. (1968). *Biochem. Biophys. Res. Commun.* **30**, 166.
Tidwell, T. (1970). *J. Cell Biol.* **46**, 370.
Tidwell, T., and Stubblefield, E. (1971). *Exp. Cell Res.* **64**, 350.
Tiedemann, H. (1968). *J. Cell. Physiol.* **72**, Suppl. 1, 129.
Tiollais, P., Galibert, F., and Boiron, M. (1971). *Proc. Nat. Acad. Sci. U. S.* **68**, 1117.
Todaro, G., Matsuga, Y., Bloom, S., Robbins, A., and Green, H. (1967). *In* "Growth Regulatory Substances for Animal Cells in Culture" (V. Defendi and M. Stocker, eds.), Wistar Inst. Symp. Monogr. No. 7, p. 87. Wistar Inst. Press, Philadelphia, Pennsylvania.
Tomkins, G. M., Gelehrter, T. D., Granner, D., Martin, D., Samuels, H. H., and Thompson, E. B. (1969). *Science* **166**, 1474.
Tomlinson, R. V., and Tener, G. M. (1962). *J. Amer. Chem. Soc.* **84**, 2644.
Tonoue, T., Eaton, J., and Frieden, E. (1969). *Biochem. Biophys. Res. Commun.* **37**, 81.
Travaglini, E. C., Petrovic, J., and Schultz, J. (1968). *J. Cell Biol.* **39**, 136a.
Travers, A. A. (1969). *Nature (London)* **223**, 1107.
Travers, A. A., Kamen, R. I., and Schleif, R. F. (1970). *Nature (London)* **228**, 748.
Trávníček, M., (1969). *Biochim. Biophys. Acta* **182**, 427.
T'so, R. O. P., Melvin, I. S., and Olson, A. C. (1963). *J. Amer. Chem. Soc.* **85**, 1289.
Turkington, R. W. (1969). *J. Biol. Chem.* **244**, 5140.
Uhlenbeck, O. C., Baller, J., and Doty, P. (1970). *Nature (London)* **225**, 508.
Ullmann, A., and Monod, J. (1968). *FEBS Lett.* **2**, 57.
Van. Etten, J. L., and Brambl, R. M. (1968). *J. Bacteriol.* **96**, 1042.
Van Iterson, W. (1969). *In* "Handbook of Molecular Cytology" (A. Lima-de-Faria, ed.), pp. 149, 174, and 197. North-Holland Publ., Amsterdam.
Vanyushin, B. F., Tkacheva, S. G., and Belozersky, A. N. (1970). *Nature (London)* **225**, 948.
Vaughan, M. H., Soiero, R., Warner, J. R., and Darnell, J. E. (1967). *Proc. Nat. Acad. Sci. U. S.* **58**, 1527.
Venetianer, P. (1969). *J. Mol. Biol.* **45**, 375.
Vesco, C., and Penman, S. (1969). *Nature (London)* **224**, 1021.
Veselý, J., and Cihak, A. (1970). *Biochim. Biophys. Acta* **204**, 614.
Vidali, G., Gershey, E. L., and Allfrey, V. G. (1968). *J. Biol. Chem.* **243**, 6361.
Vincent, W. S., Baltus, E., Løvlie, A., and Mundell, R. C. (1966). *Nat. Cancer Inst. Monogr.* **23**, 235.
Vincent, W. S., Halvorson, H. O., Chen, H. R., and Shin, D. (1969). *Exp. Cell Res.* **57**, 240.
Vinograd, J. (1970). *Rep. 7th Annu. Pathobiol. Conf.,* Aspen, Colorado.
Vittorelli, M. L., Caffarelli-Mornino, I., and Monroy, A. (1969). *Biochim. Biophys. Acta* **186**, 608.
Vogel, Z., Zamir, A., and Elson, D. (1968). *Proc. Nat. Acad. Sci. U. S.* **61**, 701.
Vold, B. S., and Sypherd, P. S. (1968). *Proc. Nat. Acad. Sci. U. S.* **59**, 453.
Volkin, E., and Astrachan, L. (1956). *Virology* **2**, 149.
von der Decke, A. (1969). *J. Cell Biol.* **43**, 138.
von Ehrenstein, G. (1966). *Cold Spring Harbor Symp. Quant. Biol.* **31**, 705.
von Ehrenstein, G. (1970). *In* "Aspects of Protein Biosynthesis" (C. B. Anfinsen, ed.), Part A, p. 139. Academic Press, New York.
von Ehrenstein, G., Weisblum, B., and Benzer, S. (1963). *Proc. Nat. Acad. Sci. U. S.* **49**, 669.
Vorobyev, V. I., Gineitis, A. A., and Vinogradova, I. A. (1969). *Exp. Cell Res.* **57**, 1.

Wada, K., Shiokawa, K., and Yamana, K. (1968). *Exp. Cell Res.* **52**, 252.
Waddington, C. H. (1956). "Principles of Embryology." Macmillan, New York.
Waddington, C. H. (1962). "New Patterns in Genetics and Development." Columbia Univ. Press, New York.
Waddington, C. H. (1966). *In* "Major Problems in Developmental Biology" (M. Locke, ed.), p. 105. Academic Press, New York.
Waddington, C. H., and Robertson, E. (1969). *Nature (London)* **221**, 933.
Waddington, C. H., and Sirlin, J. L. (1959). *Exp. Cell Res.* **17**, 582.
Wagner, E. K., and Roizman, B. (1969). *Proc. Nat. Acad. Sci. U. S.* **64**, 626.
Wainfan, E. (1968). *Virology* **35**, 282.
Wainfan, E., Srinivasan, P. R., and Borek, E. (1965). *Biochemistry* **4**, 2845.
Wainfan, E., Srinivasan, P. R., and Borek, E. (1966). *Cold Spring Harbor Symp. Quant. Biol.* **31**, 525.
Wainwright, S. D., and Wainwright, L. K. (1967). *Canad. J. Biochem.* **45**, 255.
Walker, P. M. B. (1968). *Nature (London)* **219**, 228.
Walker, P. M. B. (1969). *Progr. Nucleic Acid Res. Mol. Biol.* **9**, 301.
Walker, P. M. B. (1971). *Nature (London)* **229**, 306.
Walker, P. M. B., and Hennig, W. (1970). *Nature (London)* **225**, 915.
Walker, P. M. B., Flamm, W. G., and McLaren, A. (1969). *In* "Handbook of Molecular Cytology" (A. Lima-de-Faria, ed.), p. 52. North-Holland Publ., Amsterdam.
Wallace, H. (1966). *Nat. Cancer Inst. Monogr.* **23**, 425.
Wallace, H., Morray, J., and Langridge, W. H. R. (1971). *Nature (New Biol.)* **230**, 201.
Walter, G., Seifert, W., and Zillig, W. (1968). *Biochem. Biophys. Res. Commun.* **30**, 240.
Warner, J. R. (1966). *J. Mol. Biol.* **19**, 383.
Warner, J. R. (1971). *J. Biol. Chem.* **246**, 447.
Warner, J. R., Rich, A., and Hall. C. E. (1962). *Science* **138**, 1399.
Warocquier, R., and Scherrer, K. (1969). *Europ. J. Biochem.* **10**, 362.
Warren, K. B., ed. (1969). "Differentiation and Immunology," Symp. Int. Soc. Cell Biol. Vol. 7. Academic Press. New York.
Waters, L. C. (1969). *Biochem. Biophys. Res. Commun.* **37**, 296.
Waters, L. C., and Novelli, G. D. (1967). *Proc. Nat. Acad. Sci. U. S.* **57**, 979.
Waters, L. C., and Novelli, G. D. (1968). *Biochem. Biophys. Res. Commun.* **32**, 971.
Watson, J. D. (1968). "The Double Helix." Atheneum, New York.
Watson, J. D. (1969). *Fed. Proc. Fed. Amer. Soc. Exp. Biol.* **28**, 907.
Watson, J. D. (1970). "Molecular Biology of the Gene," 2nd Ed. Benjamin, New York.
Watson, J. D., and Crick, F. H. C. (1953). *Nature (London)* **171**, 737.
Watts, D. C. (1968). *Advan. Comp. Physiol. Biochem.* **3**, 1.
Weatherall, D. J. (1965). "The Thalassaemia Syndromes." Blackwell, Oxford.
Wehrli, W., and Staehelin, M. (1971). *Biochemistry* **10**, 1878.
Weinberg, R. A., and Penman, S. (1969). *Biochim. Biophys. Acta* **190**, 10.
Weinberg, R. A., and Penman, S. (1970). *J. Mol. Biol.* **47**, 169.
Weinstein, I. B., Friedman, S. M., and Ochoa, M. (1966). *Cold Spring Harbor Symp. Quant. Biol.* **31**, 671.
Weisblum, B., and Davies, J. (1968). *Bacteriol. Rev.* **32**, 493.
Weiss, L. (1969). *Int. Rev. Cytol.* **26**, 63.
Weiss, S. B., and Nakamoto, T. (1961). *Proc. Nat. Acad. Sci. U. S.* **47**, 1400.
Weiss, S. B., Hsu, W.-T., Foft, J. W., and Scherberg, N. H. (1968). *Proc. Nat. Acad. Sci. U. S.* **61**, 114.
Wefle, H., Bielka, H., and Böttger, M. (1969). *Mol. Gen. Genet.* **104**, 165.
Wells, R., and Birnstiel, M. L. (1969). *Biochem. J.* **112**, 777.

Werner, R. (1971). *Nature (London)* **230**, 570.
Wessells, N. K. (1968). In "Epithelial-Mesenchymal Interactions" (R. Fleischmajer and R. E. Billingham, eds.), p. 132. Williams & Wilkins, Baltimore, Maryland.
Wetmur, J. G., and Davidson, N. (1968). *J. Mol. Biol.* **31**, 349.
Wettstein. F. O. (1966). *Cold Spring Harbor Symp. Quant. Biol.* **31**, 595.
Wettstein, F. O., and Stent, G. S. (1968). *J. Mol. Biol.* **38**, 25.
Wettstein, F. O., Staehelin, T., and Noll, H. (1963). *Nature (London)* **197**, 430.
White, H. B., Laux, B. E., and Dennis, D. (1972). *Science* **175**, 1264.
Whitehouse, H. C. K. (1967). *J. Cell Sci.* **2**, 9.
Whiteley, A. H., McCarthy, B. J., and Whiteley, H. R. (1966). *Proc. Nat. Acad. Sci. U. S.* **55**, 519.
Whiteley, H. R., McCarthy, B. J., and Whiteley, A. H. (1970). *Develop. Biol.* **21**, 216.
Whitten, J. M. (1965). *Nature (London)* **208**, 1019.
Wilcox, M. (1969). *Eur. J. Biochem.* **11**, 405.
Wilde, C. E. (1961). *Advan. Morphog.* **1**, 267.
Wilde, C. E., and Crawford, R. B. (1968). In "Epithelial-Mesenchymal Interactions" (R. Fleischmajer and R. E. Billingham, eds.), p. 98. Williams & Wilkins, Baltimore, Maryland.
Willems, M., Musilova, H. A., and Malt, R. A. (1969a). *Proc. Nat. Acad. Sci. U. S.* **62**, 1189.
Willems, M., Penman, M., and Penman, S. (1969b). *J. Cell Biol.* **41**, 177.
Williams, L. S., and Freundlich, M. (1969). *Biochim. Biophys. Acta* **186**, 305.
Williams, L. S., and Neidhardt, F. C. (1969). *J. Mol. Biol.* **43**, 529.
Williamson, R., and Brownlee, G. G. (1969). *Biochem. J.* **114**, 29P.
Williamson, R., Morrison, M., Lanyon, G., Eason, R., and Paul, J. (1971). *Biochemistry* **10**, 3014.
Wilson, R. G., Russo, J. F., and Steck, T. L. (1970). *Biochim. Biophys. Acta* **204**, 42.
Wimber, D. E., and Steffensen, D. M. (1970). *Science* **170**, 639.
Winslow, R. M., and Lazzarini, R. A. (1969a). *J. Biol. Chem.* **244**, 1128.
Winslow, R. M., and Lazzarini, R. A. (1969b). *J. Biol. Chem.* **244**, 3387.
Wittliff, J. L., and Keeney, F. T. (1969). In "Problems in Biology: RNA in Development" (E. W. Hanly, ed.), p. 140. Univ. of Utah Press, Salt Lake City, Utah.
Wittman, J. S., Lee, K.-L., and Miller, O. N. (1969). *Biochim. Biophys. Acta* **174**, 536.
Wittmann, H. G. (1970). In "Organization and Control in Prokaryotic and Eukaryotic Cells" (H. P. Charles and B. C. J. G. Knight, eds.), p. 55. Cambridge Univ. Press, Cambridge.
Wittmann, H. G., and Wittmann-Liebold, B. (1966). *Cold Spring Harbor Symp. Quant. Biol.* **31**, 163.
Wobus, U., Panitz, R., and Serfling, E. (1970). *Mol. Gen. Genet.* **107**, 215.
Woese, C. R. (1967a). *Progr. Nucleic Acid Res. Mol. Biol.* **7**, 107.
Woese, C. R. (1967b). "The Genetic Code." Harper & Row, New York.
Woese, C. R. (1968a). *Nature (London)* **220**, 923.
Woese, C. R. (1968b). *Proc. Nat. Acad. Sci. U. S.* **59**, 110.
Woese, C. R. (1969). *J. Mol. Biol.* **43**, 235.
Woese, C. R. (1970a). *J. Theoret. Biol.* **26**, 83.
Woese, C. R. (1970b). In "Organization and Control in Prokaryotic and Eukaryotic Cells" (H. P. Charles and B. C. J. G. Knight, eds.), p. 39. Cambridge Univ. Press, Cambridge.
Woese, C. R., Dugre, D. H., Dugre, S. A., Kondo, M., and Saxinger, W. C. (1966). *Cold Spring Harbor Symp. Quant. Biol.* **31**, 723.
Wolfe, S. L. (1969). In "The Biological Basis of Medicine" (E. E. Bittar and N. Bittar, eds.), Vol. 4, p. 3. Academic Press, New York.
Wolff, S. (1969). *Int. Rev. Cytol.* **25**, 279.

Wolpert, L. (1969). *J. Theoret. Biol.* **25**, 1.
Wolpert, L., and Gingell, D. (1969). In "Homeostatic Regulators" (G. E. W. Wolstenholme and J. Knight, eds.), p. 241. Churchill, London.
Wong, J. T.-F., and Nazar, R. N. (1970). *J. Biol. Chem.* **245**, 4591.
Wong, J. T.-F., Mustard, M., and Herbert, E. (1969). *Biochim. Biophys. Acta* **174**, 513.
Wood, D. D., and Luck, D. J. L. (1969). *J. Mol. Biol.* **41**, 211.
Wood, W. G., Irvin, J. L., and Holbrook, D. J. (1968). *Biochemistry* **7**, 2256.
Woodland, H. R., and Graham, C. F. (1969). *Nature (London)* **221**, 327.
Woodland, H. R., and Gurdon, J. B. (1968). *J. Embryol. Exp. Morphol.* **19**, 363.
Wool, I. G., Martın, T. E., and Low, R. B. (1968). In "Regulatory Mechanisms for Protein Synthesis in Mammalian Cells" (A. San Pietro, M. R. Lamborg, and F. T. Kenney, eds.), p. 323. Academic Press, New York.
Wyatt, G. R. (1968). In "Metamorphosis" (W. Etkin and L. I. Gilbert, eds.), p. 143. Appleton, New York.
Yamada, T. (1967). In "Comprehensive Biochemistry" (M. Florkin and E. H. Stotz, eds.), Vol. 28, p. 113. Elsevier, Amsterdam.
Yang, S. S., and Comb, D. G. (1968). *J. Mol. Biol.* **31**, 139.
Yang, S. S., and Sanadi, S. R. (1969). *J. Biol. Chem.* **244**, 5081.
Yang, W.-K., and Novelli, G. D. (1968a). *Proc. Nat. Acad. Sci. U. S.* **59**, 208.
Yang, W.-K., and Novelli, G. D. (1968b). *Biochem. Biophys. Res. Commun.* **31**, 534.
Yang, W.-K., and Novelli, G. D. (1971). In "Methods in Enzymology" (S. P. Colowick and N. O. Kaplan, eds.), Vol. 20, part C, p. 44. Academic Press, New York.
Yang, W.-K., Hellman, A., Martin, D. H., Hellman, K. B., and Novelli, G. D. (1969a). *Proc. Nat. Acad. Sci. U. S.* **64**, 1411.
Yang, W.-K., Hilse, K. M., and Popp, R. A. (1969b). *Fed. Proc. Fed. Amer. Soc. Exp. Biol.* **28**, 349.
Yankofsky, S. A., and Spiegelman, S. (1963). *Proc. Nat. Acad. Sci. U. S.* **49**, 538.
Yanofsky, C. (1965). *Biochem. Biophys. Res. Commun.* **18**, 898.
Yanofsky, C., Carlton, B., Guest, J. R., Helinsky, D. R., and Henning, U. (1964). *Proc. Nat. Acad. Sci. U. S.* **51**, 266.
Yanofsky, C., Ito, J., and Horn, V. (1966). *Cold Spring Harbor Symp. Quant. Biol.* **31**, 151.
Yarus, M. (1969). *Annu. Rev. Microbiol.* **38**, 841.
Yarus, M., and Berg, P. (1969). *J. Mol. Biol.* **42**, 171.
Yasmineh, W. G., and Yunis, J. J. (1969). *Biochem. Biophys. Res. Commun.* **35**, 779.
Ycas, M. (1969). "The Biological Code." Wiley, New York.
Ycas, M., and Vincent, W. S. (1960). *Proc. Nat. Acad. Sci. U. S.* **46**, 804.
Yegian, C. D., and Stent, G. S. (1969a). *J. Mol. Biol.* **39**, 45.
Yegian, C. D., and Stent, G. S. (1969b). *J. Mol. Biol.* **39**, 59.
Yegian, C. D., Stent, G. S., and Martin, E. M. (1966). *Proc. Nat. Acad. Sci. U. S.* **55**, 839.
Yot, P., Pinck, M., Haenni, A.-L., Duranton, H. M., and Chapeville, F. (1970). *Proc. Nat. Acad. Sci. U. S.* **67**, 1345.
Yu, M. T., Vermeulen, C. W., and Atwood, K. C. (1970). *Proc. Nat. Acad. Sci. U. S.* **67**, 26.
Yudelevich, A., Ginsberg, B., and Hurwitz, J. (1968). *Proc. Nat. Acad. Sci. U. S.* **61**, 1129.
Yudkin, M. D. (1969). *Biochem. J.* **114**, 307.
Yura, T., and Igarashi, K. (1968). *Proc. Nat. Acad. Sci. U. S.* **61**, 1313.
Zachau, H. G. (1969). *Angew. Chem. Int. Ed. Engl.* **8**, 711.
Zamecnik, P. C. (1962). *Biochem. J.* **85**, 257.
Zamecnik, P. C. (1969). *Cold Spring Harbor Symp. Quant. Biol.* **34**, 1.

Zamecnik, P. C. (1971). *Cancer Res.* **31,** 716.
Zamecnik, P. C., and Stephenson, M. L. (1968). *In* "Regulatory Mechanisms for Protein Synthesis in Mammalian Cells" (A. San Pietro, M. R. Lamborg, and F. T. Kenney, eds.), p. 3. Academic Press, New York.
Zamecnik, P. C., and Stephenson, M. L. (1970). *In* "The Role of Nucleotides for the Function and Conformation of Enzymes" (H. M. Kalckar, H. Klenow, A. Munch-Petersen, M. Ottesen, and J. H. Thaysen, eds.), p. 276. Academic Press, New York.
Zehavi-Willner, T. (1970). *Biochem. Biophys. Res. Commun.* **39,** 161.
Zehavi-Willner, T., and Comb, D. G. (1966). *J. Mol. Biol.* **16,** 250.
Zeikus, J. G., Taylor, M. W., and Buck, C. A. (1969). *Exp. Cell Res.* **57,** 74.
Zeuthen, E. (1964). *In* "Synchrony in Cell Division and Growth" (E. Zeuthen, ed.), p 99. Interscience, New York.
Zilliken, F. (1967). *Exp. Biol. Med.* **1,** 199.
Zimmerman, E. F. (1968). *Biochemistry,* **7,** 3156.
Zinder, N. D. (1965). *Annu. Rev. Microbiol.* **19,** 455.
Zipser, D. (1969). *Nature (London)* **221,** 21.
Zubay, G. (1964). *In* "The Nucleohistones" (J. Bonner and P. T'so, eds.), p. 95. Holden-Day, San Francisco, California.
Zubay, G., Schwartz, D., and Beckwith, J. (1970). *Proc. Nat. Acad. Sci. U. S.* **66,** 104.
Zuckerkandl, E. (1968). *In* "Structural Chemistry and Molecular Biology" (A. Rich and N. Davidson, eds.), p. 256. Freeman, San Francisco, California.
Zuckerkandl, E., and Pauling, L. (1965). *In* "Evolving Genes and Proteins" (V. Bryson and H. J. Vogel, eds.), p. 97. Academic Press, New York.
Zwar, J. A., and Brown, R. (1968). *Nature (London)* **220,** 500.
Zylber, E. C., Vesco, C., and Penman, S. (1969). *J. Mol. Biol.* **44,** 195.

Index

A

Acidic (nonhistone) proteins, see Protein(s)
Adaptive regulation of translation, see Translation
Amino acid metabolism
  bacterial transfer RNA and, 239-241

B

Base pairing
  Hoogsteen pairs, 17-18
  non-Watson-Crick pairs, 38
  Watson-Crick pairs, 16-18

C

Cell aging and death, 343-344
Cell cycle
  acidic protein synthesis, 335
  bacterial, 104-105, 114-117
  cell division, 328
  centriole function, 337
  chalones and, 342
  chromosome condensation, 339
  chromosome replication, 334, 338
  in development, amphibian, 354-355
  DNA replication and, 333-337
  eukaryotic, 327-344
  factors, metabolic, and, 342
  G_2 arrest, 338
  G_0 period, 332
  G_1 period, 331-333
  G_2 period, 337-338
  histone synthesis, 334-335
  hormonal control, 342
  meiosis DNA, 336
  mitosis, 338-340
    adaptive regulation for, 338
    DNA, 336
    messenger RNA, 338, 340
    transcription and, 339-340
    translation and, 340
  nucleolus and, 332
  S period, 333-337
  sequential gene activation, 341
  spindle RNA, 337
  transcription and, 328-330
  translation and, 330-331
Centriole, 337
Chloroplastal ribosomal RNAs, see RNA
Chloroplasts
  genome, 218
  nucleic acid synthesis, 219-220
  synthetase, 241-242
  transfer RNA, 241-242
Chromatin, see Chromosome(s), Euchromatin, Heterochromatin
Chromosomal (derepressor) RNA, see RNA
Chromosome(s)
  B, 161
  diminution, 159
  eukaryotic program, transcriptional, 132, 139-141
  evolution, 405-410, 431-435, 472-473
  interphase, 158-162
  lampbrush, see Lampbrush chromosome
  lower eukaryotic, 408-409, 472
  organization, 88-90, 431-434
  polytene, see Polytene chromosome
  prokaryotic program, metabolic, 406, 434, 471

517

proteination, significance of, 406, 432, 473
pulverization, 339
Complementarity, principle of, 32, 405
Continuity, principle of, 430
Cyclic AMP
   hormonal effects on chromosomes and, 136-137, 447
   in transcription
      bacterial, 99-100, 116
      eukaryotic, 136-137
   in translation, 282

D

Degeneracy, *see* specific types
Development
   cell communication and, 345, 350, 397
   cell cortex and, 350-351
   cell division, rate of, 355-357
   cell membranes and, 346, 350, 397
   cellular processes in, 344-345, 349-350, 398-399
   cleavage arrest, 379
   complexity, informational, 377
   determination, 394-396
   differentiation
      molecular basis of, 394-396
      as overall process, 399-401
      phases of, 393-394
   DNA
      cytoplasmic, 355-356
      directional change, 352-353
   DNA constancy, 353
   DNA replication, 352-357
      oogenesis, 354
      rate of, 355
   dosage regulation, 361-362
   epigenesis in, 376
   epigenetics, 382
   factors, metabolic, and, 349-350, 356
   gene composition, differential, 358
   gene content, differential, 353, 358, 360
   gene expression, differential, 353, 357
   general aspects, 344-352
   germ plasm theory, 348
   hormonal control, 350
   household information, 378
   induction, 393-399
   informosome in, 385
   maternal precession, 374-375, 380
   meroistic eggs, 348-349
   methylase and, 391-392
   morphogenetic fields, 348
   morphogenetic information, 378, 380
   mosaic eggs, 347
   nucleoli in, 361, 363
   oogenesis, 346-349
   ooplasmic localization, 346-347, 357-359
   ooplasmic program, 347, 358-359, 394-395
   open-closed systems, 348
   organogenetic information, 379
   positional information, 345
   prefertilization lesion, 384-385
   preformism in, 376
   processes by categories, 344-346
   regulation of
      adaptive, 382
      polysomal, 382
      postpolysomal, 382-383
      prepolysomal, 381-382
   regulative eggs, 347
   ribosome classes, 383
   ribosome synthesis, 361
   RNA
      DNA-like, 369-381
         conservatism of, 374
         messenger, and, 375-381
            maternal, 377-379
      ribosomal, 360-365
         synthesis, rate of, 361
      5 S, 362-363
      transfer, 365-369
         isoaccepting, 390-391
         in modulation, 389-393
         status of nucleolar, 367-369
   supracellular processes in, 345, 350-351, 398-399
   supramolecular processes in, 345, 350-351, 398-399
   synthetase and, 391
   transcription in, 357-381
      informational, 369-370, 375-377, 395-396
      regulatory, 369-370, 375-377, 395-396
      stage and regional specificity, 359-381
   transcriptional program, 347, 358-359, 367, 394-395
   translation and, 381-393
      capacity, 380, 384

Index 519

household, 386
morphogenetic, 386
organogenetic, 386
potential, 380, 384
translational modulation, 389-393
ultrastructure and, 351
DNA
C-paradox, 426
complexity, 142
composition, evolution of, 427-429
compositional constraints, 428-429
crosslinks, 27-28, 429
degeneracy, 38-39, 409-410, 431
doublet pattern, 429
eukaryotic satellites, 142-145, 147-148
evolution
increase during, 425-426, 459
loss during, 425-426
rate of, 426-427
evolutionary parameters, 422-431
excess of, 141, 152-153, 168, 425-426
nuclear, and cell size, 424-425
physical structure, 16-19
postreplicative modification, 15, 353, 411
redundant eukaryotic
codon restriction and, 146, 413
development and, 369-375
evolution of, 410-431, 458-460, 466-472
extent transcribed, 149-150
functions, 146-148, 414-422
saltation, 146, 413-414
satellite, 142-145, 147-148
structure and distribution, 142-146
typical, 145-147, 411-412
repair enzyme, 22-23
replication and transcription, dual basis for, 31
sequences
eukaryotic
redundant, 141-148, 410-431
structure of, 141-148, 410-431
unique, 141-148, 410-431
evolution of, 410-437
transcription, 29-31, 32-33
extent of
eukaryotic, 149-152
prokaryotic, 112-114
unique eukaryotic
development and, 375-381

evolution of, 410-431, 458-460
extent transcribed, 150-151
informational function, 148, 416-417
DNA replication
eukaryotic, 26-29
cell cycle and, 333-337
development and, 352-357
rate of, 28
Taylor model, 27-28
prokaryotic, 20-26
Cairns model, 24-25
DNA polymerase, 23
error, level of, 33
Kornberg enzyme, 21-23
Okazaki fragments, 20, 23, 25-26
rate of, 21
replication complex, 26
rolling circle model, 24-25
DNA viruses
cell cycle and, 335
messenger RNA, 109-112
translational modulation, 292-294

E

Embryonic induction, *see* Induction
Endoplasmic reticulum, 259-261, 283-285
Enzyme regulation, *see also* Translation, adaptive regulation of
functional replication, 274
gene dosage model, 104-105, 336
linear reading model, 104-105, 337
oscillation model, 104-105
Euchromatin, 121-141, 161
Eukaryotic regulation
at cellular level, 117-119
differences with prokaryote, 93-95
noninformational RNA and, 180-183

G

Genetic code
basic nature, 46-49
code within code, 34, 446
decoding, evolution of, 49-50, 443, 446
evolution of , 46-49, 404, 430-431
codon-amino acid order, 46-49
codon-amino acid pairing model, 48, 50
interamino acid order, 46-49
intercodon order, 46-49
translation error model, 47-48
primitive code, 46-49, 404

wobble pattern, evolution of, 442
wobble theory, 38-39
Genetic information, 9-13
  flow of, 10
  second order, 13
Genetic organization
  Beermann's model, 153
  Callan's master-slave model, 153-154
  Edström's model, 154
  Whitehouse's model, 154
Genetic regulation
  Britten and Davidson's model, 155-156, 175-177, 468
  evolution of, 466-474
  Georgiev's model, 156-157, 174
Genetic system
  differences between prokaryote and eukaryote, 94, 119, 431-434
Genetic unit in eukaryotes, 163-164, 167
Genome(s)
  constitutive (informational), prokaryotic, 106-108, 114-117
  evolution of, 406-431, 458-460
  facultative, eukaryotic
    cell cycle and, 330
    evolution of, 474
    informational, 182-184, 367
    transfer RNA, 234-235, 289, 367
  inducible-repressible, prokaryotic, 95-99 106-108, 114-117
  informational, 412-413, 416-417
    evolution of, 458-460, 473
  obligatory, eukaryotic
    cell cycle and, 330
    evolution of, 458, 474
    informational, 182-184, 458
    transfer RNA, 234-235
  ploidization, 415-416, 425
  regulatory, 412-413, 420
    evolution of, 458-460, 466-474
  tandem duplication, 415-417, 425
Genotropic substance, 108
$GT_4 X$ derivative
  informational transcription and, 116
  ribosomal RNA and, 214-217

H

HeLa cell
  nucleolus, 317
  ribosomal pathway, 194-196

RNA species of, 52, 81
Heterochromatin
  constitutive, 121, 159-161, 325-327
    highly redundant DNA in, 144, 148, 326-327, 411-412, 423-424
    underreplication, 423-424
  facultative, 121-127, 158-161
Heterogeneous RNA, see RNA, labile
Histones
  biochemistry, 122-126
  cell cycle and, 334-335
  cistrons, 318
  evolution of, 432-433, 472
  functions, 122-127
  as general repressors, 122-127
  mitotic chromosomes and, 339
  posttranslational modification, 124-125
  regulation of, 318-319
  as regulators, 126, 182
Homopolymers, see Poly A
Hormones
  cyclic AMP effects on chromosomes, 136-137
  receptor proteins, 135-136
  selective transcription and, 132-135, 434-435

I

Induction, embryonic, 393-399
  competence, 393
  determination, 394
    molecular basis of, 395-396
  inducers
    hormones and, 397-398
    molecular nature of, 396-397
  macromolecular complexing, 399
  orders of, 393
  permissiveness of, 396, 434-435
  self-organization of, 399
Information, see specific type
Informosome, 57, 277-279, 463-464

L

Lampbrush chromosome, 162-164, 370-371
  loops, informational status of, 163-164

M

Macromolecular degeneracy
  evolution of, 409-410, 445, 457, 472
Macromolecular turnover, 93-94

Index 521

Messenger RNA
  characterization, 51-55
  degradation, 55-56, 102
  eukaryotic
    cell cycle and, 330
    development and, 375-381
    higher order structure, 463-465
    evolution of, 458-466
      conservatism in, 458
    molecular maturation, 172-178,
      460-461
    precursor, postulated, 172-178, 460-465
    promotion by ribosomal RNA, 175,
      286-287
    regulation
      adaptive, 184-186
      general, 179-184
      participants in, 119-139
      stable, 127-141
    specification
      monocistronic, 176
      posttranscriptional, 172-178, 460-465
    translation, 272-288
    transport to cytoplasm, 57, 170-171,
      278-279
  evolution of, 457-466
  metabolic parameters, 55-57
  prokaryotic
    coordination
      of genome, 106
      with stable RNA, 107-108
    general aspects, 100-106
    genomic representation, 112-114
    higher order structure, 462
    operon model, 95-99
    precursor, possible, 460-462
    regulation
      general, 106-108, 114-117
      specific, 95-99
      unspecific, 99-100
    regulon, 98
    translation, 266-272
    tryptophan operon, 101-103, 255
    viral, 109-112, 462
Methylase
  development and, 391-392
  translation modulation and, 300-302
Mitochondria
  evolution of, 435-437
  functional autonomy, 219, 436
  genome, 218, 436-437
  informational content, 221-222, 437
  nucleic and synthesis, 219-220
  prokaryotic ancestry, 218-222, 435-437
  protein synthesis, 219-220
  RNA
    5 S, status of, 69-70
    transfer, 241-242
  synthetase, 241-242
Mitochondrial ribosomal RNAs, *see* RNA
Mitosis, and cell cycle, 338-340
Molecular maturation, *see* Messenger RNA,
  eukaryotic; RNA, major ribosomal;
  Transfer RNA
Molecular redundancy
  in DNA, 142-146
  evolution of, 412-414, 430, 449, 451,
    468, 473
  in RNA
    ribosomal, 190, 449
    transfer, 78, 449
  use in RNA regulation, 469-470

N

Nongenetic information, 13, 345, 350-351
Nucleic acids
  structure
    orders of, 14, 20
    physical, 13-20
  synthesis, 20-33
Nucleolus
  acidic proteins, 316-317
  associated chromatin, 324-327
  cell cycle and, 332
  chromosomal contribution to, 322-323
  constitutive heterochromatin and,
    325-327
  description and function, 308-327
  evolution of, 455-456
  genesis, 321-323
  genetic control of, 321-322
  hidden ribosomal cistrons, 311-312
  micronucleoli, 310
  mineral pools, 320-321
  nucleolar nasses, 310
  organizer, 310-327
    dominant organizers, 313
  ribosomal protein pool, 316-318
  ribosomal subunit pool, 204, 321
  RNA

messenger, promotion of, 175, 286-287
nonribosomal, 314-316
5 S, pool, 318
synaptonemal complex and, 326-327
ultrastructure and organization, 319-324
Nucleus, organization of, 88-90

O

Oligopeptide synthesis, nontranslational, 264-266

P

Plasmotropic substance, 108
Poly A
  description and distribution, 82-83
  in eukaryotic messenger, 278-279, 463-464
Polyamines
  DNA regulation and, 99
  DNA structure and, 18
  regulation of synthesis, 319
  transfer RNA and, 239
Polydisperse RNA, see RNA, labile (eukaryotic)
Polysome
  aggregation, cycle of, 283-284
  formation, 260-261, 279, 283-285, 340
  free and bound, 259-261
  higher order structure, 463-465
  inactive, 279-280
  regulation, 280-282, 464-465
  segregation, 285
  specification, 260-261, 463-465
Polytene chromosome, 164-169
  puffs, informational status of, 167
Posttranscriptional modification, see also RNA, major ribosomal, Transfer RNA RNA, 14, 45
Protein(s)
  acidic (nonhistone)
    cell cycle and, 335
    description, 128-130
    evolution of, 432-433, 472
    phosphorylation, 129
    in regulation, 180-182
  degradation, 93-94, 287, 419
  effects on translation, 271-272
  evolution of sequences, 430, 458
  evolutionary divergence, 415-422
  evolutionary emergence, 415-422
  household, 331, 394
  luxury, 331, 394
  nucleolar, 316-319
  regulatory, 129-130, 180-181, 470
  ribosomal
    description, 203, 216-217, 250, 317-318
    regulation, 217, 317-318
  structure, orders of, 12

R

Redundancy, see specific types
Redundant eukaryotic DNA, see DNA
Regulation, see specific types
Replicon, 24-29, 335-336
Ribosomal protein, see Proteins
Ribosomal RNAs, see RNA, major ribosomal, 5 S, 5.5 S, mitochondrial ribosomal, chloroplastal ribosomal
Ribosome
  antibiotic effects, 263-264
  binding factors, 42
  classification, 249, 383
  evolution of, 457
  free and bound, 259-261
  inactive subunits, 270, 279-280
  membranes and, 259-261
  screening by, 44
  structure of, 250-252
  structure-function relationships, 257-259
  in termination of translation, 255-256
  topographical segregation, 269, 280
RNA
  categories in relation to genome, 441
  in cell surface, 343
  chloroplastal ribosomal
    coding genome, 220-221
    description, 69-70
  chromosomal (derepressor)
    description, 79-80
    postulated function, 130-131, 467
    putative status, 79, 131, 467
  in eukaryotic regulation, 180-183, 466-474
  evolution of, 437-474
  heterogeneous, see RNA, labile
  labile (eukaryotic)
    biological and molecular aspects, 169-172
    cytoplasmic, 60-61, 471

Index

development and, 369-375
evolution of sequence, 171-172, 470-471
mitochondrial, 61
physical and metabolic characteristics, 58-61
as precursor to messenger RNA, 172-175
as precursor-regulator to messenger RNA, 178
as regulatory RNA, 175-177
major ribosomal
  adaptive equilibria, 206
  basic design, 448, 450
  cell cycle and, 329
  cistronic multiplicity, eukaryotic, 190, 448-449
  cistronic regulation, eukaryotic, 200-202
  cistrons
    in eukaryote, 188-193
    natural history of, 309-314
    number transcribed, 209-210, 213
    in prokaryote, 193-194
  description, 62-65
  in development, 360-365
  discarded RNA, evolution of, 452
  eukaryotic regulation, 198-200
    dosage, 204-208, 314
    pathway, 202-204
    quantitative, 208-211
  eukaryotic repressor, 200-201, 364
  eukaryotic synthesis, pathways of, 194-197
  evolution of, 447-457
  genetic polymorphism, 206-207
  germinal amplification, eukaryotic, 190-193, 207, 312, 360, 422-423
  GT_4X derivative, prokaryotic, 214-217
  human dwarfism and, 207
  initiation
    factors, eukaryotic, 201-202
    rate of, 208, 212
  magnification, 206, 423
  as messenger RNA, 217
  promotion of, 175, 286-187
  minimum genome, 448
  molecular maturation, 202-204, 212, 362, 451
  partial conservatism, 450-451, 471
  precursor
    eukaryotic, 189, 194, 362

    evolution of, 452-453
    prokaryotic, 194, 197-198
  Perry model of transcription, 209
  posttranscriptional modification, 62, 202-204
  prokaryotic regulation, 211-217
  prokaryotic synthesis, pathways of, 197
  propagation, rate of, 209, 212
  relaxed (RC) control, prokaryotic, 214-217
  RNP particle formation, 194-197
  sequence divergence, eukaryotic, 189-190, 449-451
  ψ factor, 214-217
  structural complication, evolutionary, 448, 450
messenger, see Messenger RNA
mitochondrial ribosomal
  coding genome, 220
  description, 69-70
  structural simplification of, 69
molecular maturation, 172-178, 202-204, 230-231, 440, 460
polydisperse, see RNA, labile
proteination, 20
regulation by
  evolution of, 466-474
  implications of, 469-470
replication, 31-32
4 to 7 S nuclear (nucleolar)
  cell cycle and, 329-330
  description, 80-81
  function, 316
5 S
  cistronic multiplicity, 223
  cistrons in eukaryote, 222-223
    in prokaryote, 223
  description, 65-68
  development and, 362-363
  eukaryotic regulation, 224-226
  evolution of, 453-454
  mitochondrial, 69-70
  nucleolus and, 315
  ribosomal pathway and, 194-198
  structure, secondary, hypothetical, 66-68
  synthetic pathways, 223-224
5.5 S
  description, 68-69
  7 S, bacterial, 81-82, 467

small (stable) nuclear, *see* RNA, 4 to 7 S nuclear (nucleolar)
species in prokaryote and eukaryote, 438-439
stable, *see* RNA, major ribosomal, Transfer RNA
structure
 complication of, general, 438
 physical, 19-20
 simplification of organellar, 69
transfer, *see* Transfer RNA
RNA polymerases
 bacterial, 91-92, 103
 eukaryotic, 91-92
 mitochondrial, 90-91
 T7 phage, 90-91
RNA viruses
 higher order structure, RNA, 270-272, 462
 mammalian, 111-112
 replication, 31-32, 111
 translation, 270-272, 462

S

Synthetase
 bacterial amino acid metabolism and, 239-241
 in chloroplasts, 241-242
 congruency with genetic code, 446
 description, 248-249
 development and, 391
 evolution of, 445
 in mitochondria, 241-242
 translational modulation and, 299-300

T

Template hygienics, principle of, 416-417
Transcriptase, reverse, 4, 11
Transcription
 asymmetry of, 30, 91-92
 cell cycle and, 328-330
 coupling with replication, 105-106
  with translation, 101-103
 development and, 357-381
 initiation, 107-108, 110
  internal, 42
 initiation factor, 91-92, 115-116, 201-202, 332-333
 ion effects on, 137-138
 metabolic factors and, 138-139
 mitosis and, 339-340

polarity, 104
rate of, 30
structural basis of, 88-93
termination, 110
termination factors, 91-92
Transcriptional program, 139-141, 351-352, 433-434, 472
 in development, 347, 351-352, 393-395
 hypothetical evolution, 471-474
Transcriptional testing, principle of, 147, 418, 459-460, 473
Transfer RNA
 adaptor in translation, 252-255, 261-263
 allosterism, 77-78, 443
 aminoacylation, 245-248
 aminoacylation barrier, 443, 446
 ancestry, 440-442
 bacterial amino acid and, 239-241
 bacterial cell wall synthesis and, 264
 bacterial metabolism and, 238-239
 cell cycle and, 329
 cell population of, 263
 in chloroplasts, 241-242
 cistrons
  chromosomal, 227-228
  in eukaryote, 226-229, 442
  nucleolar, potential, 228-229, 316
  in prokaryote, 229-230
 degeneracy, 78, 449
 description, 70-79
 in development, 365-369
 developmental modulation and, 389-393
 economy of function, 444
 eukaryotic cistronic multiplicity, 227, 440-442
 evolution of, 440-447
  genetic code and, 46-49, 443-444
 formative pathway, 230-231
 functional differentiation, 231-234, 238, 440-441
 hormonal stimulation, 232-233
 inactive states, 76-77
 initiator, 39-42, 443
 minimum genome, 448
 in mitochondria, 241-242
 molecular maturation, 79, 230-231, 235-236, 451
 in oligopeptide synthesis, 264-265
 Perry model of transcription, 232
 plurality of function, 70, 444-445

Index

polysome aggregation and, 283-284
posttranscriptional modification, 71-72, 76, 295, 300-302, 444
precursor, 230-231, 235-236
redundancy, 78
regulation
  dosage, 233-234
  eukaryotic, 231-237
  prokaryotic, 237-239
as regulator of translation, 288-303
relaxed (RC) control, 233, 237-238
structure
  conservation of, 442
  secondary, 72-73
  tertiary, 74-75
suppressor, 43-44, 234
transcription, rate of, 235
in translation, 252-255, 261-263
structural relationships, 259
Translation
adaptive regulation of, 184-186, 338, 382
  description, 184-186
  in development, 382
  functional implications, 186
antibiotics on, 263-264
basic mechanism, 245-248, 252-256
cell architecture and, 284-286
coupling with transcription, 101-103
in development, 381-393
elongation, 253-255
elongation factors, 253-254
eukaryotic characteristics, 256-257
evolution of, 446-447, 457, 465-466
GTP and, 253-254, 447
higher order structure, 270-272, 462-466
initiation, 39-42, 253, 256, 267
  "ticket theory" of, 281, 464
initiation complex, 40-42

initiation factor, 40, 253
mistranslation, 44-46
modulation
  animal, 295-297
  bacterial, 290-292
  development and, 389-393
  ethylase and, 302
  hypothesis, 289
  methylase and, 300-302
  neoplastic cell, 297-298
  plant, 295
  synthetase and, 299-300
  viral, 292-294
peptidyl transferase, 254
polarity, 101
polysomes, inactive, 279-280
rate of, 257
regulation
  eukaryotic, 272-288
    integral, 275-277, 284
    polysomal, 280-282
    postpolysomal, 282-283
    prepolysomal, 277-278
  prokaryotic, 266-272
ribosomal subunits, free, 279-280
ribosome binding sites, 42, 268, 270, 462
symmetry relationships, 274-275
termination, 42-44, 255
  suppression of, 43-44
termination factors, 43-44, 255
transfer RNA as regulator, 288-303
"translation package", 279
translocase, 254-255
transpeptidation, 266
viral initiation factors, 267

U

Unique eukaryotic DNA, see DNA